INSTRUMENTATION AND AUTOMATION FOR MANUFACTURING

INSTRUMENTATION AND AUTOMATION FOR MANUFACTURING

(An Overview for Manufacturing Students, Supervisors, and Managers)

JACK W. CHAPLIN Ed.D

Professor Emeritus of Industrial Technology, San Jose State University

Delmar Publishers Inc.®

NOTICE TO THE READER

Publisher does not warrant or guarantee any of the products described herein or perform any independent analysis in connection with any of the product information contained herein. Publisher does not assume, and expressly disclaims, any obligation to obtain and include information other than that provided to it by the manufacturer.

The reader is expressly warned to consider and adopt all safety precautions that might be indicated by the activities described herein and to avoid all potential hazards. By following the instructions contained herein, the reader willingly assumes all risks in connection with such instructions.

The publisher makes no representations or warranties of any kind, including but not limited to, the warranties of fitness for particular purpose or merchantability, nor are any such representations implied with respect to the material set forth herein, and the publisher takes no responsibility with respect to such material. The publisher shall not be liable for any special, consequential or exemplary damages resulting, in whole or in part, from the reader's use of, or reliance upon, this material.

Cover illustration courtesy of Allen-Bradley Co.
Cover design by Juanita Brown

Delmar Staff

Executive Editor: Michael McDermott
Project Editor: Carol Micheli
Art and Design Supervisor: Rita Stevens
Production Coordinator: Sandra Woods
Art Coordinator: John Lent
Design Supervisor: Susan Mathews

For information, address Delmar Publishers Inc.
Two Computer Drive West, Box 15-015
Albany, New York 12212

Printed in the United States of America
published simultaneously in Canada
by Nelson Canda,
a division of The Thomson Corporation

10 9 8 7 6 5 4 3 2 1

Library of Congress Cataloging in Publication Data

Chaplin, Jack W.
Instrumentation and automation for manufacturing: an overview for manufacturing students, supervisors, and managers / Jack W. Chaplin
p. cm.
Includes bibliographical references (p.) and index.
ISBN 0-8273-4713-8 (textbook)
1. Manufacturing processes—Automation. 2. Industrial electronics. 3. Engineering instruments. I. Title.
TS183.C48 1991
670.42'7—dc20

91-18700
CIP

TABLE OF CONTENTS

PREFACE ix

ACKNOWLEDGEMENTS xiii

1 Introduction to Instrumentation and Automation for Manufacturing 1
- Objectives 1
- Introduction—Text Overview, History, and Definitions. 1
- Manufacturing Process Control Measurement 14
- Summary/Facts 27
- Review Questions 27

2 Electronics Fundamentals 29
- Objectives 29
- Introduction 29
- Instrumentation Electronics Fundamentals 30
- Summary/Facts 79
- Review Questions 80

3 Instrumentation Concepts 82
- Objectives 82
- Introduction 82
- Physical Phenomena Used for Measuring 83
- Summary/Facts 108
- Review Questions 108

4 Sensing Physical Variables: Temperature and Pressure Sensor Instruments 110
- Objectives 110
- Introduction 110
- Sensors Used to Acquire Temperature Data 110
- Sensors Used to Acquire Pressure Data 139

Summary/Facts 157
Review Questions 158

5 **Sensors of Flow and Level** 160
Objectives 160
Introduction 160
Transducers Used to Sense Flow Data 160
Sensors Used to Obtain Level Data 183
Summary/Facts 201
Review Questions 201

6 **Density, Specific Gravity, Viscosity, Humidity, Acidity and Alkalinity, Weight, Force, Rotation, Position, Motion, Acceleration, Vibration, and Dimension Sensors** 203
Objectives 203
Introduction 203
Density and Specific Gravity 204
Viscosity Sensors 209
Humidity 213
Acidity and Alkalinity 221
Weight and Force Sensors 223
Rotational Speed Sensors 228
Position Displacement and Motion Transducers 234
Acceleration and Vibration Transducers 242
Dimensional Transducers 249
Summary/Facts 260
Review Questions 261

7 **Decision and Control of the System** 263
Objectives 263
Introduction 263
The Process Control Loop and Response Characteristics 264
Microprocessor 286
Microcomputer 286
Direct Digital Control Technology 287
Digital Computer Control 290
Distributive Control Systems 292
Summary/Facts 293
Review Questions 294

8 **Control Actuators and Control Valves** 296
Objectives 296
Introduction: Actuators 296
Final Control Elements - Control Valves 325
Summary/Facts 332
Review Questions 333

9 **Automated Material Handling** 335
Objectives 335
Introduction 335
Automated Material-Handling Machinery 336
Handling Bulk Materials 336
Handling Piece Parts 343
Assembly and Part Handling 351
Automated Quality and Position Control 354
Finished Product Handling Instrumentation 355
Summary/Facts 358
Review Questions 358

10 **Instrumentation Applied to Manufacturing Machinery** 360
Objectives 360
Introduction 360
Mechanical Instrumentation 360
Transfer Line Production Instrumentation 378
Manufacturing Instrumentation with Numerical Control 386
Computer Controlled Machines 390
Summary/Facts 392
Review Questions 394

11 **Strategies of Production Automation** 396
Objectives 396
Introduction 396
Manufacturing Cells 397
Flexible Manufacturing 398
Robot Classifications 406
Controllers for Robots 412
Power Supply 421
Summary/Facts 431
Review Questions 432

12 **A Strategy of Automation with Computers** 434
Objectives 434
Introduction 434
Computer Integrated Manufacturing 435
Identification of Inventory and Storage 454
The Automated Factory 459
Summary/Facts 469
Review Questions 471

Acronyms 473

Glossary 477

Bibliography 484

PREFACE

A word about the objectives and intent of *Instrumentation and Automation for Manufacturing* is needed. This text is not designed as an Engineering Instrument Design or an Automation Engineering textbook. It is designed for use in a class for the two-year technology technician, four-year technologist, or business major who desires a technology background. Upon graduation these people will enter manufacturing at beginning managerial positions within an organization. They will often carry the payroll title of Floor Engineer, Manufacturing Engineer, Production Planner, Associate Systems Analyst, Process Engineer, Associate Engineer, Associate Q.A. Engineer, Inventory Control Manager, OSHA Coordinator, and others.

These people are sought after because of their technical and managerial training studied at the university. They will be managing rather than designing. They will apply the principles, concepts, and application of data gathering, controlling processes and automation that effect efficient manufacturing. Engineering design and mathematical application and formula derivations are not stressed because they are for the engineer's use in analysis and design rather than understanding and management of the application of a total system.

The first chapter in the text is an overview of the text and includes a brief history of manufacturing and instrumentation with the addition of important definitions throughout the chapter.

Electronics is a key element in the understanding of instrumentation and automation. Therefore, practical and fundamental electronic information is needed. Chapter 2 introduces these elements of electronics. When basic electronics has been presented in other course work, chapter 2 may be reserved for students without such knowledge.

"Instrument Concepts," Chapter 3, introduces the basis of how variables are measured. The physical, chemical, and electronic principles that are necessary for instrumentation sensors to use in measuring changes in physical phenomena are presented. In addition, manufacturing process variable concepts and a group of the common process variables that are frequently controlled are described.

Chapter 4 is a presentation of the major sensors utilized in the measurement of the variables of temperature and pressure. These variables were selected for the first group to be studied because they are commonly used in a great number of industrial processes and are the basis for so many control systems.

Flow and level measurement are industrial variables that appear in the manufacturing process as well as in daily living. It is important to understand the application of these sensors because they appear so frequently.

The content of chapter 6 is a group of variables to be sensed that are more closely associated with manufacturing and processing. In the past, many of these variables were measured in laboratories and the data passed to production control for the correction of the manufacturing process. Thus many of these variables would appear under a heading of "Analysis" in other textbooks. At the present, variables such as density, specific gravity, viscosity, acidity, and alkalinity are carried out as continuous production-process modes rather than in a laboratory. For this reason these variables are grouped together and treated as any other industrial variable to be measured and controlled.

Chapter 7 "Decision and Control of the System," makes use of the data from the sensor and puts it in a usable form to control the system. It introduces the concepts of the control loop, process response characteristics, modes of control, and control technologies, including distributive control systems.

Chapter 8 is concerned with what is to be done with the data from the sensor and with decision and control from the instrumentation system. What is still needed is something to make a physical change in the total system: the actuator and the final element. They are either a control valve, solenoid, relay, or a motor. These devices come in many forms and are responsible for making a physical change in the variable within the production system. These actuators will add or reduce the controlled variable that is entering the manufacturing processing system, thus providing control.

Material handling is concerned with the machinery for handling materials in a manufacturing setting. The handling of automated bulk materials, piece parts, assembly parts, automated positioning control, and instrumentation applied to finished product handling are the content of chapter 9.

Chapter 10 is concerned with the mechanical aspects of instrumentation of manufacturing machinery. It relates to the application of devices used to carry out the delivery of data and control of the machinery. The application of kinematic mechanisms, counters, and sequencing and coding for numerical control to provide instrumentation for manufacturing machinery is presented.

"Strategies of Production Automation," chapter 11, focuses on flexible manufacturing. This chapter discusses the various units required to build a flexible manufacturing system. Direct numerical control, a communication protocol system, modular manufacturing systems, industrial robots, position sensing systems, robot controllers, various power supplies, and robotic applications and inspection with computer measuring machines are designed to provide a flexible manufacturing system.

Chapter 12, "A Strategy of Automation with Computers," has as its central theme computer integrated manufacturing (CIM). This chapter brings instrumentation and manufacturing into a complete whole. The sensing, control, and manufacturing culminates in a total automated factory.

The sources of material for this text are many. The foundation for this book has been more than twenty years of teaching a course by the same name at the university. The materials have come from many sources over the years. Textbooks, magazines, journals, and an instrumentation apprenticeship and experience at the Sacramento Air Depot are among them. The most important source of textual materials has been the current literature and materials furnished by the manufacturers of the products illustrated in the text. The vendors of numerous products have been a great source of knowledge and help.

Included at the end of the text is a list of acronyms that will help in understanding the notations commonly used within the instrumentation and manufacturing industries. A glossary of terms is also provided in the back of the book for aid in defining the italicized terms in the text.

About the author: I graduated from Sacramento Junior College, San Jose State University, and Stanford University. My advanced degrees are from Stanford University. My teaching experiences have been at Fremont Union High School (Machine Shop), and San Jose State University (wide range of courses including classes in graduate professional courses, Machine Shop, Manufacturing, Instrumentation and Automation). I have also written a previous text, *Metal Manufacturing Technology*.

ACKNOWLEDGEMENTS

I would like to acknowledge and thank the many people who supported me in the preparation of this book. First I would thank the Instrumentation, Manufacturing, and Electronic Faculty of the Division of Technology at San Jose State University and the students who passed through my classes of Instrumentation and Automation for Manufacturing. The students were enthusiastic and supportive in their lecture and laboratory classes. I learned much in preparing for these classes and I learned a great deal from the students as they worked out their laboratory problems.

In addition I would like to thank Dr. Mohan Kim for reading the sensor chapters of the early manuscript, as well as Mr. Clyde N. Herrick for reading the early electronic chapter. I am also grateful to Mr. Charles Lichtenstein for his many suggestions and final readings of chapters dealing with electronics and automation.

Acknowledgements are in order for the many companies and organizations that freely furnished materials, photographs, technical manuals, and information. Their contributions are greatly appreciated and the text would not be possible without their generous contributions and help. The photographs are cited in the text giving recognition to their organizations.

Finally, I want to thank my wife Arlene for her continued encouragement and cooperation, and the Chaplin family for their interest, understanding, patience, and help during the writing and completion of this book.

1

INTRODUCTION TO INSTRUMENTATION AND AUTOMATION FOR MANUFACTURING

OBJECTIVES

Upon completing this chapter, you will have the following:

- An understanding of instrumentation and its relationship to automation in manufacturing
- Knowledge of the history of the involvement of manufacturing since ancient times
- Recognition of the improvement of manufacturing with the advent of the computer
- Knowledge of the development, principles, and operation of instrumentation
- Understanding and application of the stages of controller development
- Knowledge of manufacturing process control measurement
- Understanding of the application of the control loop
- Basic application of instrumentation to automation

INTRODUCTION — TEXT OVERVIEW, HISTORY, AND DEFINITIONS

Instrumentation was invented by man in order to be able to measure and make corrections and changes in his manufacturing processes. Through measurement, he gained a method of evaluating and controlling his manufacturing processes.

This added control increased his reliability and effectiveness over his manufacturing procedures.

As instrumentation improved, it has made it possible to supply better information and data about the ongoing processes of manufacturing. This improvement of information on product data has provided the opportunity for better control for the monitoring of manufacturing. As the ability to take quick and accurate measurements improved, corrective action was taken on the item under control while the item was still in the manufacturing process. These real time corrections in the process provided for excellent production and greatly reduced the number of rejected components and products.

The Improvement of Manufacturing Since Ancient Times

Manufacturing technology has constantly evolved since ancient times. In the stone age, man learned to direct pressure on a piece of flint or obsidian that would flake to produce a conchoidal or curved fracture. When a fragment of volcanic glass was chipped from both sides of the piece where the intersecting fractures met, an extremely sharp cutting edge was formed. These cutting edges were highly desired by the members of the tribe and thus a method of production was perfected.

Through *observation* and *comparison*, the only means of measurement available during the stone age, the chipper was able to improve his manufacturing ability. This work resulted in early tools such as knives, hand axes, awls, and weapon points used by the tribes. In like manner other ancient manufacturing technologies evolved — always utilizing measurement through observation and comparison.

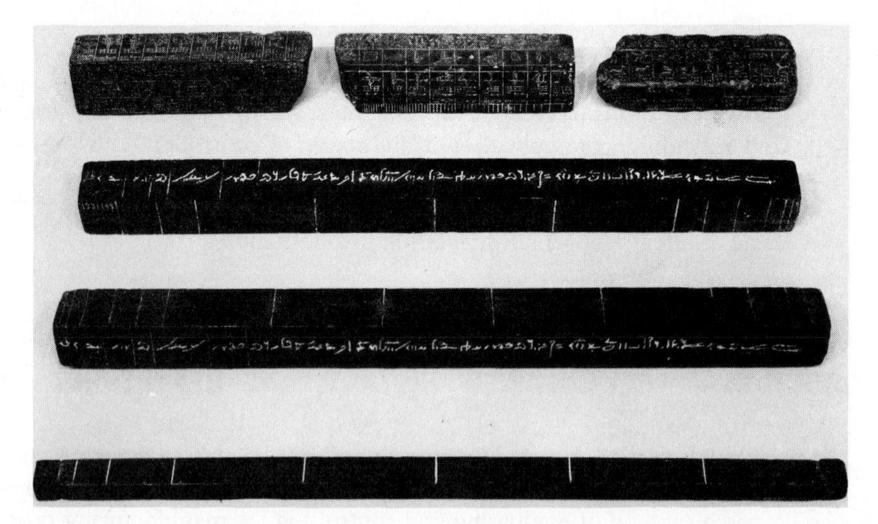

Figure 1-1 Stone Egyptian cubit measures. *(Courtesy of the Trustees of the Science Museum, London)*

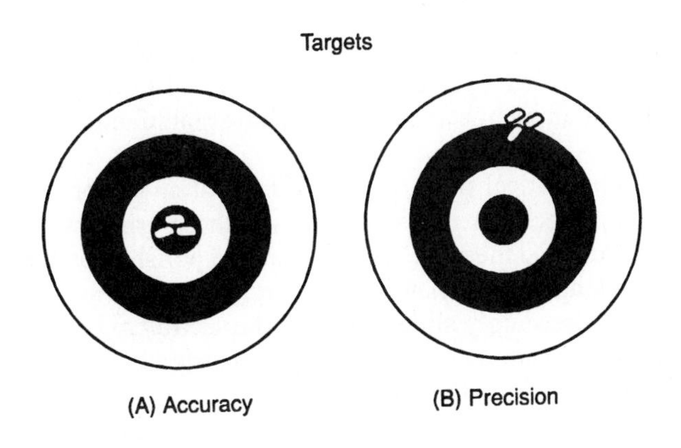

Figure 1- 2 Accuracy versus precision: (A) A bull's eye group; (B) A tight group

Handcrafted Production

Early handcrafted manufactured products were built to the specifications of the customer. Very often only one object was built to be used in a specific way or place. Measurement for these products was still through observation or comparison. Comparisons were made to the length of a stick or body parts, e.g., a digit, finger, or thumb; a hand span; or a cubit, which is the length from the elbow to the tip of the index finger, Figure 1-1.

During the Middle Ages, each product was designed for an individual person. Thus the *accuracy* of the measurements was relative to the customer's requirements. Because of these requirements, relatively coarse measurements were applied. However, precision was available in these products because each article had its parts individually fitted into the whole, resulting in a highly crafted and beautiful individual product.

Precision has always been the craftsman's hallmark, while accuracy has evolved with the improvement of measurement. Exactness is accuracy and is dependent upon measurement. In modern instrumentation both accuracy and precision are stressed, Figure 1-2.

Mechanized Production

The English Industrial Revolution brought great strides in measurement. A large number of industrial materials were produced in one area and assembled in another. The products of iron rolling mills, cut timber, bricks, stone, and textiles were materials for industry. Because these were items used by any number of manufacturers, these products were standardized as to size and weight. Measurement standards that were most often selected were fractions of the foot and the inch. For weight, the pound and ton were applied.

For most manufacturing applications at this time, accuracy to the fraction of an inch was satisfactory. Accuracy in the railroad and ship building industries was one-eighth of an inch; accuracy in machine manufacturing was one-sixteenth of an inch.

In special critical parts, precision measurement for bearing and shaft sizes were fitted. Precision measurement for these parts was established by end measures. An "end measure" is a rod cut and polished on the ends so that the rod is the same length or diameter as the part to be measured. The part is fitted until the end measure snugly slides between the section to be measured. With careful use, parts so measured can be delivered consistently within three-thousandths of an inch variation from the rod's length.

The mechanization during the industrial revolution lead to rapid development of measurement. A group of English tool builders invented machine tools that would manufacture accurate surfaces and screw threads. These improvements accelerated the parallel growth in instrument manufacturing with machine tool manufacture. Navigational and astronomical instruments also expanded the ability to measure.

Among these English tool builders, Henry Maudslay designed a screw cutting lathe with gearing that could be changed to produce different screw pitches. Joseph Whitworth designed a number of efficient machine tools that greatly reduced the time necessary to produce an accurate machined surface. He was very concerned about measurement and screw threads; he did much to aid in the standardization of screw threads. By 1856, Whitworth had built a bench micrometer which he used in the machine shop. He claimed that he could discriminate to one-millionth of an inch on a measuring machine that he had designed with a combination of a screw thread and a worm and gear. He delighted in the settling of arguments among his men by measuring a disputed workpiece's size on the measuring instrument. He called it "The Lord Chancellor" because it ended the discussion. Measuring instruments provided the consistency in machine tools that was necessary for the mass production tools used in industrial manufacturing during and after the English industrial revolution.

Mass Production of the 1920s

Mass produced manufacturing of precision products became a reality in the early twentieth century. Measuring instruments and standards were refined. Commercial, interchangeable, mass produced parts with a common industrial manufacturing requirement of one one-thousandth of an inch accuracy were expected. Starting in the 1920s Henry Ford introduced the mass production assembly line that set the climate for industrial manufacturing and automation. For measuring, the company adopted a mathematical system of eighty-one precision gage blocks. With a combination of these steel blocks most measurements could be obtained. The Johansson gage blocks ranged in size from 0.1001 inches to 4.0 inches and when these blocks were properly wrung together, they delivered accurate measurements to millionths of an inch. These blocks were used to check fixed gages and to set comparison gages.

Fluid Processing Industries

The chemical and petroleum industries applied mechanical and pneumatic instrumentation to the monitoring and control of their manufacturing process. Pressure, temperature, and flow instruments were developed employing the physical properties of metals and fluids. Pneumatic technology evolved so that most industrial process variables were measured as pressure measurements. These measurements became the input to a pneumatic controller to make the decisions of process control. The control signal was sent out to a pneumatic control valve in the form of a pressure. Process measurements-whether the variables of pressure, temperature, flow, level, or others-are all easily sensed and transmitted as a variable pressure. This technology provided improvements in instrumentation during the early half of the 1900s.

Transfer Machines and Lines

Transfer machines and lines were the beginnings of what is now referred to as "hard tooling automation." These were very long machines built from a series of machines containing modular machine tool units. The machine basic modules were the base, column, way, and tool heads. Other units that could be added were angular units and wing units. The transfer machines were many machines interlocked into one machine, Figure 1-3. Large workpieces were moved in and positioned from workstation to workstation automatically. These machines produced massive quantities of the same product. They were very efficient, but in order to a change the product, a massive rebuilding of the machine was required.

During the 1940s, instrumentation of these machines was improved rapidly and changed to electrical, mechanical control with the aid of many hydraulic

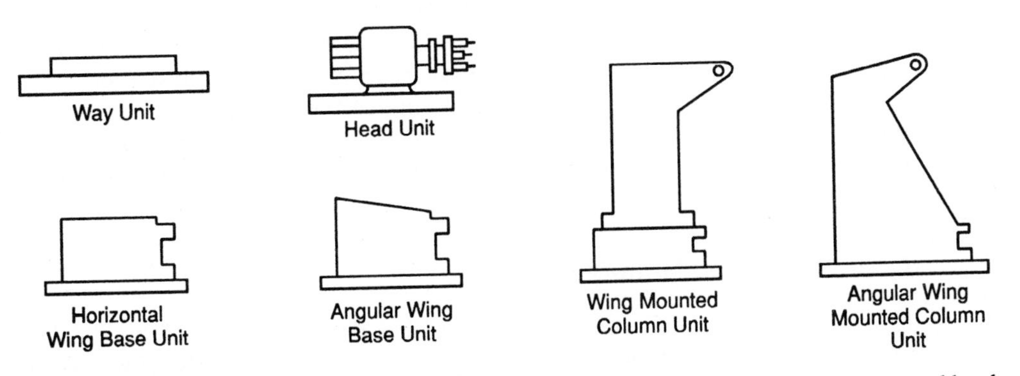

Figure 1-3 Basic building block units for modular production machines are the base, column, and heads. Other units are added such as angular and wing bases.

valves and *actuators* for tools and for the transferring slides. The machine was controlled by the application of electrical relay logic, warning lights, and interrupt circuits. Instrumentation provided a count of the work cycles of cutting tools and the inspection of the workpiece for chips or broken tools lodged in the workpiece.

Numerical Control

Instrumentation and control in the 1960s made rapid changes. A major contribution occurred with the application of numerical control to manufacturing tools and processes. Through numerical control a new range of flexibility was added. This flexibility was provided by an ability to program the machine control systems. With the added capability of programming the machines, it became possible to produce a large number of different workpieces with the same tools and machines. A vast variety of similar parts with varying shapes and dimensions could be accurately manufactured. Numerical control helped solve the problems of the inflexibility of rigid or hard tools and tooling of the transfer machines by being able to reprogram the movements and dimensions of the machine's cutter paths. This made the machine available to perform work on an unlimited number of different parts. The ability to change a machine's action by the use of software modification added great adaptability to the machines. The application of this new flexibility concept has been given the name of "Soft Automation."

The 1960s was a period when electronics was applied to many old and newly invented sensors. The application of transistor technology to electronic instrumentation of all types resulted in an exponential expansion and flexibility of industrial controls. This new sensing and controlling technology was applied to most industrial process variables by including electronic sensors, comparators, controllers, amplifiers, and actuators-all of which rapidly moved to electronic control.

The Advent of the Computer

The evolution from individual *solid-state* component electronic circuit technology to large-scale integrated circuit technology (or chips) occurred at a very rapid pace. With improvements in large-scale integrated circuit technology, the cost of microprocessors and computers was dramatically reduced. From the 1970s to the present, we have seen the wide application of this technology to instrumentation, automation, and manufacturing.

Computer Aided Manufacturing

Computer aided manufacturing resulted to a large extent from applying the data generated by computer aided drafting. Progress proceeded rapidly with applications in engineering and planning. Computer aided drafting had been successful in designing large steel structures, such as office buildings, with their lists of specifications that amounted to a great deal of data. This class of work had developed the essential geometry for representing various components into a computer data base system. The system and method had been developed that could be applied to smaller parts, and the geometry of these parts could be electronically stored.

The computer is very adept at performing the arithmetic operations that can be applied to solving algebraic formulae. The computer is excellent at accepting massive amounts of data, performing mathematical functions, and quickly delivering the data in a mathematical or statistical result. The computer can generate and output data points that can be drawn by a plotter to produce an accurate plan or drawing. These same data points can also result in a numerical control position of a machine tool or the position of robot.

Computer aided manufacturing has resulted from the integration of these technologies. The computer provides the necessary communication and control for computer numerical control. A system in which a machine is directly under the control of a computer provides programmed instructions to make the machine's operation error free on complex or very dull repetitive work, once the machine's program and system have been debugged.

Robot Units

Robot units can be controlled by the same type of system that is used to control numerical control machine tools. The difference is that the work frequently is transported by the robot or a tool attached to the end effector/gripper of the robot arm. The instrumentation again is largely found in the *feedback* sensors and the signal conditioning for the microprocessor or minicomputer. Robots have been successful in manufacturing because they have reliability and have flexibility in that they can be reprogrammed and retooled much more rapidly than is possible with the transfer machine or other designs of manufacturing production.

Flexible Manufacturing

Flexible manufacturing is a broad term that includes other broad production concepts within it. Flexible manufacturing facilities and machines are designed so that the production system can handle batches of parts that are not identical but belong to a family of parts or a group of parts with similar basic geometric configurations. An example of this concept could be a series of hydraulic valves with different functions that can be machined from the same stock body casting. The instrumentation can sense a code on the casting and provide the correct holding fixture and location. The code also requests the machining program, the needed tools, and the machinability data for the workpiece. The valve receives the machining of holes and surfaces required for that particular model and function of valve. It is possible that the next valve to be machined by the same machine will be done with a different program and tools. The first valve may have been a hydraulic *sequence valve*, while the second valve may have been a hydraulic pressure *relief valve*, both being machined from the same basic castings. These valves are part of a family of products capable of being machined from the same basic castings. The flexibility to produce this family of valves is achieved by a code that requests the reprogramming and retooling of the machine as necessary.

Manufacturing Cells

The *manufacturing cell* consists of a group of machines that may machine or assemble manufactured products. The cell may contain a robot or robots to load and unload machines or to supply workpieces and tools for assembly by the production machines. Frequently the manufacturing cell will consist of a machining center, robot and material handling systems with software technology, all under computerized control.

Integration of Manufacturing Cells

The integration of a group of manufacturing cells and their material transportation system makes up a computer integrated manufacturing system. Computer integrated manufacturing (CIM) is the result of linking the above systems together with computers to include the control of the production machines or assembly tools, the total material transportation system, and an inspection system. The computer provides real time corrections back to the machines and controls transportation conveyors and the automated warehousing system. The computer integrated manufacturing system has all the critical processes under control at all times as well as supplying the data and information for the generation of management reports.

Factory Automation

Factory automation is a method of manufacturing for long periods of time with very minimal human supervision. In many cases the equipment is specially designed and has little resemblance to our present generic tools. Within automation a whole factory will be integrated into one large machine. This type of automation has been referred to as the automatic factory. This level of automation is approached by the petroleum industry and some space projects, but is in the developmental stages in most manufacturing facilities.

Instrument Development

One of the most important factors needed in instrumentation is the development and application of sensors for the control of manufacturing. The role of the sensor is to supply reliable information. This input is necessary for accurate control of the various processes of manufacturing and is supplied by measuring the industrial variables to be controlled. Measuring instruments have evolved over long periods of time, slowly improving as technology and science progressed.

Within the past few generations, instrumentation has developed because of the need to control the various manufacturing processes more adequately. This has resulted in the exploitation of the physical and chemical properties of selected materials for the construction of measuring sensors.

Like manufacturing, instrument development has had a very long history. In speculating about ancient times a case could be made that very early measurements were concerned with level, flow, and length. Simple instruments, such as a stick pushed into the mud of a river bank, would determine the level of a river and whether or not it was rising. Ancient stone devices were housed along the Nile River and were observed by the priest to determine the rising of the Nile. Flow was observed by providing a restricted opening in a channel thus delivering a controlled rate of flow of water to a field. Length was measured by the cubit and various lengths of cord. After a flood of the Nile, field boundaries were located and reestablished using reference points, geometry, and distances measured by cord lengths.

Simple Instruments

The early industrial instruments were designed to sense the condition of the materials under process. The conditions of the material to be controlled are referred to as variables. Temperature, pressure, level, flow, and size are considered as variables that will be measured and controlled. Simple instruments measure by employing basic physics and properties of materials, such as expansion or deflection. These sensors in turn produce a mechanical motion sufficient to drive an indicator or actuator.

Improved Pneumatic Instruments

Instrument measurement technology quickly expanded by applying pneumatic devices to the increasing need for more accurate and automatic control. The application of the dynamics of fluids was employed in the building of controllers and actuators. These devices were designed to operate pneumatic control valves or electrical devices.

Electrical Instruments

Electromechanical and electrical instruments provided expansion of sensing and controlling. These instruments were designed around the various characteristics of electrical circuits: resistance, capacitance, induction, voltage change, amperage change, or frequency change. These devices were built with smaller sensors and control equipment into a large system of *automatic control* and were centralized into a control room. As new instruments were added, they were incorporated into a large system and staffed by specially trained people who could control a whole plant.

Electronic Computer Sequencing and Process Control

Electronics provided a true revolution in instrument development. With the application of the transistor, and later, large-scale integrated circuits, the size of the total instrumentation requirements was drastically reduced. Large-scale integrated circuits have evolved that contain microprocessor and sensor within the same chip. These sensors sense a signal and condition the signal for process control. These devices are referred to as "smart sensors." These sensors have become more efficient and provide greater flexibility to industrial controls. These smart sensors have resulted in distributed instrument systems, where video-based operations work-centers replace large control panels in control rooms.

For local control sensing, the control parameters are programmed into a microprocessor on a single chip. These control requirements are directly transmitted to the control valve or other actuator. When a group of sensors require monitoring and controlling, a microcomputer can process a thousand or so of these control loops. A minicomputer has the capability of controlling all the necessary information for the control of a petrochemical production system, cement manufacturing processes, or a manufacturing cell. The manufacturing cell computer will include the numerical control programs for all the machine tools and robots within the cell. Very frequently the minicomputer will be interfaced with the corporate main frame computer. These computers scan and monitor the various instrument loops, controlling the processes in each loop as well as sequencing the work to the next process.

The main frame computer is the chief data storage of programs and is the data base for the corporation's computer aided design and computer aided manufacturing data as well as for all the business and management functions of the corporation.

Stages of Instrument and Controller Development

Early control systems were mechanical systems based on the applications of levers, springs, cams, balance beams, gear trains, and valves.

Mechanical Instrument Controllers

With simple mechanical components applied to temperature and pressure sensing units, a considerable number of industrial variables could be controlled. To illustrate: a pressure signal was used to control the industrial variables of level and flow; springs were applied to control weight and force; gear trains were used to control position and speed; variations of levers and cams were applied to other ingenious mechanical systems to control additional variables.

Pneumatic Instrument Controllers

Pneumatic devices provided some of the early remote control instrumentation. With the employment of diaphragm, bellows, Bourdon tubes, and cylinders, systems could be designed that would provide sensing, control, and actuation of systems. These components could be installed with the sensors, the controller, and the actuators separated by several hundred feet. This was accomplished by a small diameter pipe or tube that was used to transport a variable air pressure signal to a diaphragm motor that in turn positioned a control valve. The valve usually controlled the amount of energy passing into the system-whether from steam, natural gas, or oil. The instrument often employed to do this type of work was the Bourdon tube in a controller or the application of a balance beam built of a series of pneumatic bellows. These devices delivered an air pressure that would correct the error in the system and provide control.

This instrument was very reliable and accurate for the work required and would function in severe weather conditions. Another advantage of the pneumatic devices was that they had no electrical connections and thus were spark free. They

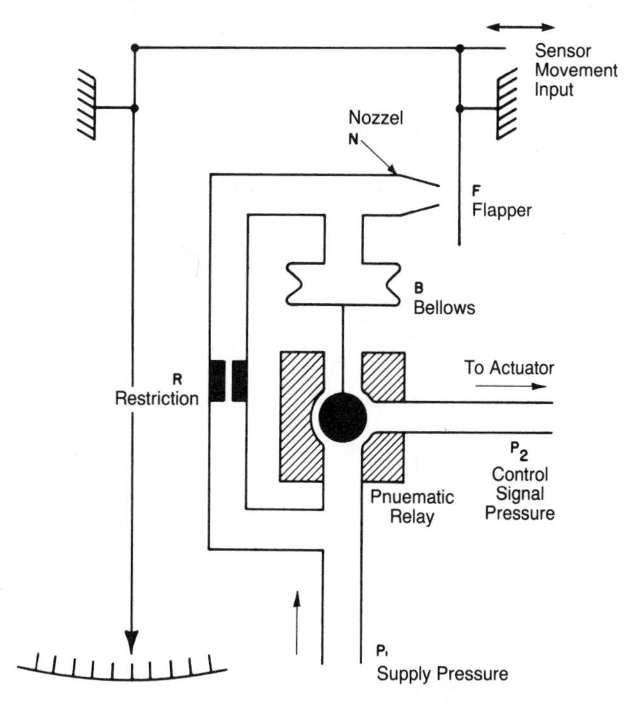

Figure 1-4 Schematic of a pneumatic control system

could perform in flammable conditions of petrochemical plants and other explosively hazardous atmospheres. These controls were designed using bellows, flappers, restrictions, and air-bleeds to provide the various modes of control, Figure 1-4.

Electrical Instrument Controllers

The electrical instrument controllers are based upon signals provided by a change in one of the characteristics of electricity. The application of Ohm's law (involving the relationships of voltage, amperage, and resistance) is fundamental for sensing and making control measurements.

A basic electrical controller is a device that senses a voltage deviation from a constant voltage source. The constant voltage source represents the *set-point* or what value is desired. The difference between the voltage coming from the sensor and that of the set-point is the error signal or correction that needs to be made. There are a number of variations in the physical circuits that can perform this control. Potentiometers balance an unknown or changing voltage to provide a correction. *Wheatstone bridge* circuits evaluate a change in resistance. Transformer circuits are used to control amperage, Figure 1-5.

The output of these control circuits is used to activate relays, solenoids, switches, and motors. These electromechanical devices make a corrective change in the process.

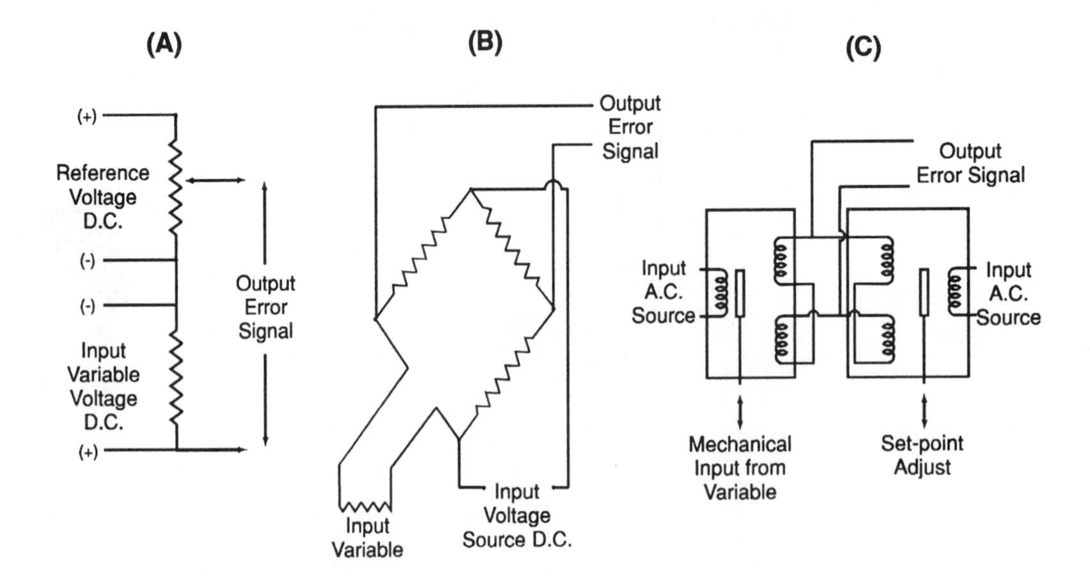

Figure 1-5 (A) Variable potentiometer circuit; (B) Wheatstone bridge circuit; (C) Variable differential transtomer circuit

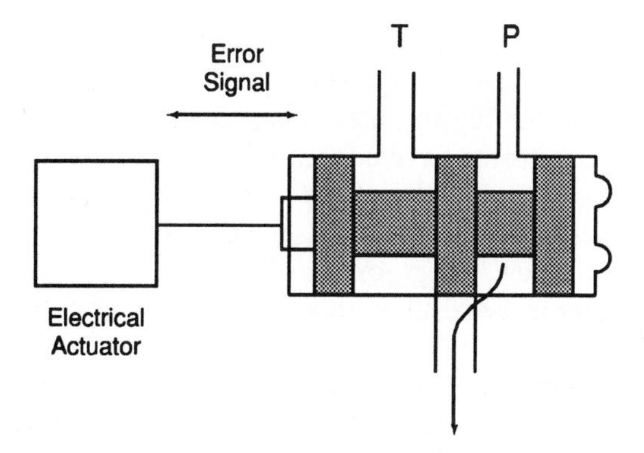

Figure 1- 6 Spool valve

Hydraulic Instrument Controllers

Hydraulic controllers are designed to receive an *error* signal from the variable being controlled, resulting in the movement of the position of a directional flow control valve. The error signal may be a mechanical displacement, a pressure, an electrical voltage, a digital pulse, or an optical or laser beam. The error signal is transmitted to a hydraulic relay (most frequently, a *spool valve*) to supply the energy to move the actuator and correct the manufactured product, Figure 1-6.

Electronic Instrument Controllers

Electronic instrument controllers depend on diodes, transistors, resistors, and capacitors (each appearing as individual components or integrated circuits) to make up circuits. A most valuable component is the operational amplifier. This device can be designed into a large number of circuits and perform many different functions in an electronic circuit. Some applications of operational amplifiers in electronic instrument controllers are voltage comparators, voltage *amplifiers*, and waveform generators. Each classification of operational amplifiers is used in a number of different ways to provide electronic solutions to control problems.

Digital Instrument Controllers - Digital instrument controllers are in reality a continuation of electronic controllers. A special variation of the operational amplifier has been designed to receive an analog or continuous signal and convert it into a digital signal or a pulsing signal representing the values of the analog signal. The new signal is now a digital signal and is in a pulse mode. This device is called an analog-to-digital converter. There is also another device called a digital-to-analog converter that reverses the process described. The advantage of the A/D converter is that a computer can receive and perform various operations on the signal data at very high speeds.

To provide for digital interfacing to a computer, the above electrical and electronic components have been applied to a silicon chip to produce integrated circuits. These integrated circuits have been further condensed resulting in a complete microprocessor being placed on a single silicon chip. With additional circuitry and a few integrated circuit chips mounted onto a circuit board, a microcomputer is the result with a great amount of control power and flexibility to adjust a system.

Digital electronics has become the major growth area in the field of controllers. With the application of large-scale integrated circuits as smart sensors, real time instrument control over all the sensors within the process is under the surveillance of the computer. The processes under control may be in a flexible manufacturing cell, a petroleum cracking plant, or an entire automatic factory.

MANUFACTURING PROCESS CONTROL MEASUREMENT

Process control measurement instruments have operational characteristics that need to be understood. The instruments have responses caused by deviations within the instrument during the measurement of the manufacturing process variables.

Measurement of Variation in Process Instruments

Error can be present in an instrument because of the characteristics of its design or materials used to build the instrument. There can also be an unexpected interaction between different materials used in the construction of the instrument.

Characteristics of Instruments

The characteristics of instruments fall into two major classifications: static and dynamic. The static characteristics are changes that take place when the variable being measured is not in reality changing. These characteristics refer to accuracy, reproducibility, and sensitivity.

Static Characteristics of Instruments - Accuracy in instrumentation is concerned with the actual value of the variable being reported by the instrumentation, being the same as the real or true value of the variable. The accuracy of an instrument is reported by the manufacturer as a plus or minus percentage of the total range of the instrument. If a difference does occur between the actual and the reported value of the measured variable, it is referred to as the static error. This static error is the amount that the controller deviates from the desired value or the set-point.

Sensors and sensor installation can be a contributor to the error. Items such as variations of air temperature surrounding the sensor and its extensions, variations of air pressure, or variations in humidity may introduce instrumentation problems. Sensors and instruments that contain moving parts are subject to friction between these parts. The parts will resist movement (because of inertia, hysteresis, and friction) resulting in error. Metal and plastic instrument parts deteriorate over periods of time, leading to the breakdown of insulation, worn or damaged parts, resulting in unresponsive components. The aging of instruments is a factor in the *calibration* and indicating characteristics of instruments. With the extreme sensitivity of electronic instrumentation, electrical interference can be a serious factor. This problem is referred to as *noise*. Noise is voltage fluctuations superimposed on the input voltage. Noise is capable of triggering high gain instruments unless special provisions are installed to filter out the unwanted and extraneous signals. Noise can be introduced from any area of the instrumentation circuit: sensors, transmission, or controller section.

Another static characteristic of instruments is *reproductibility*. This is the instrument's ability to deliver identical values of the measured variable each time identical conditions occur. Repeated measurements over a period of time of the same value of the variable provide a measure of reproducibitity of an instrument. A slow undesired change in this repeated measurement is referred to as *drift* and can be the result of hysteresis or the result of a dead band within the system.

The *sensitivity* of an instrument is the smallest change in the variable to which the system's instrumentation will respond. Sensitivity is a ratio of a change in the amount of output to the change in the amount of input causing the change. This ratio is measured after the system has stabilized and in a steady-state condition. This ratio is reported in the same units as those being measured.

Dynamic Characteristics of Instruments - There are a number of dynamic characteristics of instrumentation. Most of these characteristics are concerned with how fast an instrumentation system responds, or after its response, how stable the system remains. A few of the dynamic characteristics cluster around the responsiveness and fidelity of an instrument system.

Responsiveness is the ability of an instrument system to follow the change in the value of the measured variable in the production process. If an instrument does not respond to a change in the variable for a period of time, this is referred to as *dead time*. If the controller does not respond following an input change, the inability to completely respond is referred to as controller lag, Figure 1-7.

Fidelity is the characteristic of instruments to correctly indicate precision in following a change in the measured variable. It is the degree of ability of the output signal to follow the changes in the input signal of the sensor.

Cycling is a dynamic error above or below the set-point. It is an oscillation of the system that may occur in some modes of control. In the case where the cycling occurs moderately and continuously, the system is considered in a condition of steady-state cycling. In the instrumentation system the primary concern relates to

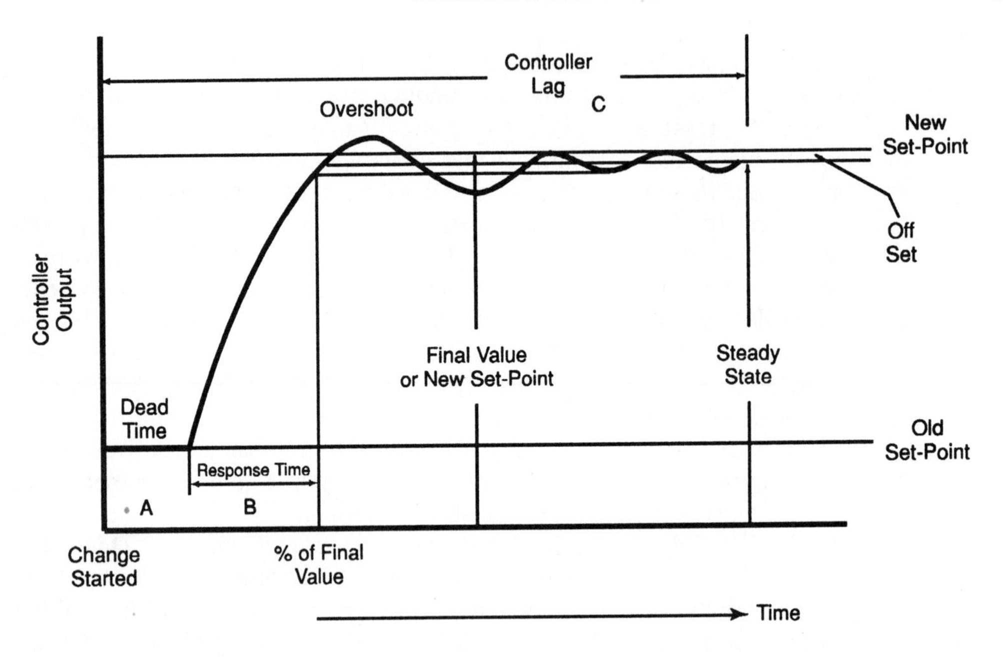

Figure 1-7 Process response characteristics: (A) Dead time; (B) Response time; (C) Controller lag

the initial error, the period of the *cycle*, and the time needed to decay to zero or the new set-point.

Calibration of Instruments

The process of calibrating instruments is carried out by applying an accurate and known input to the instrument under calibration and comparing the measured output to the appropriate standard. The calibration is performed by adjusting the instrument to the specified value and test conditions.

Calibration can also be carried out on electronic or fluid circuits, electronic components, as well as on mechanical devices.

The Control Loop

"Control loop" is a term that is applied to a method of controlling a condition. It may relate to a process, position, procedure, or atmosphere. The open control loop means that a measurement is made of the condition, and it is reported so that monitoring or a manual correction may be made. A closed loop exists when a controller with a set-point is added, so that a decision can be made to determine if the measured condition is above, below, or at the set-point or desired value. The output of the controller's decision is an error signal that results in the correction of the condition until it changes to the set-point value. This action is a control loop.

The Open Loop

The *open loop* is a system that measures a variable in the process and indicates what is happening. The loop does not correct the system but records the results and in some cases may sound an alarm if the system has reached a predetermined value. The open loop system is usually a reporting system, and changes or adjustments are done manually.

The Closed Loop

The *closed loop* measures the variable as in the open system, but it provides a device for evaluating the condition of the process or system. This evaluation takes place by a comparison device that may be mechanical, hydraulic, pneumatic, *fluidic*, electric, electronic, or digital. Whatever system is employed, it will compare the condition that exists with what is desired. The condition that exists is the measured variable and what is desired is the set-point or that which is requested for the condition. The decision or result of the controller is a control signal referred to as an error signal. The error signal is amplified and converted by a number of different amplifiers (depending on which system is employed) into enough energy to power an actuator. The actuator could be a control valve, motor, solenoid, hydraulic *servo valve*, stepper motor, or a number of different actuators

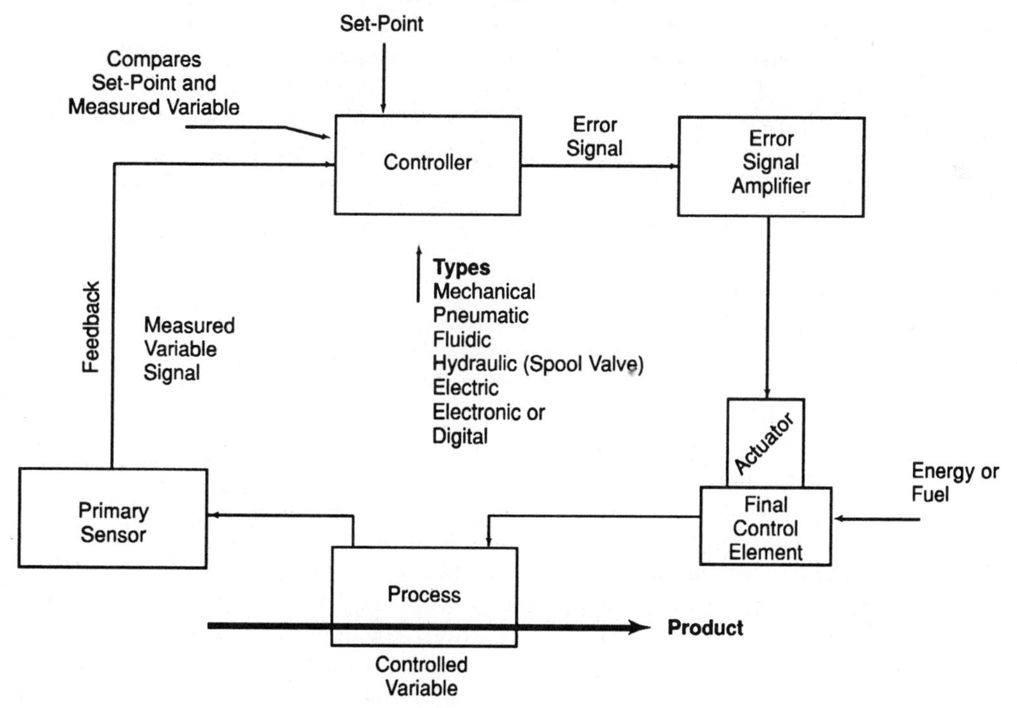

Figure 1-8 A closed loop control system

providing the change necessary to align the system with the set-point or what is desired, Figure 1-8.

Process Response Characteristics

Response is the reaction of the system under control to the changes introduced by an error signal related to time. Manufacturing process systems vary as to the amount of resistance or capacity they contain. These factors control the time the systems require to respond to controllers' changes. The systems' factors of resistance, capacity, and time result in a series of characteristics and generate a process reaction curve for the system.

Process response characteristics include resistance, capacity, dead time, transfer lag, gain, *overshoot*, *offset*, and others. These characteristics interact to produce a typical response curve to a step increase of a system. The process reaction rate is the time it takes for the system to achieve its new steady-state after a step change has been introduced.

Modes of Process Control

Different process requirements and the types of controllers applied to the processes make it necessary to select different modes of process controllers. The major types of control modes are these: on/off, proportional, proportional with integral, proportional with integral and derivative. These modes of control are found in pneumatic, electronic, and digital types of controllers.

The controllers deliver the signals to the actuators that provide the energy to move the final control element of the system.

Actuators and Final Control Elements

The actuator is a machine that takes a control signal and converts it into a mechanical movement. Actuators are designed using most of the energy conversion applications of mechanical, pneumatic, electrical, hydraulic, or digital technologies. These machines follow the signals of the controller and move the final control element.

The *final control element* is the unit that is in contact with the product being controlled. Often the final control element is a control valve that alters the flow of a fuel to some type of manufacturing process or controls the flow of product in a fluid measuring and mixing operation. The actuator and the final control element are often combined into one unit called a control valve, stepping motor, pneumatic or hydraulic motor, diaphragm, or *cylinder*. A servo valve is also an actuator but it has a unique feedback function.

Instrumentation Applied to Material Handling

Engineers have seen that production can be rapidly improved by increasing the efficiency of material handling. Often materials have been on the floor too long waiting processing or lost in a corner when needed. Organized material handling greatly improves the efficiency of production.

Automated Material Handling

With the application of the computer to receiving, storing, processing, assembling, and shipping, it was logical to use the computer for inventory and control of materials and products within the production facility. Automated material-handling systems are designed to inventory and transport bulk materials, piece parts, subassemblies, and finished products within the plant so that manufacturing can proceed at a rational pace with efficiency.

Instrumentation Applied to Machinery

Instrumentation that is applied in mass production involves the use of many mechanisms. The application of linkage systems, rotary motion converters, and other *kinematic* devices are a part of mechanical instrumentation.

Mechanical Instrumentation

Kinematic linkage systems are motion conversion devices. They are employed in the transmitting of linear and rotary data to many sensors and therefore become a part of the instrument system. Among the linear devices are a number of lever systems that produce straight line movements, ratio movements, as well as balance beam movements — all as parts of instrument systems. Rotary motion converters transform mechanical movements or positions into data that can be stored, integrated, or calculated to provide an instant analog readout. The cam, screw, and ratchet are common data acquisition units for other analog sensor and actuator systems, Figure 1-9.

Instrumentation in Transfer Lines

Mass production transfer lines employing pallets, jigs, and/or fixtures have the functions of loading and counting the parts entering and within the transfer machines. The parts under manufacture within the transfer machine are sequenced as to the required operation. The machine units are controlled to maintain these operational counts and sequences.

The workpieces are inspected as the parts are being manufactured in the transfer machine. This instrumentation measures work dimensions, tool wear, or damaged or broken tools remaining within the machine or workpieces.

Figure 1-9 Kinematic linkage and rotary motion converter

Strategies of Production Automation

Early automation was applied to production by using hard tooling methods of manufacture. With the advent of instrumentation, numerical control, and the computer, unlimited new opportunities of production have been and are rapidly emerging.

Flexible Manufacturing

Flexible manufacturing systems (FMS) have introduced new methods of applying mass production techniques to batch manufacturing. The key to this methodology is — with the aid of the computer controlled production tool — the capability of being reprogrammed so that parts with different configurations can be processed with the same machines.

Production with Reprogrammable Tools - Computer-numerically-controlled (CNC) machining centers with their supporting equipment (such as automated warehouses, conveyors, and robots) are capable of being readjusted to accommodate the batch product being manufactured. This manufacturing tool flexibility increases the production efficiency of the manufacturing system because of better utilization of the system's tools.

Modular tools and assembly machines are designed so that they can be assembled and reassembled in different configurations, Figure 1-10. This innovation adds flexibility to manufacturing if model or other changes require product modification.

Part feeders, conveyors, and robots and material-handling equipment can be controlled by a minicomputer. These systems are capable of being reprogrammed just as other units in the manufacturing environment, providing a new aspect of flexibility in production.

Manufacturing Cells - Manufacturing cells are made up of a group of processing modules needed to process a family of parts. The machine centers or assembly centers employing high production with the aid of computer numerical control (CNC) provide flexibility. A major part of the high productivity is achieved with the application of robots. Robots feed the machines of the machine centers that

Figure 1-10 A computer controlled manufacturing system *(Courtesy of K. T. Swasey)*

are built in configurations best adapted to the requirements of the workpiece.

The control of the robots is performed at a number of levels of technology depending on the demand of the machines that are interfaced. The robots are energized by electrical, pneumatic, or hydraulic power. Their control for positioning depends upon encoders, tachometers, servo devices, or potentiometer feedback equipment frequently controlled by the available computer. The cell is designed so that raw industrial materials, such as castings or bar stock, enter one end of the cell and finished parts leave the machines.

The cells are the core manufacturing units that are often interfaced with other cells with automated part transferring and assembly equipment to represent a flexible manufacturing system. This application of manufacturing is helping to bring batch-manufactured products economically back into a competitive world market, Figure 1-11.

Figure 1-11 Cellular systems, FMS large cells, and Modular automation *(Courtesy of Kingsbury Machine Tool Corporation)*

The Industrial Robot

The robot is primarily used as a manipulator to handle material or tools. With this application it has been programmed to load, position, and unload workpieces for a number of production machine centers. The robot may be stationary, or servicing machines in a circle. A line of workpieces and fixtures may move past the robot as in the case of spot welding auto bodies. In other cases, the robot moves along a track to provide work on a line. It may apply a tool, such as in drilling holes in large sheet metal sections. Robots also ride on gantry cranes, which allows work on surfaces and locations above the workpiece. Robots are designed to work with almost any production configuration required.

Robot End Effectors or Grippers

The part of the robot that is in actual contact with the workpiece or the tool is the gripper. The gripper holds the work during transportation or work. The gripper or end effector — when fitted with a tool such as a drill, spot welder, spray gun, or other device — provides a complete new category of work open to the robotics.

Robot Controls

The control of robots is carried out using a number of different technologies. A very important requirement for the robot is that of position. Position and location can be delivered to a robot by pneumatic, hydraulic, or electrical technologies. The feedback information crucial to position is provided by electrical or electronic applications. Instrumentation sensors employ encoders, tachometers, resolvers, synchros, and inductosyns to deliver position information. In addition, servos supply feedback for pneumatic and hydraulic systems. These instruments commonly interact with a minicomputer to supply programs and to sequence the system.

Power Supply

Energy for the manufacturing system is received from the local electrical company and is considered as unregulated power. This power is available in a number of different voltages but all at the same frequency. This power is used for large energy devices such as motors and common electrical applications. For control applications and instrumentation utilization, this unregulated power is converted into regulated power. This power may be direct current, special frequencies, or a very stable source of regulated power. These power applications are necessary for instrumentation and computer use.

Electrical energy is also converted into pneumatic energy for instrumentation and operational devices by compressing and carefully regulating air. Pumps deliver hydraulic oil at high pressure for the hydraulic control and actuator systems. All these secondary power supply systems are carefully controlled because any variation in the source affects the accuracy of the instrumentation.

Robotic Application in Automation

Flexibility and adaptability are important concepts in batch manufacturing. The ability to reprogram manufacturing machines and equipment has provided flexibility for quality control. To provide control over the products, the quality control also required automation. The inclusion of the coordinated measuring machine has supplied a way of having a 100 percent inspection and feedback through a computer to the manufacturing cell. The coordinated measuring machine is programmed and becomes an automated robot that measures and supplies control data over production, Figure 1-12.

Figure 1-12 Measuring robot for in-line coordinate inspection *(Courtesy of Digital Equipment Automation Inc.)*

A Strategy of Automation with Computers

The application of the computer to instrumentation and automation for manufacturing has provided a new rejuvenation for production. This rejuvenation has resulted in a flexibility unknown previously. The flexibility is provided by the opportunity of being able to reprogram for new products as well as to integrate the operations of many machines and processes within a total factory.

Computer Integrated Manufacturing

Computer integrated manufacturing (CIM) includes the interrelating of a data base with all the functions of manufacturing. It is a full spectrum approach. The system in the initial stages - finance, market research, product design, product development, engineering, production engineering, production machine control, and inspection - interacts through a system data base. Other services within manufacturing including material handling, storage, inventory control, and shipping are all part of the integration.

Some of the objectives of CIM are to increase productivity by increasing the amount of actual machine work time. Efficiency is increased by having material and tools at the proper place just in time for work. Costs are reduced by having only the inventory on hand that is needed for that batch in manufacture. Accurate knowledge is obtained of where manufacturing tools (including their tooling data) and special materials are stored.

Instrumentation

The computer is a very important part of the instrumentation system. In a CIM application it has taken over the responsibilities of the older controllers and recorders and has become the new center of process and manufacturing control. To provide this data base for CIM control, instrumentation is the essential component. A great variety of sensors are utilized to count, measure, and verify the many variables for the computer and are the bases of control.

Because of the vast amount of instrument and machine collected data, design data, engineering and management data, a problem is presented for the organization and retrieval of the data, Figure 1-13. This problem has resulted in a new discipline — data base management — a part of the software industry.

The integration of many sensors and the ability to select different options for reprogramming of machines, processes, and material-handling systems provide the flexibility for optimizing the production strategy.

The Automated Factory

The application of CIM to production facilities brings the concept of the automated factory close to a reality. The engineering workstation is a computer

Figure 1-13 An engineering workstation *(Courtesy of Hewlett-Packard)*

workstation where thinking, planning, and execution of computer programming systems are created. The thinking and data generated in the application of computer aided design, computer aided manufacturing, and computer aided quality assurance and management data, provide a powerful data base that is an available resource for supervision and management as well as direction at the floor level of production.

A computer controlled tool center provides management control over cutting tools, jigs and fixtures, machines and hand tools, storage and retrieval, inventory, condition of the tools, tool setting for size and length programs, purchase, and alternate applications.

In earlier times, most tool cribs were in a condition of disorder and cost inefficiency. A computer managed system greatly improves the location and utilization of tools employed within the manufacturing system.

In the automated factory, management reports are available from any level of the facility. These provide information for rational decisions where the information did not exist before. This information supplies management with more control over the factory. The expectation is that with this knowledge, productivity will increase and a successful competitive enterprise will result.

SUMMARY/FACTS

- Measurement is a key factor in industrial processes and manufacturing control.
- Many old and newly invented sensors are applied to electronic instruments to provide multiple functions.
- The role of the sensor is to supply reliable information about a process variable.
- Gage blocks are a system of eighty-one blocks that can be used to measure dimensions as small as a few millionths of an inch.
- Large-scale integrated circuit technology made microprocessors and computers more affordable.
- The computer is very adept in the performance of arithmetic operations used in solving algebraic formulae.
- Flexible manufacturing makes it possible to manufacture a family of parts of similar configurations.
- The accuracy of an instrument is concerned with the actual or true value of the variable being measured by the instrument.
- Kinematic linkage systems perform motion conversions.
- The key to flexible manufacturing is the capability of the tools to be reprogrammed.
- For computer integrated manufacturing to be truly successful, a holistic approach to the total factory integration must be planned and carried out.

REVIEW QUESTIONS

1. What is the chief application of instrumentation?
2. Why is quick and accurate measurement critical in the manufacturing process?
3. In what way does real time correction in production processes increase production?
4. Why can we say that the concept of technology is very old?
5. What were the means of measurement in ancient times?
6. Name some early comparison measuring units.

7. How does accuracy and precision differ?
8. What particular aspect of manufacturing has increased the need for measurement?
9. What measuring tools were employed to fit early precision parts?
10. How does a screw thread provide measurement?
11. What are Johansson gage blocks?
12. List some process variables that are measured.
13. How were early transfer machines controlled?
14. Why was the emergence of numerical control important to manufacturing?
15. How did electronic and transistor technology change industrial controls?
16. Why has the application of the computer been so helpful in complex part shape production?
17. Why is the concept of flexible manufacturing so interesting to producers?
18. What is a manufacturing cell?
19. What is computer integrated manufacturing?
20. What was the primary purpose for the development and application of sensors?
21. Early mechanical devices were control systems applications of what?
22. How did early electrical instrument controllers receive the variable's signal?
23. What is one of the very common functions of an electronic instrument controller?
24. In what form is the signal of a digital controller received?
25. What are the two major classifications of instrument characterstics?
26. How is calibration performed?
27. What is a control loop?
28. What are the various modes of control?
29. What does an actuator do?
30. Name some final control elements.
31. What are activities associated with automated material handling?
32. What does a kinematic linkage system usually perform?
33. What are some procedures employed within flexible manufacturing?
34. What is a manufacturing cell?
35. How has the coordinated measuring machine evolved in the application of manufacturing automation?
36. What is necessary for computer integrated manufacturing?

2

ELECTRONICS FUNDAMENTALS

OBJECTIVES

Upon completing this chapter, you will learn the following:

- The fundamentals of electronic principles and components function
- The interpretation and application of the measurement of electrical values
- An application of semiconductor principles
- The understanding of semiconductor logic
- The application of digital technology
- The characteristics and application of operational amplifiers
- The interpretation and application of integrated circuits
- The use of the microprocessor
- A description of the microcomputer

INTRODUCTION

The manufacturing supervisor will find a basic knowledge of electronics and its application valuable assets in applying instrumentation to manufacturing. Electronics is utilized because of its speed, flexibility, reliability, and small space requirements that provide a highly successful controlling system.

INSTRUMENTATION ELECTRONICS FUNDAMENTALS

Electronics fundamentals are included in this text because the newer industrial control systems employ electronics as the technology in manufacturing to measure and control. To understand this technology, some electronics fundamentals are presented that provide a few building blocks that lead to the understanding of the computer. Some elements such as resistance, voltage, amperage, and capacitance are basic to construct circuits and provide a base for other later devices. Circuits are used to apply the elements, diodes, and transistors to build operational amplifiers. In turn all the earlier devices are used to build integrated circuits. Integrated circuits are of many types and are used to build microprocessors and smart sensors. With a microprocessor and the addition of memory circuits and in/out devices, a microcomputer is completed. The sensor, microprocessor, and the microcomputer are the powerful tools of instrumentation.

Physical Properties in Electronics

Electronics is based on three concepts of energy: voltage, current flow, and resistance. These three become the building blocks from which develop the concepts and devices of electronics. The concepts may be analogous to the phenomena occurring to the water in a fire hose while a fireman is fighting a fire.

The fire system must have a pump or reservoir to supply water pressure to the hose. This pressure can be thought of in electronics as voltage. The fire hose will have a difference in pressure from the pump end of the hose to the nozzle end of the hose. This difference in pressure in the hose will cause a rate of flow of water through the hose. This current or rate of flow in electronics is called amperage. The interior of the hose has roughness, size variations, irregularities, and length, all of which cause more turbulence or resistance to the rate of flow. This slowing of the flow or current in electronics is called resistance. These physical properties of hydraulics provide a conceptual model for the understanding of invisible properties of basic electrical principles, Figure 2-1.

Electronic Basics

The measurement of electrical values is an essential requirement of an instrumentation system. The primary measurements are for voltage, amperage, and resistance. Many instruments are based on the measurements of these electrical variables.

Voltage Concepts

Potential difference may be described as an electrical "pressure" that is capable of causing a flow of current. A unit of one volt is the difference in the voltage

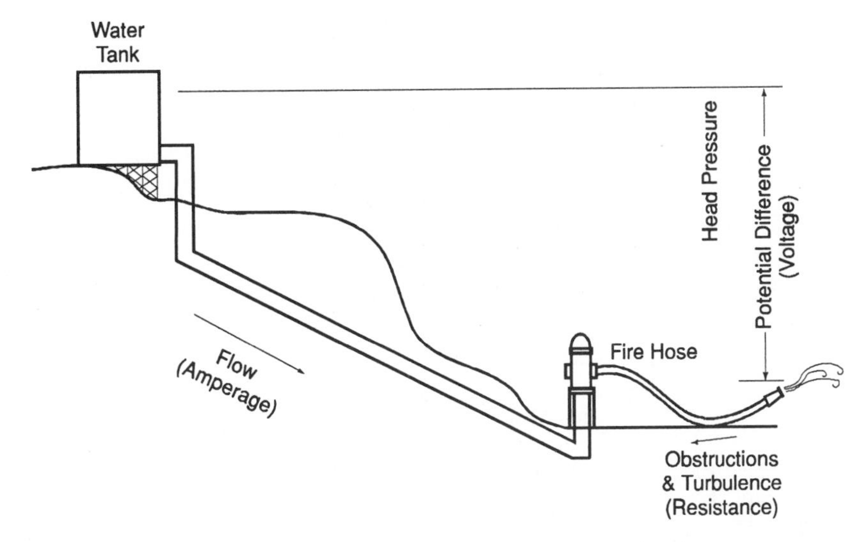

Figure 2-1 (A) Hydraulic electronic analogy of voltage; (B) amperage and resistance.

between two points of a conducting wire carrying a constant current of one ampere through a resistance of one ohm.

A voltage results from this potential difference. It is the measurement of the amount of difference in polarity between two points, as between the connections of a battery or two different points in an electrical circuit. The term "electromotive force" is also used to express the same concept, i.e., that a potential difference provides a force that causes the current to flow in a circuit, Figure 2-2. The more common name for potential difference is voltage, thus the name given to its measuring instrument is the voltmeter.

The electrical units used in manufacturing are amperes, volts, and kilovolts — which are large units. In instrumentation and electronic controls, the units are small — volts, millivolts, milliamps, and microamps — applied to electronic circuitry.

Amperage Concepts

Amperage is the rate of flow of electrical current. This flow is also considered as electrical current. It is the flow of current that provides the electrical energy that is capable of doing work. Our early electrical experimenter, Mr. Benjamin Franklin, expounded an electrical theory based on the concept of an invisible fluid: a sufficient amount of fluid represented a positive charge and a lesser amount of fluid represented a negative charge. His concept of flow was that the greater amount of fluid would flow to the lesser amount, therefore yielding a flow that was from positive to negative. This reasoning lead to the establishment of current flow

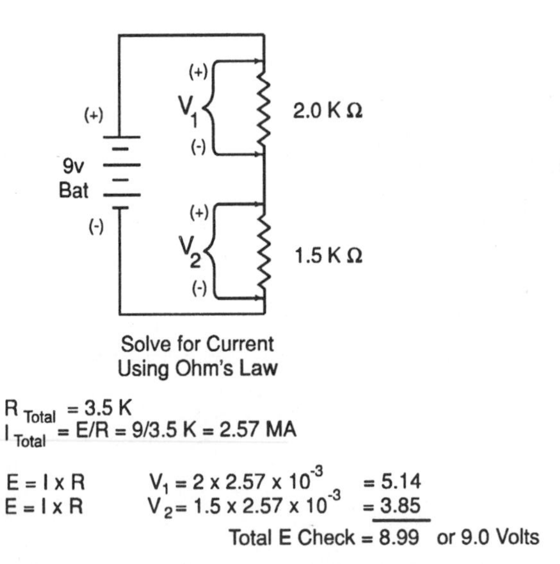

Figure 2-2 Circuit diagrams involving voltages and current

as being from positive to negative. This direction of current flow is referred to as conventional current theory.

The second concept of the flow of electrical current is based upon the discovery of the electron in association with atomic structure. The theory yields a nucleus having a positive charge and the electron having a negative charge. The electrons are arranged in orbits around the nucleus with any weakly attracted electrons on the exterior of the system. This weakly attracted electron or free electron may be attracted to another atom and travel from atom to atom. With the movement of vast amounts of electrons a current is established. This theory is known as the electron flow theory; the current flows from negative to positive polarity.

Today both theories of current flow are used but are not usually specified on drawings. The polarity of the circuitry is given and the proper theory applied to the device under study.

Electrical Resistance

Electrical resistance is one of the fundamental concepts that underlies the operation of all electrical and electronic devices. This is important because electrical resistance determines the amount of current that can flow through an electrical circuit. The physical and chemical properties of materials determine their resistance. Resistance is a property of material that is a function of its resistivity (rho). Resistivity is a proportional constant derived by comparing a cube of the material under study to a like cube of annealed copper at the same temperature, thus yielding a resistive constant for the material under study. Materials with a high value of rho are insulators, while materials with a low value of rho are excellent conductors. Materials between these values are semiconductors.

These materials with very high resistance restrict the movement of electrons and are the insulators. Examples of such materials are wood, paper, wax, glass, air, porcelain, hard rubber, epoxies, some plastics, etc.

The property of resistance opposes the flow of current through electronic devices. This property is the governing factor that controls the magnitude of the electric current that flows through a conductor. This restrictive phenomenon is referred to as electrical resistance. Resistance changes in a conductor as the length of conductor, cross-sectional area of the conductor, or its temperature varies.

Resistant Wire - Resistors are applied to limit the flow of electrical current and control how much current will flow in a circuit. Resistors are fundamental devices for the construction of electronic circuits.

In instrumentation circuitry, a wide *range* of resistor values are employed. They vary from less than 1 (ohm) to 1,000 (kilohm) to more than 1,000,000 (megohm). Resistors are affected by temperature. Devices made of metals will generally increase in resistance as the temperature is increased. However, other materials, such as carbon and other nonmetallic substances, will decrease in resistance as the temperature increases. This phenomenon is referred to as an inverse temperature relationship or negative temperature coefficient. It offers a lower resistance at a higher temperature.

Electrical wire has resistance that is a function of the wire's diameter, length, and its material. Conductor wire is available in standard sizes, referred to as wire gauges. The American Wire Gauge is most frequently used and is designated for a solid, round copper wire. Sizes range from 0000 gauge (0.4600 in) to 40 gauge (0.003145 in). A common size used in electronics is 24 gauge (0.0201 in). As the length of a wire is increased or the wire diameter reduced in size, the resistance of the total wire is increased. If a smaller gauge number of wire is used (the diameter of the actual wire increased), or the wire shortened, the resistance of the wire is reduced. In electrical applications, the gauge of the wire is very important. The safe electrical current carrying capacity is specified for each gauge of wire.

Resistors - Resistors are constructed in two forms, a fixed resistor that has a set amount of resistance and a variable resistor that has adjustable resistance throughout the range of the device. The fixed resistor is manufactured with three characteristics: the value of resistance, the power rating, and the material that is used in its manufacture. The resistance value is indicated in ohms with a specified tolerance. The power rating — the unit of power (P=EI) — is indicated in watts of output at which the component will operate without a permanent change of value caused by power dissipation (heat). The characteristic is controlled by the materials and manufacturing techniques used.

The carbon composition resistor is a commonly employed resistor in electronics and manufactured in many sizes. The larger sized resistors provide higher power ratings and are larger than the lower wattage resistors. Resistors that require high precision are manufactured from metal or metal oxide films that are deposited on insulator materials. For heavy power applications, wire is wound on a ceramic form that will dissipate the heat generated by the power passing through the circuit, Figure 2-3.

The identification of the value of fixed resistors is determined by a code. The code is marked on the resistors with a number of colored bands. These bands are of different colors to represent the value of the resistor. Starting with the band closest to the end, the first two band colors represent the value of the resistance. The third band represents the number of zeros by which to multiply the digits or to multiply in the power of ten. The fourth band represents the tolerance of the resistor or error that can be expected from the resistor's true value. A gold band in the fourth position reports a plus or minus 5 percent resistor; a silver band reports a plus or minus 10 percent resistor; and no band reports a plus or minus 20 percent resistor.

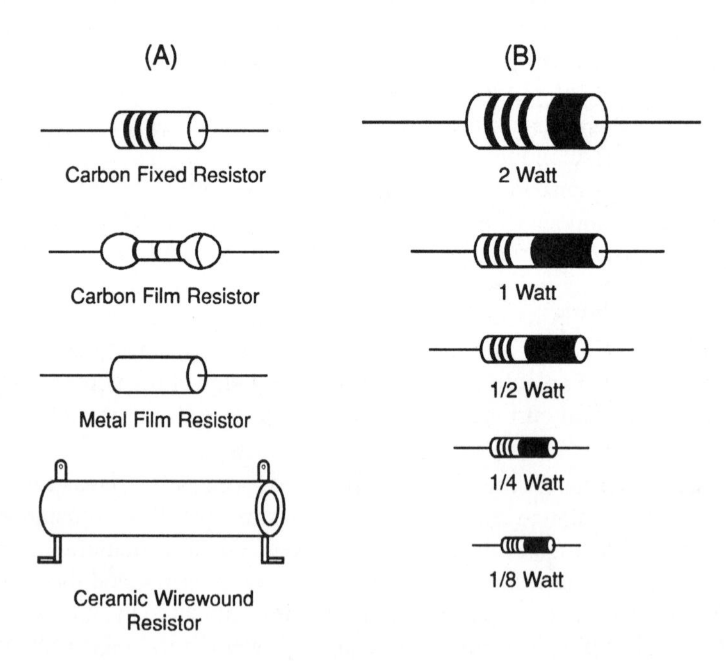

Figure 2-3 (A) Types of fixed resistors; (B) Carbon resistors of different wattages

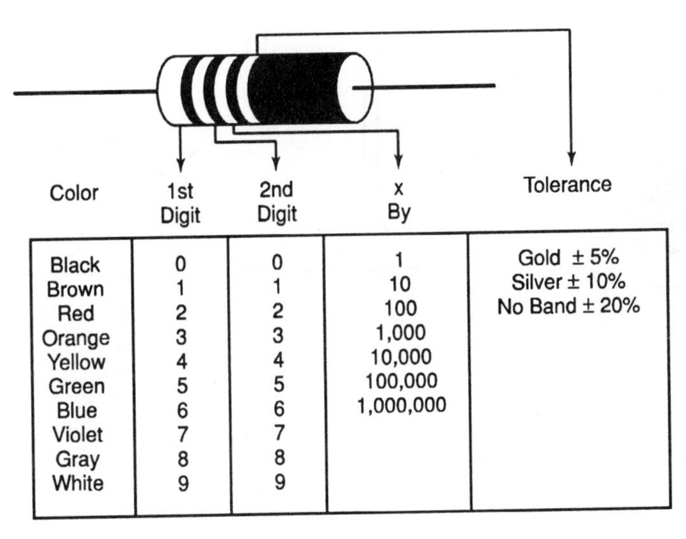

Color	1st Digit	2nd Digit	x By	Tolerance
Black	0	0	1	Gold ± 5%
Brown	1	1	10	Silver ± 10%
Red	2	2	100	No Band ± 20%
Orange	3	3	1,000	
Yellow	4	4	10,000	
Green	5	5	100,000	
Blue	6	6	1,000,000	
Violet	7	7		
Gray	8	8		
White	9	9		

Table 2-1 Color Code for Carbon Resistor Values

An example of carbon resistor value reading:

First Band	Second Band	Third Band	Fourth Band
Yellow 4	Blue 6	Orange 1,000	Silver 10%

The resistance of this carbon resistor is 46,000 ohms, with a 10 percent tolerance.

Other types of resistors may be constructed from materials such as ceramics, wire, and conductive plastics. These are generally marked with the actual value or a coded representation of the value.

Variable Resistors - Variable resistors are also called "potentiometers." They are used to trim or to control the current flowing through an electronic circuit. Being able to vary the resistance in a critical part of a circuit allows it to be calibrated. In some applications, adjustment of the system's set-point is made. This type of resistor is constructed so that it receives its movement from the rotation of a shaft moving a contactor along a cylindrical resistance. The contactor or wiper moves smoothly over the resistor; the smoother the change of resistance the better the resolution of the potentiometer. The rotation of the shaft is about three hundred degrees, and in this rotation the resistance varies from near zero resistance to a position at the other end to the maximum value usually marked on the back of the potentiometer. When the potentiometer wiper is in the center position, 50 percent of the value of the potentiometer should be added to the circuit.

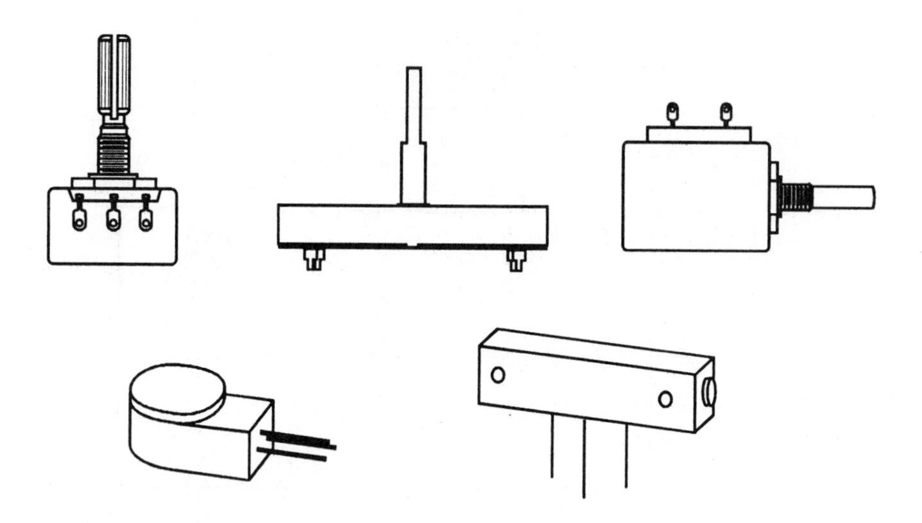

Figure 2-4 Various types of variable resistors (potentiometers)

The coil of the resistor may be wound so that the resistor may have different changes or resistance proportional to the amount of shaft rotation. The spacing of the coil's winding determines whether it results in a linear or a logarithmic change of resistance in the potentiometer, Figure 2-4.

In instrumentation for data acquisition of closed loop operations, a sliding or linear potentiometer is employed. The linear variable resistance is constructed with a wiper moving along the axis of the coil rather than on the rotary action of a shaft. Because many of the analog instrumentation systems provide a straight-line movement, it is easy to sense these with a linear potentiometer.

In the event that a very precise variable resistance device is required, a multiturn potentiometer is utilized. This variable resistance component requires ten turns of the shaft to move the wiper from the low value of the potentiometer to the high value. It allows very high resolution and precise resistance adjustments. These multiturn potentiometers may be constructed of carbon, ceramics, plastics, or wire-wound resistance components.

Conductors - Conductors are materials that permit the free flow or movement of electrons through the substance. They have a low resistance and transport electrical energy with a low loss. When losses occur, they appear as heat in the conductor. Typical conductors are gold, silver, copper, aluminum, iron, and lead. In instrumentation, the conductors may take on many shapes and forms: wire, ribbons, rivets, bolts, bands, foils, shaped forms, circuit boards, integrated circuits, and others. Material of any shape or form that will allow a current to flow through it is an electrical conductor.

Conductivity is the reciprocal of resistance. Not all conductors contain the same resistance value per unit of length. In some materials (like nichrome, used in heating units), the wire has a higher resistance and when a current is passed through the unit, heat is generated. Because of the resistance in the conductor a voltage drop results between the voltage applied at one end of the conductor and that which is measured at the other end. In many cases this is a power loss in the conductor, an I^2R drop. In extreme cases the heat generated will be enough to destroy the conductor. The purpose of the design of the conductor and equipment is to supply the electrical energy to the next component with as little energy loss as possible, thus the use of the metals listed above.

Insulators - *Insulators* are materials that have very few free electrons available to carry an electrical current. Insulating materials such as air, porcelain, rubber, glass, mica, and some plastics are used to protect conductive devices from shorting to ground. These materials are used on power lines, electronic circuit boards, and integrated circuits with the objective of keeping the current moving through the designed circuits.

Insulators do have a breakdown voltage at which point electrons are freed from the atomic structure. This allows the passage of a current. This is why each insulating product has a voltage specification that should not be exceeded. When the insulator is used with the proper care and at or under the specified voltage, the insulator will provide excellent and reliable service.

Material	Breakdown Voltage (kV/cm)
Air	30
Porcelain	70
Rubber	270
Glass	1200
Mica	2000

Table 2-2 Examples of Breakdown Voltage of Materials

Electrical Capacitors

A capacitor is an electrical component that is made of conductive surfaces separated by a dielectric material. The surfaces take on an electrical charge that is directly proportional to the voltage applied. If a capacitor is attached to a direct current power source, the positive attached plates take on a positive charge (+Q) and the negatively connected plates receive a negative charge (-Q). The charge builds until it reaches the charge equal to the source.

Electrical Charge Storage - When a capacitor is charged and the connections from the power source are removed, the charge is stored in the capacitor. If the capacitor is reconnected to an electrical system, the stored energy is released back to the system on demand.

If the voltage is doubled, the charge in the capacitor also doubles. This proportionality of storage is the device's capacitance at that voltage: the larger the surface area in the capacitor, the greater the charge that can be stored in the capacitor per volt, Figure 2-5.

Time Constants - In addition to storage, capacitors can be attached to other components and provide a function of timing. When a capacitor is connected in a series with resistors, an electronic time delay is created in the circuit. The size of the resistor establishes the time it will take to fully charge the capacitor. This delay is referred to as a "time constant." In one time constant the capacitor will receive 63.2 percent of its full charge; after five time constants the capacitor will have received 99.3 percent of a full charge. This charging of the capacitor is the basis of an RC circuit that results in a time delay that provides timing in an electronic circuit. This is useful for timing control for electronic components in electronic circuits, Figure 2-6.

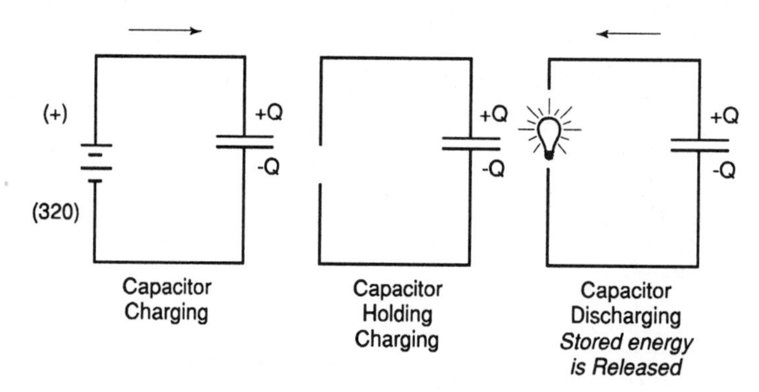

Figure 2-5 Capacitor operation: The charge is proportional to the voltage applied.

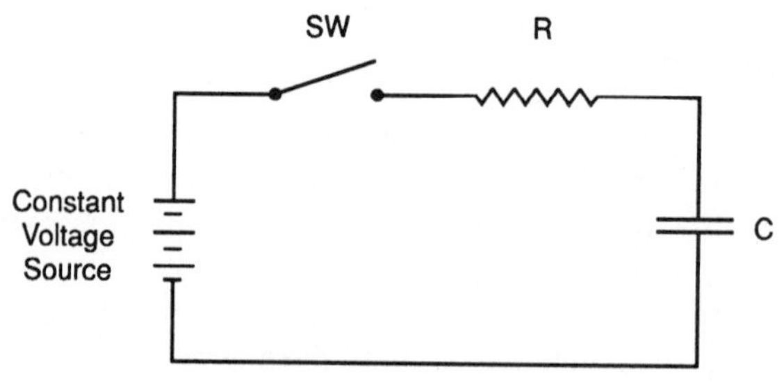

Figure 2-6 Diagram of a time constant circuit

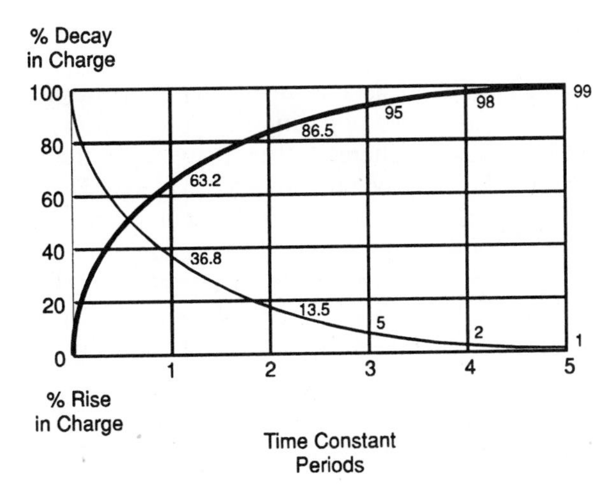

Figure 2-7 Diagram of the charge and discharge curves

When a charged capacitor is discharged through a resistor, the capacitor's charge will result in an exponential discharge curve that is simply an inversion of the original charging curve, Figure 2-7

Electrical Inductance

Inductance is the effect of either opposing or restraining the flow of current, or opposing a change in the flow of current. This property of energy tends to maintain the flow of energy even after the source has been shut off. The inductor usually consists of a coil of wire that, when energized, possesses electromagnetic properties.

Electromagnetic Inductance - Inductance is that property of a circuit which opposes a change in current in the circuit. It may also be described as the property of a circuit where energy is stored in a magnetic field. When a magnetic field is established around an energized coil, a counter voltage and current is produced in the circuit if the energy source is removed. This counter voltage and current is the result of the collapse of the magnetic field that was established around the wire coil. The induced voltage that results is a current flow that acts to oppose any change in the original voltage. This counter flow is *impedance*. Impedance reduces the original flow in alternating currents. The electromagnetic property of the coil characterizes its alternating current output behavior, which is a *frequency* dependent device. Because these inductive components are large and expensive, they are not commonly employed in instrumentation devices. They have been largely replaced by an operational amplifier circuit called a "gyrator" that simulates an inductor at below radio frequencies.

Electrical Impedance

Impedance functions as resistance in an alternating or pulsating direct current circuit. Impedance provides opposition to current flow; it limits the flow of current. The impedance is altered by the frequency of the energy being applied, whether it is an AC voltage or a plusating DC voltage. The higher the frequency, the greater the impedance. In alternating current circuits, the frequency can also be used to control current flow. Because impedance resists the flow of current, devices with a high impedance (such as an operational amplifier) provide a very accurate and reliable buffer in sensing measurements. Thus, because of the high impedance of the device, it does not draw current from the circuit being measured.

Measurement of Electrical Values

The measurement of electrical value is an essential requirement of an instrumentation system. Most electrical instruments are based upon the measurements of the electrical variables of voltage, amperage, and resistance.

A voltage measurement results from a potential difference. It is the measurement of the amount of difference in polarity between two points, as between the connections of a battery or two different points in an electrical circuit. The term "electromotive force" is also used to express the same concept — a potential difference provides a force that causes the electrons to move within a circuit. The more common name for this energy is voltage. Thus the name given to the measuring device is the voltmeter.

Electronic Digital Multimeters

To overcome the energy consumption during measurement of the passive-type voltmeters, the electronic voltmeter was developed. The D'Arsonval type of voltmeter draws the energy from the circuit under test to activate the meter movement. This does have some reduction of the system's accuracy. Very sensitive and precise meters with higher internal resistances are fragile and are not suited to manufacturing level testing and troubleshooting applications.

The electronic digital multimeter utilizes an external power supply that greatly reduces the energy drawn from the circuit under test. The electronic voltmeter also incorporates amplifier circuits, which greatly increases the sensitivity of the measurements. Voltages are measured with a sensitivity as low as one millivolt and resistance values of ten megohms or greater at full scale deflection.

Digitally designed electronic multirange multimeters incorporate integrated circuits to produce instruments with accuracies of 0.05 percent. These instruments employ a power source, an amplifier, an analog-to-digital converter, and a digital display. In addition they may contain automatic ranging circuits to select the range that displays the greatest accuracy automatically and also includes an overload protection function. One of the reasons for the digital readout multimeters being so successful is that they are easy to read without an error. They allow discrimination to the smallest digit. These instruments are available in a competitive price range, Figures 2-8 and 2-9.

These measuring instruments are portable and have become very popular for testing and circuit troubleshooting and calibration. They are not appropriate for continuous measurements found in an automated control system.

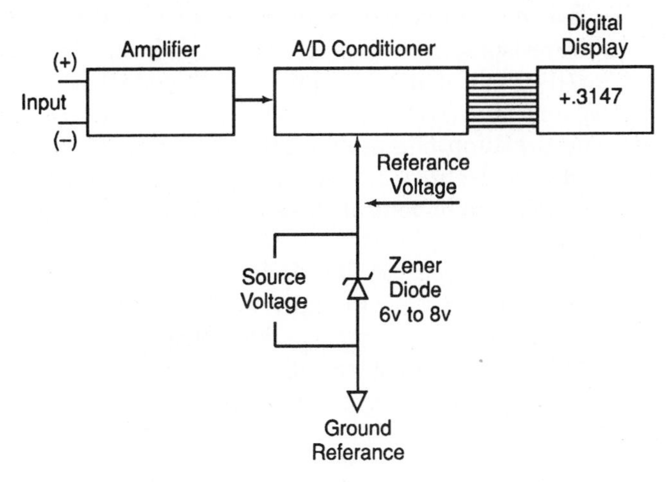

Figure 2-8 A schematic of a digital display of a multimeter circuit. A comparison is made between the input voltage and a stable reference voltage.

Figure 2-9 An electronic digital multimeter *(Courtesy of Simpson Electric Company)*

Relationship Between Resistance, Amperage, and Voltage

To measure an electrical circuit with a constant voltage, the resistance in the circuit determines the current flow or amperage in the circuit. This provides an opportunity for controlling the amperage in a circuit and to establish a relationship with the voltage. The mathematical relationship is that voltage (E) is equal to the amperage times the resistance, (E = I x R) — Ohm's law. The flow of current through a resistance is directly proportional to the voltage and inversely proportional to the resistance. To illustrate, if the voltage in the circuit were increased by five, the current would increase by five if the resistance remained the same. If the resistance were increased by five, the current would be reduced to 0.5 ampere (five volts divided by ten ohms). Notice that when the resistance (R) increases, amperage (I) must decrease in order to keep a constant voltage in the circuit. These relationships provide information needed to calculate many of the values needed in electronics, Figure 2-10. With this knowledge of circuits, sensors can be designed to control temperature, pressure, level, flow, or other industrial variables.

OHM'S LAW

To determine voltage: multiply amperage times resistance
To determine amperage: divide voltage by resistance
To determine resistance: divide voltage by amperage
(For circuits consisting of resistances only)

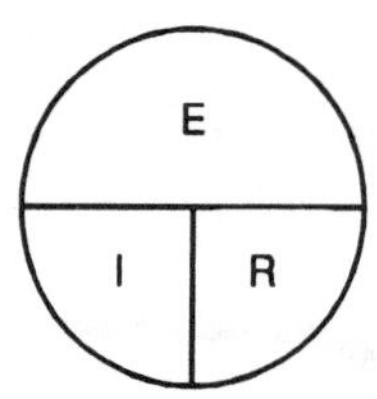

Figure 2-10 Ohm's law circle. Find the property desired and cover it with your finger. The visible letters remaining provide the mathematical operation required to solve your equation.

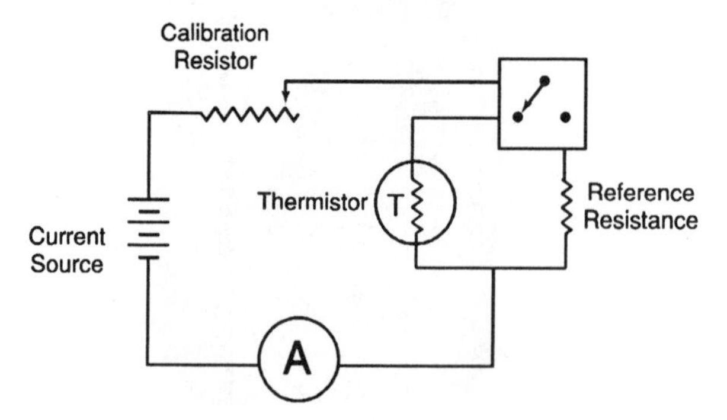

Figure 2-11 Schematic of a temperature instrument using an ammeter and thermistor to report degrees Fahrenheit or Celsius.

In designing a circuit the values of the components in the circuit should be calculated so that when the instrument is operating normally, the indicating device should be at midrange, Figure 2-11. The midpoint of the instrument is usually its most accurate position. Ohm's law can be used to calculate the values of the components in the instrument so that the instrument will operate in midrange.

Semiconductors

Semiconductors are materials with properties that are between the characteristics of the insulator and the conductor. Under activation, they have the property of being conductive or resistive. They have a unique covalent bonding structure that allows an electrical phenomenon to occur. The properties of material such as silicon and germanium provide a large number of solid-state components that utilize the electrical behavior of these materials.

Semiconductor Principles

Elements of various materials are held together by different atomic bonding. Materials like copper have a metallic bonding whereby free electrons are available in the metal and are thus available for use in conducting a current. The electrons that are free provide easy movement of electrons and thus the materials are known as conductors. Other materials, such as silicon, have a different atomic binding called "covalent bonding." In this bonding theory, the electrons are bonded to other silicon atoms sharing valence electrons, Figure 2-12. At low temperatures all electrons are covalent bonded and the silicon material responds to an electric current as an insulator. There are no free electrons available to provide a flow of current. At higher temperatures or with the addition of a different kind of atom, the material can be made to carry a current. This chemical process is called *doping* the silicon.

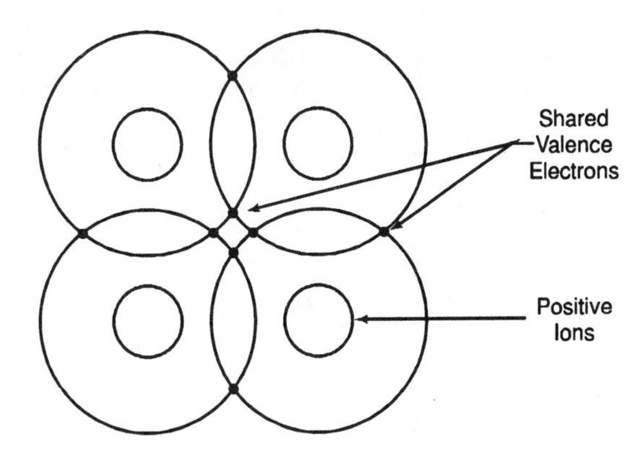

Figure 2-12 Covalent atomic bonding

Both silicon and germanium are doped with different materials so that when a certain voltage is impressed on the silicon, electrons become available and the total device becomes conductive. This becomes the basis for many of the solid-state devices. They are applied as *transistor* to produce very fast switches, amplifiers, oscillators, memory circuits, and numerous other solid-state devices. This is a valuable employment of solid-state semiconductors in the field of electronics.

The Diode

Semiconductor devices are based on covalent bond. In application the diode is a fundamental component. The diode has a different voltage-to-amperage relationship than that of a resistor. In the resistor, as the voltage is increased, the amperage increases at a linear rate. Semiconductor materials such as those used in a solid-state diode have a low voltage so that very little current is conducted until a transition voltage is reached. At that time the material becomes very conductive. This transition voltage will vary with the material used in the semiconductor. If the diode is made of germanium as its semiconductor material, the transition voltage when it becomes highly conductive is approximately 0.2 volts. Below this level so little current will flow that it can be ignored. If a silicon material is used in the semiconductor, the response is the same, except that the transition voltage is approximately 0.6 volts. In either material the voltages required to activate the diode into a heavy current-carrying capacity are very low control voltages, Figure 2-13.

Because the current-carrying capacity below the transition voltage is near zero, the diode responds as if it were an open circuit. This becomes a device that will allow current to flow only in one direction. This one-way current flow is a very important characteristic for electronic devices used in many circuits.

The identification of the direction of current flow on a diode is given by a white line that corresponds to the bar the symbol points toward and indicates the cathode terminal. A diode used in this fashion is referred to as "forward biased." If the diode is used in the opposite fashion it is "back biased" and allows only a small leakage current to pass through, so small that it is not considered, Figure 2-14.

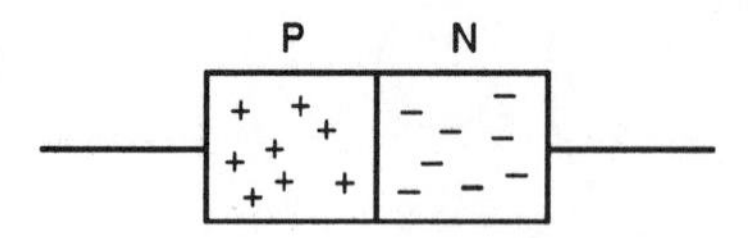

Figure 2-13 PN crystal diode

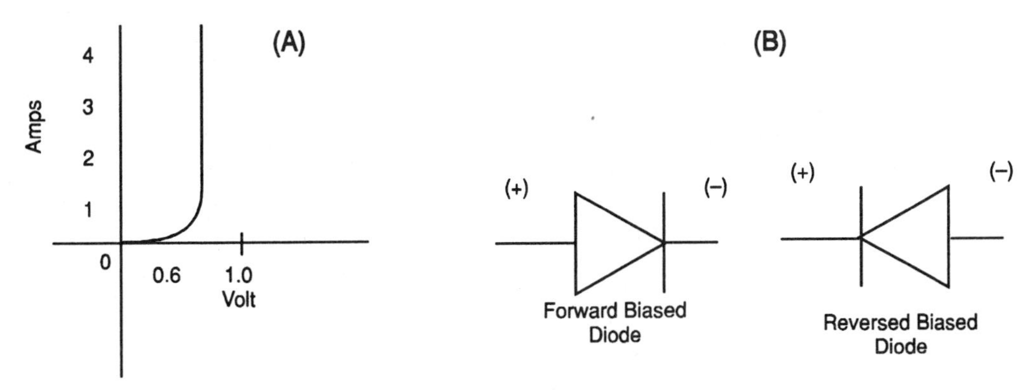

Figure 2-14 (A) Diode transition voltage diagram; (B) A diode symbol

Light-Emitting Diode

The light-emitting diode (LED) is manufactured with a special characteristic: it emits light when it is conducting a current. This is helpful because it provides a quick visual check of a circuit's operation. The diode also performs other applications, such as the activation of optical relays and of many display functions. These LEDs are provided in a number of colors: red, green, and yellow, Figure 2-15. They are seen on clock segments, test panels, computers, battery chargers, power supplies, instruments, and countless other applications.

Zener Diode

The zener diode finds its application in constant voltage control. It is based on the volt-amp characteristic of a very steep drop in its curve at a specific negative voltage. They are manufactured with breakdown voltages from two to two hundred volts. A dramatic change occurs at the designed voltage, which is referred to as "reverse breakdown" or "zener voltage." At this voltage, the current will increase with very small additional increases in reverse voltages. The zener diode has a varying current range with a very constant voltage change. A voltage regulator diode normally will employ a reverse-biased zener diode. This constant

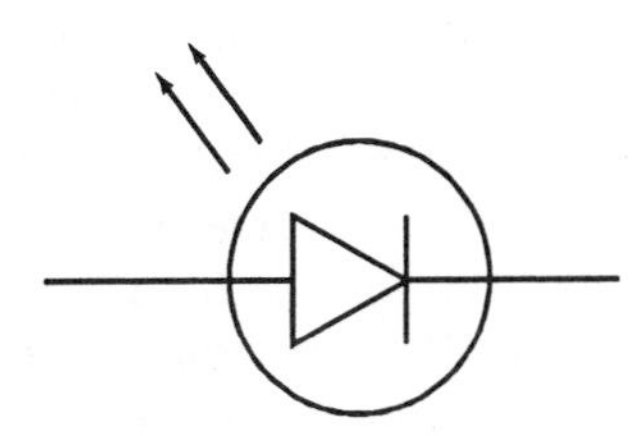

Figure 2-15 Symbol of the light emitting diode - (LED)

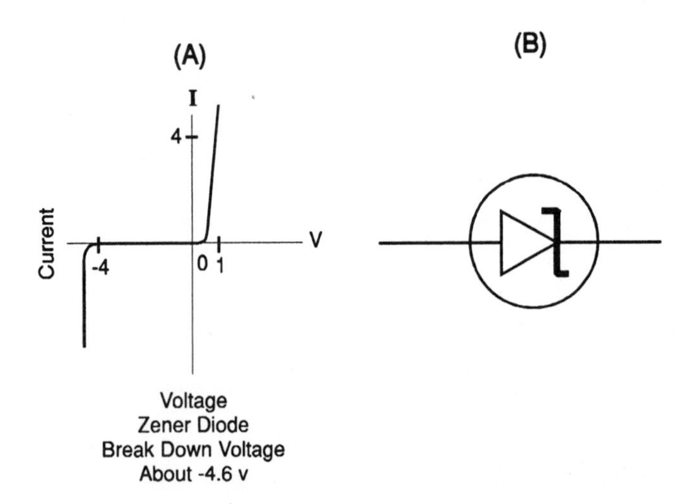

Figure 2-16 (A) The reverse breakdown voltage of a zener diode; (B)Symbol of a zener diode. The reverse breakdown of a zener diode will provide a wide range of current at a constant voltage.

voltage characteristic becomes very important in power supplies used for solid-state electronic devices, Figure 2-16.

The Transistor

Transistors are made from a number of combinations of diodes. The NPN crystal is designed with a heavily doped emitter that injects electrons into the base. The base is a thin, lightly doped section designed to pass emitter-received electrons on to a medium doped collector. It collects the electrons from the base and passes the flow on to the next device. Because a collector receives more heat when in service, it is usually made larger than the emitter section, Figure 2-17.

The transistors are applied in instrumentation and manufacturing systems as switches, amplifiers, and oscillators. The transistor is a voltage and current controlled switch, activated with a small voltage and current applied to its base terminal. When a voltage to the base reaches the threshold of 0.6 millivolts, the silicon transistor becomes saturated and the device becomes conductive. Current flows from the collector to the emitter terminals. This flow of current is controlled by a low voltage and current on the base terminal providing for a high-speed switch that can turn on and off higher voltages that duplicate the pattern of the low control voltages. This duplication of high-speed switching will also function as an amplifier. A very low variation in voltage and current can cause faithful replication and control over a high voltage and current.

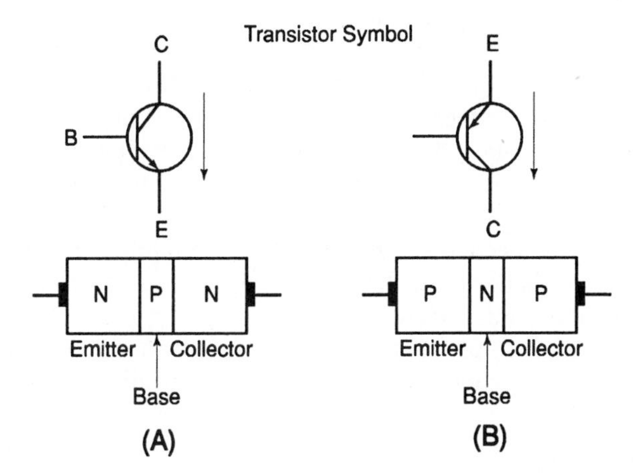

Figure 2-17 Transistor symbol and lead positions. (A) The three sections of an NPN transistor; (B)The three sections of a PNP transistor

Biasing the Transistor

Transistors are manufactured in two main types: NPN and PNP. Their applications depend upon whether a forward biased or reverse biased transistor is desired. In a forward biased NPN transistor, the conduction will be toward the emitter and base junction. With a reverse biased voltage on the base, little voltage will flow between the base and the collector junction, but a large current will flow between the collector and the emitter, Figure 2-18.

The transistor has a broad range of applications — from that of a switch and relay to that of the control element in operational amplifiers. It has become a basic building block of electronics.

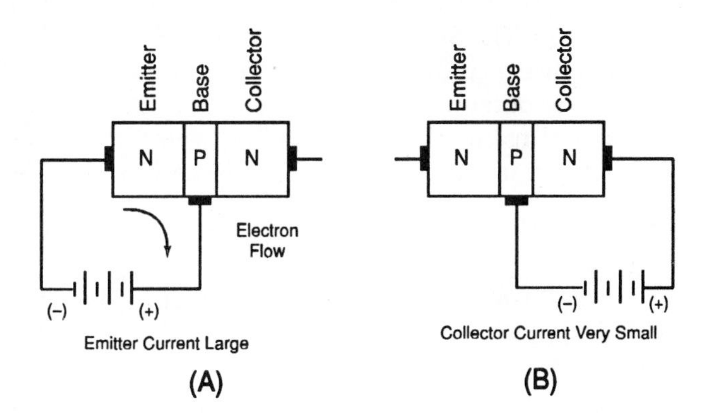

Figure 2-18 (A) Forward biased transistor; (B) Reverse biased transistor

The Field Effect Transistor

The transistor is normally a current control device. However, the field effect transistor (FET) is an exception. It is a voltage controlled device. The FET has three terminals: a gate, a source, and a *drain.* A P type of material is diffused into an N type of substrate. The major current carriers are between the source and drain. However, the conduction between the source and the drain current can be substantially reduced with a negative voltage applied to the gate. This relatively small voltage change ("pinch off voltage") at the gate will provide a large change in the available drain current. For this device, the gate is not forward biased, Figure 2-19. The FET is employed as an amplifier; with the addition of other components of resistors and capacitor, it provides a stable oscillator.

The Thyristor

The thyristor is a four-layer semiconductor with internal feedback that provides a latching action. The thyristor is primarily applied to the switching of heavy current carrying devices such as motors, heaters, lighting, and power supplies.

Silicon Controlled Rectifier - The silicon controlled rectifier (SCR) is formed by combining a diode and a transistor into a single solid-state device. It is a four-layer semiconductor device that blocks the flow of direct current through it until it is triggered or turned on by an electrical signal to its gate. It is a latching switch with very high gain. Gain is the ratio of output voltage to input voltage. Only a small current through the gate to the cathode is needed to cause a forward main current to flow from the anode to the cathode. Once the SCR is gated, it will remain on, even with the gating current removed. The device can be stopped by opening an additional switch that lowers its holding current below the transistor's saturated state and the latching current drops, opening the SCR. Another method of opening the latched SCR is to provide reversed bias triggering which opens the latch, Figure 2-20. The SCR is used in controlling large industrial currents. The currents controlled may be lower than one ampere to as high as twenty-five hundred amperes or more.

When an alternating current is supplied to an SCR, the diode section of the SCR will rectify it into a direct current; and a half cycle of the current will pass

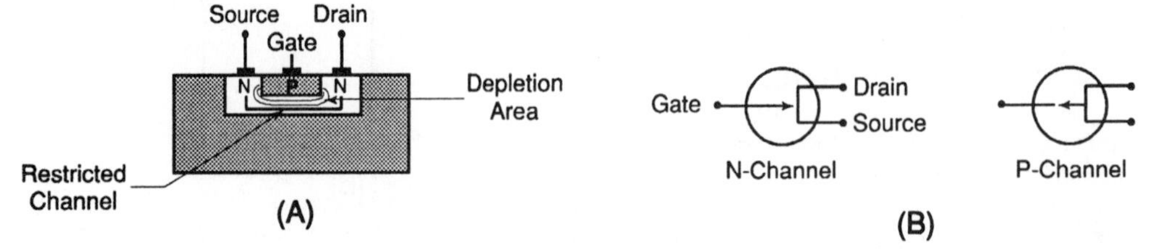

Figure 2-19 (A) Field effect transistor; (B) FET symbols

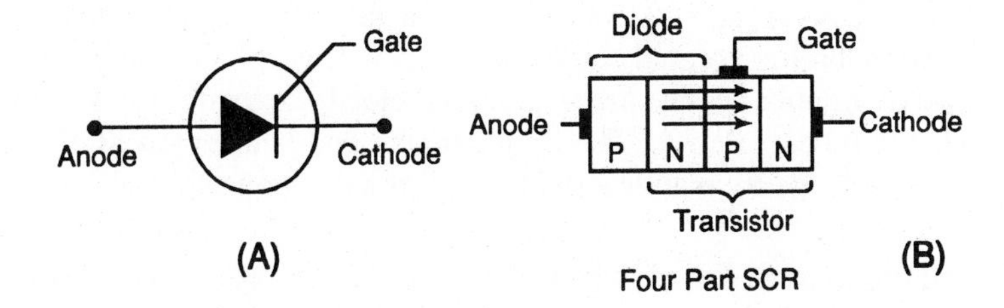

Figure 2-20 (A) Symbol of an SCR; (B) Operation of an SCR

through the device if the gate is energized. The SCR does not allow the reverse cycle to pass through the diode, thus the SCR is shut off and will restart only when gated by the cathode current. To state it another way, the gating signal starts the forward current, anode to cathode, while the reverse current from the cathode shuts off the flow for each cycle. This characteristic of the SCR makes it valuable because a high-speed transistor switch can be used to control the gate and other frequencies can be produced. With the employment of this solid-state device, the power frequency can precisely control the speed of electric motors and other actuators, Figure 2-21.

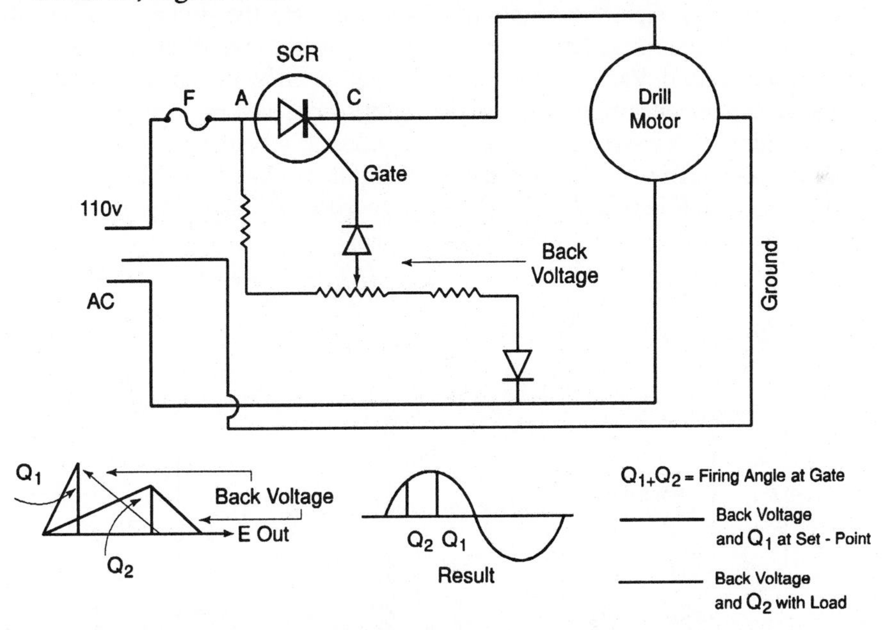

Figure 2-21 Schematic of an SCR motor speed control for an electric drill. A positive voltage o the gate will cause the SCR to conduct. A small negative voltage will cut off the conduction. As the drill slows down, the back voltage in series with the SCR cathode and gate decreases. The SCR voltage — actually the firing angle — increases. The extra gate voltage causes the SCR to conduct over a larger angle and more current is supplied to the drill.

Photo-SCR - A light-activated SCR is designed with a window in the transistor area so that a light signal can pass through. The light signal is strong enough to dislodge valence electrons from a collector to activate a base thus causing the device to be conductive. The light triggers and latches the SCR. When the light is removed, the base is deactivated and the device is shut off.

The Unijunction Transistor - The unijuction transistor (UJT) is used as a very rapidly acting switch. It is a three-terminal device with two base terminals and an emitter terminal. A bar of N-type crystal material with a section of P emitter crystal is alloyed to the larger crystal to make up a unijunction transistor. When the emitter voltage reaches its peak voltage (typically 0.6 to 0.8), the junction quickly closes to form a short circuit across from the emitter to one of the bases and the switch is closed. When the voltage falls below a valley voltage, the other circuit base is complete and the switch is open. The unijuction transistor acts like a voltage-sensitive switch, Figure 2-22.

Voltage Divider

A voltage divider is a network of resistors. A number of types of voltage dividers are applied to electronic components to perform the division and control of voltages. In instrumentation and control hardware, a resistance series network is commonly used. Various voltages can be delivered by placing a series of resistors between a power supply and ground. In this case, each resistor drops the voltage in relation with the resistor's resistance value. If a voltage is tapped off a power supply between two resistors, the new output will be different from a tap above the resistors, Figure 2-23. It will be a proportional voltage of a resistance series.

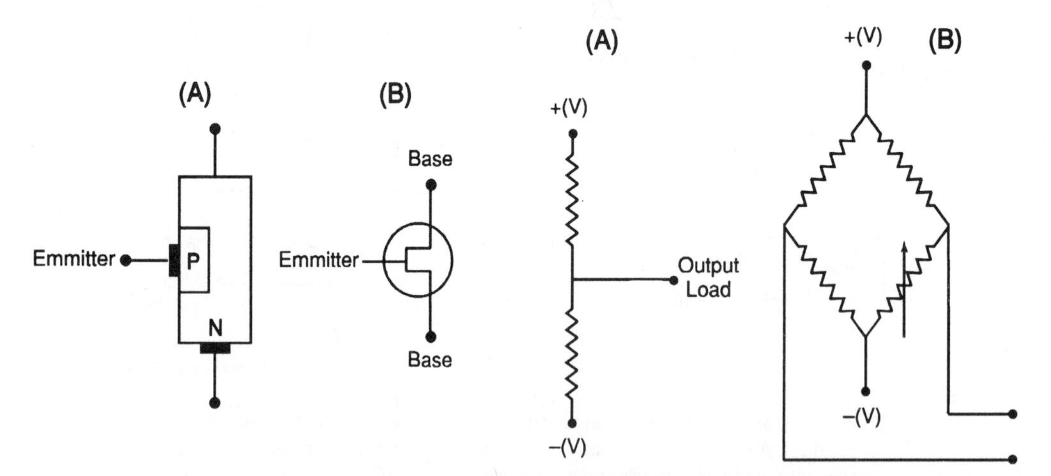

Figure 2-22 (A) A unijuction transistor (UJT) circuit; (B) UJT symbol

Figure 2-23 (A) A simple voltage divider with resistors; (B) A bridge voltage divider

The voltage varies directly with the resistance. Thus a useful proportion may be set up to determine the voltage drop across any series of resistances. Ohm's law is applied to calculate the value to deliver the required voltages.

Digital Technology

Digital technology is based on any device that operates on discrete or discontinuous signals and utilizes binary or two-state information — binary (0 and 1) — in the application of code systems.

Binary Communication

People often use the concept of a two-condition information system without being aware of doing so: an object can be present or not be present; something can be on or off; an object can move or not move; a voltage may be high or low. Information can be based on one or another state or condition. This is the concept behind the binary system that is applied in the digital electronic computer and electronic control technology.

These two conditions are accurately represented by two voltages in electronic communications — usually zero volts and five volts for positive logic or five volts and zero volts for negative logic. (Other voltages may be used but these are the common voltages used.) These high or low levels of voltages can be represented by a very short duration pulse and still accurately represent the data communicated. Each level represents a bit of information, Figure 2-24. These pulses have voltage values far enough apart so that electronic contamination, referred to as noise, does not provide false information superimposed on top of the true signal.

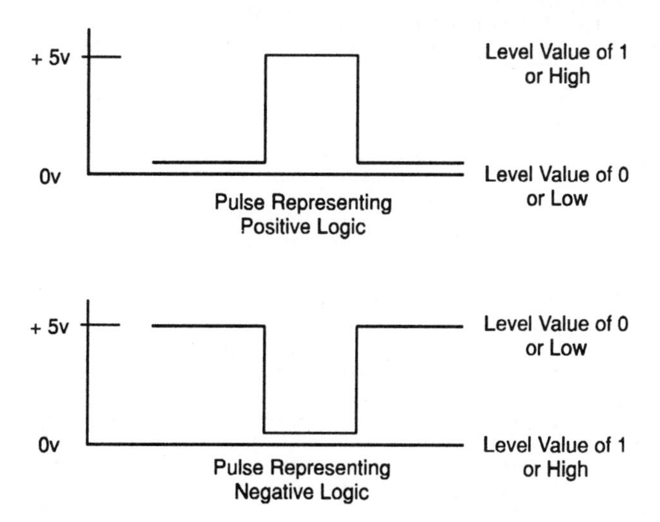

Figure 2-24 A voltage fuse representing binary logic

Being able to communicate a condition is a requirement of a sensor. That data can be manipulated through further arithmetic and algebric computations. To provide these arithmetic functions, a numbering or counting system is needed.

Binary Numbering System - A binary numbering system based on the on/off condition or high/low voltage or amperage value is represented by a base system of two. The traditional decimal system is represented by the base ten system of numbers. The binary system uses a base two number system and like the decimal system, the least significant digit is the digit furthest to right. A system of columnar notation is used: starting with the least significant digit the value of the digit is the digit multiplied the base raised to an exponent starting with zero.

DECIMAL:					
Exponent	10^4	10^3	10^2	10^1	10^0
Number	2	3	4	5	6
Value	20000+	3000+	400+	50+	6
BINARY:					
Exponent	2^4	2^3	2^2	2^1	2^0
Number	1	0	1	0	1
Value	16 +	0 +	4 +	0 +	1

Table 2-3

(A)

Power Value (of 2)	2^8	2^7	2^6	2^5	2^4	2^3	2^2	2^1	2^0
Positive Value (Common Number)	256	128	64	32	16	8	4	2	1

(B)

Decimal Number (Base 10)	Positive Value Decimally Added	Power Values 32 16 8 4 2 1	Binary Number
5	4+1	101	101
8	8	1000	1000
9	8+1	1001	1001
12	8+4	1100	1100
27	16+8+2+1	11011	11011

Figure 2-25 (A) Coding a binary number ; (B) Power value in a common number yielding a binary number

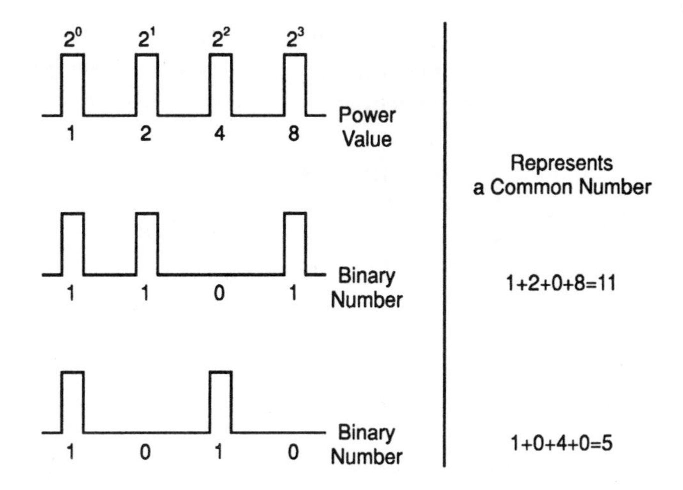

Figure 2-26 A train of pulses delivers a binary number

This system applies to any number system in any base, each number from left to right is multiplied by the power of that value, Figure 2-25.

With the use of a series of pulses, each representing a bit of information, a whole number or value of a variable can be communicated to an electronic machine. The pulses come in a train with each clock count representing either a high/low or 1/0. The pattern results in a four-bit binary number, Figure 2-26.

The one's position (or least significant digit) in the pulse train is on the left and proceeds to the right. The most significant digit is on the right of the pulse train. In everyday printed or written form, the one's position occurs on the right and proceeds to the left. This binary number procedure is the reverse of our natural number notation.

Digital Arithmetic

Boolean algebra is a mathematical technique of analysis involving two conditions. This algebra provides a method of computation for two-state devices that allows adding, subtraction, multiplication, and division. With these basic mathematical techniques, algebraic expressions can be developed that will solve most mathematical equations.

Addition - Logic gates can perform the addition of binary numbers. These additions are based on three rules:

Rule 1: 0 + 0 = 0
Rule 2: 0 + 1 or 1 + 0 either = 1
Rule 3: 1 + 1 = 0 and a carry bit of one (1-0, not 10)

In the addition of binary numbers, the pattern is the same as in decimal addition, except that in binary a carry to the next position is not generated after the sum reaches ten (as in decimal), but is generated when the sum reaches two.

	Decimal	5	=	101	Binary
	Decimal	2	=	010	Binary
Add		—		—	
		7		111	No carry

				11	Carries
	Decimal	7	=	111	Binary number
Add	Decimal	5	=	101	Binary number
		—		—	
		12		1100	

The carry concept is explained in the second example where 7 and 5 are added together. When 111 is added to 101, in the first column 1 + 1 = 0, resulting in a carry of a 1. In the second column, 1 plus the carry 1 = 0 and another carry results. Then in the third column is 1 + 1 = 0 with a carry plus the previous carry, or 1 + 1 + 1 = 11. The binary answer is 1100, or in decimal, 5 + 7 = 12.

Subtraction by Complement - Subtraction by the complement is obtained through the application of a number's complement. A complement is the result of changing all zeros of the binary number into ones and all the ones into zeros. Subtraction can be executed by converting the larger number to be subtracted from to its complement and performing the subtraction by adding the two numbers together and then complementing the answer to provide the subtracted answer.

Decimal	50	=	110010	Binary number
(Subtract) Decimal	45	=	101101	Binary number

The first item in subtraction by complement is to complement the larger number. The binary number 110010 becomes 001101. Add the 50 complement to the 45 binary number.

	Decimal	50 complement	=	001101	Binary number
Add	Decimal	45	=	101101	Binary number
				—	
				111010	

Again complement the addition answer 111010 = 000101.

The binary number 00101 = the decimal number 5.

One difficulty can occur: it may be difficult for a computer to determine which of the two numbers is larger and thus which number to complement. Because of this problem, other schemes of calculation may be employed.

With the application of subtraction circuits and addition circuits employing the shift and carry processes, addition, subtraction, multiplication, and division processes are performed. With these processes, these arithmetic calculations can be performed at very high speeds within digital computer logic circuits.

Binary Based Codes

A digital code is a system of symbols that represents data values and makes up a language that digital circuits can understand and use to perform, store, manipulate, and communicate information. Binary based codes include letters, symbols, punctuation marks, and decimal digits as well as binary numbers.

The Binary Numbering System

The binary code is the simplest code. It is a two-state code consisting of zero (off) and one (on) states. A series of four binary digits, or bits, can represent any of sixteen different decimal numbers ranging from zero to fifteen. Numbers larger than sixteen require an eight-bit binary digit and can encode up to two hundred fifty-six different decimal numbers, or encoded "things." With a sixteen-bit binary digit, 65,536 different memory locations can be selected, Figure 2-27.

A bit is a binary digit. Four bits equals a nibble; eight bits equals a byte; sixteen bits equals a word or two bytes or a small word; thirty-two bits can equal four bytes, two small words or a (large) word, depending on the size of the system. You can also point out the maximum number associated with each group of bytes or words, but be careful about mixing up bits and bytes.

The Octal Code Numbers

Because binary numbers require a large number of bits to express a value, an octal code was developed to reduce the number of bits required in the code. The octal code is based on a radix of eight rather than the binary of two. Each octal number digit is equivalent to three digits in a binary number, Figure 2-28.

The octal counting system contains eight different symbols: 0, 1, 2, 3, 4, 5, 6, and 7. To convert a binary number into an octal number, three steps are required. Write an eight-bit binary number: 10011101. First — divide the binary number into three groups: 10 011 101. Second — add a logic 0 in front of the group of two bits: 010 011 101. Third — substitute the octal digit for each of the groups: 2 3 5. These steps convert the binary number 10011101 into the octal number 235.

Decimal Number	Binary Number
0	0000
1	0001
2	0010
3	0011
4	0100
5	0101
6	0110
7	0111
8	1000
9	1001
10	1010
11	1011
12	1100
13	1101
14	1110
15	1111
16	10000

Decimal Number	Binary Number	
0	0	
1	1	
2	10	
3	11	
4	100	
7	111	
8	1000	= 4 bits
15	1111	= 4 bits
16	10000	
31	11111	
32	100000	
63	111111	
64	1000000	
127	1111111	
128	10000000	= 8 bits
255	11111111	= 8 bits
256	100000000	
511	111111111	
512	1000000000	= 10 bits
1023	1111111111	= 10 bits
1024	10000000000	
2047	11111111111	
2048	100000000000	
4095	111111111111	
4096	1000000000000	
8191	1111111111111	
8192	10000000000000	
16,383	11111111111111	
16,384	100000000000000	
32,767	111111111111111	
32,768	1000000000000000	= 16 bits
65,535	1111111111111111	= 16 bits

Figure 2-27 A decimal number to a binary number

Octal Number	Binary Number
0	000
1	001
2	010
3	011
4	100
5	101
6	110
7	111
10	001 000
11	001 001
12	001 010
13	001 011
14	001 100
15	001 101
16	001 110
17	001 111
20	010 000
21	010 001
22	010 010
23	010 011
24	010 100
25	010 101
26	010 110

Octal Number	Binary Number
27	010 111
30	011 000
40	100 000
50	101 000
60	110 000
70	111 000
77	111 111
100	01 000 000
110	01 001 000
111	01 001 001
120	01 010 000
140	01 100 000
170	01 111 000
177	01 111 111
200	10 000 000
240	10 100 000
270	10 111 000
277	10 111 111
300	11 000 000
340	11 100 000
370	11 111 000
377	11 111 111

Figure 2-28 Binary to octal code numbers

Hexadecimal Number	Binary Number
0	0000
1	0001
2	0010
3	0011
4	0100
5	0101
6	0110
7	0111
8	1000
9	1001
A	1010
B	1011
C	1100
D	1101
E	1110
F	1111

Figure 2-29 Binary to hexadecimal numbers

The Hexadecimal Code Numbers

The hexadecimal byte code is designed to reduce a binary code of sixteen bits to a code of two hex digits. To provide this conversion, a change is performed similar to that performed in the octal code except that the four bits are chosen from the binary number rather than the three of the octal code, Figure 2-29.

The hexadecimal code counting system contains sixteen different symbols: 0, 1, 2, 3, 4, 5, 6, 7, 8, 9, A, B, C, D, E, and F. To convert a binary number into a hexadecimal number, three steps are required. Write the binary number 1001110. First — divide the binary number into groups of four bits: 100 1110. Second — add a logic zero in front of the first group: 0100 1110. Third — substitute a hexadecimal symbol for each of the groups. The most significant bit is 0100, which is 4. The least significant bit is 1110, which is E. Therefore, 01001110 binary is equal to 4E in a hexadecimal code.

The Encoder

The *encoder* transforms information from real world data, such as a position or motion of a device, and converts that information into binary logic signals. This new information is in a form the computer can manipulate and determine if the incoming information is equal to, greater than, or less than that which is required by the controlling system. This information may be supplied by a code wheel or shaft encoder. The code wheel may be a disc with printed lines at its edge, punched slots or holes in the disc, or insulated sections between contact areas. The disc is mounted on the shaft or part to be encoded and is contacted by brushes, reflected light, or directed light through the disc. The disc is divided in sectors and bands from the outside edge to the center. Sensors can be set so each band or track will deliver a bit of information for the code depending on the position and thus the on or off data represented by its surface. A high energy or reflection would represent a logic HIGH and no or little reflection represent a logic LOW. In this fashion, a phototransistor detector can be used to provide binary code data to a control system.

The Gray Code - A binary encoding wheel will deliver an accurate reading of where a mechanism is located, except when the location is between wheel sectors or on the sector edges. This error may cause a misreading of the correct number of bits and produce a false readout. To eliminate these crossover problems, the Gray code was devised.

The Gray code solves the crossover problem in that the code wheel is designed so that only one bit of the code wheel changes as the wheel progressively turns. One problem is solved but another is evident. Digital circuits do not communicate with Gray code. Therefore the code is converted into binary with a Gray-to-binary converter made up of exclusive OR gates resulting in a binary communication system, Figure 2-30A and Figure 2-30B.

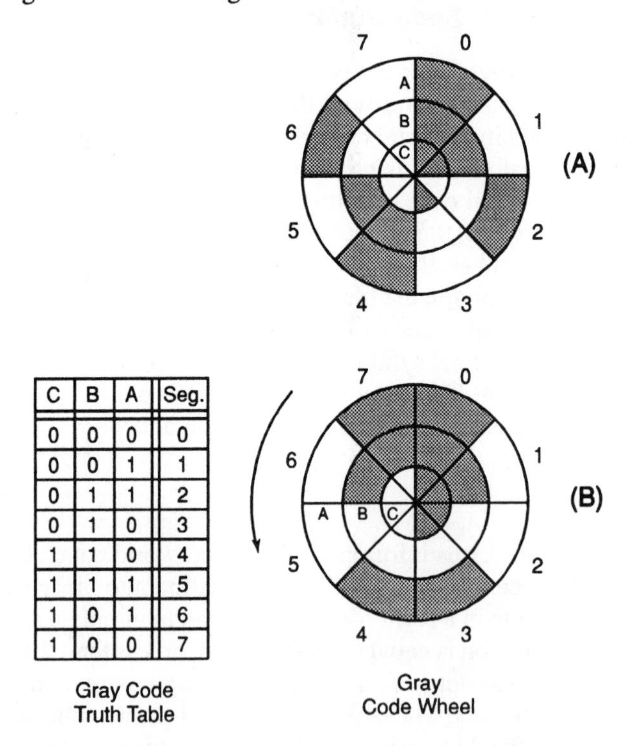

C	B	A	Seg.
0	0	0	0
0	0	1	1
0	1	1	2
0	1	0	3
1	1	0	4
1	1	1	5
1	0	1	6
1	0	0	7

Figure 2-30 Encoder wheels: (A) Binary code wheel; (B) Gray code wheel

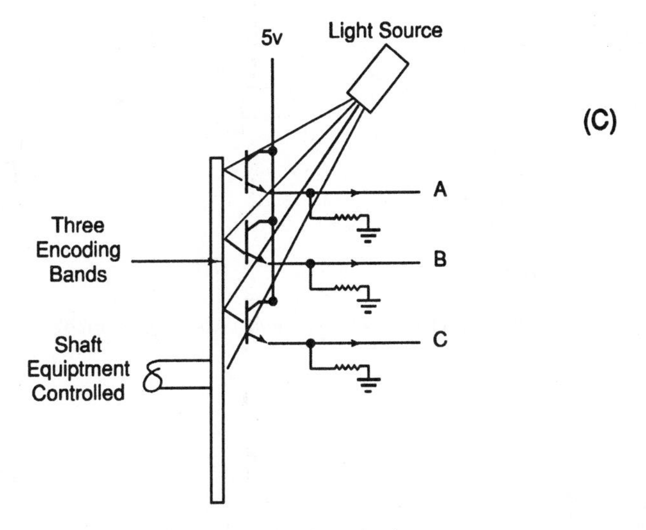

Figure 2-30 (Continued) (C) Phototransistor reflective wheel

Semiconductor Logic

George Boole devised a mathematical logic for the solving of problems by posing a series of questions that could be responded to as true or false. He represented the true or false statements as independent variables and the answers as dependent variables into algebraic equations. These equations could be solved by using a systems of rules. The rules are the basis of Boolean algebra.

The binary numbering system was applied to a source data system employing the digits of 1 and 0. The data system represented true or false statements expressed in mathematical notation that could be solved by Boolean algebra.

Logic Gates

Logic gates within the circuits provide a mathematical analysis of the inputs into the various gates and deliver an output of either a high (HI) or low (LO) voltage. The HI voltage is usually a +5 volts and indicates a binary bit with a value of 1. The LO voltage is usually 0 or ground potential and indicates a binary bit with the value of 0.

The AND Gate - The AND gate is a semiconductor logic control device that applies Boolean algebra and analysis resulting in a logical choice involving two conditions: a HI or a LO output. The device is like switches in series that provide a single output. All of the sources are required to be HI to have an output of HI. The possible combinations of the device are presented by a truth table, Figure 2-31.

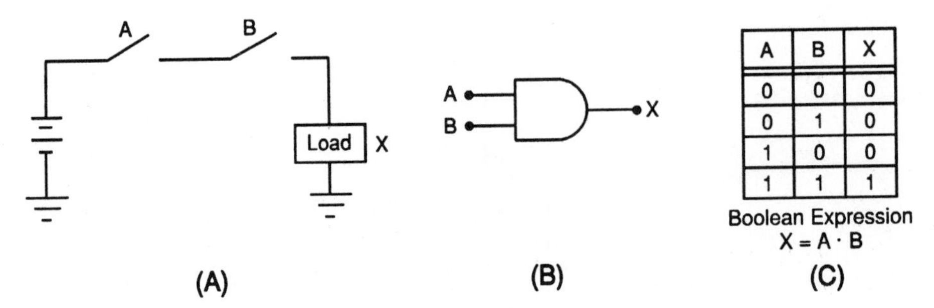

A	B	X
0	0	0
0	1	0
1	0	0
1	1	1

Figure 2-31 AND logic circuit: (A) AND gate diagram; (B) AND gate symbol; (C) AND gate truth table

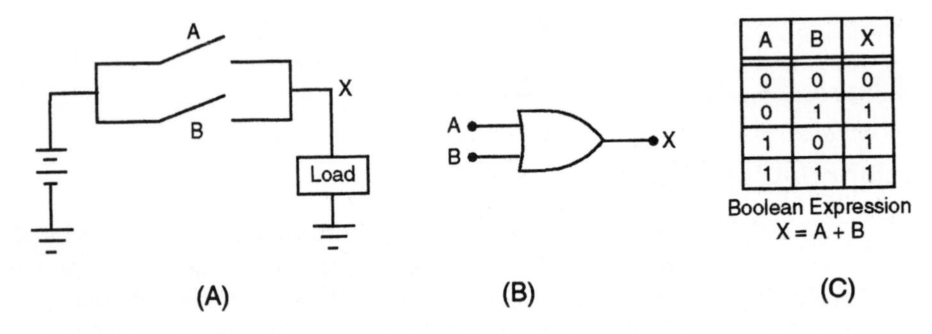

A	B	X
0	0	0
0	1	1
1	0	1
1	1	1

Figure 2-32 OR logic gate circuit: (A) OR gate diagram; (B) OR gate symbol; (C) OR gate truth table

The OR Gate - The OR gate analyzes the condition of the inputs and if either A or B or both are HI, the output is HI. In this logic gate, the "+", as shown in the Boolean expression X = A + B, is read as A or B. The plus sign is read as "OR." In the OR gate, the output is 0 only if all its inputs are 0. In other cases the output will be 1, Figure 2-32. An exclusive OR gate is defined only for two inputs. The output of this gate is 1 if — and only if — one input is 1 and the other is 0.

The NOT Operator - An important logic function for use in solid-state circuitry is a NOT operator. The NOT operator inverts the function of the device. If applied to the output, the NOT operator will provide a function opposite of what it would normally be. Applied to the input, it would require a negative logic input instead of a positive logic input. The NOT operator is a small circle placed on either end of a logic symbol. The NOT operator is never employed alone. They are always applied to the input or output of another logic element, Figure 2-33. This inverter circuit is made up of a transistor and resistor that is connected to other circuits.

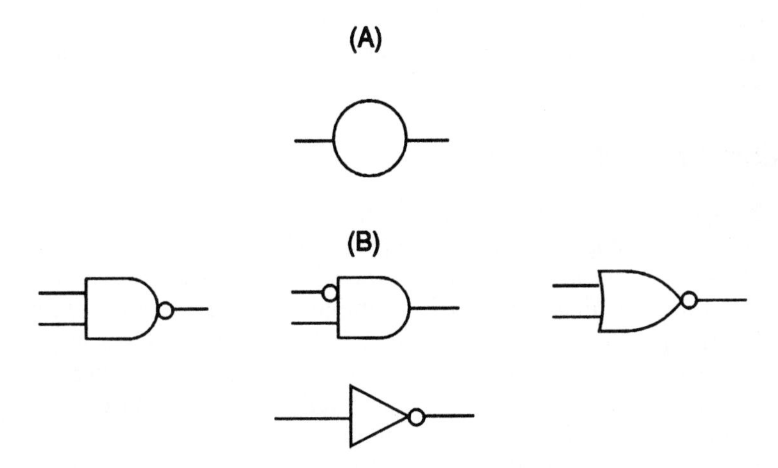

Figure 2-33 (A) NOT operator; (B) symbols

NAND and NOR Gates - The NAND and NOR gates are AND and OR gates with a NOT operator placed in series. In the NAND application, the AND portion of the gate must receive a high to both inputs to deliver a high to the NOT operator portion of the gate. The NOT operator reverses the results of these two high inputs of the AND gate and results in a low output of the NAND gate.

The NOR gate is an OR gate followed by a NOT operator. When either terminal of the OR gate is high the output of the OR gate is a high; again the NOT operator reverses the output and a low from the NOR gate results. This application provides two more gate possibilities, Figure 2-34. The chief importance is that in integrated circuits these gates are easier to manufacture. When starting from a beginning design, computer logic can be carried out using NOR and NAND gates for construction.

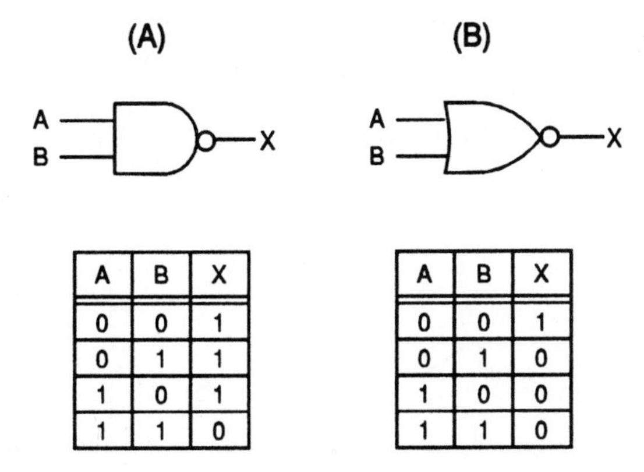

(A)

A	B	X
0	0	1
0	1	1
1	0	1
1	1	0

(B)

A	B	X
0	0	1
0	1	0
1	0	0
1	1	0

Figure 2-34 (A) NAND and (B) NOR symbols and gates

Multiple-Inputs Gates - Logic gates can have multipleinput on AND, OR, NAND, and NOR gates. Three and more inputs are found on digital circuit designs, Figure 2-35.

Operational Amplifiers

The operational amplifier is a solid-state device employing integrated technology that has as its chief objective the increase of energy of the signal coming into the unit. The operational amplifier has the advantage of being able to perform many more functions than gain or the multiplication of an original signal. They were designed primarily to solve mathematical computations in computers, but their usage has been extended to many industrial control applications.

The operational amplifier is essentially made up of three amplifier units: a differential amplifier, a voltage amplifier, and an output amplifier. The advantage of the operational amplifiers over other circuits is their small size, high input impedence, low output impedence, and high voltage gain. They have many of the same characteristics of other electronic amplifiers in that they have impedance, offset, and drift; they utilize electrical power and develop heat as well as provide gain, Figure 2-36.

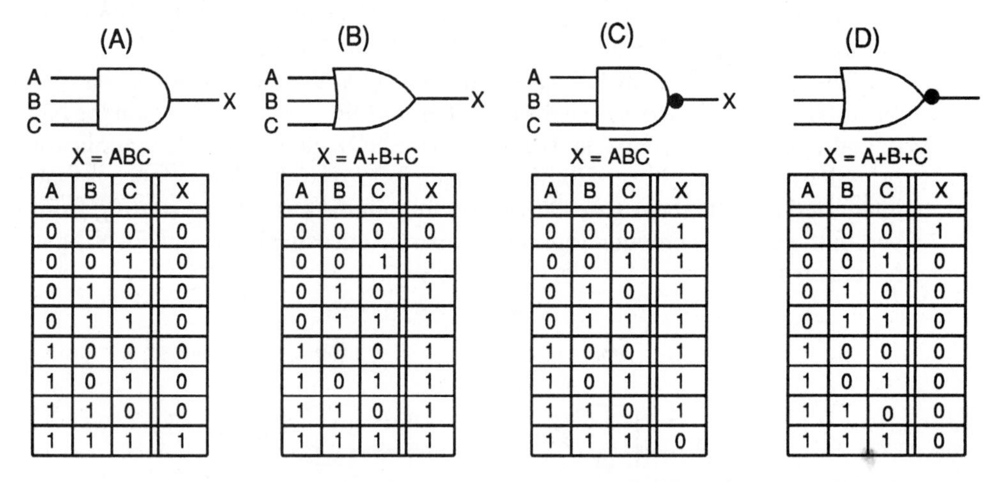

(A) $X = ABC$

A	B	C	X
0	0	0	0
0	0	1	0
0	1	0	0
0	1	1	0
1	0	0	0
1	0	1	0
1	1	0	0
1	1	1	1

(B) $X = A+B+C$

A	B	C	X
0	0	0	0
0	0	1	1
0	1	0	1
0	1	1	1
1	0	0	1
1	0	1	1
1	1	0	1
1	1	1	1

(C) $X = \overline{ABC}$

A	B	C	X
0	0	0	1
0	0	1	1
0	1	0	1
0	1	1	1
1	0	0	1
1	0	1	1
1	1	0	1
1	1	1	0

(D) $X = \overline{A+B+C}$

A	B	C	X
0	0	0	1
0	0	1	0
0	1	0	0
0	1	1	0
1	0	0	0
1	0	1	0
1	1	0	0
1	1	1	0

Figure 2-35 Multiple input gates (A) AND, (B) OR, (C) NAND, and (D) NOR with truth tables

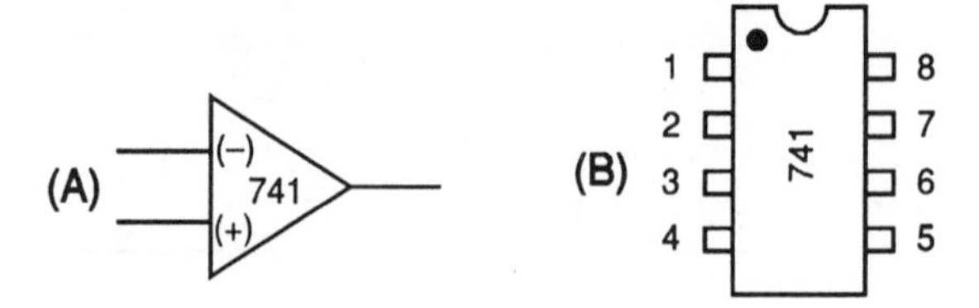

Figure 2-36 (A) Op amp symbol and (B) pin numbers

Construction of the Operational Amplifier

This device is more commonly called the "op amp." It is produced with integrated circuit technology that includes resistors, capacitors, conductors, transistors, and completing circuitry, containing twenty to thirty active electronic elements. All of these elements are on a silicon chip about one-eighth of an inch square. These devices are mounted in cases that are either round or rectangular with attachments for leads. The case and leads require considerably more space than the chip itself. Some op amps are supplied in round metal case packages with eight, ten, or sixteen leads coming from the container. Another container more commonly seen in commercial use is one designed with plastic or ceramic enclosures. The ceramic packages are used where higher heat dissipation is required. The dual-in-line (DIP) packages are available in eight or fourteen pin (terminal) packages, most frequently employed in industrial applications.

For application in this text, the discussion will center around the operational amplifier 741, a device widely used in electronics and instrumentation. These integrated circuits are used as building blocks for electronic circuits. Other components are assembled around them to provide the requirements. Resistors, capacitors, voltage dividers, and diodes all may be externally added to provide a specific instrumentation function.

Op amps are designed with a number of stages. A basic op amp will consist of three stages: an input stage that receives the signal and increases the current or the gain of the amplifier; an isolation stage that prevents unwanted electronic noise from entering the output stage; and the output stage that provides gain and feedback to the input depending upon the application of the op amp. These stages are connected together by direct coupling for applications for DC signals, or with capacitors between if they are used for AC signals. For many instrument

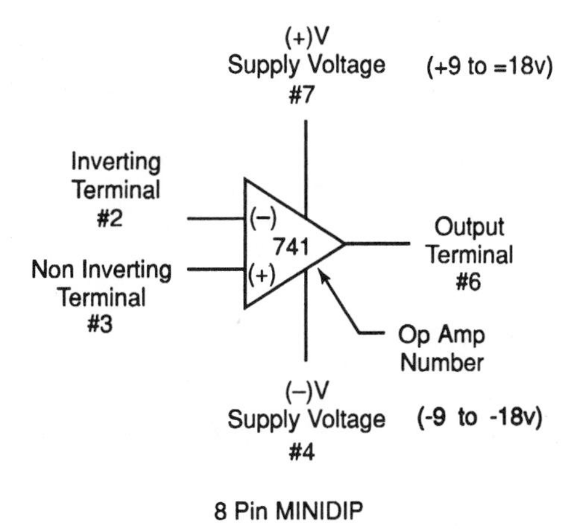

Figure 2-37 Op Amp — some pin locations and numbers

applications, DC signals are received from the sensors and therefore the direct-coupled amplifier will be utilized.

Identification of the pins on an op amp rectangular case is made by locating the notch on the end of the case and counting the pins counterclockwise. The first pin to the left is the number one pin. Frequently there is a dot near the number one pin position, Figure 2-37.

Op Amp Gain

Open Loop Gain - The 741 op amp has an open loop gain of approximately two hundred thousand — which would indicate that an input voltage of one volt would deliver an output voltage of two hundred thousand volts. However in a practical application with a power supply voltage of plus or minus fifteen volts delivered to the op amp, an output voltage signal of about thirteen volts is expected. The saturation voltage does not produce a voltage gain greater than the supply voltage, even when the gain is considered infinite. The difference between the input voltages when they are referred to ground is the value that is used to divide into the output value to find the open loop gain for an amplifier, Figure 2-38A.

Closed Loop Gain - Closed loop gain is the overall gain when a feedback connection is provided. The op amp output is connected through a voltage divider back to the negative polarity input terminal of the op amp. This negative feedback decreases the gain slightly but greatly stabilizes the closed loop output. When changes occur in the op amp output voltage, a fraction of this error voltage is fed back to the negative input terminal that automatically compensates for the original change in the voltage of the op amp output, thus providing its designed gain, Figure 2-38B.

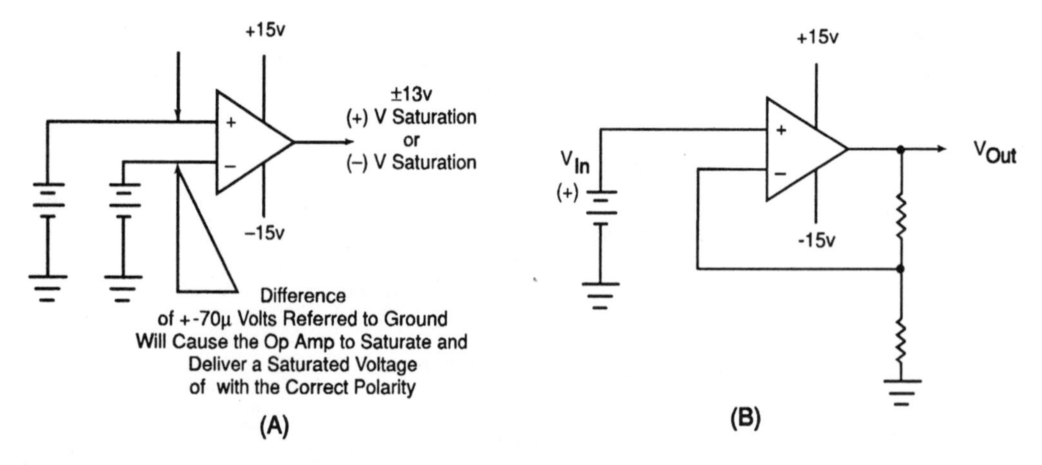

Figure 2-38 (A) Open loop operational amplifier gain. Voltage input (+) - voltage output (-) = voltage difference and polarity referenced to ground at that instant. (B) Closed loop operational amplifier gain

Op Amp as a Voltage Comparator

The amount of voltage difference between the two inputs necessary to cause the op amp to deliver nearly the voltage of the supply voltage is defined as the saturation voltage for the op amp. In the case of the 741, it goes to a saturation voltage with about plus or minus seventy microvolts and delivers plus or minus thirteen volts of a plus or minus fifteen volt supply. The seventy microvolts difference is so small that for practical purposes, any plus or minus voltage differences will cause the op amp to provide a saturated condition and deliver the correct polarity voltage at the output terminal.

The characteristic of high gain at low saturation values makes the op amp very valuable as a voltage comparator. The comparator can determine which of two voltages is higher as they enter the input of an op amp. In instrumentation applications, a reference voltage may be supplied to one input and an unknown voltage to the other input. The higher voltage saturates the op amp to produce an inverted op amp output.

The Inverting Operational Amplifier

The operational amplifier can be used to build an inverting amplifier. In this case, the (+) op amp input is grounded and an input signal is applied from the op amp output to a voltage divider and fed back into the (-) op amp input. Delivering a feedback signal to the (-) op amp input inverts the output voltage providing a voltage opposite in polarity to the input voltage, Figure 2-39.

When following circuit diagrams in the application of op amps, the + and - indicated on the op amp do not refer to the polarity of the input voltage but rather to the inverting and noninverting input terminals. A input into the (-) terminal of an op amp will provide an inverted output.

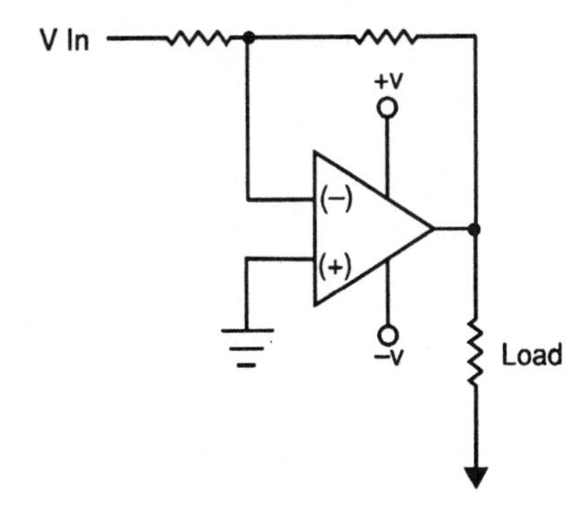

Figure 2-39 An inverting operational amplifier. A closed loop is provided by feedback to the noninverting (-) terminal of the op amp.

A Zero-Crossing Voltage Level Detector

The zero-crossing voltage detectors can be set to detect voltages either above zero (positive) or below zero (negative). With a positive detector, the reference voltage is above zero and connected to a positive voltage input that triggers an output change when the reference level is reached. With a negative voltage detector, the output voltage polarity changes whenever the input exceeds a negative reference level.

An Inverting Zero-Crossing Voltage Detector

In the case where the reference voltage is set at ground potential, an inverting zero-crossing detector is employed. Every time the unknown voltage crosses this zero potential, the output voltage will change polarity at a voltage near to the voltage of the reference voltage. The required plus or minus seventy microvolts for the op amp saturation is so small it can be ignored in the voltage effecting the polarity changes of the output of the device, Figure 2-40.

A Noninverting Zero-Crossing Detector

In the event that a noninverting zero-crossing detector is desired, the only change necessary is to ground the negative input terminal of the op amp. With this completed, the output terminal will have the same polarity as the input signal voltage.

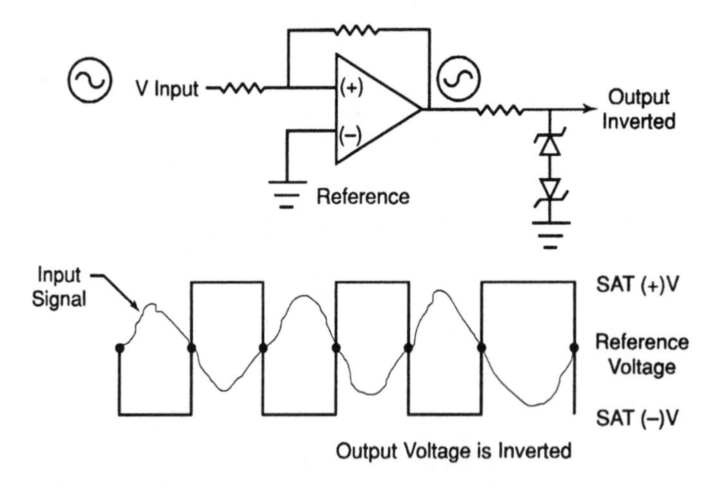

Figure 2-40 An inverting zero-crossing detector diagram

A Nonzero-Crossing Detector

Another op amp, called a nonzero-crossing detector, functions so that the saturation voltage is not at ground potential or zero volts. In this case, the reference voltage can be established at a positive or negative voltage level either above or below ground or zero potential. Since an op amp will not saturate until the reference voltage level is reached, being able to set the reference point gives an added advantage to this voltage level detector, Figure 2-41. This is very important to instrumentation because it can be used to establish a set-point in a control system. The voltage level detector can be set up for either a noninverted or an inverted output voltage.

The op amp saturates at the set plus one volt reference level and delivers a positive maximum source voltage polarity to the output. When the input voltage drops to the minus one volt reference level, the op amp saturates and delivers the negative maximum source voltage to the output of the op amp. The sensor in an instrumentation application will activate a control signal whenever its voltage exceeds the two set levels. The setting of the levels can be varied but they must be high or low enough to cause the op amp to saturate and produce a polarity change.

A Voltage Range Detector

A voltage range detector can be designed with the application of two comparators. The system is built so that input voltage remains within a specified voltage range. These voltage limits are referred to as threshold voltages. When the input voltage exceeds the set threshold voltage of that op amp, it will saturate and provide an output signal. One op amp's input is set for an upper threshold voltage and is

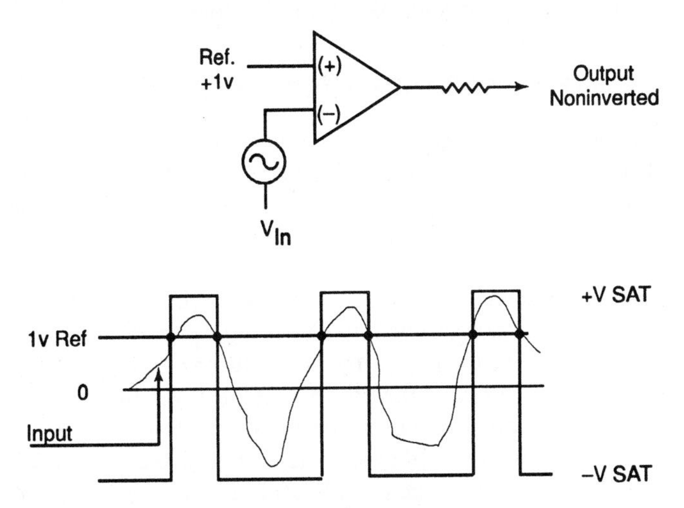

Figure 2-41 A (+) one volt voltage level detector (noninverting)

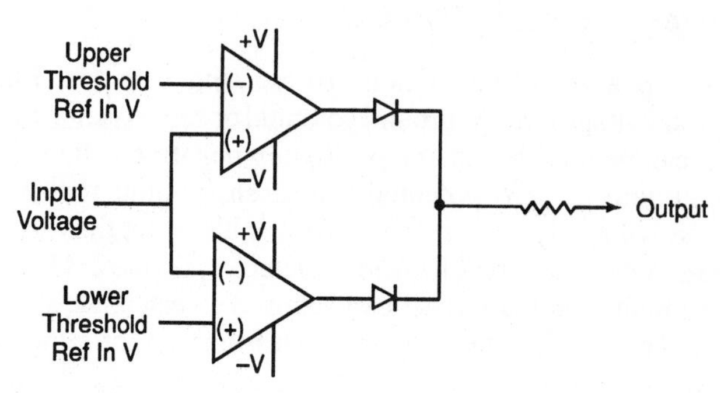

Figure 2-42 A voltage range detector

attached to the (-) terminal of the other op amp. The lower threshold input reference is attached to the (+) terminal of the second op amp that is set at the low threshold voltage. The input voltage to be measured is attached to the (+) of the upper threshold terminal as well as to the (-) terminal of the low threshold op amp, Figure 2-42.

When the voltage entering the input is less than the lower threshold voltage or greater than the upper threshold voltage, the output of the circuit will be "on," with a small voltage drop caused by the diodes. The op amp will deliver a voltage close to the source voltage indicating that the voltage variation is larger than the range of the required voltage. The output voltage is "off" when the input voltage is between the two threshold voltages. This condition exists because the two diodes are both biased and the output from the op amps are at voltage saturation; each diode blocks the current flow to the op amp and the output voltage is zero.

This range detecting circuit can be utilized in a number of instrumentation protecting, testing, or controlling applications.

Hysteresis Added to Operational Amplifiers

Undesired extraneous electrical signals called noise may be added to the signal that is being measured. These voltages are frequently generated from composite electrical environmental contamination. These slight fluctuations of induced voltages can cause an op amp to go into a saturated condition when it should not be saturated. The polarity output is changed by this unrelated signal, thus providing incorrect data. To prevent this condition, "hysteresis" is added to the op amp. This is accomplished by providing a positive feedback connection from the op amp output through a resistor and back to the op amp (+) input. This provides a higher upper threshold voltage level that must be reached before saturation in the op amp will occur and the device changes its polarity. A lower threshold voltage level is likewise established so that the input voltage must reach that level before the op amp will saturate and change its polarity. The upper threshold voltage and the lower threshold voltage are established so that they are greater than the noise voltages and less than the op amp saturation level. Thus the op amp does not saturate until a signal is above or below the threshold saturation level of the op amp. This procedure does not allow the electrical noise to pass through the op amp. The difference between the triggering voltage thresholds and op amp saturation eliminate the noise signal, Figure 2-43. The op amp responds like an inverting zero-crossing detector that changes the range in which saturation occurs.

A portion of the op amp's output is introduced into the positive input terminal of the op amp. Another resistor is added to create a voltage divider. It is placed between the resistor and ground to help calibrate the feedback voltage. Establishing an upper trigger point and a lower trigger point within the op amp via the positive feedback result is a voltage transfer characteristic which eliminates or reduces electrical noise signals.

A Voltage Follower

A voltage follower (also called an isolation amplifier, a unity gain amplifier, or a buffer) is made up of a circuit with a feedback. A major advantage of a voltage follower circuit is that it has a very high input impedance input value. This is important because this high resistance protects the source by not allowing a high current to be drawn from the source. But the op amp will respond to a voltage change. This is accomplished by a feedback connection directly from the op amp output back to the (-) input terminal for 100 percent feedback.

The original input enters the op amp at the (+) input terminal, while the output of the op amp feedback is connected to the (-) input terminal and is opposite in polarity. This provides a gain of one where the output tracks the input and has the same voltage value and polarity. With this application of the op amp, the values will be equal but the two voltages will be isolated from each other.

A proportional feedback can be achieved providing a voltage divider between the op amp output and the (-) input. As mentioned earlier, the ratio of the two resistors in the voltage divider provides the gain desired. This ratio of resistors

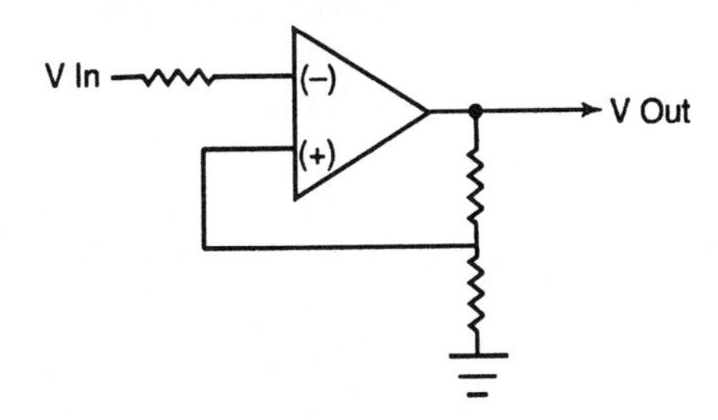

Figure 2-43 Op amp with hysteresis circuit

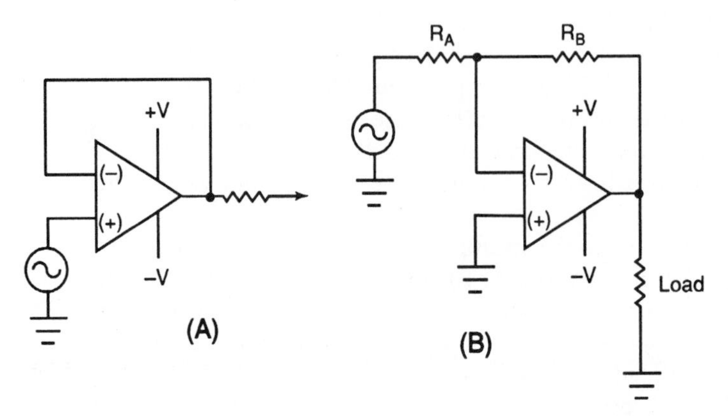

Figure 2-44 (A) Diagram of a voltage follower circuit; (B) Diagram of an inverting amplifier circuit with gain established by the ratio between Ra and Rb

can provide gains from one to one thousand depending upon output requirements. Also, the voltage divider can provide the function of producing a noninverting or inverting amplifier depending upon the input connections.

The chief advantage of this circuit is its high input resistance. The voltage follower draws a negligible amount of current from the input source, making possible accurate measurements or reproductions of the original voltage values, Figure 2-44.

Voltage-To-Current Conversion

Voltage-to-current conversion amplifier circuits are employed where an instrumentation sensor is a considerable distance from a central computer or instrument controller. The data is transmitted via a variable current rather than a variable voltage. A voltage-to-current operational amplifier circuit is used to perform the conversion.

This op amp is of a noninverting type. The current in the feedback loop passes through a feed component (a resistor, diode, or capacitor depending upon the circuit requirements) that is connected to the (-) input terminal. The (-) input is directly proportional to the voltage input and the gain responds as 1/Ri. As Ri decreases, 1/Ri increases and the feedback of amperage feedback increases proportionally to provide the current conversion. The output is provided between V in (-) and Vo, not to ground; this is used only in an electrical "floating" connection.

The common instrumentation amperages converted for sensor signals range from four to twenty mA. This current range will transport instrumentation signals over distances where voltage signals would suffer considerable voltage drop,

Figure 2-45. The instrument amperages use four to twenty mA rather than zero to twenty mA because a zero mA indicates a nonoperative circuit. Starting with four mA indicates that the circuit is functioning.

Bridge Amplifiers

Bridge amplifiers are based on the voltage dividers concept of the Wheatstone bridge coupled with an op amp. The four resistor network is the basis for making very accurate resistance measurements. The Wheatstone bridge is two voltage dividers connected in parallel whereby a voltage variation due to resistance can be measured between the two dividers. This concept is widely used to obtain the resistance value of an unknown resistor by comparing it in a circuit with three resistors of known resistance value.

The voltage in the Wheatstone bridge is not important, but it must be constant. The function of the voltage in the Wheatstone bridge is to excite the circuit so that a voltage drop in the unknown resistor may be measured. If R4 resistor is replaced by an unknown resistance, the input (+ and - voltage to the op amp) is not equal to zero as it would be if all four of the resistors were equal. The voltage output of the op amp is proportional to the differences of the voltage at - and + inputs. The amplified voltage output is calibrated on a readout instrument as the resistance of the unknown resistor. The circuit can be converted into a very accurate thermometer by replacing the unknown resistor with a thermistor whose resistance changes with a change in temperature. This instrument output is capable of measuring or controlling a temperature to one one-thousanth of a degree Fahrenheit, Figure 2-46.

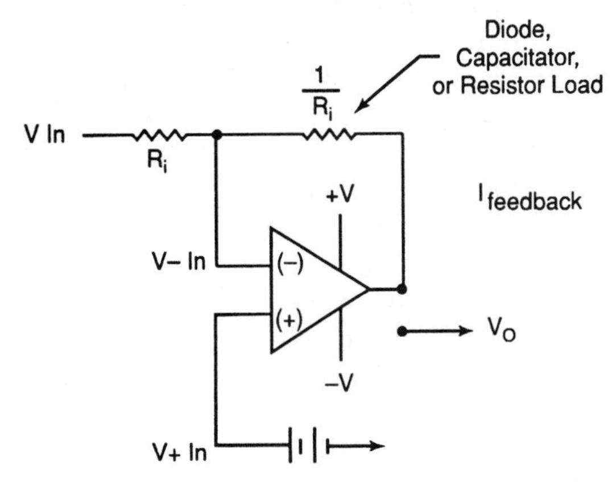

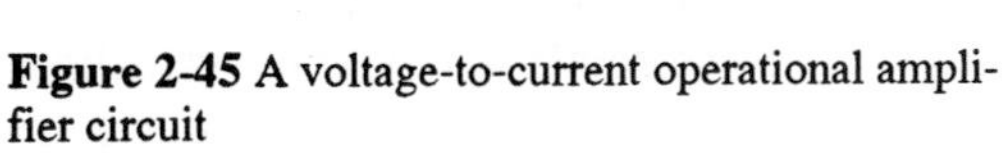

Figure 2-45 A voltage-to-current operational amplifier circuit

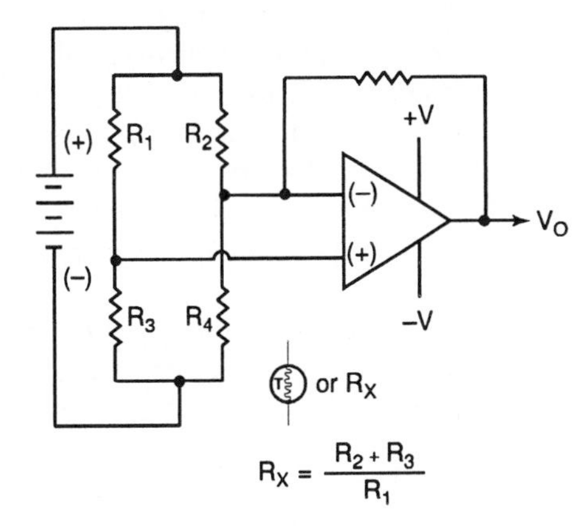

Figure 2-46 An operational amplifier circuit applied to a Wheatstone bridge circuit

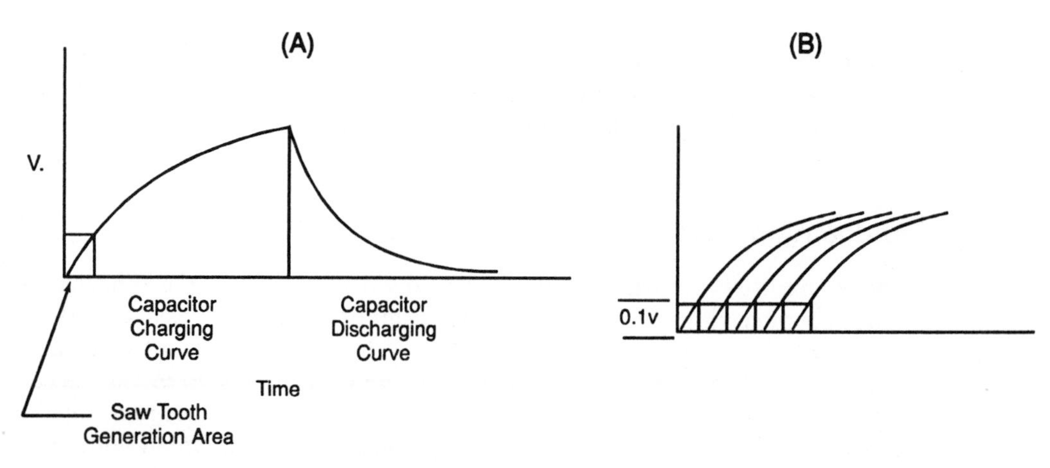

Figure 2-47 (A) Portion of the RC circuit; (B) Diagram of a sawtooth generator output

Sawtooth Frequency Generator

A sawtooth frequency generator is designed by charging a capacitor at a rate controlled by a resistor of one size and discharging it through a resistor of a different value, or by shorting the capacitor directly to ground. This is an application of a time constant in a discharging circuit, Figure 2-47. One design of this device produces a very accurate clock.

The Unijunction Transistor High Speed Switch - When a high-speed switch is added to an RC circuit, the charging time constant curve and discharge curve are controlled by an electronic voltage sensitive switch called a unijunction transistor. When open, this transistor will charge the capacitor until it reaches a predetermined voltage value; this voltage then closes the switch and discharges the capacitor. Typically, voltage values of 0.6 to 0.7 microvolts will cause the transistor to switch. The switching is set so that only the linear portion of the charging curve is used, which results in a sawtooth-shaped output signal. These source voltage oscillations, combined with a high-speed switch, provide the basic elements for a solid-state timer that can provide a wide range of applications. As an example, the unijuction transistor is often used to trigger an SCR or triac, Figure 2-48.

Using the unijunction transistor as a high-speed voltage sensitive switch, the capacitor's voltage oscillates with a continuous timing between a peak voltage and a valley voltage. However, the capacitor voltage does not go all the way to zero voltage.

The 555 Timer - By combining capacitors with resistors and voltage sensitive switching devices, a variety of signal-generating circuits with voltage threshold detectors are designed. One application is the 555 timer. A 555 integrated circuit

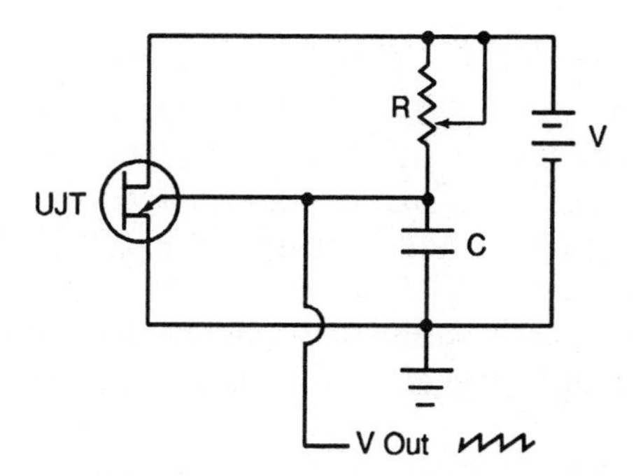

Figure 2-48 The unijunction transistor circuit as a high-speed voltage sensitive switch

is a solid-state device with a large amount of circuitry inside, but it is considered wholly as a single component. External capacitors and resistors are added to the integrated circuit in order to have it deliver the output characteristics desired. A dual-in-line (DIP) package integrated circuit has eight terminals, Figure 2-49.

Pin 1 and pin 8 are connected to an external DC power supply and vary as demanded by the other external components. Pin 3 is the output voltage used to control external components and the voltage is either zero or equal to the voltage of pin 8. Pin 2 and pin 6 are important because they control the behavior of the integrated circuit and thus the output. Pin 2 is the trigger and pin 6 the threshold -- these pins are voltage level detectors. Either pin will trigger a voltage output from pin 3. Pin 5 is the control voltage pin and is not always used; frequently it will be grounded through an external capacitor. Pin 4 is the reset pin and attached to the power supply. Pin 7 is the discharge and provides a conductor to ground when the capacitor is discharged; thus its voltage will go to zero. Finally, to repeat,the voltage of pin 3 is the output that goes to the component to be controlled.

A Square Wave Generator - A square wave generator is constructed with an RC circuit, an op amp with a voltage divider, a feedback circuit to provide an astable multivibrator circuit. This square wave device vibrates between two voltage levels. The voltage levels are usually zero volts and plus five volts. The device is built around a 555 integrated circuit with external capacitors and

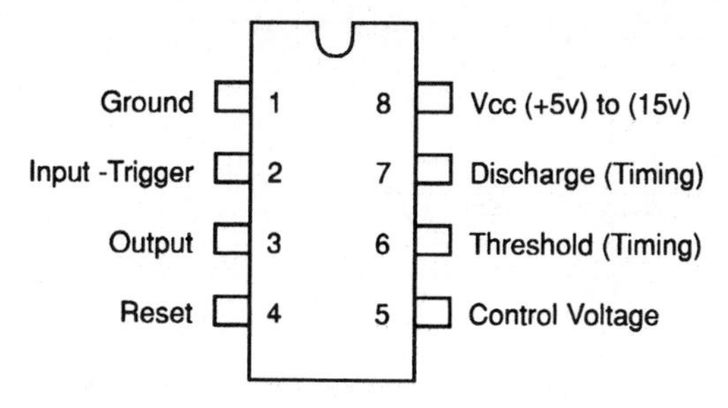

Figure 2-49 Eight pin dual-in-line (DIP) package

resistors added to provide an RC circuit to make it an astable multivibrator with an extended duty cycle. The extended duty cycle means that its minimum can be extended below 50 percent by having different resistors R1 and R2 charge the capacitor and using only R2. A diode that is forward biased provides a short circuit for R2 when discharging the capacitor to ground. The diode will not pass reverse current through itself, thus the discharge time constant and the width of the low output remain as before. Resistors R1 and R2 are varied independently until the device delivers a full range of duty cycles from 0 percent to 100 percent, Figure 2-50.

To explain how this circuit functions, it is assumed that pin 2 and pin 6 are low, and the output pin 3 is high. At this time pin 7 is open — in the nonconducting state — therefore the energy from the voltage supply enters R1 and R2 and starts charging the capacitor (C). Charging the capacitor through both resistors provides the time constant or delay. When pin 6 voltage reaches two-thirds of the charge on the capacitor, the internal threshold switches the circuit and changes the output to a low by shorting pin 7 to ground through resistor R2. The capacitor discharges until the voltage at the trigger pin 2 drops below one-third of the capacitor's voltage. When this point is reached, an internal switch opens at pin 7 and the capacitor repeats its charging. This cycle continues to produce a pulsed frequency with an output pulse width that is determined by the time required to charge the capacitor through both resistors R1 and R2. The time between the pulses is determined by the discharge of the capacitor through resistance R2. The period of the square wave is the sum of the charging and discharging time of the capacitor through the various resistor combinations.

An example of the use of the 555 timer in a RC circuit is an audible temperature alarm. As a thermistor changes in its resistance because of a change in temperature, the 555 timer generates a square wave that changes its frequency

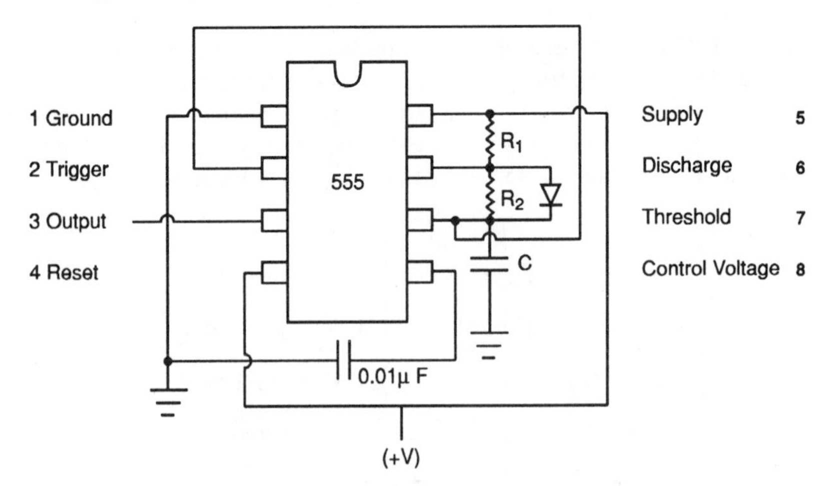

Figure 2-50 Schematic of a 555 IC modified to produce a square wave generator

with the temperature change. The resistance change of the thermistor is set to provide an audible signal at a specific temperature.

In this application, changes are made to develop a square wave generator (astable multivibrator) that will deliver nearly a full range of output duty cycles from 0 percent up to 100 percent. The change in frequency is brought about by a thermistor being placed in a square wave generator circuit by replacing resistor R1. Because it is a thermistor, it varies its resistance with a varying temperature. As the temperature of an industrial process changes, the square wave generator's frequency varies with the temperature. The frequency of the square wave generator is set so that at a designated temperature, the frequency is heard in a human's audible range, and the alarm is heard through a loudspeaker.

Alarm systems can be designed using a square wave generator to report most industrial control variables such as the level of fluids, the rate of flow, acid or alkaline levels (pH), pressures, etc. The basic ingredient in such warning devices is a sensor that will change its resistance as the industrial variable changes.

Solid-State Switching

An integrated circuit, the 555 IC, is referred to as a clock because it produces a very accurately pulsed output. To build a solid-state switch, additional external resistors and capacitors are attached to the 555 IC so that it can be operated as a timed switch. This new configuration of the circuit can be used to turn on or off a variety of different actuators such as solenoids, relays, motors, lights, pumps, valves, or other devices.

Time Delayed Switch-An RC circuit attached to the IC circuit creates a time delayed circuit that is based upon the time constant of the circuit. The time constant is the result of the charging and discharging of the capacitor. This produces a monostable multivibrator and is controlled by a voltage on pin 2 on the IC, changing from a high to a low condition.

An application of this type of circuit is for timers used in various production processes. For example, in adding acids to a chemical batch, the volume of acid is dropped by an interval of time rather than by a volume measurement. A button is pushed to activate the action, and the acid solution is added until the required amount is dropped. This solid-state switching device is capable of timing the duration of the drop to a microsecond.

This circuit is named a Schmitt trigger or one-shot multivibrator, because to start the timing sequence, pin 2 receives a low pulse and the unit will complete its time cycle without any other signal. This could be started manually by inserting a spring-loaded switch to provide a ground or low to pin 2. In the application of automated equipment, the low is provided by a voltage comparator (op amp) to start the sequence.

The timing sequence starts when pin 2 is grounded or receives a low pulse — a zero voltage pulse — that internally produces a high in the IC at pin 3 equal to the supply voltage of the IC source. The capacitor starts to charge by receiving energy from the supply through resistor R2. The output (pin 3) is high and remains in that condition while the capacitor is charging and until it reaches two-thirds of the supply voltage. Upon reaching two-thirds of charging voltage, pin 7 closes an internal switch and discharges the capacitor to ground. At that point pin 3 drops to the low state of zero volts, and the timed cycle is complete. The time that pin 3 was high was determined by the time constant of the resistor R2 and the capacitor. The length of the time delay is adjusted with the values of the resistor and the capacitor.

The voltage in the capacitor oscillates between one-third of the supply voltage to two-thirds of the supply voltage, and the length of time it takes to charge the capacitor determines the amount of time delay. Because any low or zero voltage on pin 2 will start the time delay that always delivers a pulse with the same height and width, any input *transducer* that will produce a low can interface with this timed switch. This time delayed switch can produce a delay of a fraction of a second to an hour, Figure 2-51.

The switch is closed by a transistor or push button and the output goes to the supply voltage for as long as the time constant of the RC. In the general field of electronics, capacitors have other uses in addition to storing energy, time-delay switching, and frequency generation. They are used in other circuit applications, e.g., tuning and trimming; by employing variable capacitors, filtering to remove AC voltages from DC circuits; or for coupling, used to pass or couple a particular frequency to another stage in a circuit.

Operational amplifiers are very versatile devices. In addition to the principles discussed, there are other amplifiers that have been designed to solve many other problems of electronics, science, and industry.

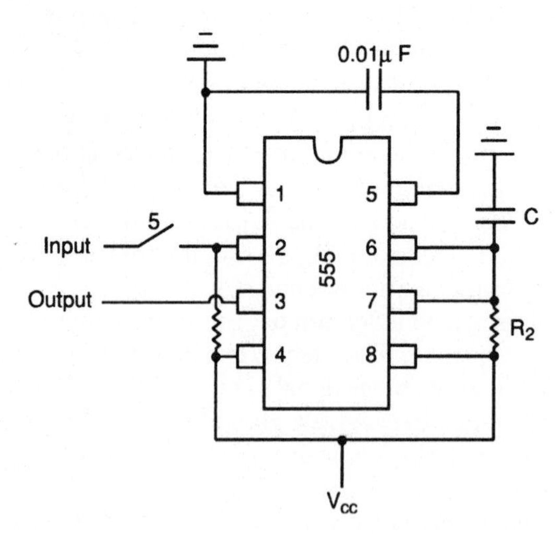

Figure 2-51 A time delayed switch

The Integrated Circuit

Integrated circuits (ICS) and large-scale integrated circuits (LSI) (or microchips) consist of complex arrays of circuits imprinted on to silicon chips. These circuits are made up of the elements mentioned above. They consist of resistors, capacitors, diodes, transistors, and a large number of operational amplifiers and memory circuits — all on a chip that will fit on your fingernail. These chips today have hundreds of thousands of circuits packed into this small space, thus allowing the circuits to function very rapidly and inexpensively in a compact area.

With the application of integrated circuit measuring instruments, not only do they make measurements but they are also capable of analyzing data and conditioning the signals received from a number of sensor inputs. This type of instrumentation is referred to as "smart sensor" instrumentation.

The Microprocessor

The microprocesser is frequently referred to as a computer on a chip because it contains all the essentials for a computer's central processing unit. There are circuits within the microprocessor that provide nearly all digital functions required. The microprocessor is programmable. With suitable programming, it can activate these circuits and sequence and perform the functions that other external digital systems previously performed. This capability makes it possible to replace many circuit boards, hardware, and gates with a programmed microprocessor, Figure 2-52.

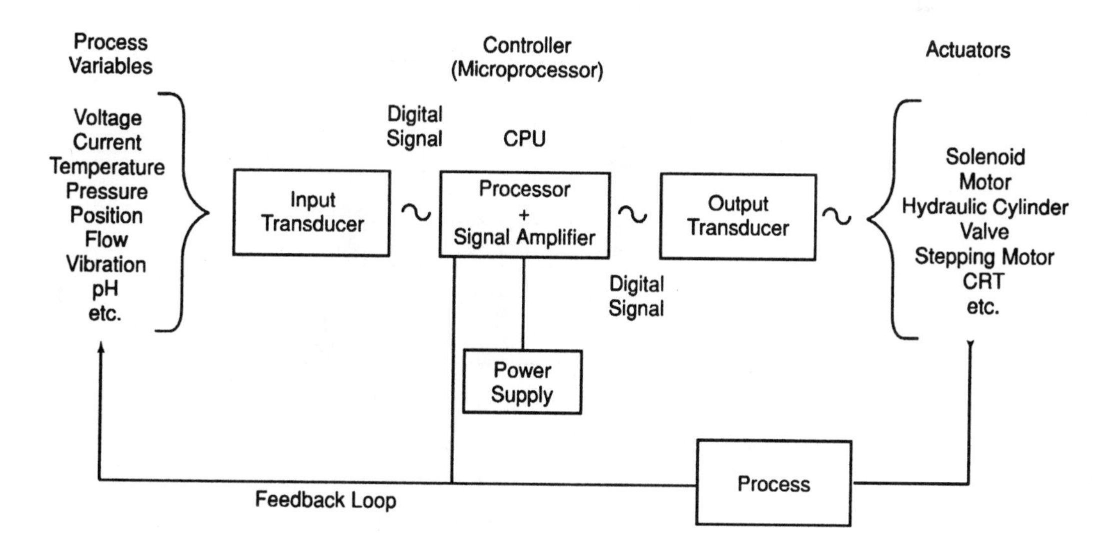

Figure 2-52 A diagram of a microprocessor manufacturing control

Microprocessor Construction

A typical microprocessor is a single device that contains the following units: a decode and control unit that interprets and proceeds on the instructions from the stored program, an arithmetic and logic unit that performs these functions, and registers that provide easily accessible memory to store frequently manipulated data. This is the interpreting section of the microprocessor. This section is the central processing unit (CPU) — it interprets the input and performs actions that are based on the memory's instructions.

Addition or extra memory may be accessed outside in a microcomputer. The microprocessor contains an accumulator, which is a special register associated with the arithmetic logic unit. Address buffers read the instructions or data coming into or out of the microprocessor and provide the control memory with the address from which to fetch the next instruction that is a part of the controlling unit.

Microcomputer Construction

The microcomputer system is constructed with three basic parts: a microprocessor, a memory, and an input/output section. The input part is made up of a semiconductor memory. The memory depends on the microcomputer's design: it may be a random access memory (RAM), read only memory (ROM), programmable read only memory (PROM), or erasable progragrammable read only memory (EPROM or EROM). Instructions are stored within the memory for the interpreting section of the microprocessor. The microcomputer will always follow these instructions and sequence.

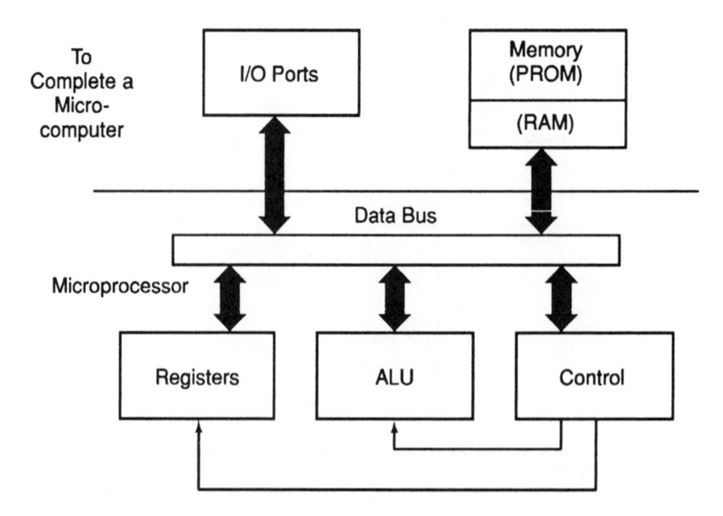

Figure 2-53 The microprocessor expanded to form a microcomputer

The third or final part of the microcomputer is the input/output (I/O). The output is a digital signal that activates the display device (CRT) or the actuator (relays, solenoid, motors, valves), making a physical change to control the manufacturing process or product, Figure 2-53.

Today, microprocessors can be obtained on a single chip with additional memory and I/O. They are also referred to as microcomputers.

SUMMARY/FACTS

- Electronics is used in instrumentation because of its speed, flexibility, reliability, and the small space it requires.
- The larger the electrical charge difference between two dissimilar charges, the greater the potential difference or voltage.
- Amperage is the rate of flow of an electrical current. It provides the electrical energy that is capable of doing work.
- Electrical resistance is measured in ohms and is a fundamental element because it determines the amount of current that can flow through an electrical circuit.
- Conductors are materials that permit the free flow of electrons through the substance. They have a low resistance and transport electrical energy with a low loss.
- When a capacitor is connected in series with a resistor, an electronic time delay circuit is created. This circuit has a time constant. One time constant is the time required to charge the capacitor to 63.2 percent of its full charge.
- Ohm's law reveals that the flow of current through a resistance is directly proportional to the voltage and inversely proportional to the resistance.
- Both silicon and germaninum crystals can have other material added to their surfaces so that when a certain voltage is impressed on them, electrons become available and the total device becomes conductive. This is the basis for many solid-state devices.
- A voltage divider is a network of resistors designed to perform the division and control of voltages.
- Digital technology, based on devices that operate on discrete or discontinuous signals, utilizes a binary or two-state system of information. The information, in the form of the digits 0 and 1, is applied to a code system.
- Logic gates within a circuit provide a mathematical analysis of all the inputs into the various gates of the circuit and deliver an output of the circuit analysis.
- The operational amplifier is essentially made up of three amplifier units: a differential amplifier, a voltage amplifier, and an output amplifier.

- The unijunction transistor is a high-speed switch that is often used to trigger an SCR or triac.
- The parts of a chip microcomputer are the CPU, memory, and input/output sections.

REVIEW QUESTIONS

1. What are the three building block concepts of energy applied in electronics?
2. When there is a large electrical charge difference between two dissimilar charges, what is the result?
3. What is amperage?
4. What is resistivity?
5. Define resistance.
6. List some factors that cause resistance to vary in a conductor.
7. How is the value of a fixed resistor determined?
8. In instruments, in what shapes and forms may conductors appear?
9. What are insulators made of?
10. What is an electrical capacitor?
11. How is a time constant established?
12. What is a time constant?
13. Where is the energy stored in an electromagnetic inductance device?
14. What is the role of impedance in measuring with electrical instruments?
15. What test instrument is employed for troubleshooting electrical circuits?
16. What are the basic relationships of Ohm's law?
17. How does a semiconductor work?
18. What is the function of the transition voltage in a diode?
19. Why is the transistor a voltage controlled switch?
20. Why is the operational amplifier so important to electronic instrumentation?
21. What is an application of the field effect transistor?
22. What name is given to a four-layer semiconductor that has internal feedback and provides a latching action?
23. The SCR is another four-layer semiconductor. Why is it valuable?
24. What is the application of the unijunction transistor?

25. What is the function of a voltage divider?
26. What is a binary numbering system based upon?
27. Describe why the binary numbering system is important to the electronic industry.
28. Why is Boolean algebra used with digital equipment?
29. Why is an octal code used?
30. What is the function of the logic gate?
31. What is the chief function performed by operational amplifiers?
32. What is the 555 timer?
33. What function does a Schmidt trigger perform?
34. Why is the microprocessor frequently referred to as a computer on a chip?
35. What are the three basic parts of a microcomputer?

3

INSTRUMENTATION CONCEPTS

OBJECTIVES

Upon completing this chapter, you will know and be able to describe and apply the following concepts:

- How data is acquired through change in physical phenomenon — such as changes in solids, liquids, gases, vapor pressure, resistance, voltage, semiconductors, logic states, etc.
- How these concepts are applied to the common manufacturing process variables — such as temperature, pressure, flow, level, density, viscosity, force, humidity, position, etc.

INTRODUCTION

This chapter is concerned with the concepts that underlie the basic phenomena necessary to generate the sensor data. This is brought about by some sort of change in the sensor material. One of the primary uses of instrumentation is to gather the information about this change in the phenomena and use it as a data signal. The rate of flow, temperature, pressure, or position are examples of what are commonly referred to as variables in the system. Any change in the variable may be measured by the instrumentation. The instrumentation system is designed to control the variable thereby controlling the total system.

PHYSICAL PHENOMENA USED FOR MEASURING

The characteristics of various materials are used to accomplish the measurements that are required. The materials used are found in three forms: solids, liquids, and gases. These material forms may be subjected to heat, pressure, or voltage, for example, to change their physical size or electrical properties. These changes are sensed and provide a measurement. Measurement in instrumentation is the extraction of a signal that is proportional to the physical, chemical, or electrical change in the material as the result of a change in the variable.

A change may be provided by heat causing a metal sensor to expand. A change in the length or size of the metal sensor results in the signal of a change of measurement. This new data will be used to correct the variable and thus the system.

Measurement in instrumentation therefore is the extraction of a signal that is proportional to the physical, chemical, electronic system or processes that represent the control variable. A change of condition in the variable will usually be sensed by a change in the materials, such as the expansion or contraction of metals, fluids, or gases, or a change in electrical properties. These changes result in an electrical output or a mechanical linkage in the form of a number or value that is relative to a calibrated reference position. These numbers or values appear on a scale, display, recorder, or controller that has been calibrated against physical standards to confirm the accuracy of the measurements.

To refer these measuring concepts to human senses, we may measure in broad terms, such as "This soup tastes too salty." In this case we have sampled the product in process (soup) and compared it to a standard in the mind which made an analysis resulting in a decision that the soup was too salty. The question of the soup should be, "saltier than what?" Good instrumentation would give a scale that has been standardized to answer that question. The answer may be reported in specific gravity of the solution or as to the number of parts of salt per million parts of water. The answer above is subjective because of the lack of a repeatable scientific standard. Industry requires accurate and repeatable process instruments referred to as standards — traceable back to the National Bureau of Standards (NBS) of the United States.

In any case, measurement is the sampling of the variable under consideration, supplying a standardized signal that reliably represents the condition of the variable at the time of measurement.

Measuring by Changes in Materials

Some changes in the physical characteristics of materials are obvious, e.g., length and volume. There are other not-so-apparent changes, including electrical characteristics such as resistance, voltage, amperage, and frequency. Whatever variable is considered, a measurable change in the sensor is required for an accurate measurement of that variable.

Changes in Solids

With the addition of heat, solid materials expand their volume. This is the result of increased molecular activity from the applied heat energy. Likewise, the cooling of solids reduces their volume. This change in thermal expansion may be applied as the basis for a sensor and used with many temperature sensitive instruments. The selection of the material and the designing of its configuration or cross section of the sensor are used to create many different instrument applications. Various metals are rolled into ribbons, stamped into discs, or formed into many different shapes and made into many inexpensive heat sensors. These devices are called transducers because they measure the phenomenon of heat by its change in volume or length. These transducers change heat energy to mechanical movement that can be observed and recorded.

Changes in Liquids

Liquids also expand when heated because of the increase in molecular activity. When a liquid is enclosed in a container such as a sensing bulb, bourdon tube, or bellows and sealed, it can be used as a heat sensing device. In this case, the liquids expand and produce a pressure in the confined system. The new pressure can be transmitted to a number of different types of devices that convert the new pressure to a signal that represents the change in temperature. A large number of transducers have been designed using this phenomenon.

Changes in Gases

Gases also expand when they are heated and follow the dictates of the ideal gas laws regarding temperature, pressure, and volume. Various gases can be confined in a system made up of a bulb, capillary tube, and bourdon tube to build a very temperature-sensitive instrument. In this case, as the heat increases, the pressure increases in the confined volume; the bourdon tube responds to the pressure and changes its configuration, producing mechanical movement that can be scaled.

Changes in Vapor Pressure

Vapor pressures exist above confined volatile liquids that are subject to temperature variation. The vapor pressure of specific liquids generates a pressure at a specific temperature. When an equilibrium has been reached above the liquid, a vapor pressure is established. This vapor pressure can be used to provide a signal to various devices. A small bulb, capillary tube, and bourdon tube are again employed to measure temperature by converting temperature into a pressure.

Changes in Resistance

Different materials vary in their ability to conduct electricity. In addition, most conductors increase in resistance with the increase of temperature. Resistance is also inversely proportional to a conductor's size and length. Factors that effect a change in the resistance of a device are these: the temperature of the resistor, the material from which it is made, the wire diameter of the resistor, and the length of the resistor. Any changes in any or all of these parameters will result in a different resistance value, thus making it possible to sense a change that will provide a usable signal. With the advent of superconductors, the responses of these characteristics may be different.

Changes in Inductance

Inductance is the property of a conductor that opposes a change in the current flow in the conductor because of the opposing energy stored in the magnetic flux around the conductor. Transducer devices produce an inductance change as a change in a voltage output. A sensor can utilize this voltage change as a signal to control a production process. Inductance devices accomplish this by repositioning the core material within a coil, increasing or decreasing the number of conductor turns in the coil, increasing or decreasing the diameter of the coil, or increasing or decreasing the length of the coil. In each case, a change will produce a voltage change in the output.

Changes in Conductivity

Conductivity is the transmission of electrical energy through a material. It is the reciprocal of resistance. Conductivity is a measure of the ease with which electricity flows. An excellent conductor requires little energy to free electrons to allow the transfer of electrical energy. A sensor that utilizes a change in conductivity is governed by the same factors that produce a change in resistance, namely, a change in the device's temperature, selection of material, the diameter of the conductor, or the length of the conductor. The output will be given as a voltage change and be the reciprocal of the resistance of the device.

Change in Voltage

Voltage is the amount of energy transferred to a charge as it moves through a conductor. Voltage defines the size of the potential difference and determines the rate at which a charge will flow in a conductor. Voltage devices sense a voltage or the change in an existing voltage. Voltages can be generated by sensors by utilizing thermoelectric, electromagnetic, photoelectric, electrochemical, *piezo-electric*, or electrostatic principles. Voltage devices may sense by resistance, impedance, or other electrical signal conditioning.

Change in Amperage

Amperage is referred to as an electric current caused by electrons in motion. The electrical rate of flow, i.e., the current, is measured in coulombs past a point in a conductor per second. A device that senses a change in amperage is referred to as an ammeter. It is calibrated to measure a broad range of current values. In addition, a number of electronic circuits have been designed to measure and provide an amperage signal.

Change in Capacitance

Capacitance is the property of a material to have instantaneous storage of an electrical charge. The unit of measurement used for capacitance is the farad. The factors that control capacitance in the device are the size and number of plates in the capacitor, the distance between the plates, and the permittivity or the dielectric constant of the insulating material. The electrical charge stored in a capacitor is directly proportional to the voltage applied to the capacitor. Any device that alters the relationship between these factors provides a signal that can be calibrated and used to measure a variable. In capacitive and inductive transducers with an alternating current applied, a change in either the capacitance or inductance will alter the current output. A change in current is measured that provides a signal of the process variable being measured and controlled.

Change in Frequency

Electrical frequency is the number of vibrations or cycles per second. It is reported as hertz per second. Data acquisition sensors are capable of counting the number of vibrations or cycles or of measuring the amount of time required for a specific number of cycles to occur. The frequency is changed commonly with the function of an RC circuit.

Electronically, an impedance bridge (with arms of known impedance and frequency) can be balanced against a frequency to be measured to produce a measurement. Another procedure involves an electronic circuit using resistances and capacitors so that a standard or set-point frequency is produced. This output frequency is compared to the unknown frequency coming from a transducer, and the unknown frequency is tuned to a condition of resonance. In this condition, the unknown frequency of the variable has been controlled, and the set-point has been represented by a frequency.

Change in a Semiconductor

Semiconductors can be either insulators or conductors, depending upon their electrical condition. They are usually made of germanium or silicon materials. These materials share electrons with other atoms to form a covalent bonded crystal. This material may be formulated so that an electrical current will flow in only one direction and can be activated by applying only a very small voltage. A small increase in the voltage will suddenly cause a change to occur. The device becomes conductive, that is, able to conduct a large current through the device. It becomes a one-way electronic valve.

The devices may be assembled in different configurations to produce transistors. Transistors are able to control a large current with a very small voltage. These are solid-state devices.

Change in a Logic State

Solid-state devices have revolutionized the sensor and control industry in data acquisition and analysis. The application of digital systems uses a binary number system based on the Boolean technique of analyzing data with a decision procedure of yes or no, true or false, on or off, high or low, 1 or 0 to indicate the condition of a sensor or basic data.

Solid-state electronics has been employed to create a series of devices that can integrate digital circuits into a decision-making system. These logic devices are designed to take sensor data and analyze it, to make a decision as to whether to start or to stop a process based on a change in logic state.

Table 3-1
Common Process Variables

Acceleration	Impedance	Temperature
Capacitance	Inductance	Thickness
Color	Level	Time
Conductivity	Light	Vacuum
Current	Mass	Vibration
Density	Moisture	Viscosity
Distance	pH concentration	Voltage
Flow	Position	Volume
Force	Pressure	Weight
Frequency	Resistance	
Hardness	Sound	
Humidity	Specific gravity	
Hydrostatic head	Speed	

Manufacturing Process Variable Concepts

The process variable is the parameter in manufacturing that is manipulated to provide control so that a consistent, uniform, and satisfactory product within required specifications is produced. As an example, in the manufacturing process of heat treating, it is very important to control the process variable of temperature. A specific temperature must be maintained. For example, even when a load of cold steel is added to the furnace, that temperature must be held. Because of the change of conditions, the control system must compensate for the new thermal *load*.

Temperature

Temperature measurement and control is one of the most important variables used in a very large number of industrial processes. Temperature is a measure of the relative hotness or coldness of a material. It is the result of molecular activity within the substance. Heat is a form of energy that is increased or decreased in relationship to the substance's molecular activity. Heat is the amount of internal energy that is available to flow from one substance to another substance. Temperature, then, is a measure of the average velocity of the molecules in a substance.

Temperature Scales - Temperature can be reported on a number of scales. The four common scales are these:

Fahrenheit	F
Celsius	C
Kelvin	K
Rankine	R

Temperatures can be converted from one scale to another with the application of these formulas:

$F = 9/5\ (C + 32)$ $\qquad$ $C = 5/9\ (F - 32)$

$K = C + 273.15$ $\qquad$ $R = F + 459.67$

Temperature Transfer - The measurement of temperature in manufacturing is important because the variable of temperature is used to control so many processes. Heat energy transfers from one medium to another by three methods: conduction, convection, and radiation.

Conduction heat transfer takes place when one surface is in contact with another, e.g., when one metal object is contact with another metal object. Metals like aluminum and copper transfer heat quickly by conduction, Figure 3-1.

In convection, heat is transferred by a medium such as a gas or liquid absorbing heat energy from the source and circulating that energy in the medium. These currents and the heated medium transfer the heat. The gulf stream that transfers the heat energy from the Caribbean Sea to Northern Europe is an example of heat transfer by convection, Figure 3-2.

In radiation, heat energy is transferred by electromagnetic waves. This type of heat transfer can occur over vast distances. A heated source radiates in all directions and its rays transmit its energy through a *vacuum*, liquids, or solids.

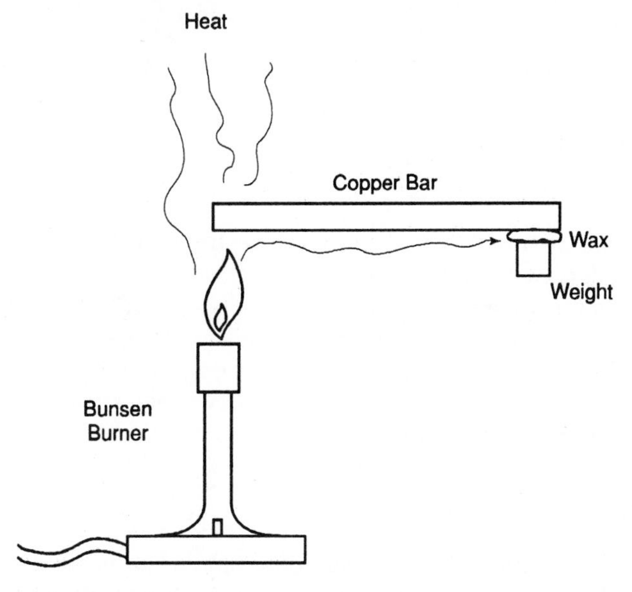

Figure 3-1 Heat transfer by conduction. The heat is transferred through the solid copper bar, melting the wax and allowing the weight to fall.

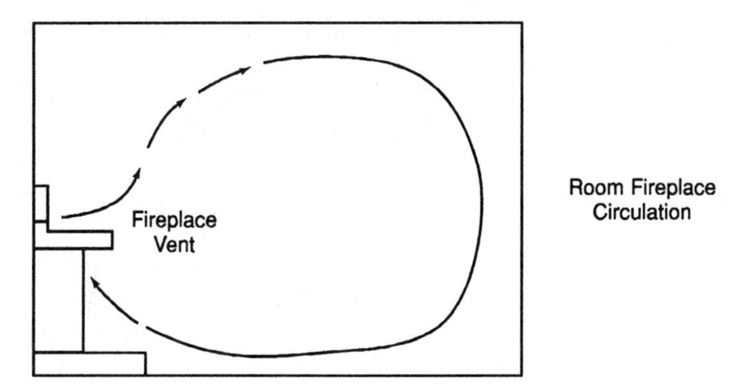

Figure 3-2 Heat transfer by convection. The heat is distributed around the room by the movement of the warm air.

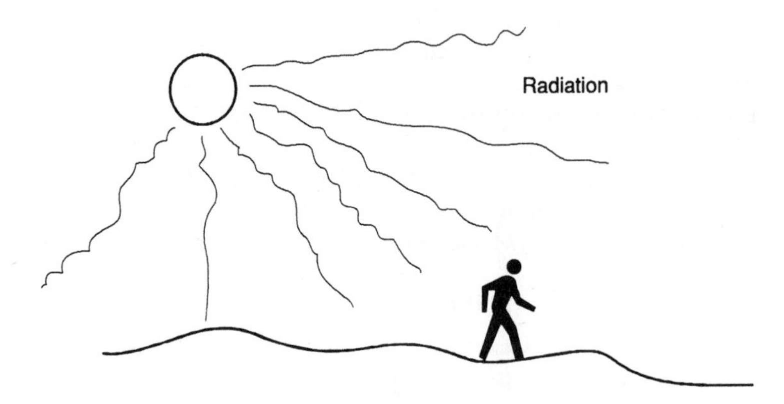

Figure 3-3 Heat transfer by radiation

Radiant energy traveling in the form of electromagnetic waves is retransformed into heat energy when it comes in contact with an object. The object will gain heat, and as its temperature rises, it itself will radiate energy. The sun's heat energy is transformed into radiant energy and travels to earth whereupon by striking the earth's objects, retransforms the radiant energy to heat, Figure 3-3.

All of these methods of heat transfer have been applied to a large group of instruments that measure temperature.

Pressure

Pressure is a variable that is commonly applied to process measurement and control. Pressure is a force applied to an area. The area may be in square inches, square feet, square yards, or square millimeters, square centimeters, square meters. In United States industrial plants, pressure is most frequently measured in pounds per square inch. Pressure is therefore defined as force divided by area.

Pressure measurements apply to both liquids and gases; both media fall under the common term "fluids." In order to understand the instrumentation applications of liquids under pressure another concept is needed. Liquids for practical purposes are incompressible and when they are confined to a restricted area such as a chamber or pipe system, Pascal's principle applies. His principle states that a pressure applied to this system will be transmitted undiminished to all of the walls and branches of the container and system, Figure 3-4.

Pressure is a variable that is measured directly as well as being used as a medium for transmitting signals and analog data from other variables. To illustrate, a pressure may be measured directly by placing a gage in a hydraulic system and measuring the pressure in the hydraulic system. Or, a pressure may be used to transmit a signal from a temperature bulb to an indicating instrument, as in a bourdon tube. Thus the variable of pressure has a dual role in measurement.

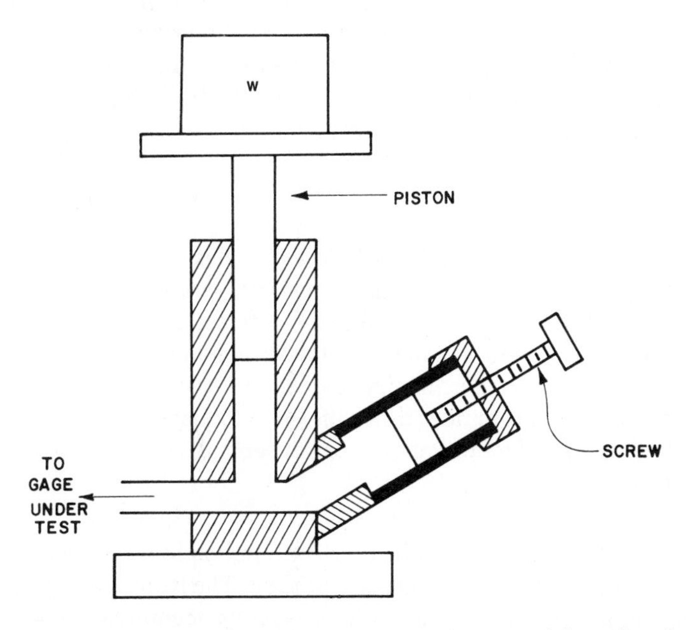

Figure 3-4 Pascal's principle. The test pressure is delivered by known weight against the gage under test. The secondary piston and screw provide elevation for the weight.

Hydrostatic Head

Liquid height or hydrostatic head is frequently measured by industrial instrumentation. Here again pressure is measured, but it is caused by the weight of the water above the instrument tap point. The higher a column of fluid above the reference point, the greater the pressure that will appear at the base of the fluid. If the fluid is water, it weighs 0.0361 pounds per square inch. If this number is multiplied by the height in inches, the pressure generated by the water column may be obtained. This pressure may be applied to a pressure gage, and it will indicate pressure; or the dial of the gage may be calibrated to indicate the level, how full the tank is, or the number of inches of water in the tank, Figure 3-5.

Gage and Absolute Pressure - Gases have the properties of fluids but they are compressible and therefore have different relationships between pressure, volume, and temperature. Gases behave according to Boyle's law: the *absolute pressure* of a stated mass of gas changes inversely as the volume changes when the temperature of the gas is held constant. A series of instruments apply this law by holding the volume constant in the system, exposing a gas temperature bulb to the medium and allowing the change in pressure to represent the signal of the temperature change.

When making calculations that are applied to gases it is important not to neglect the effects of atmospheric pressure (the *ambient condition*) on the system under consideration. Pressures that appear on a bourdon tube and like gages report *"gage pressure."* Before calculations are made, these pressures must be converted to "absolute pressure" measurements. Absolute pressure measurements are made from a vacuum base reference, Figure 3-6.

Flow

Fluid in motion is a definition of flow. This is a very important concept to industrial plants that utilize steam, water, and various other fluids and gases.

The common process variable of flow refers to a *flow rate*, that is, the volume of fluid passing a point in a measured time in some type of conduit, pipe, duct, or channel. When considering pipe, the flow rate (volume per minute) is equal to the velocity (feet per minute) times the area (square feet). The flow rate result is reported as cubic feet per minute or per second, or cubic meters per second in the SI system.

Flow in Confined Conduits - In industrial applications, the fluid flow movement in pipes is not as ideal as calculations indicate. The insides of pipes are frequently rough, fittings have burrs, and various deposits accumulate within the pipe and fittings. As a result there are other factors to consider in studying flow.

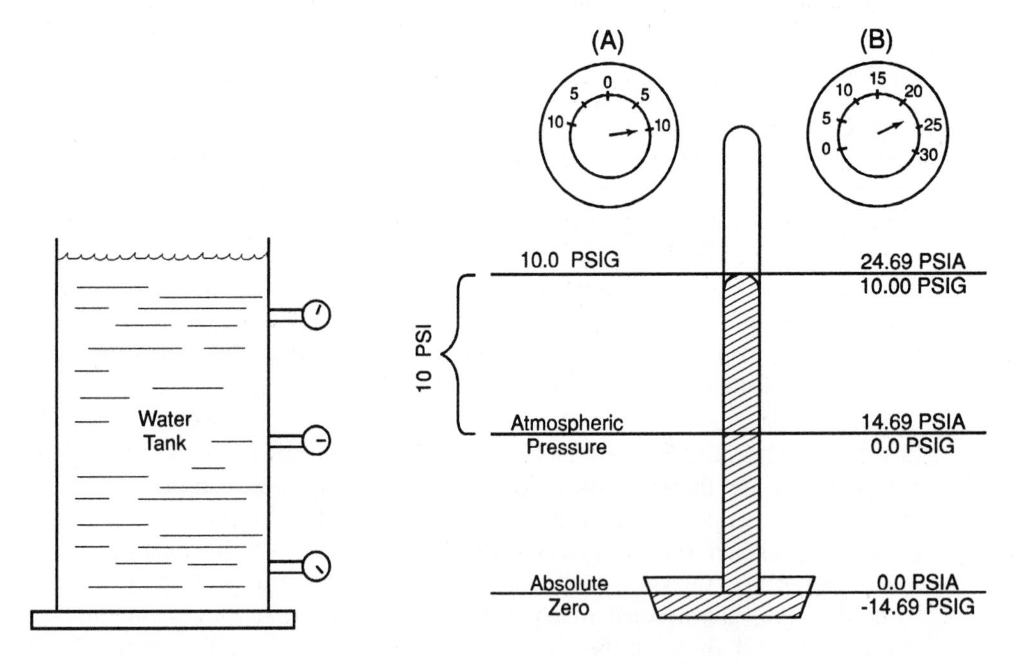

Figure 3-5 Hydrostatic head

Figure 3-6 (A) Compound gage (PSIG); (B) Absolute gage (PSIA)

Laminar and Turbulent Flow - Friction and viscosity are factors that enter into the rate of flow. When the *velocity* of a fluid flowing in a pipe is low, the fluid in the center of the pipe moves nearly as fast as the fluid near or in contact with the pipe. The total flow moves in parallel lines along the pipe. This range of flow is considered viscous or laminar. As the velocity of the fluid increases, a critical flow rate will be reached and the fluid in the center of the pipe will travel at a faster rate. The walls of the pipe create a fluid friction and retard the flow at the area of contact. At this point, eddy currents form within the fluid and a breakup of the smooth flow occurs. The flow in the pipe becomes more and more turbulent. Engineers have described this change over from laminar to turbulent flow by calculation and have assigned a Reynold's number to the phenomenon. A Reynold's number is used to describe the condition of flow within a pipe. The formula considers the diameter of the pipe, the mass density, and the viscosity of the fluids. If the calculated number is less than two thousand, the flow is considered laminar or viscous flow. When the Reynold's number is greater than four thousand, the flow is considered turbulent. Reynold's numbers between two thousand and four thousand are considered in transition and may be changing from one phase to the other and back. Engineers frequently consider this zone of flow turbulent, Figure 3-7.

Pressure Drop - In addition, pressure loss is due to friction between fluid and the pipe wall, so measurement of the flow at two different locations along an extended piece of pipe is different. This is expressed as head loss or pressure drop. To illustrate, a pipe may have a pressure applied at the source end with a given flow rate, but at the other end of the long pipe the pressure will be different — it will have dropped and, as a result, the rate of flow will have lessened. Friction will cause a loss of energy as the fluid flows through the pipe. These losses are expressed in tables dealing with the mechanics of fluids. It must be noted that valves and various fittings also add to the loss in the flow of fluids. The pressure drop for these components is reported as losses equal to the like number of feet of straight pipe, Figure 3-8.

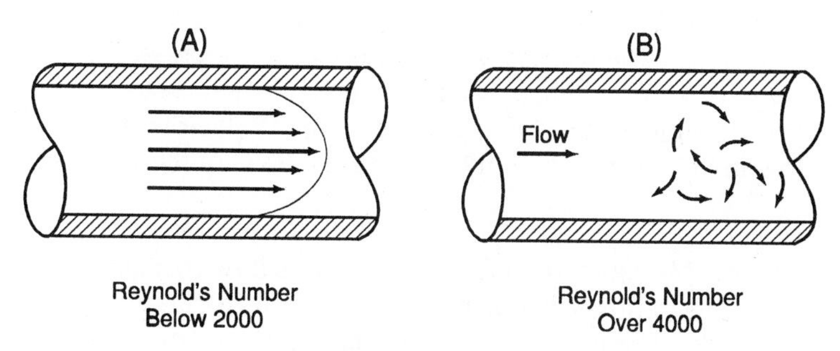

Figure 3-7 (A) Laminar flow; (B) Turbulent flow

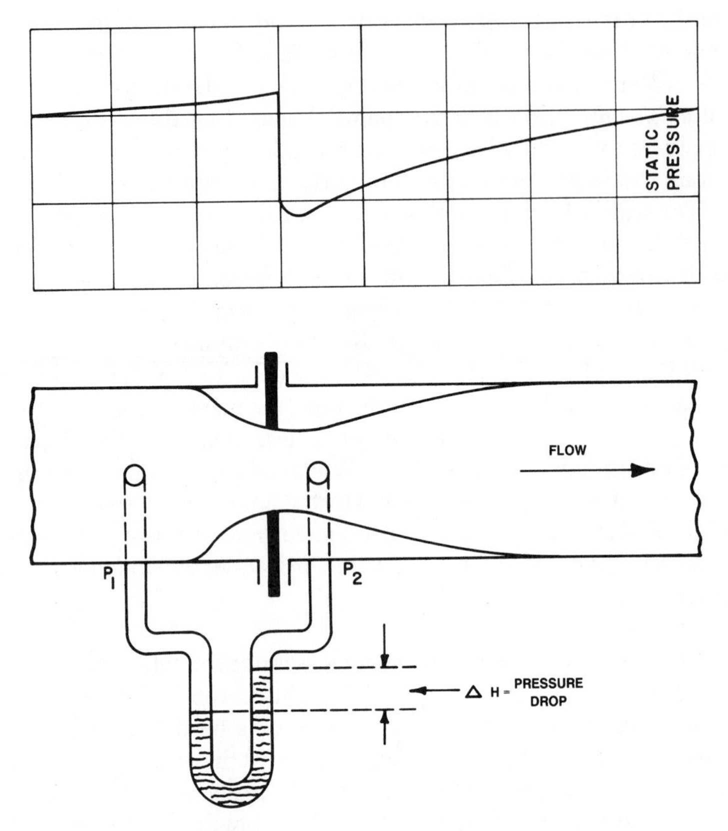

Figure 3-8 The flow through a restriction provides a pressure drop that indicates a flow rate.

Flow in Open Channels - Large open flow channel applications such as on rivers, canals, ditches, and sewerage systems are made. The flow can be measured according to the concept that a specific shaped restriction, placed in the path of flow, will cause a change in the hydrostatic head behind the restriction.

These restrictions are called "weirs" and "flumes." They cause the flow to back up behind the restriction, thus providing a rise in the fluid's level on the upstream side. The restrictions generally have a cross-sectional area that is in the shape of a "V" notch, a rectangular notch, or a trapezoidal notch. When the canal is flowing, the level of the water behind the weir will stabilize and the rate of flow in the ditch will be indicated by the level of the water behind the weir, Figure 3-9.

Mass Flow Measurement Concepts - Mass flow changes are indicated by the measurement of flow without the use of physical volumetric measuring displacement devices. There are a number of mass flow instruments manufactured for industry using an application of physics — the Coriolis (or gyroscopic) effect. The

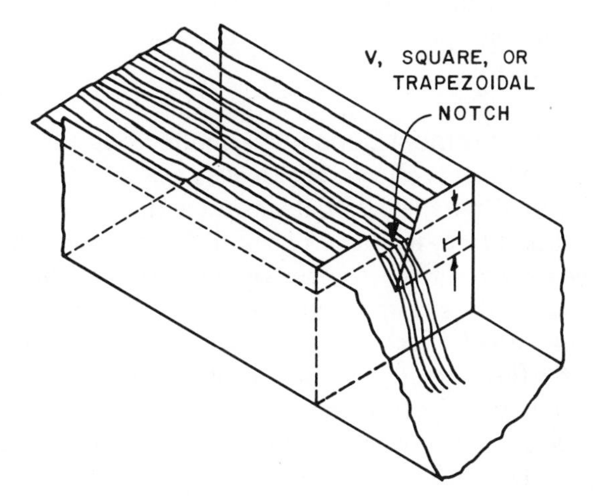

Figure 3-9 The weir and hydrostatic head. Any obstruction in an open channel causes a backup depending on flow rate. This can be calibrated in head versus flow.

mass flow meter applies a flow-sensing tube that deflects a small amount in response to the mass-flow-induced Coriolis forces. A combination of the flow in the tube elements and vibration of the tube results in a twisting or Coriolis force that is measured in a curved area by a drive and sense coil, Figure 3-10. These forces act upon the fluid mass within the instrument's vibrating elements and

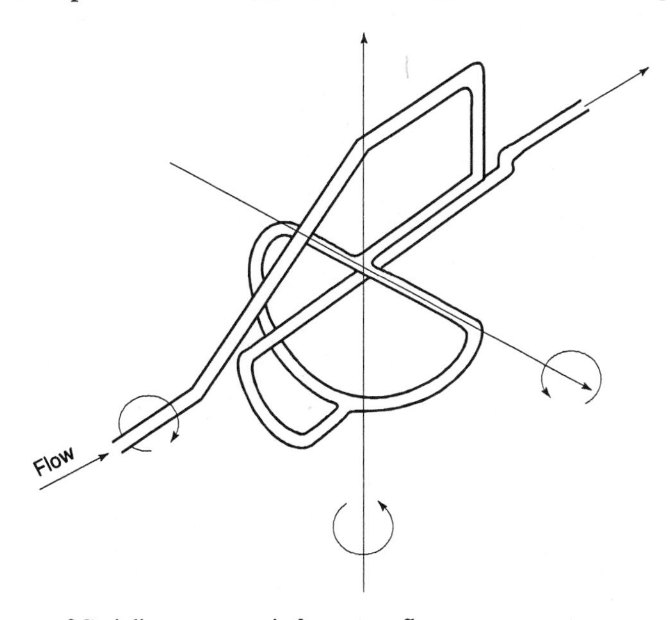

Figure 3-10 Diagram of Coriolis or gyroscopic force mass flow measurement

provide mass flow data that is independent of the physical properties of the fluid. Solid-state electronics provide digital readout data that is convenient for mass flow measurements.

Fluidic Flow Measurement Concepts - Another indicator of flow change is the application of the Coanda effect. The Coanda effect is demonstrated by passing a jet stream of fluid parallel to a flat vertical surface. When the fluid comes in contact with the surface, it attaches itself to the surface. Once the stream has attached to the vertical surface it remains in contact even if the surface is bent away from the stream. If the stream is interrupted, however, it disengages itself. This is referred to as the *wall effect*. The wall effect is used in the design of instruments where part of the flow is redirected back at ninety degrees to the main flow stream and is used to interrupt the flow from one channel to another. This effect can be used as a switch or to set up an oscillation where the frequency is a function of the flow, Figure 3-11A and Figure 3-11B.

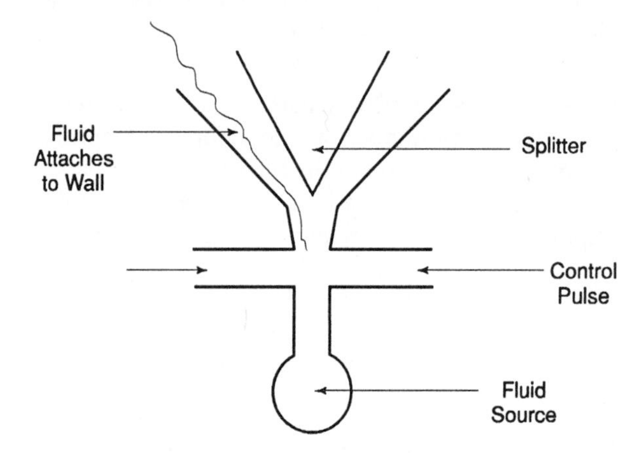

Figure 3-11A Diagram indicating the wall effect — binary switching

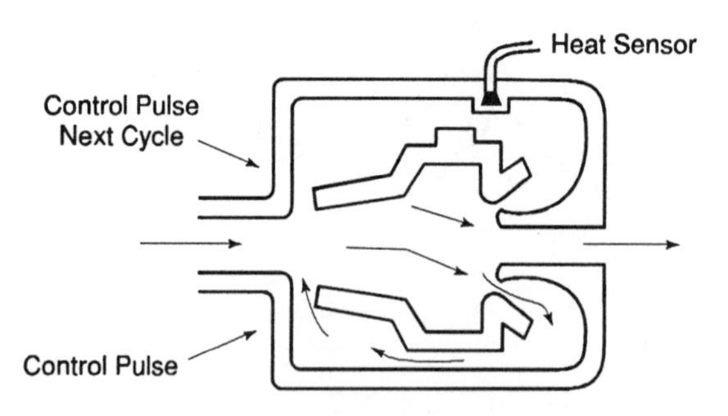

Figure 3-11B Fluid oscillator — each binary switch provides a count to a heat sensor — related to flow

Level

The determination of level is an important variable to measure in manufacturing. Level change information is necessary in the measurement of fluids, slurries, and powdered dry materials. Level frequently is a measuring factor in the control of the feeding of fluids into mixing and blending operations in many industrial processes and products. Levels in dry products stored in silos and bins are frequently measured to gather inventory data for accounting and ordering purposes, Figure 3-12. The control system is designed to respond to changes in the demand for materials. Level actually changes by the process of consumption of the material, whether it be fuel oil, milk, wheat, plastic granules, gasoline, solvents, paints, water, coal, acid, or many other industrial materials used in production.

Levels of materials can be sensed by a number of methods that will usually activate either a pump, conveyor belt, or hopper to replenish the material level that is required by the industrial process.

Density and Specific Gravity

Density and specific gravity are manufacturing variables that are frequently measured to control the quality of a product. Density is the weight per volume of a material. Specific gravity is the comparison of the density of a material to the density of water. It answers the question "Is the compared substance more or less dense than water?" The variable of specific gravity is applied to liquids such as salt solutions, petroleum products, acids, sugar solutions, and others that become ingredients in manufactured products. The density and specific gravity change with the addition or elimination of other materials.

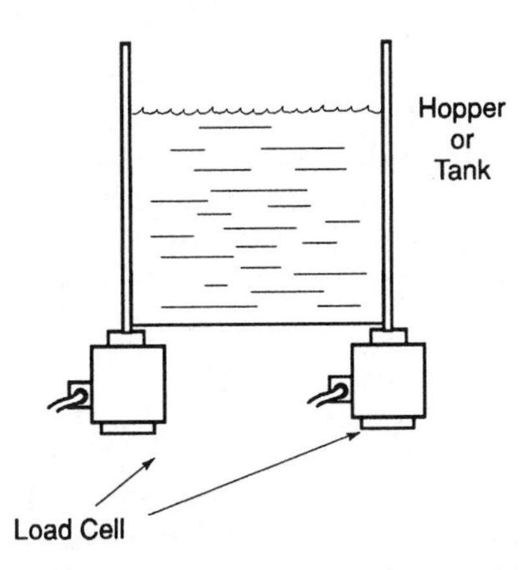

Figure 3-12 Level measurements for liquids and solids. Level of materials can be measured by weighing the total container. Load cells are commonly constructed with strain gages.

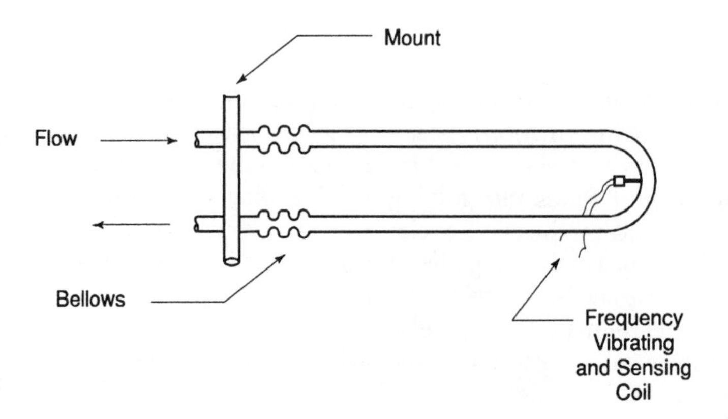

Figure 3-13 Density and specific gravity can be monitored by a change in a vibrating U-tube's frequency (Coriolis measurement). Density — weight per unit volume specific gravity — compared to water

The measurements of the variables are frequently designed around buoyancy, weight comparison, Coriolis forces, differential pressure measuring of the weight of fluids, gamma ray and ultrasonic technology applications, and others. Specific gravity provides an excellent means to measure a product's quality, Figure 3-13.

Viscosity

Viscosity is the property of a fluid that resists the force that causes it to flow. It can be measured by a timed flow through a nozzle at a specific temperature. This resistance is partly due to the friction between the boundary surface (such as a pipe) and the moving fluid area. The fluid's resistance to movement or flow is also caused by friction between the liquid particles or the fluid's shearing force.

When fluids are subject to shear stress, they flow in two fashions: constant flow and nonconstant flow. The constant flow materials are referred to as Newtonian substances. They flow with a ratio of flow-to-force as in materials such as water, gasoline, salt solutions, etc. Non-Newtonian materials such as catsup, chewing gum, most paints, peanut butter, and other products have a nonconstant flow, Figure 3-14. Viscosity change is a very important measurement variable in the chemical and petroleum industries.

Force, Weight, Stress, and Strain

Force is the action of energy applied from a source to a body; it has a magnitude and direction. An unbalanced force on a body will cause it to accelerate in the direction of the force. Force is stress multiplied by the dimensions of the area under consideration. Stress is the compressive or tensile load on a body and is measured in pounds per square inch or pounds per square meter.

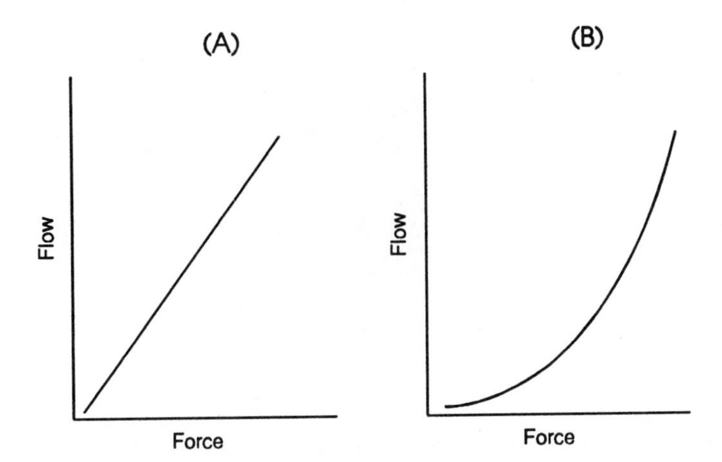

Figure 3-14 (A) Newtonian materials flow forces: Newtonian substances have a constant deformation ratio of shear rate (flow) to shear stress (force); (B) Non-Newtonian materials flow forces: Non-Newtonian substances have a nonconstant ratio of shear rate (flow) to shear stress (force).

Weight is the force produced by gravity between two bodies or masses. Weight is measured by the units of the pound or the newton. Weight is the interaction of mass and gravity.

Stress is the applied force to a body, and the change in the body is referred to as "deformation." This deformation is the result of strain. Strain is a derived measurement from the above variables and is very useful in measuring process variables such as flow, pressure, and acceleration. This is done with the application of strain gages, Figure 3-15A-C.

These variables are all subject to measurable change and can be sensed for many industrial processes.

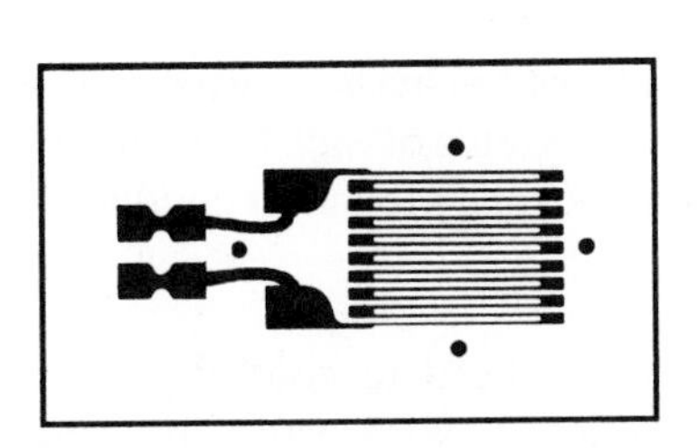

Figure 3-15A The concepts of measuring weight, mass, gravity, force, stress, and strain with strain gages. (A) Foil strain gage.

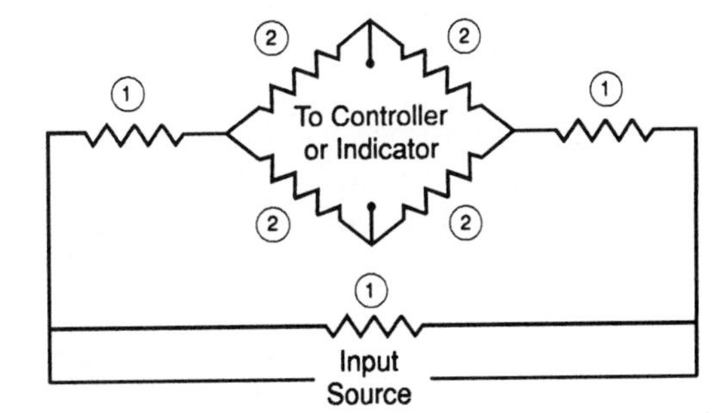

① Calibration Resistors
② Strain Gages

Figure 3-15B Wheatstone measuring circuit.

Figure 3-15C Center loaded, shear beam, load cell, and mounted *(Courtesy of Transducers, Inc.)*

Humidity and Dew Point

The process variable of humidity is concerned with the measurement of the water content in the air or the liquids in a specific gaseous environment. Humidity is measured by applying one of these three definitions, depending upon the objective of the measurement: relative humidity, absolute humidity, or specific humidity.

Relative Humidity - Relative humidity is the ratio of the moisture content of the air compared to what the air would have at the same temperature if the air were saturated. When the relative humidity is 100 percent, the air is saturated at that temperature.

Absolute Humidity - Absolute humidity is the actual amount of water in a volume of air. It is measured in grains of water per cubic foot or as grams of water per cubic centimeter.

Specific Humidity - Specific humidity is the ratio of the mass of water vapor to the mass of dry air in the sample. It is measured as grains of moist air per pound of dry air. This definition of humidity is applied frequently to the air conditioning industry.

Dew Point - Dew point is a variable commonly used to control industrial processes. The dew point is the temperature at which the water vapor in a sample of air begins to condense. Moisture will form on the coolest surfaces. The dew point can be an important warning of danger in many industries such as textiles, paper, fertilizer, cement, and most chemical powders, flour, and wood products.

A major concept in the measurement of humidity is a change in conductivity. A number of instruments utilize the concept by employing hydroscopic salts that

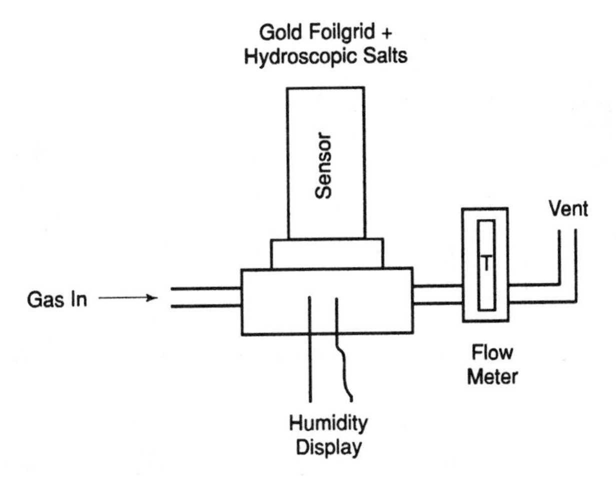

Figure 3-16 A simplified electrolytic hygrometer using conductivity as the humidity measuring principle.

change their water content with a change of humidity. The change of water content provides a change in the salt's electrical conductivity, easily measured as a voltage change, Figure 3-16.

Position, Rotation, and Speed

A desired position is the exact place with reference to a location or situation. *Displacement* is the error between the reference and the actual position desired.

Position - Position is concerned with locating a point or place — the point may be a linear position, a distance from a reference position. These distances and locations are variables that are very important to all types of manufacturing. In manufacturing processes, this data is constantly monitored so that quality products may be produced. Automation utilizes this technology. It ranges from placing parts on a conveyor belt to positioning a robot arm used to place component parts into precisely located holes to be drilled into an elaborate casting that will contain a host of interacting precision parts, Figure 3-17.

Rotation - Position measurements are also concerned with rotary positions. Rotary positions provide information concerning the angular displacement of shafts, screws, and various types of motors. The data can be used to control the precise movement of a large number of machine applications in manufacturing, assembly, and/or control of the many positions required in robot operation. The devices that perform this type of work are referred to as encoders, resolvers, and digitizers. They provide the information that verifies that the desired radial position has been achieved or if not, what correction is necessary.

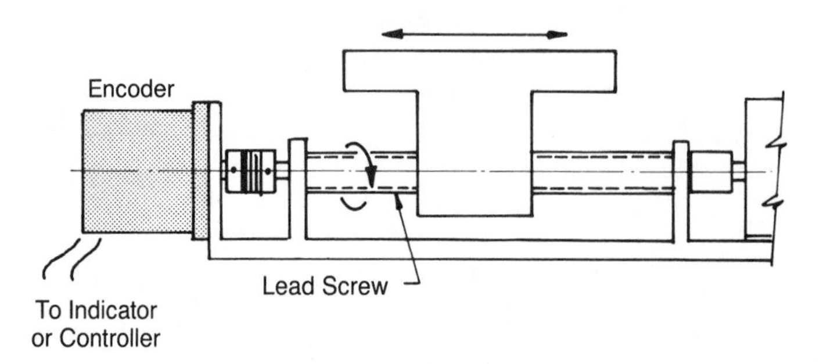

Figure 3-17 A position can be reported by an encoder *(Courtesy of BEI Motion Systems Company)*

Speed - Speed is the time rate of change of distance. Speed is reported in miles per hour, feet per minute, or feet per second, but does not indicate a direction. Linear motion indicates a direction of motion and angular speed in the case of rotation. The class of instrument used to measure linear motion is the encoder that reports movement in inches. The greater portion of industrial machinery measurement is rotational in movement and therefore its speed is measured in revolutions per minute. The class of instrument used for these measurements is the tachometer or electronic variations of this instrument, Figure 3-18.

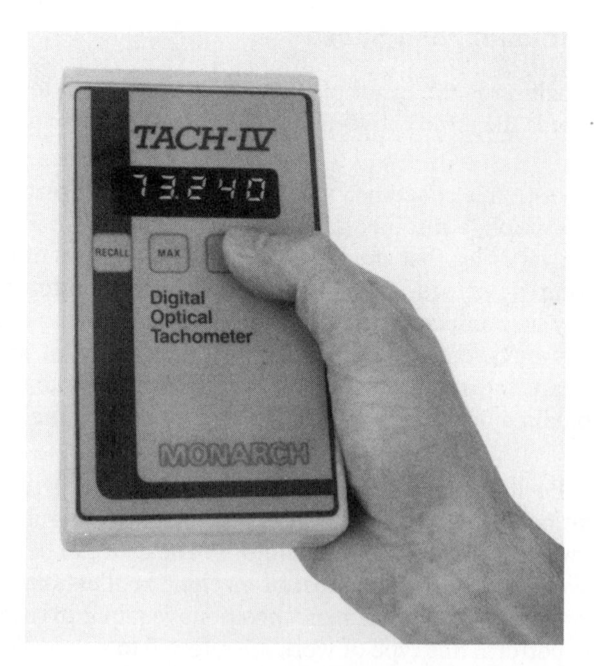

Figure 3-18 Speed measured with a digital optical tachometer *(Courtesy of Monarch Instrument, A Division of Monarch International Inc.)*

Acceleration

Acceleration is the process variable measured for the collection of data about a rate of change. Acceleration is the rate of change in the velocity of an object during an interval of time, divided by that interval of time. Acceleration is thus a rate variable in which time is the common component of the measurement. An acceleration sensor primarily perceives movement through the resistance of the movement of mass. In addition, this type of instrument is frequently used to measure wavelike or sinusoidal vibrations to maintain various machines within safe operation parameters, Figure 3-19A and Figure 19-B.

Thickness

The process variable of thickness is applied to the measurement of materials or coatings on continuous sheetlike materials. Measurements are frequently made of the support-base material and compared to the total thickness and therefore yield a thickness of the coating, Figure 3-20.

Thickness changes are measured by using the concepts of pneumatics, ultrasonics, X rays, nuclear radiation, and mechanics.

Radioactivity - Radioactivity is the breaking away from the nucleus of alpha rays, beta rays, and gamma rays resulting in nuclear radiation. This radiation results in or from atoms that have like numbers of protons but unlike numbers of neutrons. These materials are called isotopes. Isotopes decay at different rates from lengths of half-life measuring in microseconds to thousands of years. Industrially, an isotope is selected with a radiation output that will generate a signal necessary for

Figure 3-19A Resistance to the movement of a mass (A) A quartz piezo electronic general purpose accelerometer

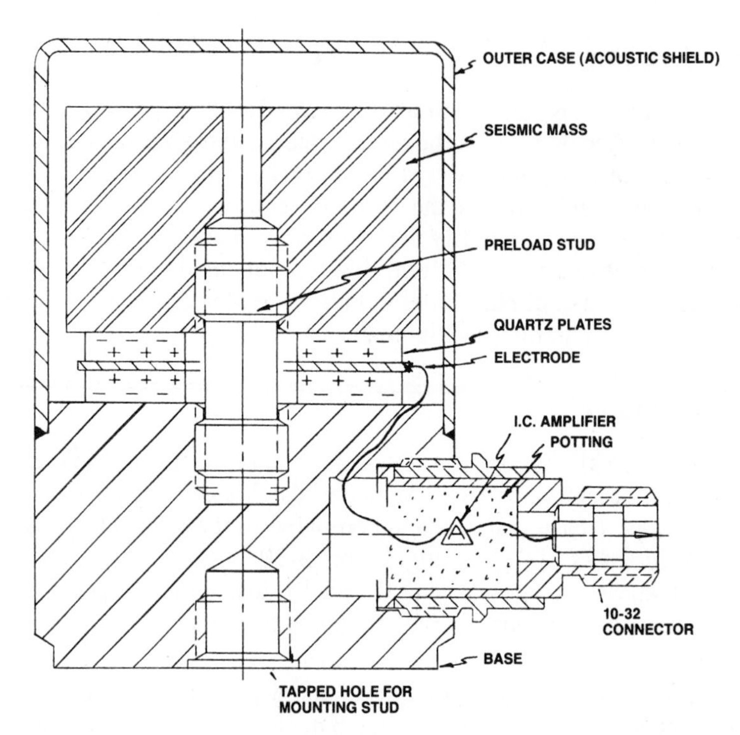

Figure 3-19B A cross section of a typical quartz accelerometer with built in integrated circuit electronics *(Courtesy of PCB Piezotronics Inc.)*

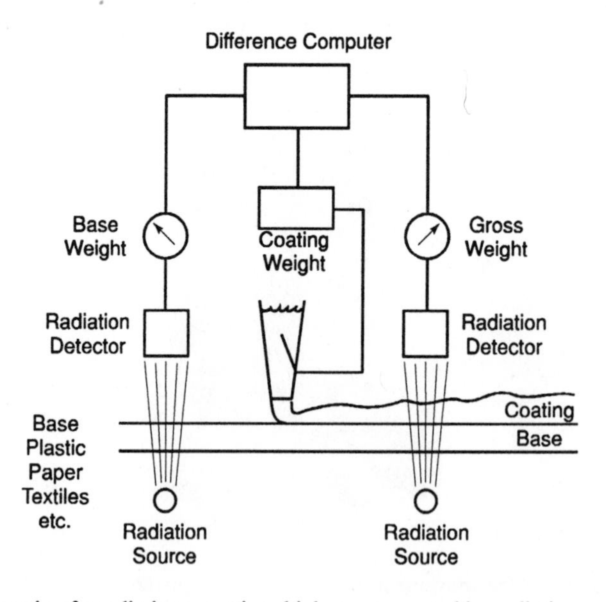

Figure 3-20 A schematic of a radiation a coating thickness measured by radiation

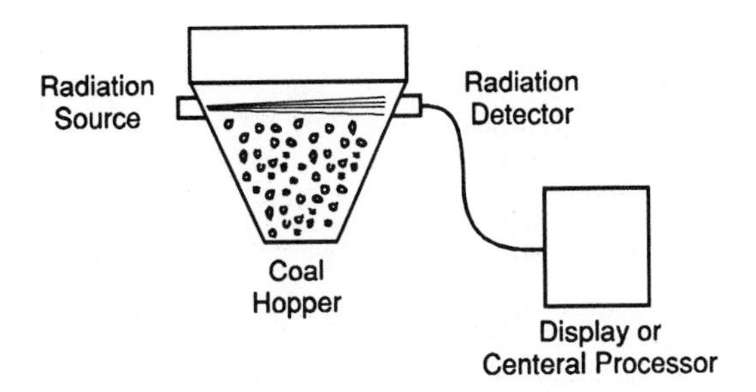

Figure 3-21 Level measured with radioisotopes. At a high level the material in the hopper attenuates the beam and less energy reaches the detector, resulting in an alarm or signal.

the requirements of the materials measured and range of the instruments employed, Figure 3-21.

A number of industrial variables may be measured by the use of radioisotopes. Measurements are made by passing radiation through the product being measured. For example, the level of cement in a tank, the thickness of a product on a conveyor belt, or the thickness of a product on a film can be measured in this way. The measurement is made by the change in energy received by the sensor, usually a Geiger counter. The change is indicated by the absorption of some of the energy by the product. Another method employed is referred to as "back scattering." In this application, the radiation strikes the product and a percentage of the energy is reflected back to the Geiger tube providing a signal.

pH — Hydrogen Ion - The measurement of the hydrogen or hydroxyl ions present in a solution offers the opportunity to determine the *acidity* or *alkalinity* of the solution. The degree of acidity or alkalinity of solution is an important variable in many industrial processes within the chemical, food technology, biological technology, cosmetic, sewage treatment, waste water treatment, and drinking water industries.

A number of pH meters exist, but commonly used industrial instruments consist of a reference electrode and an indicator electrode. These electrodes produce a small electrical potential. The two electrodes operate as two half-cells. Thus the tested solution's hydrogen ion content will vary the two electrodes by producing a millivolt change and indicate the pH of the solution tested, Figure 3-22. These instruments are chemically very sensitive and must be guarded against contamination. They must also be compensated for temperature variation.

Vibration

Vibration originates as periodic motion. When a body has insufficient inertia to remain stationary, it can be excited into rapid movement. This displacement or movement is frequently of an oscillating type and may be part of a machine. The

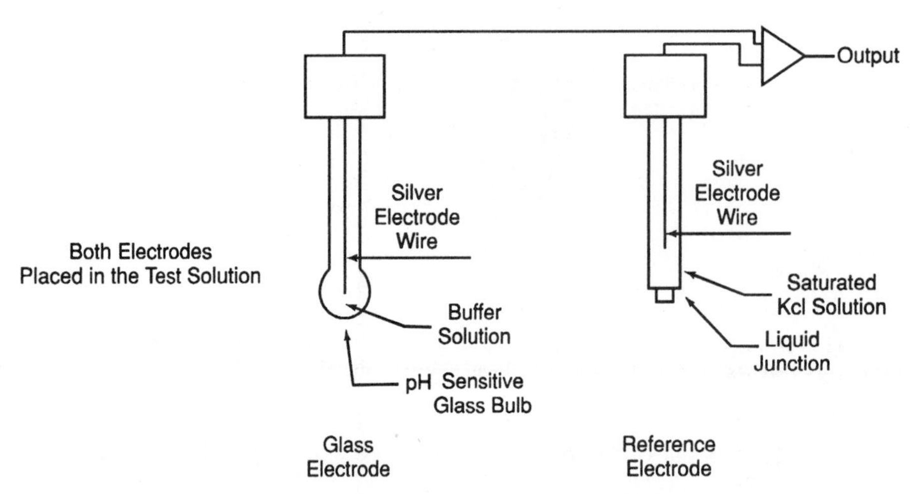

Figure 3-22 A pH cell for the measurement of alkalinity or acidity of solutions

flexibility of a machine is set by its natural frequencies, resonant vibration, or modes of vibration. With added excited movement, a rigid body will resonate as a flexible body at one or more characteristic modes. This will involve bending, twisting, or other complicated motions. Uncontrolled shaking of a machine could cause a failure of the machine or system. To improve the system, *damping* can be added to reduce or eliminate the vibrations.

To measure vibrations, the sensors convert changes of displacement into electrical signals by providing changes in resistance, capacitance, inductance, or eddy current outputs to a recording instrument. *Strain gage* and piezoelectric accelerometers are measuring sensors that employ resistance and voltage outputs for these types of instrumentation, Figure 3-23.

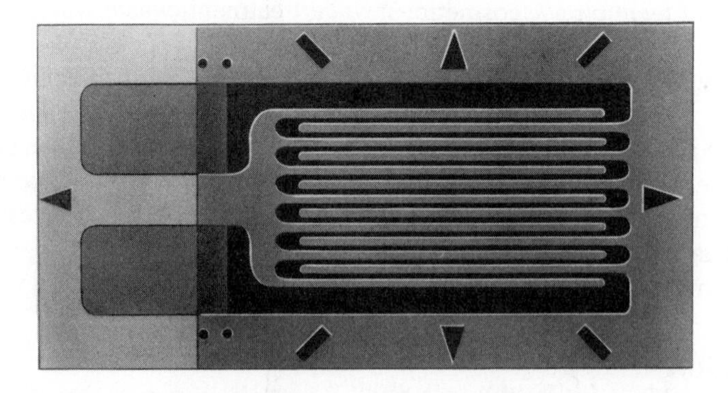

Figure 3-23 A strain gage applied to measuring vibration. The bending of the strain gage changes its resistance, thus providing an electrical signal. *(Courtesy of Micro-Measurements Division, Measurements Group Inc.)*

Distance — Length and Size

Control of the linear distances that a product or machine moves during manufacturing is critical for successful operation. The measuring of movements for locating and positioning machine tools, cutting tools, or components and materials requires accuracy as well as consistency. Mechanical measurements with scales, micrometers, lead screws and dials, stops, limit switches, and movement of stepping motors are supplemented with measurements directly on the workpiece rather than a measurement of the movement of the tool machine table. These instruments can employ measuring methods such as laser interferometry.

Interferometry - Interferometry is the application of a wavelength of coherent light that is split — half to the instrument and half to a reflecting mirror on the workpiece. Part of the beam is reflected back to a device that recombines the two light beams. The source and the reflected beams result in an "in phase" and "out of phase" light condition. The "in phase" light adds its brightness to the beam and the "out of phase" light reduces the brightness, resulting in a banded light and dark pattern. A photodetector counts and measures the light intensity difference caused by a change of movement of the reflecting mirror. The variable is distance — how far the reflected beam travels with the workpiece. The interference bands are counted by the photodetector as the mirror moves to a location and is multiplied by a factor (the wavelength of the light used), resulting in a distance measurement in microinches, Figure 3-24.

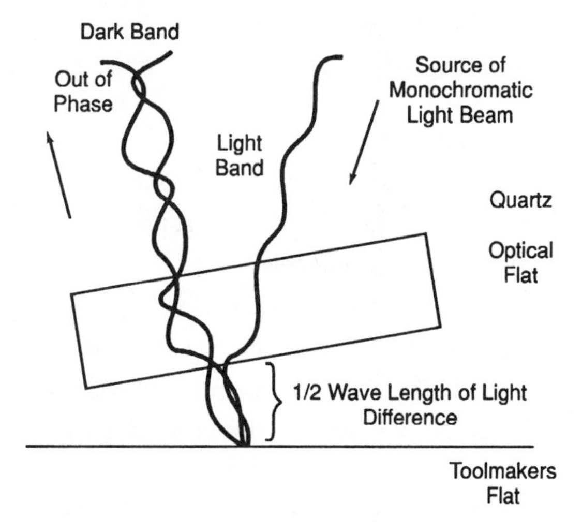

Figure 3-24 Interference band theory

SUMMARY/FACTS

- Changes in the physical characteristics of materials (such as length or volume) with a temperature change are relatively easy to measure with a contact instrument. Electrical changes (such as frequency change, resistance change, or voltage change) require more sophisticated measuring instruments.
- Manufacturing process variables are the parameters in manufacturing that are manipulated to provide control over the products being manufactured. There are many different process variables that are controlled.

REVIEW QUESTIONS

1. What is a process variable?
2. What is meant by a change in phenomenon in regard to instrumentation?
3. What is the function of a sensor in instrumentation?
4. How are gases and vapor pressures used to make temperature measuring instruments?
5. How is an industrial standard verified?
6. List some of the electrical characteristics of materials that are very important in instrumentation.
7. What are three ways that heat is transferred?
8. In calculations involving ideal gas laws, what must gage pressures (PSIG) be converted to?
9. What is defined as "fluid in motion?"
10. What type of flow is frequently found in high volume common pipe?
11. What type of flow does the Coriolis effect measure?
12. How is flow measured in open channels?
13. What is another name used for the Coanda effect?
14. In the industrial plant, what is the variable that is measured in stored products?
15. What measurement is designed around the principle of buoyancy?
16. What variable is frequently measured by timing a volume of flow through a standard nozzle?
17. What device is frequently used to measure weight, force, and stress?
18. What are the process variables that are concerned with the measurement of water content or moisture?
19. What two positions' measurements are important to manufacturing technology?

20. How does speed vary from displacement?
21. How is acceleration measured?
22. A number of industrial variables are measured by the use of radioisotopes. How do they measure the variables?
23. What are hydrogen or hydroxyl ions frequently used to measured?
24. How is the variable of vibration measured?
25. How is the variable of distance measured?
26. What term is used to describe the measurement of very accurate distances using light?

4

SENSING PHYSICAL VARIABLES: TEMPERATURE AND PRESSURE SENSOR INSTRUMENTS

OBJECTIVES

This chapter will allow you to learn about a series of temperature and pressure sensors. After completing this chapter, you will be able to explain and apply these temperature and pressure measuring devices:

- Sensors that measure the physical variable of temperature
- Sensors that measure the physical variable of pressure

INTRODUCTION

The manufacturing supervisor needs to be acquainted with data acquisition, sensing analysis, control, and carried out by instrumentation in systematizing automatic production or automation. The study of sensors provides data and insight as to how control of manufacturing processes that govern product quality is obtained.

SENSORS USED TO ACQUIRE TEMPERATURE DATA

Sensors are devices used to acquire data that will measure a change in the variable and produce a signal. The transducer may use a variety of methods to obtain and transmit the signal. There are four major divisions of temperature transducers: (1) mechanical systems employing the principles of expanding metals, liquids, and gases; (2) electrical systems utilizing data acquired from resistance changes in circuits, thermoelectricity generated by a thermocouple circuit, or total radiation measuring in the pyrometric range using the principles of the thermopile; (3) optical pyrometers using the principle of a photometric match, and fiber optics which has the ability of transmitting light energy around corners for measurement by radiation instruments; (4) solid-state integrated circuits that sense the data and condition the signal as well as amplify the signal to a level that can be applied to decision-making circuitry.

Reference Temperatures

When considering temperatures, it is necessary that reference points be established so that different groups of people can compare their temperatures and have an understanding of what those temperatures mean.

To resolve this problem, The International Practical Temperature Scale was formulated, establishing eleven reference temperatures.

Table 4-1
IPTS-68 Reference Temperature

EQUILIBRIUM POINT	Degrees K	Degrees C
Triple Point of Hydrogen	13.81	-259.34
Liquid/Vapor Phase of Hydrogen at 25/76 Std. Atmosphere	17.042	-256.108
Boiling Point of Hydrogen	20.28	-252.87
Boiling Point of Neon	27.102	-246.048
Triple Point of Oxygen	54.361	-218.789
Boiling Point of Oxygen	90.188	-182.962
Triple Point of Water	273.16	.01
Boiling Point of Water	373.15	100
Freezing Point of Zinc	692.73	419.58
Freezing Point of Silver	1235.08	961.93
Freezing Point of Gold	1337.58	1064.43

The triple point is the equilibrium between the solid, liquid, and vapor phases of equilibrium hydrogen, etc.

Mechanical Temperature Sensing Systems

Expanding Metal Systems

Temperature may be sensed by a metal, such as mercury expanding in a thermometer stem. When the mercury rises in the capillary tube of a thermometer, it can be designed to intercept a prepositioned sensor. The rising mercury causes a change in the circuitry of the sensor and activates an electrical signal starting a blower or other device needed to correct the manufacturing process.

The expansion of the mercury was the primary sensor and provided the initial signal in the system that did rapidly become very sophisticated and modern.

Liquid Filled Thermometers

The common glass stem thermometers have been in use since the time of Fahrenheit (1686-1736) and have been based primarily on the expansion of mercury. The expansion of mercury is 0.01 percent per degree Fahrenheit, and its expansion rate is very linear. The instrument is designed with a bulb filled with mercury that expands with a change in temperature into a capillary tube of a very small diameter. The capillary tube is contained within the stem which in turn has a graduated scale printed on its surface. Frequently the stem provides a visual magnification of the column of mercury. A restriction at the base of the capillary tube holds the mercury column at its highest point when the thermometer is removed from its heat source, making it easier to read accurately. In addition, there is an expansion chamber at the upper end of the capillary tube to protect the instrument in the event that an incorrect temperature range has been selected. Frequently an immersion line is scribed on the stem below the main scale; this indicates how far the thermometer is immersed into the medium to be measured, Figure 4-1.

When a thermometer nears the end of its practical range (600°F), dry inert nitrogen placed above the mercury prevents its evaporation and increases its temperature range. For low temperature ranges, alcohol, pentane, and other hydrocarbons are used. These thermometers frequently also may have colored dyes added for easy reading.

In the application of the glass filled thermometer, reproducible results can be obtained by understanding the correct thermometer immersion techniques. Total immersion means that the bulb and the scale or liquid column are immersed in the medium. Partial immersion means the bulb and the stem are immersed to the scribed line in the medium to be measured.

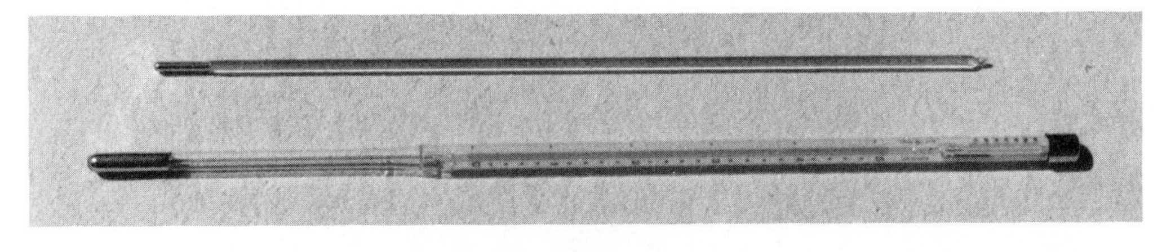

(A)

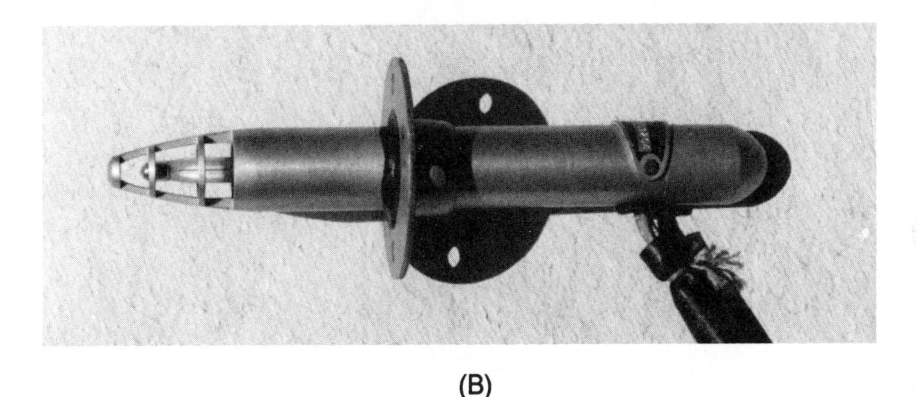

(B)

Figure 4-1 (A) Liquid metal in a glass thermometer; (B) Liquid filled thermometer with an electrical sensor

Fluid-filled Thermometers

Pressure bulbs fluid-filled thermometers have long been a successful industrial measuring device. This type was one of the earliest instruments that could provide a remote and continuously charted record. Essentially they are small bulbs that absorb heat and expand in a restricted volume. The expanding fluid in the chamber supplies a pressure to the capillary tube that correlates with the change in temperature, Figure 4-2A. This pressure is transmitted to an indicating or charting instrument. These bulbs can be charged with liquids, liquid/vapor, or gases. When the bulbs are charged with gas, the bulbs need to be considerably larger, Figure 4-2B.

Gas-filled Thermometers

Theoretically, a gas-filled thermometer can be used in a wide range of temperatures so long as the temperature is above the condensation point of the gas. If the bulb is filled with nitrogen, temperatures of -50°F to 1200°F can be measured, Figure 4-3.

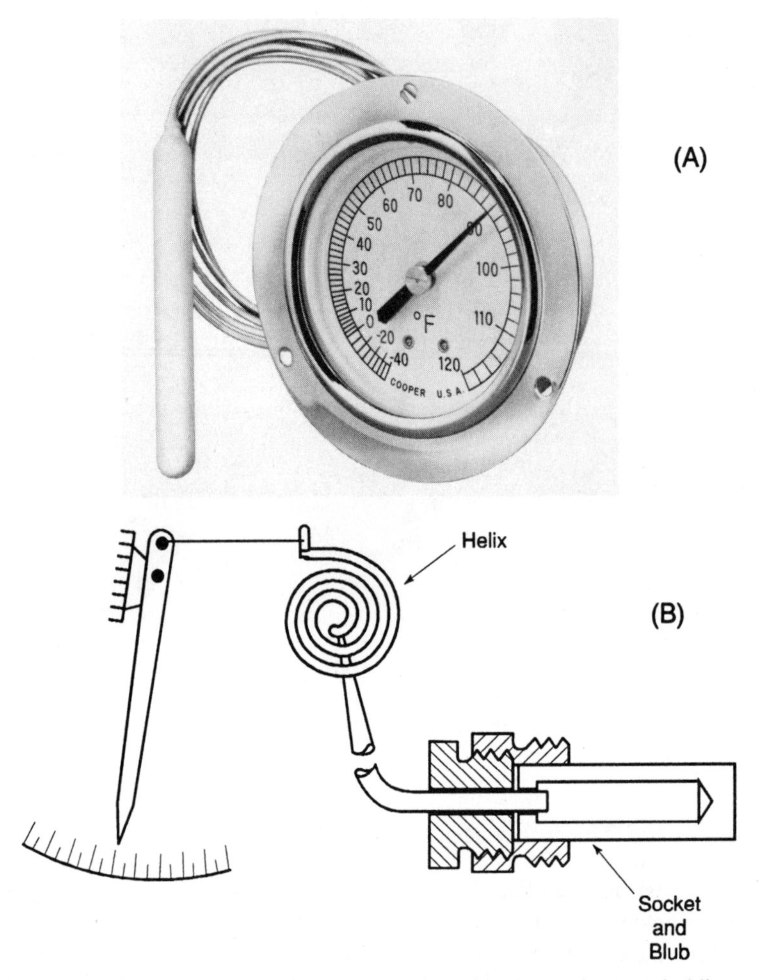

Figure 4-2 (A) Vapor tension remote reading thermometer for refrigerators, freezers, holding cabinets, etc. *(Courtesy of Cooper Thermometer Co.)* (B) Liquid-filled bulb thermometer diagram with a bulb connected by a capillary tube to transmit the pressure to a helical bourdon tube

Bimetallic Thermometers

The bimetallic thermometer is designed from two temperature sensitive metals with different coefficients of expansion. When a metal such as Invar (64 percent iron, 36 percent nickel) is rolled into a thin strip and fastened to a like strip of highly expandable nickel-iron alloy with chromium, a heat sensor is produced. When a strip of these materials is wound into a helix and one end fastened to an anchor or case, while adding a pointer fastened to the inside end of the helix, an instrument is produced. As the temperature on the helix increases the higher expansion rated material, the spiral winds up, thus moving the pointer and

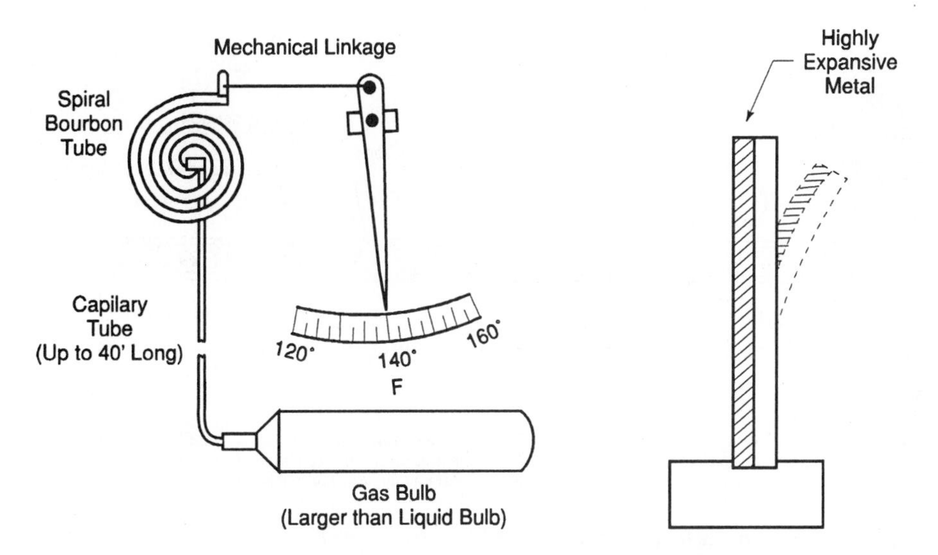

Figure 4-3 Gas-filled thermometer (uncompensated)

Figure 4-4A Bimetallic Strip

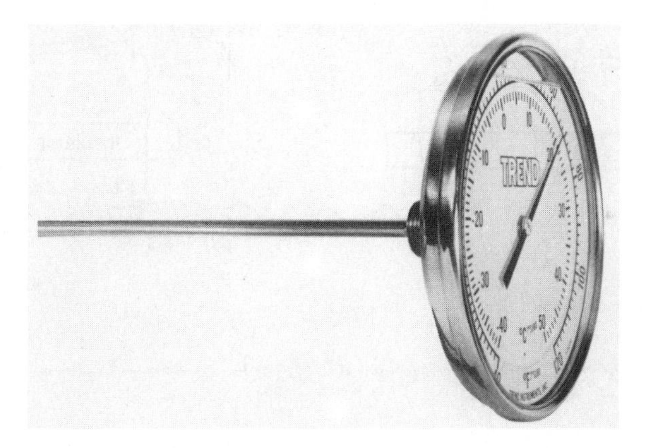

Figure 4-4B Stem thermometer

indicating the new temperature, Figure 4-4A, Figure 4-4B, and Figure 4-4C (next page).

The bimetallic strip sensor develops enough torque so that it moves a pen on a circular chart that is rotated by a clock. Because of this ability, an inexpensive twenty-four hour temperature recording instrument has been built. This type of instrument has numerous applications with an error of only a 1-2 percent.

Bimetallic materials can be produced in many different shapes (spirals, helixes, coils, cantilever strips, and discs), Figure 4-5. These devices are activated by a temperature range that provides a temperature control. In a temperature-sensitive bimetallic disc, a high-wattage electrical current flowing through a

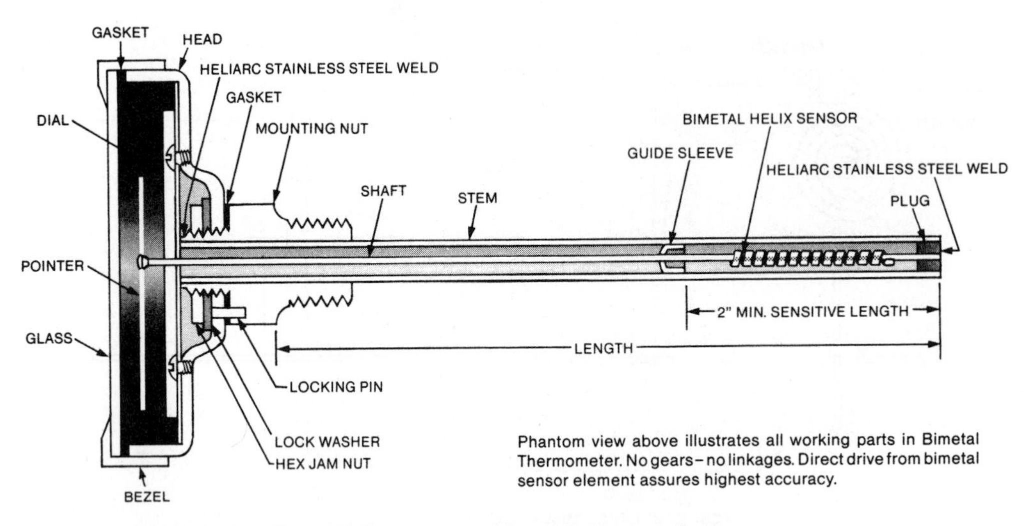

Figure 4-4C Bimetallic thermometer *(Courtesy of Trend Instrument, Inc.)*

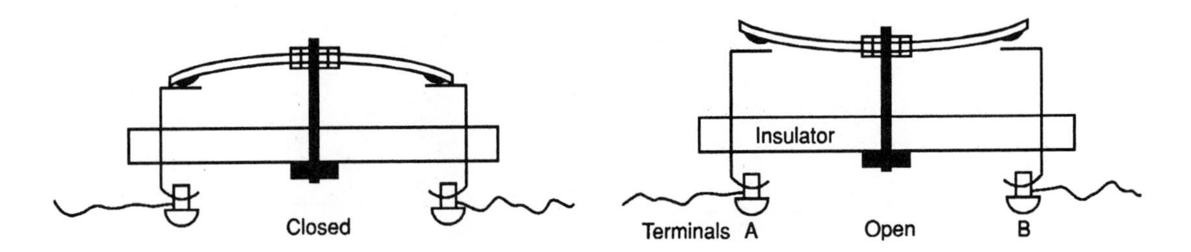

Figure 4-5A Bimetallic disk switch

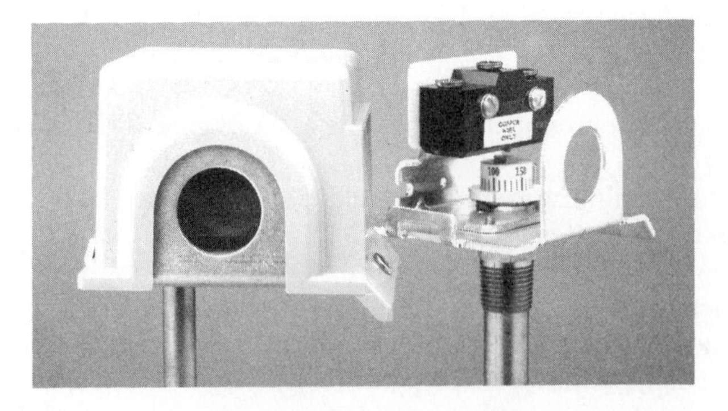

Figure 4-5B Snap action bimetallic or liquid-filled temperature switch *(Courtesy of Wahl Instruments Inc.)*

device will raise the temperature of the disc thus causing the device to be switched. This principle is also employed in a circuit breaker and other devices to protect circuits from being overloaded. The heating causes the bimetallic device to open the circuit thus providing protection from dangerously overloaded equipment.

Electrical and Electronic Temperature Sensing Systems

Resistance Thermometers

Electrical resistance thermometers are based on the principle that a metal's resistance increases with an increase in the metal's temperature. This fact allows a series of sensors to be designed that produce reliable temperature data for industrial process control.

In 1871, before the Royal Society, Sir William Siemens demonstrated the use of a platinum wire for temperature measurement of metallurgical furnaces. Platinum wire has high resistivity that changes almost linearly with temperature changes. In addition, platinum is a noble metal, which means that it is inactive with most chemical reactions. With these properties, the platinum resistance thermometer became an international standard for temperature measurement. A practical scale of resistance is available from -182.97°C to 630.74°C; and it is possible to measure within a few thousandths of a degree.

Resistance sensing elements are manufactured from platinum, nickel, copper, and tungsten. Other metals such as iridium, rhodium, silver, iron, and tantalum are less frequently used. A thermoresistant sensor is commonly manufactured by wrapping a piece of wire or thin film of platinum around a ceramic core and the whole assembly mounted inside a stainless steel mount resembling a test tube. A low electrical conductive filler is added to provide support and location of the core. Electrical leads come out of the open end of the stainless steel tube and are fastened to extension wires that lead to the instrument. This sensor is referred to as a Resistance Thermal Detector (RTD), Figure 4-6.

A Wheatstone bridge circuit is utilized to read the RTD sensor. Various circuits do exist that will compensate for errors that may be introduced by variations of temperature to the extension leads. The simplest readout is the two-wire bridge measuring circuit, Figure 4-7.

A four-wire Mueller bridge circuit is a modified Wheatstone bridge. It allows an extra two wires to nullify an error that may be introduced because of a temperature difference occurring in the extension leads, Figure 4-8.

Thin film RTD technology is utilized in two precision types: a wire-wound element and a thin film laser cut element. Both types have a wide temperature range and long-term stability unique to pure platinum metal. They are supplied in two configurations. One is a flat film platinum sensor with radial, perpendicular, and axial leads. The other configuration is a cylindrical sensor. The cylindrical

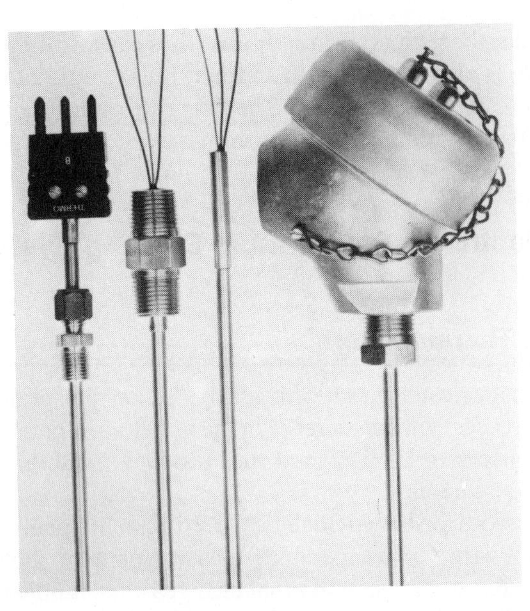

Figure 4-6 Miniature and industrial thermal resistance detector *(Courtesy of Thermo Electric Co. Inc.)*

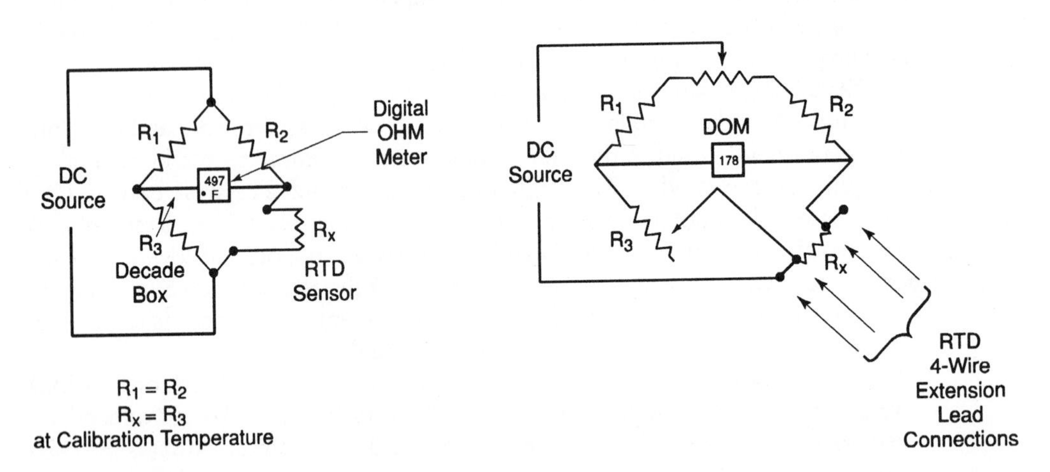

Figure 4-7 The two-wire bridge resistance measuring circuit to indicate temperature

Figure 4-8 A four-wire Mueller bridge

sensor has the capability of dual configuration that provides reliable and redundant sensing. Each sensor contains two separate units — one provides temperature data and one is a safety backup with an independent duplicate circuit, Figure 4-9A and Figure 4-9B.

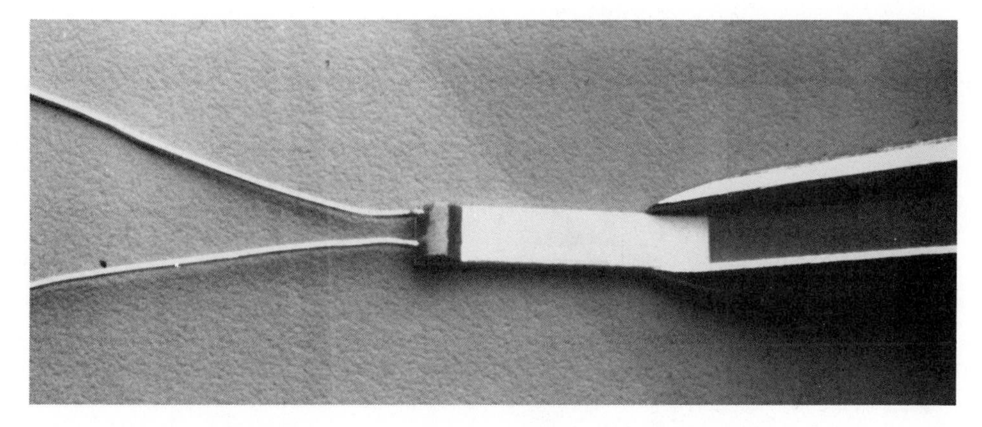

Figure 4-9A Miniature thin film platinum RDT sensors *(Courtesy of Wahl Instruments Inc.)*

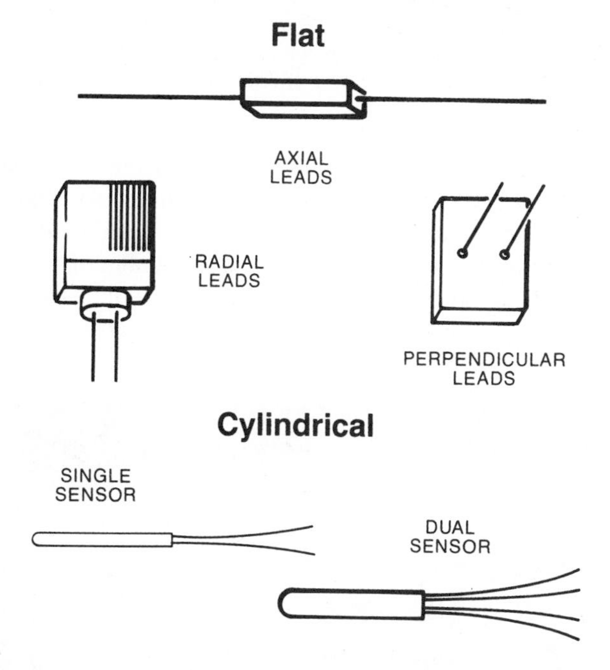

Figure 4-9B Configurations and lead desgins *(Courtesy of Wahl Instruments, Inc.)*

Digital thermometers that combine solid-state electronics with a resistance thermal detector are available as portable instruments. Their range is from -290°F to 1450°F (-188°C to 788°C). Other models are available within various ranges. A platinum RTD model will produce a resolution to 0.1°F when applied in gases, powders, liquids, or to surfaces. These thermometers incorporate the latest state-of-the-art microprocessor technology, Figure 4-10.

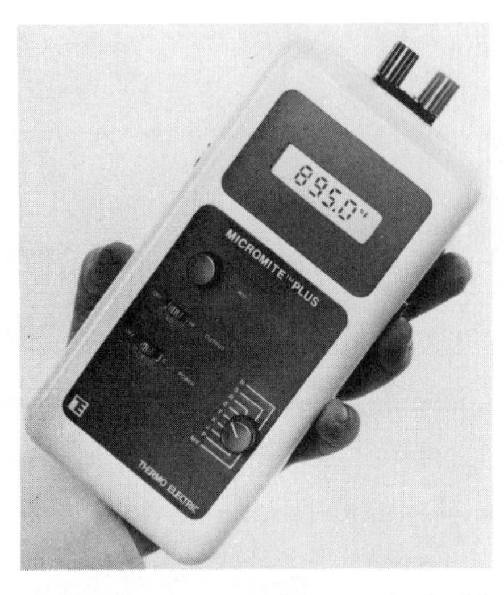

Figure 4-10 A hand held, nine-position thermometer; with seven standard thermocouple calibration ranges and an RTD ranges; microprocessor based. *(Courtesy of Thermo Electric Co. Inc.)*

A precision resistance thermometer bridge offers an added dimension of simplicity and precision for temperature measurement in the laboratory, plant, or field. Temperature changes as small as 0.001°C are made under controlled conditions. In industrial use, temperature measurement of steam lines, air ducts, waste water, and other products are made for energy studies and accurate control, Figure 4-11.

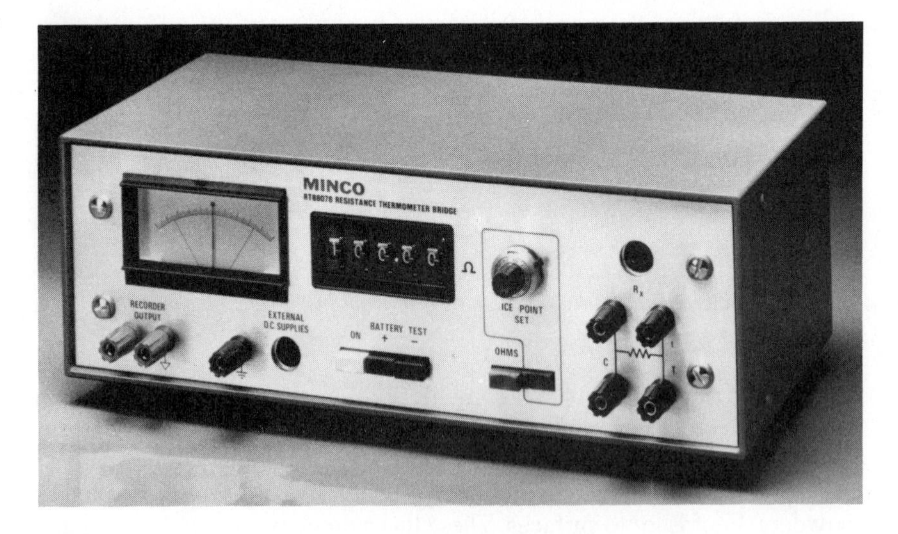

Figure 4-11 Resistance thermometer bridge *(Courtesy of Minco Products Inc.)*

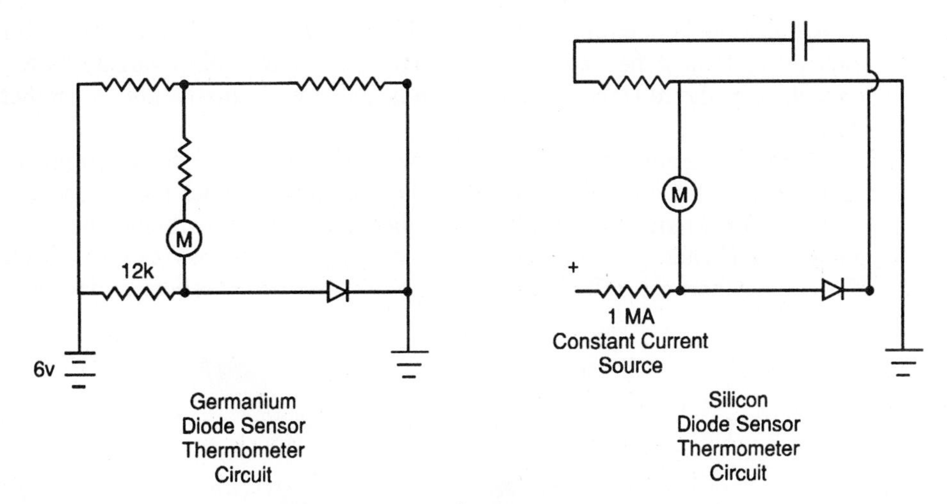

Figure 4-12 Principle of diode thermonetry circuits *(Courtesy of Analog Devices, Inc.)*

Diode Thermometry

In semiconductor devices, the junctions between the *n-type* and *p-type* doped silicon diodes change about 2.2mV/degree C over its measuring range. Diode voltage is a function of current. Therefore a constant current is used to excite the device to make use of the diode's thermal properties.

Diodes of silicon or germanium have been used as temperature sensors. Silicon diodes provide a forward voltage drop that is quite linear. The silicon diode will cover the range adequately from -60°F to 300°F. The germanium diode will cover the range from -420°F to 110°F, with sensitivities of 0.1°F. Because these devices are so inexpensive and linear they are considered a strong competitor to the resistance thermal detectors. The diodes can be placed and calibrated in control circuits that are read with a common microampere indicating meter, Figure 4-12.

Thermistor Temperature Sensors

A thermistor is a thermally sensitive resistor. It is a semiconductor of compressed metallic oxides that has a large negative temperature coefficient of resistance. As the temperature of the device is increased, the resistance of the device decreases making more current available for the sensing and indication of temperature changes.

These semiconductors are made of sintered oxides of nickel, manganese, copper, cobalt, iron, magnesium, and titanium, and include other metal oxides for special characteristics. The semiconductors usually have a negative temperature coefficient, but the oxides can be compounded so that they will have a thermistor with a positive temperature coefficient, Figure 4-13. Thermistors are nonlinear devices and must be calibrated or compensated for this characteristic. An advantage of thermistors is that they can be made very small and in a number of different shapes.

Thermistors are supplied into two broad catagories: those in which embedded beads are sintered into the thermistor body; and those in which metal contacts are applied to the thermistor surface. The embedded thermistors include bare beads, glass coated beads, glass probes, glass rods, and beads in glass (tube) enclosures. The contact thermistors include discs, chips, flakes, rods, washers, and wafers. In

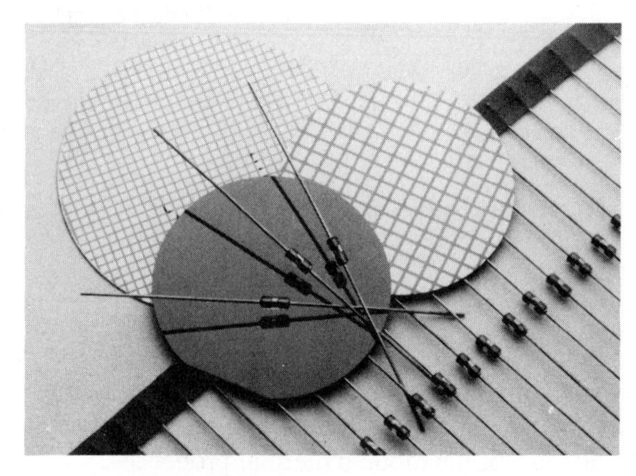

Figure 4-13A Glass encapsulated unichip thermistors, high volume for automation *(Courtesy of Fenwal Electronics)*

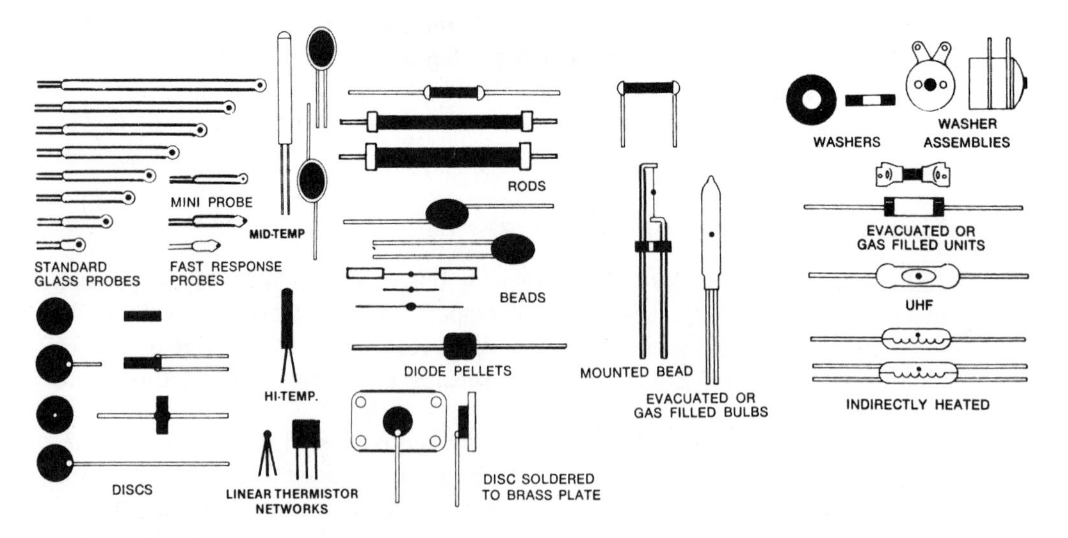

Figure 4-13B Thermistor configurations

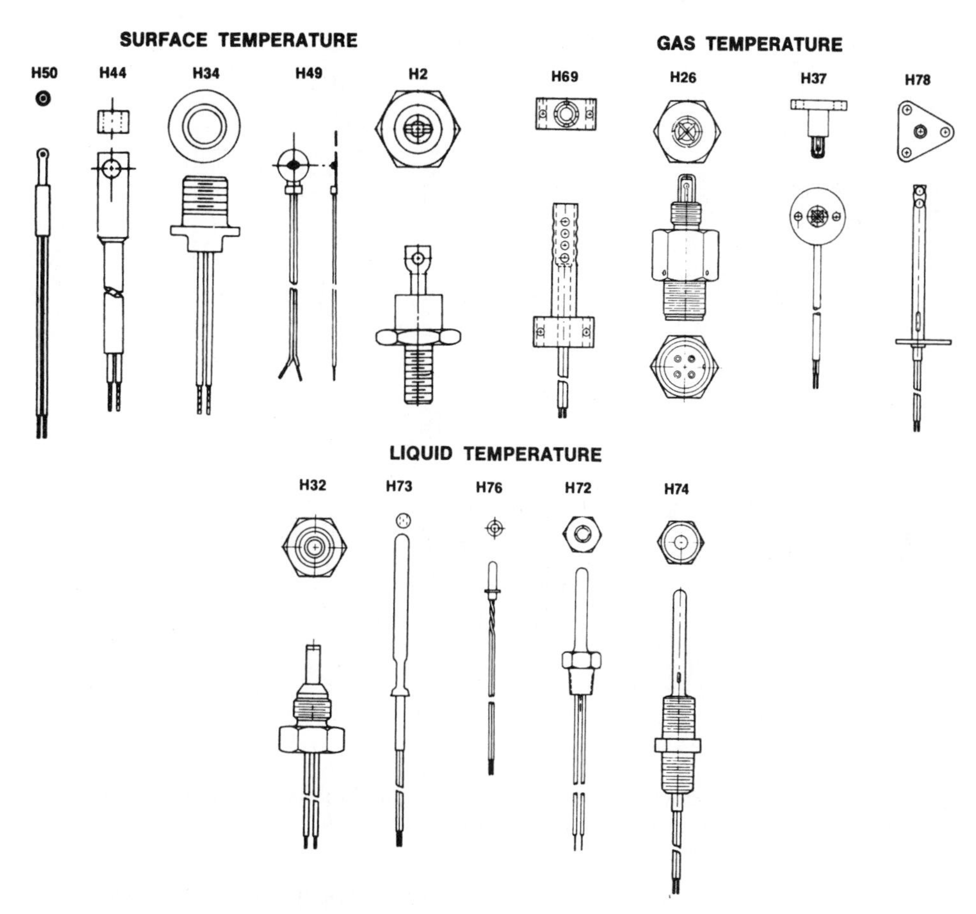

Figure 4-14 Thermistor probe assemblies *(Courtesy of Fenwal Electronics)*

addition to glass coating, the bare thermistors may be covered with plastic or cements to give them strength and protect them from various environments.

The thermistor functions with a small flow of current through the device. When a small increase of temperature is applied to the thermistor there is a change in its resistance. Most thermistors are designed so that the resistance drops as the heat increases. The resistance is proportional to the device's absolute temperature and thus the temperature may be inferred from the resistance of the circuit, Figure 4-14.

The thermistor's resistance values may be calibrated to a specific temperature range and sensitivity by adjusting the thermistor's chemical compositions until the required resistance for a specific temperature range requirement is reached. The resistance of a thermistor is usually high enough so that a considerable length of extension wire may be installed without the extension wire resistance disturbing the temperature data.

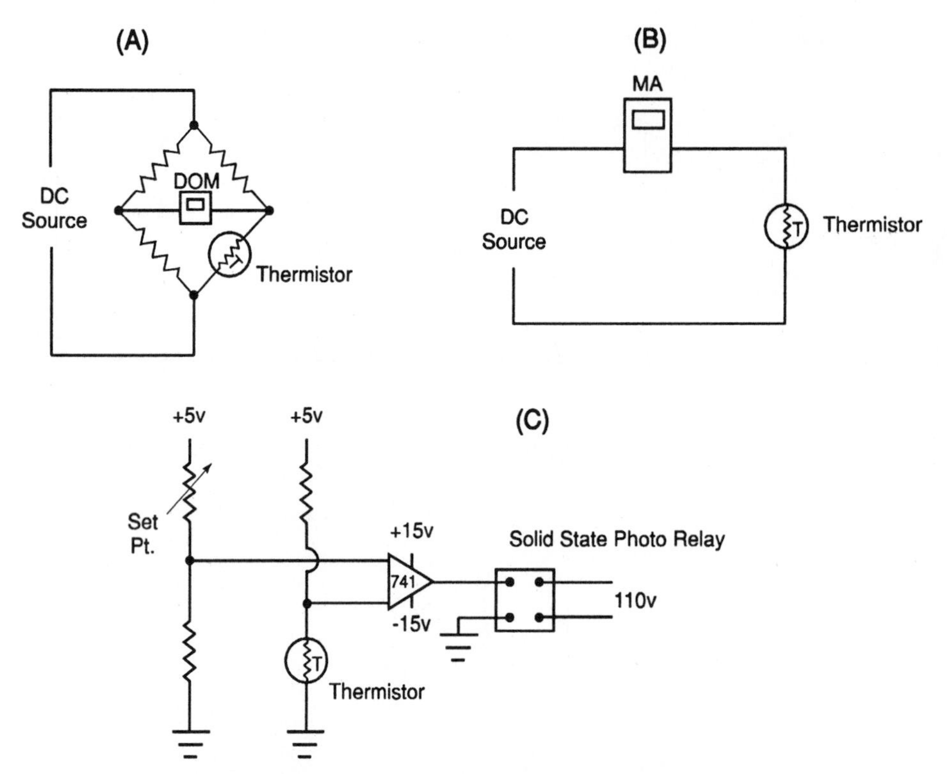

Figure 4-15 Thermistor readout methods (A) Thermistor and Wheatstone bridge; (B) Thermistor and microammeter; (C) Thermistor and solid-state comparator/controller

Thermistor data may be read by a number of methods, Figure 4-15. A common method is with the application of the Wheatstone bridge circuit using a zero center galvanometer and a regulated power supply. Another readout for indication only is with the application of a voltage divider and a voltage comparator operational amplifier. A set-point can be selected and a voltage value determined as to when an inversion will take place in the operational amplifier. This inversion produces a change in the control circuitry and triggers a relay making the change in control.

Thermoelectricity — Thermocouples

A thermocouple is produced when two dissimilar metals are fastened together and the connection is subjected to heat. A millivoltage output is generated that is proportional to the amount of heat applied. In 1821, Thomas J. Seebeck, a German scientist, discovered this phenomenon and since that time it has been widely used in temperature measurement, Figure 4-16.

The thermocouple measurement is based upon the difference between the bonded end of the wires in the heated area and the temperature of the wires at the

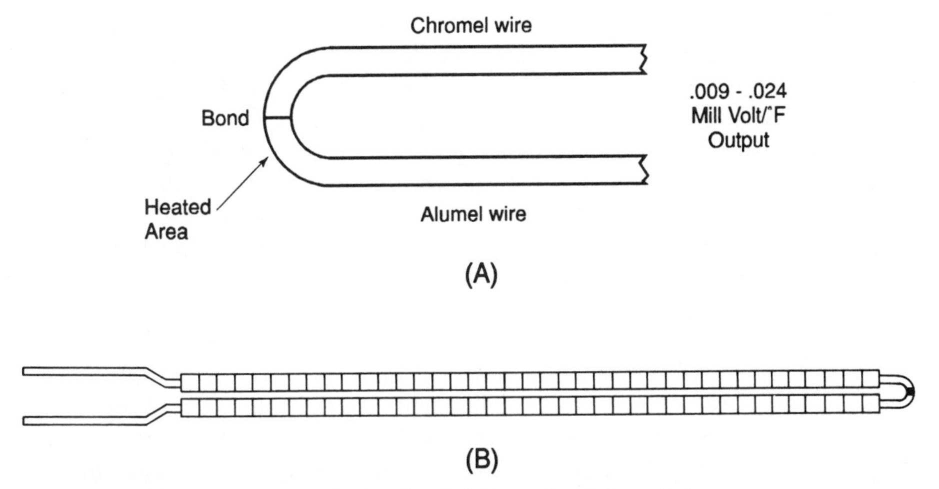

Figure 4-16 (A) K-type thermocouple; (B) Bonded chromel and alumel wire

instrument. These temperatures are referred to as the *hot (measuring) junction* and the *cold (reference) junction* respectively. The difference in temperature of these two junctions is a function of the millivoltage output. The greater the differences in temperature, the higher the millivoltage output, Figure 4-17.

When a thermocouple is placed into a measuring circuit, other phenomena may also be introduced into the measuring voltages. In 1834, Jean Peltier, a French watchmaker, discovered in his experiments the transportation of heat when a current was flowing in a wire. Heat is absorbed or liberated at either junction when a current is flowing in the circuit. The heating or cooling of the second junction will be set by the direction of the current flow, Figure 4-18.

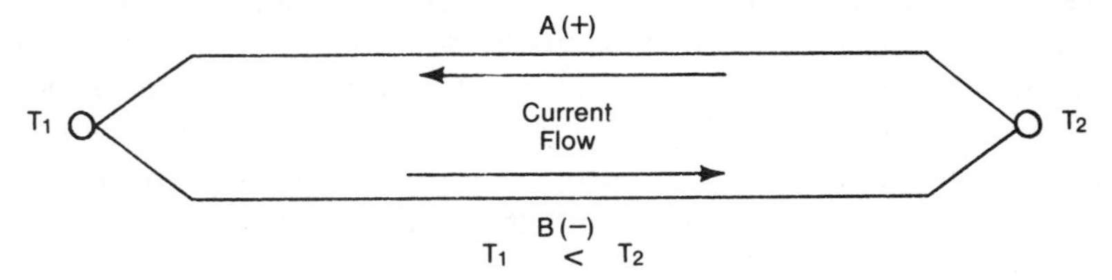

Figure 4-17 Seebeck effect thermocouple circuit *(Courtesy of Thermo Electric Company, Inc.)*

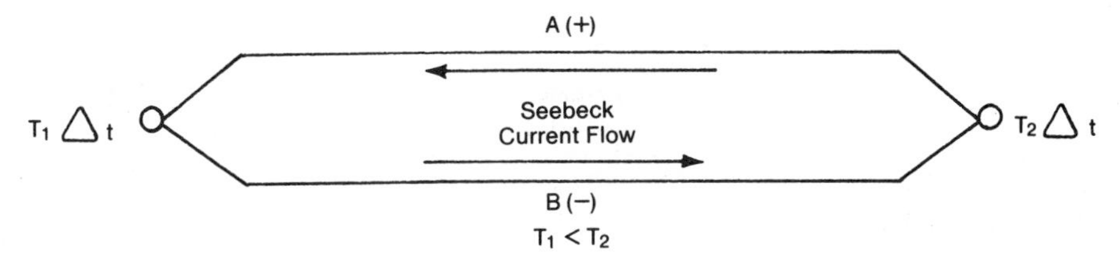

Figure 4-18 The Peltier effect. A flow of current will transfer heat *(Courtesy of Thermo Electric Co. Inc.)*

Thermocouple wire can be very small in diameter (0.005 in) but most industrial applications use a 14 gauge wire for sensitivity and response. For heavy duty and high temperatures, an 8 gauge wire is frequently selected, although the speed of response is reduced with the heavier wire.

The Peltier effect is being explored by engineers working with electrical heating and refrigeration. When the current flow is reversed, the heating or cooling end of the system is also reversed. For this reason electrical heat pump application for heating and cooling are being investigated for new products.

The application of intermediate metals in a thermocouple circuit can be detrimental to the signal being measured. The concept expressed by the phrase "Law of Intermediate Metals" is applied to installation and repair of a thermocouple system. If by accident or lack of material or knowledge, a technician introduces nuts, bolts, or wires of different materials on different branches of a thermocouple, an error would be introduced. If these different metals were used on only one side of the circuit, they would respond as secondary or tertiary thermocouples and provide additional electrical sources. For this reason, it is important that no inadvertent thermocouples be allowed to exist in the extension lines because of poor design or craftsmanship in the installation of the thermocouple. If new sources are allowed, the new polarity may add to or subtract from the millivoltage reaching the instrument. To have a *homogeneous* circuit, the principle that must be applied is that whatever material (bolt, nut, or wire) is used to fasten or connect the circuit, the same amount, type, and material must also be applied to the other branch of the circuit, Figure 4-19. It would be all right to use a steel nut on an extension connection, but the same type of nut must be used on each side of the circuit. Adding new materials, even to both sides, will affect the voltage, requiring that the system be recalibrated.

The value of microvoltages from a thermocouple is a function of the differences in temperature between the reference junction and the measuring junction. Accurate measurements are made with the reference junction held at a constant temperature. The reference junction is placed in a cracked ice bath that maintains the junction at 32°F, thus delivering a stable reference temperature and providing a large practical difference necessary to produce a maximum microvoltage for that spread of temperature, Figure 4-20.

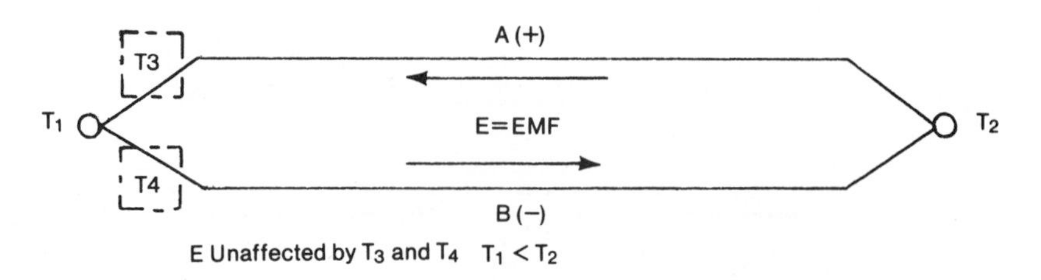

Figure 4-19 Law of homogeneous circuits *(Courtesy of Thermo Electric Co. Inc.)*

Table 4-2
Types of Thermocouples
Commonly Found in Industrial Applications

TYPE	CONDUCTOR	TEMP/RANGE DEGREES F	NOTES
E	Chrome/*constantan*	32 to 600 632 to 1600	Highest output
J	Iron/constantan	32 to 530 530 to 1600	Reducing atmosphere
T	Copper/constantan	-300 to -75 -150 to -75 -75 to +200 +200 to +700	Oxidizing atmosphere
K	*Chromel/alumel*	32 to 530 530 to 2300	Oxidizing atmosphere
S	Platinum/platinum Rhodium 10 percent	32 to 1000 1000 to 2700	Laboratory standard
R	Platinum/platinum Rhodium 13 percent	32 to 1000 1000 to 2700	Oxidizing atmosphere

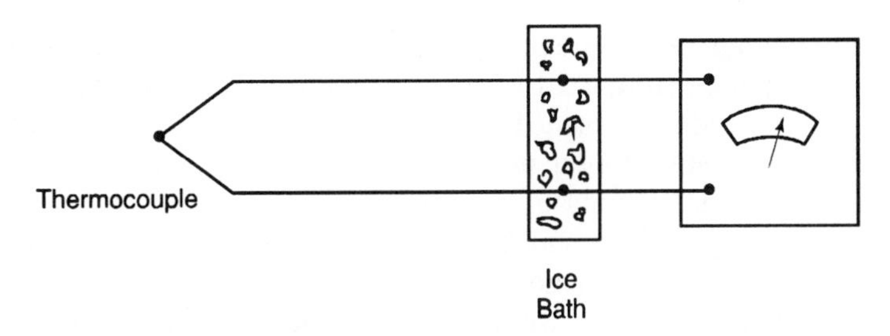

Figure 4-20 Accurate thermocouple measurements. The thermocouple wire and extension wire junctions are electrically separate and held at 32°F during the measurement.

Thermocouple output tables are published by a number of thermocouple equipment manufacturers. The tables convert a millivoltage output of the thermocouple to an equivalent temperature. These tables are very accurate and are based upon the reference junction being at 32°F (0°C) for the expressed values.

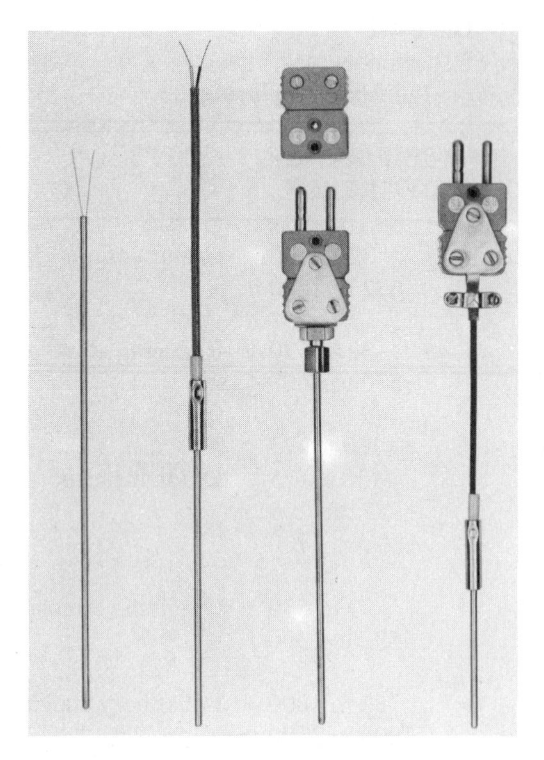

Figure 4-21 Hastelloy X sheathed thermocouples *(Courtesy of ARI Industries Inc.)*

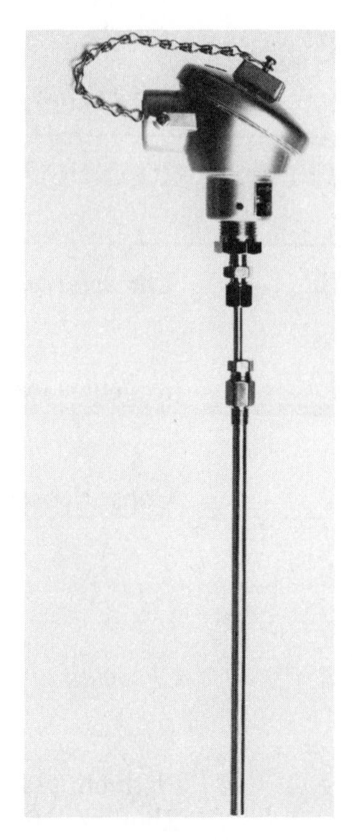

Figure 4-22 Thermocouple head construction in cast iron, die cast, or aluminum with sheathed thermocouples in 3/4 in NPT conduit openings *(Courtesy of ARI Industries Inc.)*

Commercial measurements are made with sheathed thermocouples to protect the thermocouples from oxidation, sulfur compounds, and other chemical reactions. These sheaths are made from stainless steel, hastelloy, ceramics, and other high temperature resistant materials, Figure 4-21.

Industrial installations of thermocouples require very rugged equipment because of possible physical damage that may be caused by workers or the movement of materials around the sensing equipment. The thermocouples may be mounted in plumbing-like assemblies to provide strength to resist the abuse of the every day environment, Figure 4-22.

Installations of thermocouple equipment are usually designed for long-term use and are designed so that they can be maintained. The standardized parts are used within the system so that they can be replaced if a failure occurs. These parts can be repaired without an interruption of production, Figure 4-23.

Panel digital thermometers that scan a number of thermocouples made up of different materials have been designed. Some will accept any of six types of

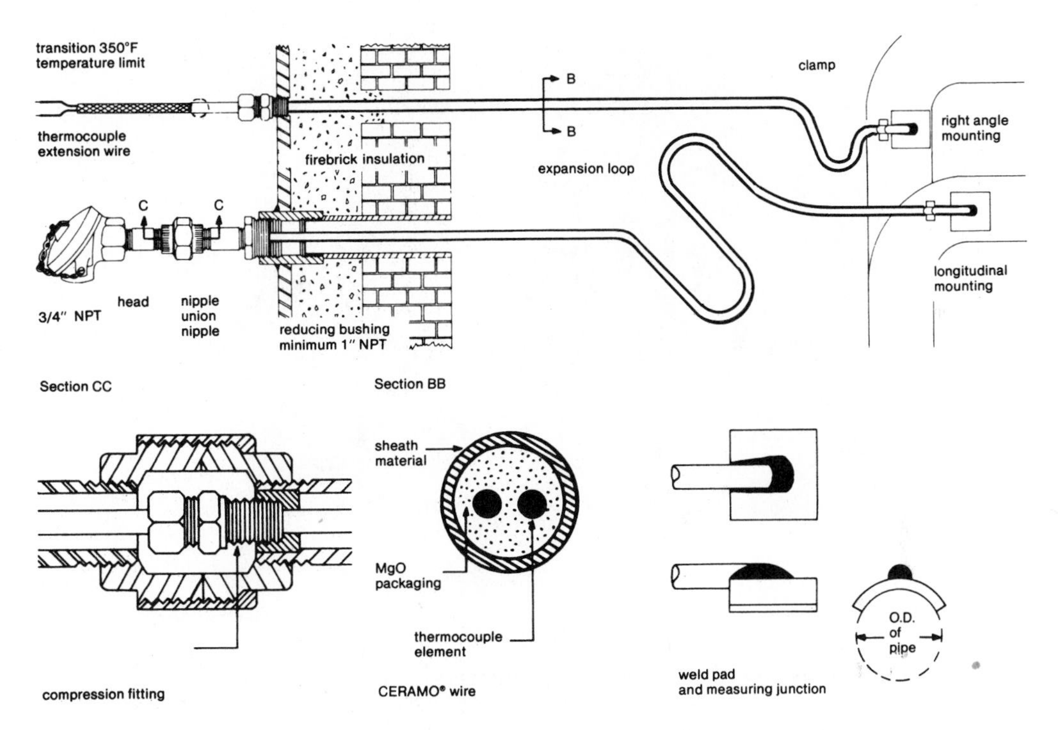

Figure 4-23 Industrial assemblies and wells *(Courtesy of Thermo Electric Co. Inc.)*

thermocouple inputs and displays and will linearize the values digitally with one degree of *resolution.* Fully isolated output options are available for either analog voltages (1 mV/deg) for strip chart recorders and other instruments, or tri-state binary coded decimals (BCD) for computers and printers, Figure 2-24.

Instruments for measuring on-line sensors with microprocessor-based multiple-function indicator/calibrators are available. These instruments will perform field or lab calibrations and will do the troubleshooting of thermocouples, RTDs, transmitters, and a wide variety of instrumentation use. The instrument is based on a single chip microprocessor that conforms to NBS thermocouple curves. This type of instrument can be used to calibrate most electronic devices, Figure 4-25.

Radiation Thermometry

All bodies or surfaces that are not at absolute zero temperature give off thermal radiation that may be used to measure their temperature.

Radiation Pyrometers

Radiant energy is always present at all times in the natural environment. The thermal energy from a hot object is transformed into electromagnetic waves and

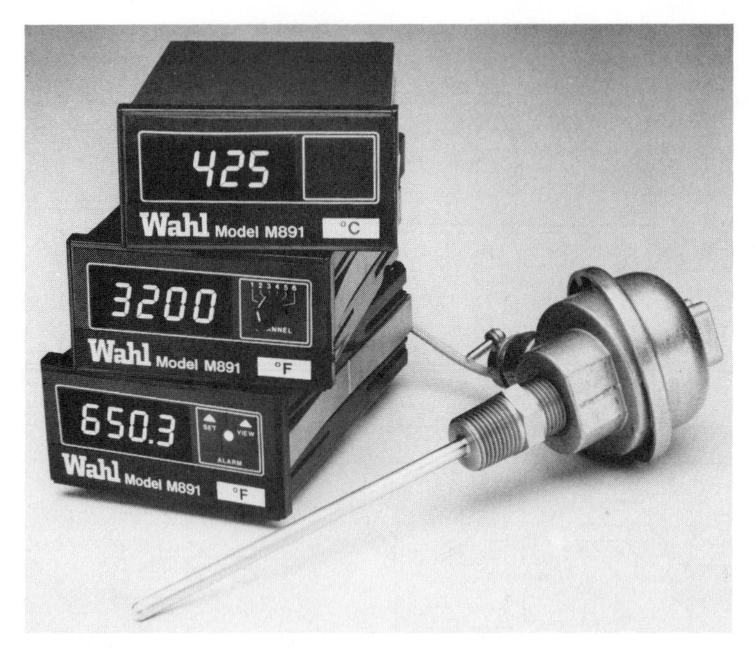

Figure 4-24 Panel thermometer *(Courtesy of Wahl Instruments Inc.)*

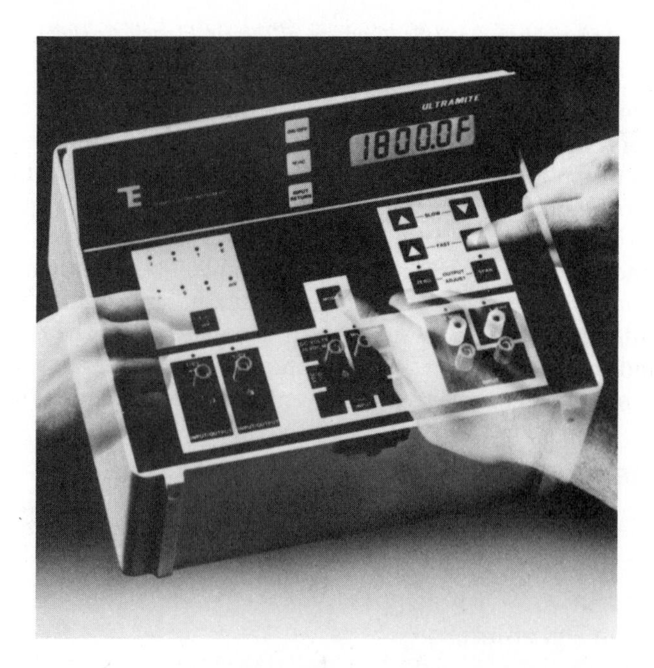

Figure 4-25 A multiple indicator/calibrator microprocessor-based, membrane switching, input, and output for J, K, T, E, R, S, and B thermocouples *(Courtesy of Thermo Electric Co. Inc.)*

is transmitted as radiant energy. When the electromagnetic waves strike an object they retransform into thermal energy. All bodies, by virtue of their temperature, emit radiant energy. The radiation energy loss is reduced from the heat energy the body contains and it becomes cooler. Radiation is emitted in accordance with a law presented by Stefan and Boltzman which states that "Total radiation per unit area per unit time is proportional to the fourth power of its absolute temperature" on any temperature scale.

Objects under radiation are considered to be either emitters, absorbers, reflectors, or a combination of all three applications. Emission is the rate that a body radiates energy and will depend upon the difference between the surrounding temperature and the object's temperature. Emission takes place at the surface of the body. A rough surface is a better emitter than a smooth surface because of the increased surface area. Emission is the same for both long wave heat radiation as for shortwave light radiation. Absorption is the reverse of emission, that is to say an excellent absorber emits little light back and appears black. A body that absorbs all light falling on it will appear totally black to the eye. A polished surface is a poor absorber and reflects most of the incident radiation. A radiation reflector functions similar to a mirror with visible light.

Radiation measurement relies upon the energy radiated by an object resulting from its temperature. The concept of a perfect radiating body is referred to as a *blackbody*. This concept means that the body absorbs all the radiation it receives and the same blackbody also radiates more thermal energy for all wavelengths than any other body of the same area at the same temperature. The ratio of the emitted energy of a blackbody to that of a material at the same temperature is referred to as the material's *emissivity*.

Radiation pyrometery instrumentation requires an emissivity standard to be used to measure a radiating body's temperature. The total emissivity of a material under measurement is required so that a calculation can be made and the temperature of the material under measurement can be determined.

$$\text{Emissivity} = \frac{\text{Total radiation from a non-blackbody}}{\text{Total radiation from a blackbody}}$$

Each industrial material during a temperature measurement will have its own emissivity value. This value is used in setting the emmisitivity compensation for the instrument.

Characteristics of Radiation Thermometers

Radiation pyrometry measurements are divided into two groups of instruments. Wide band pyrometers measure emitted energy over a broad band of wavelengths.

Narrow band pyrometers measure optical or brightness over a narrow band such as visible light and infrared radiation.

Wide Band Pyrometers - Wide band pyrometers are noncontact instruments capable of reading very high as well as low temperatures. This instrument is used to determine the temperature of such diverse temperature ranges as that of the sun or stars to that of smoke stacks of metal melting furnaces. The sensing unit is made of a series of thermocouples connected in series to become a thermopile, Figure 4-26A and Figure 4-26B. The radiant energy enters the instrument and is focused on the thermopile by using optical glass for high temperature measurement, quartz optics for most industrial applications, and mirrors for the measurement of low temperatures. The radiation pyrometers are internally temperature compensated so that the incoming radiation will not be affected by the instrument case temperature.

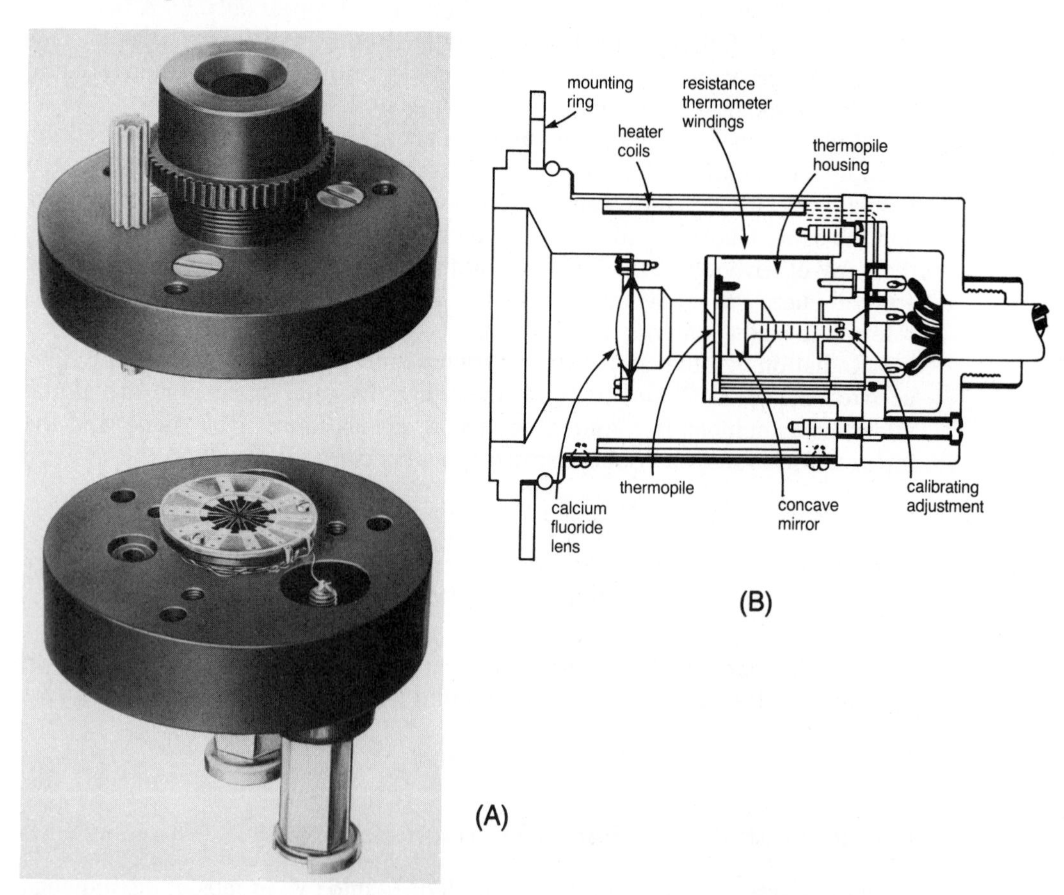

Figure 4-26 (A) A thermopile; (B) A broad band pyrometer *(Courtesy of Honeywell Industrial Products Group)*

Narrow Band — Optical/Brightness

Narrow band or optical/brightness instruments also allow industrial temperature measurements to be made without any physical contact with the object being measured. This is done by using the radiant energy coming from the heated object. In this case, the human eye is used as the detector to judge the color match between a electrically heated filament and the surface of the heated material under measurement. The light from the target material passes through an optical system to a lamp filament and on to a red filter through an optical eyepiece and into the eye. The filament is heated and emits radiant energy that is compared to the radiant energy coming from the hot target. The electrical current is adjusted until the filament disappears into the hot target background. The target and the filament at this point are emitting the same wavelength of monochromatic light. The instrument is calibrated so that the dial controlling the electrical energy is graduated in temperature symbols, reporting the temperature of the target area. Industrial temperature measurements are not usually blackbodies, thus corrections will need to be made for the emittance of the target when the instrument is being calibrated.

This instrument is frequently portable and convenient for the supervisor or technician to use. It is a reliable, versatile, and inexpensive method of gathering thermal data.

Ratio or Two-color Pyrometer

The ratio or two-color pyrometer measures radiation being received from the object as a ratio between two frequencies. This visible light is filtered into two different colors or wavelengths. These separate wavelengths are compared in amplitude to establish a ratio that represents a function of the temperature. The emmissivity ratio of most industrial materials varies slowly so that there is a minimal effect on the ratio of wavelengths; thus these are effective in industrial environments.

This type of temperature measurement instrument is very useful in an industrial setting where measurements need to be made through atmospheres of haze, smoke, dust, steam or vapors, and dirty optical systems, Figure 4-27A.

Infrared Temperature Measurement

Infrared thermometers are the only solution when the product is small, fragile, moving, or in a vacuum or other controlled atmosphere. They are very useful because they are noncontact measurements that do not add or remove heat or disturb the process in any way. These thermometers utilize various spectral ranges in the measurement of temperature, Figure 4-27A-C. Spectral wavelengths of 8 -14 microns measure temperatures from 0°F to 1000°F. These wavelengths are for low temperature applications such as print drying, food, wood, paper, textile

Figure 4-27A A portable two-color ratio pyrometer *(Courtesy of Capintec Instruments Inc.)*

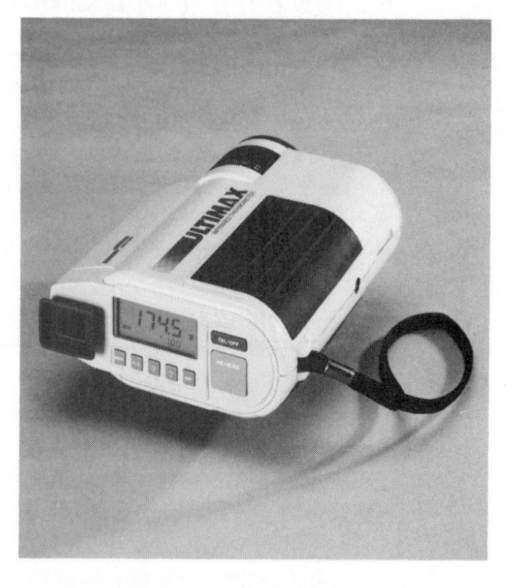

Figure 4-27B A portable infrared temperature thermometer *(Courtesy of IRCON, Inc., a Subsidiary of Square D Company)*

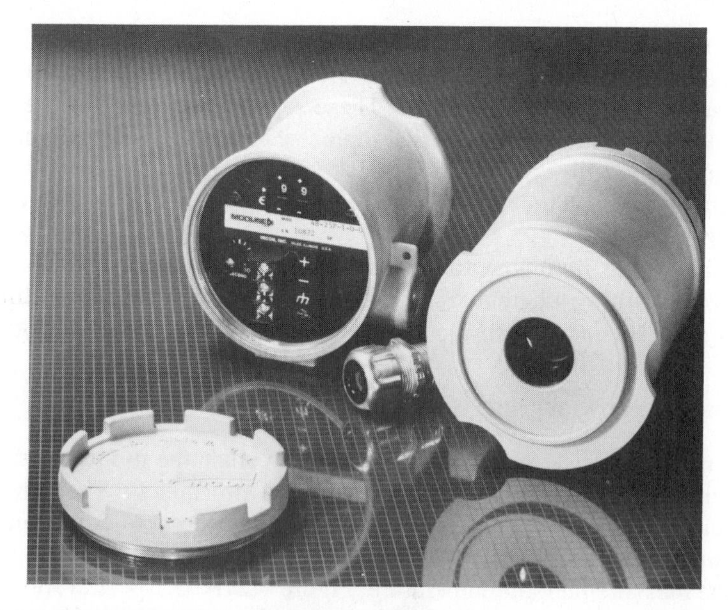

Figure 4-27C A built-in two-wire infrared thermometer transmitter with a 4 to 20 mA output *(Courtesy of IRCON, Inc., a Subsidiary of Square D Company)*

processing, and vacuum forming. Wavelengths of 3.7 - 4.0 microns measure temperatures from 500°F to 2500°F for the measurement of glass temperatures. The wavelengths of 2.0 - 2.6 microns are used to measure temperatures from 1000°F, providing measurements for all types of general applications of medium temperatures including metals. Wavelengths of 4.8 - 5.2 microns are used to measure surface temperatures in glass manufacturing operations such as forming, bending, tempering, annealing, and sealing. Finally, for a broad temperature range from 0°F to 2500°F, wavelengths from 7.5 microns to 8.5 microns are ideally suited for the temperature measurement of thin films of plastic, oils, paper, wood, and other organic materials. These wavelengths will also measure the temperature of refractory brick, painted surfaces, and glass.

Temperature Sensitive Materials

Temperature sensitive materials are a means of temperature indication. Specially compounded chemicals are applied to the surface of products under manufacture or later when these products are in shipment, working, or in storage. They indicate that a predetermined temperature has been reached by a change in physical appearance. Temperature sensitive materials have been formulated that can be applied to a product: paint, recording labels, crayons or pellets, and pyrometric cones. These temperature sensitive materials, upon reaching a designed temperature and time, will abruptly change color or phase from a solid to a liquid. The change is indicated by a chemical transformation of the material. These materials can be so carefully compounded that they will respond to a specific temperature value. They are designed to reveal temperatures from below 100°F to 2500°F.

Temperature Sensitive Paints

Paints that are temperature sensitive will change color as the result of the dislocation of water vapor or gas from the crystalline salts. These materials can be used to assure the buyer of a manufactured product that heat damage has not occurred during manufacture, storage, or shipment. This assurance frequently is a very important requirement to the electronics industry.

Temperature Recording Labels

Irreversible temperature recording labels provide an instant direct readout of the recorded temperature. These self-adhesive temperature recording labels have sensor windows that contain one to several sealed heat-sensitive elements that progressively change chemical structure when exposed to heat exceeding their calibrated temperatures. These decals help to monitor, and thus help to prevent, overheating of valuable machinery, electrical parts, electronics components, perishable goods, and other heat-sensitive materials and equipment, Figure 4-28.

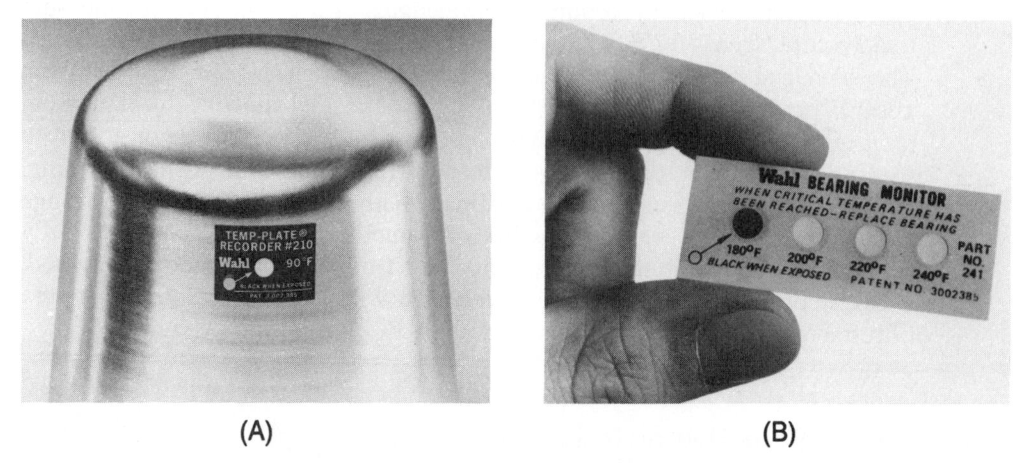

Figure 4-28 (A) Irreversible temperature recorders 80°F and 90°F decals; (B) 180°F to 240°F temp-plate recording bearing monitor *(Courtesy of Wahl Instruments, Inc.)*

Crayon and Pellet Temperature Indicators

Color crayon or pellet temperature indicators have a specific melting point at which a phase change takes place in the crayon or pellet. The crayon is stroked on the material under test and observed. When the mark begins to leave a liquid smear on the work, the piece is at the temperature indicated by the crayon's label. Crayon-type indicators are manufactured in a large number of temperature ratings, from 100°F to 2500°F. Fusible pellets are used when a large workpiece must be observed at a distance and the temperature checked at a number of different levels and distances around the work, Figure 4-29.

Pyrometric Cone Temperature Indicators

Pyrometric cones have been used for scores of years to indicate surface temperatures in the ceramic industry. These clay products are slender trihedral pyramids

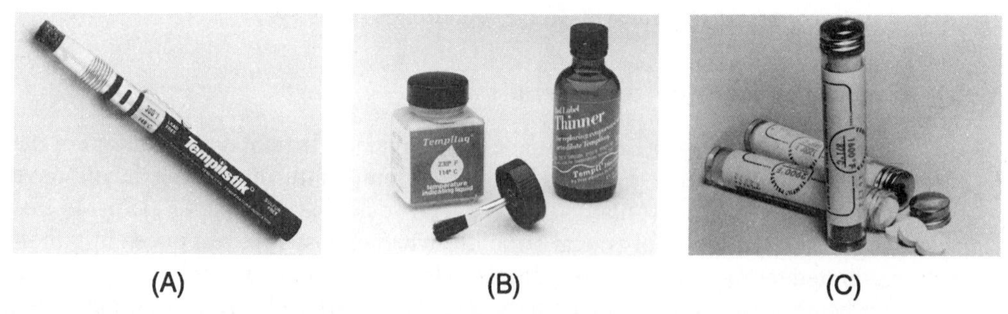

Figure 4-29 Temperature-indicating crayons, liquids, and pellets *(Courtesy of Tempil Division, Big Three Industries, Inc.)*

that deform into an arch when the cone's designed temperature is reached. These cones are usually employed over a range of temperatures in sets or three or four. Each cone has its specific temperature and time value when it deforms. The temperature is read by the position of the cone whose tip has bent into an arch and touches the base mounting. The temperatures covered by these cones range from 584°C (1085°F) to 2015°C (3659°F). These indicators are frequently used in the ceramic industry as a check on the total thermal environmental exposure of the products during manufacture and to provide a permanent temperature record, Figure 4-30. The devices must be observed directly; they do not produce a monitoring signal.

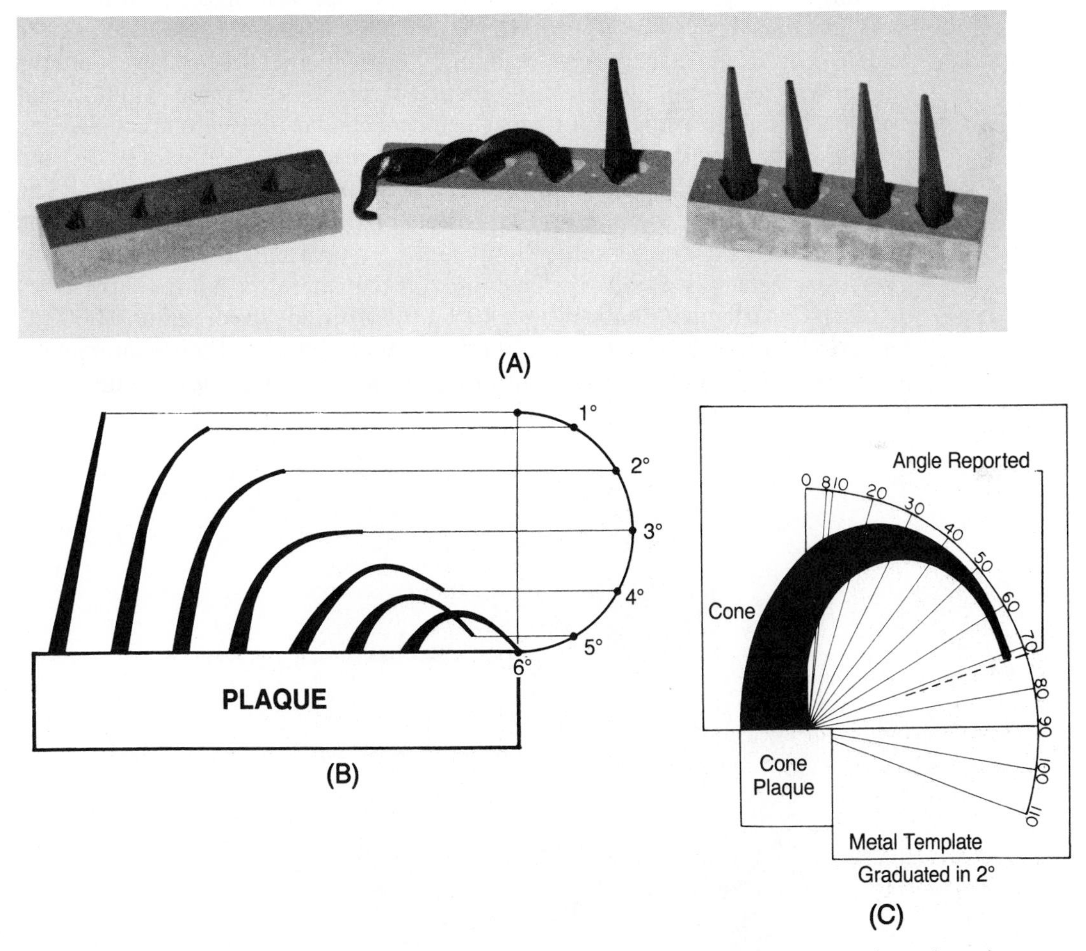

Figure 4-30 Pyrometric Cones (A) Temperature measurement with pyrometric cones determines the combined effect of temperature and time in the firing of ceramics; (B and C) When the pyrometric cone tip is five degrees from the level of the cone plaque, the time and temperature has been reached for the maturing of the fired ware. *(Courtesy of The Edward Orton Jr. Ceramic Foundation)*

Quartz Thermometer

Temperature is also measured by resonation quartz crystals. Crystals change resonant frequency when their temperature is changed. This change of frequency is employed as a temperature measuring instrument. A reference oscillator is established as the reference frequency against which the sensor crystal is compared. A change in the number of cycles per second results from a temperature change. The new frequency provides a signal that is compared to the reference frequency. The output of the comparison will be signal conditioned and calibrated for an output signal to a digital readout, strip-chart readout or be the input to a microcomputer.

The sensor probes are sealed in a stainless steel well (like a thermocouple well) and thus this device will withstand very high pressures without damage or impairment to the temperature readings. Each quartz sensor has a unique temperature response; each quartz sensor is individually calibrated. This calibration data is fitted to a curve using a regression technique. The unique coefficients for the sensor are recorded into a programmable read-only-memory (ROM) and is mounted in a calibration module that is shipped with each probe. The thermometer's microprocessor reads the data on the calibration module and the frequency of the quartz sensor to compute the temperature of the quartz sensor. This instrument, when calibrated, will provide temperature readings in the order of .0001°C in a temperature range of -40°F to 450°F, and plus or minus 0.0002°C for periods of months or more. One of the chief advantages of this equipment is that it can withstand shocks without sustaining a loss of accuracy, Figure 4-31.

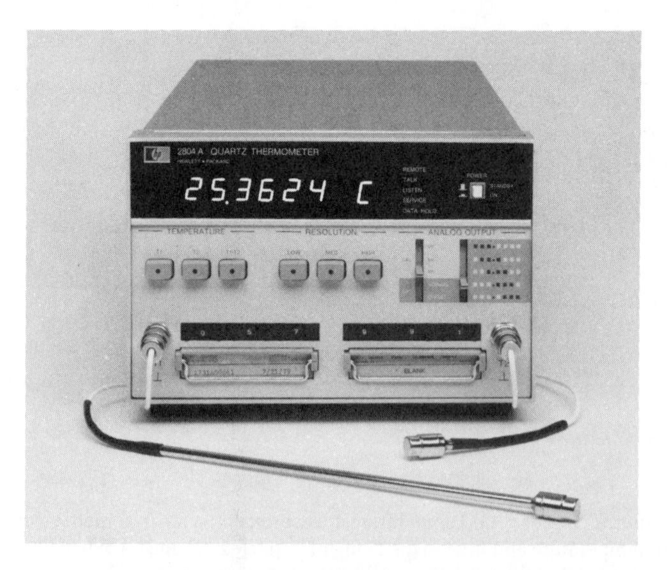

Figure 4-31 Quartz thermometer for high accuracy measurements *(Courtesy of Hewlett Packard Co.)*

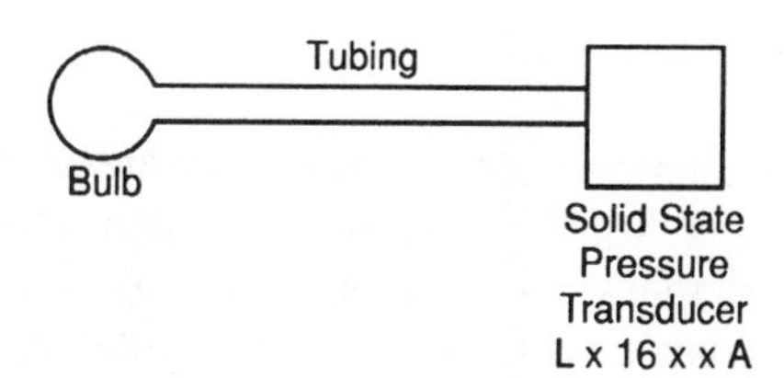

Figure 4-32 A hot bulb thermometer *(Courtesy of SenSym)*

Hot Bulb Temperature System — Solid-state

Hot bulb temperature system utilizes a solid-state silicon pressure sensor in an integrated circuit pressure transducer to sense a heated constant volume gas bulb. This transducer in turn produces an electrical output signal.

A container or bulb with a constant volume of gas is subjected to the medium being measured. This thermometer is based on Charles' law which states that for a gas at a constant volume, pressure is proportional to absolute temperature. This is a multiple-step transduction process: temperature to a pressure then pressure to a voltage which represents the temperature measurement. These IC pressure transducers are interfaced with other systems and deliver a temperature signal to a microcomputer to produce a low-cost and reliable system for control in automation technology.

This system provides an alternative choice to an electronic thermometer, because above 200°C, most junction techniques suffer from error and Seebeck effects begin to appear. In the hot bulb system, the bulb is placed in the medium and the sensor can be connected to a capillary tube with the sensor out of the heat-affected area, thus providing a reliable and stable sensor, Figure 4-32.

Cryogenic Temperature Measurement

Cryogenic temperature measurement is concerned with very cold temperatures. These temperatures occur in the liquefaction of industrial gases. Thermocouples, resistance bulbs, and thermistors are among the devices that measure this temperature range. A number of industrial organizations prefer the use of the resistance bulb as the sensor because the platinum wire in the bulb is a stable and linear sensor for these temperature measurements.

SENSORS USED TO ACQUIRE PRESSURE DATA

Pressure Sensors

Pressure is the source of many industrial control measurements. This variable is defined as force applied to an area. Force in the case of fluids is the application of energy that is transmitted through the fluid and is governed by Pascal's principles.

There are four major divisions in pressure measuring systems: (1) mechanical systems employing liquid columns; (2) elastic materials employing bourdon tubes, bellows, diaphragms, and capsules; (3) electrical systems applying the variables of resistance, current, voltage, frequency, conductance, inductance, capacitance, and impedance; (4) solid-state electronic systems utilizing hybrid integrated circuits.

Liquid Column Pressure Devices

The Barometer - The *barometer* is a fundamental liquid column instrument that measures absolute pressure. A tube sealed at one end, filled with fluid, inverted, and placed with its open end submerged beneath the surface of a container of like fluid produces a barometer. The fluid in the tube will fall until the hydrostatic head of the fluid in the tube is equal to the atmospheric or surrounding pressure. This instrument has a near vacuum above the fluid and therefore it indicates a pressure reading with reference to the vacuum above the fluid. This instrument provides an absolute pressure measurement commonly referred to as PSIA or pounds per square inch absolute. This instrument is used to measure atmospheric pressure and also can be applied to the precise calibration of other instruments, Figure 4-33.

The Manometer - The manometer is a very useful instrument for measuring variable pressures and differential pressures. The U-tube manometer is made by bending a tube into a U shape with the vertical members of the U extended to become the measuring channels. A measuring scale is mounted behind the tubes so that the heights of the fluids may be compared. The U-tube contains some liquid in the bottom third of the tube, usually mercury because of its high specific gravity.

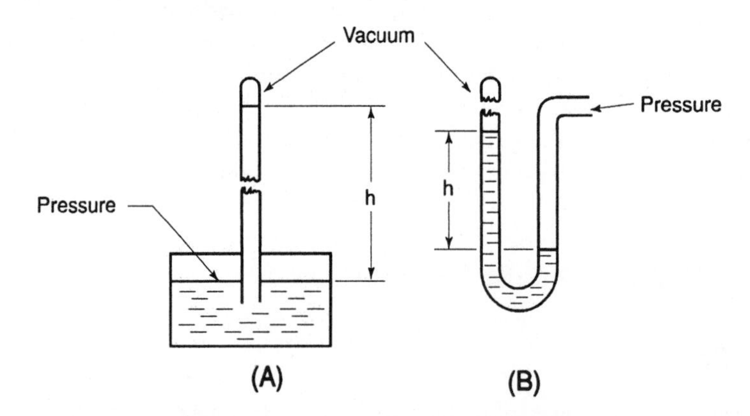

Figure 4-33 (A) Absolute pressure barometer; (B) Absolute pressure gage

During measurements, one end of the tube is fastened to the medium being measured while the other end of the U-tube is open to the atmosphere.

When the fluid is at rest (no pressure on the system) the mercury will be at the same height in the tubes. When pressure to be measured is applied to the measuring branch of the U-tube the mercury is displaced. A measurement is made of the difference between the heights of the mercury in the two branches of the U-tube. The pressure measurement is the difference between the two heights of mercury times the density of the fluid. Fluids other than mercury, with different specific gravities, may be applied. Those with a lesser specific gravity will deliver an expanded scale reading for a given pressure.

Manometer Pressure = Density of the fluid times difference of the heights of the fluids in the two branches of the U-tube.

The measurements may be made in inches or in centimeters. It is referred to as a PSIG measurement in contrast to the PSIA measurement. The reason for this is that atmospheric pressure is applied to the system through the open end of the U-tube, providing a 14.7 pound pressure to the system, Figure 4-34. The pressure is not compared to a vacuum as in the earlier case. Many industrial gages are subject to atmospheric pressure and thus report a PSIG reading. When the ideal gas laws calculations are performed using these data, the PSIG information must be converted into PSIA values.

The manometer can be also used for the measurement of the difference between two pressures. This is referred to as a *differential pressure* gage. One pressure is applied to one branch of the open end of the U-tube and a second

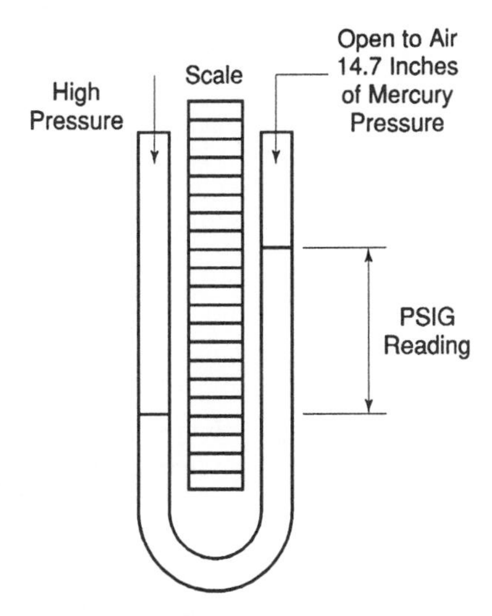

Figure 4-34 A U-tube manometer open to air pressure will yield a PSIG reading

pressure measurement to the other end of the U-tube. The two pressures are applied, and the manometer fluid will balance at the difference in height between the columns of fluid. This difference in height represents the difference between the two pressures, Figure 4-35.

In the making of precision measurements using these techniques, consideration must be given to temperature. Density and viscosity change with temperature, thus these factors must be stabilized or compensated for during accurate measurements. In some cases, the capillary action of the fluids in tubes must also receive a correction.

The Well Manometer - The well manometer is a variation of the U-tube manometer. The well manometer has a cistern or well as a base and a single vertical tube for the mercury or other fluid. The ratio between the diameter of the measuring tube and the well is important. The area of the two should not be less than a ratio of one to ten in order to minimize error. To compensate for the lowering of the fluid in the well as the fluid rises in the measuring tube is the reason for the required ratio, Figure 4-36. In some instruments, the height of the mercury can be adjusted so that a zero position for the height scale can be accurately positioned. These instruments are frequently used to calibrate other industrial instruments.

The Inclined Manometer - The inclined manometer is designed to read small pressure differentials. The measuring tube is inclined nearly into a horizontal position rather than the traditional vertical position. This slanted position expands the measuring scale as a vertical measurement expands when measured on the

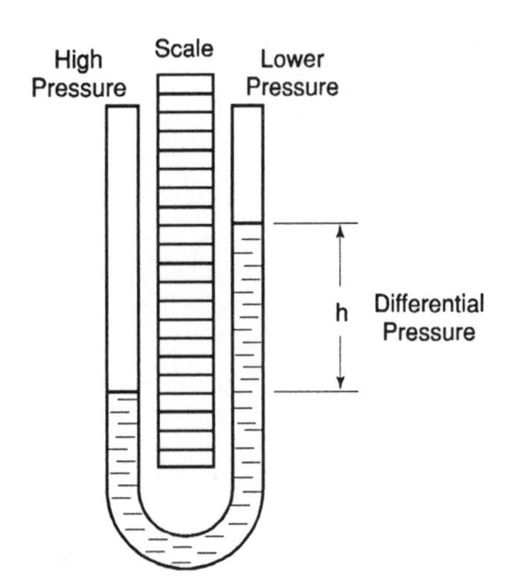

Figure 4-35 Differential U-tube manometer

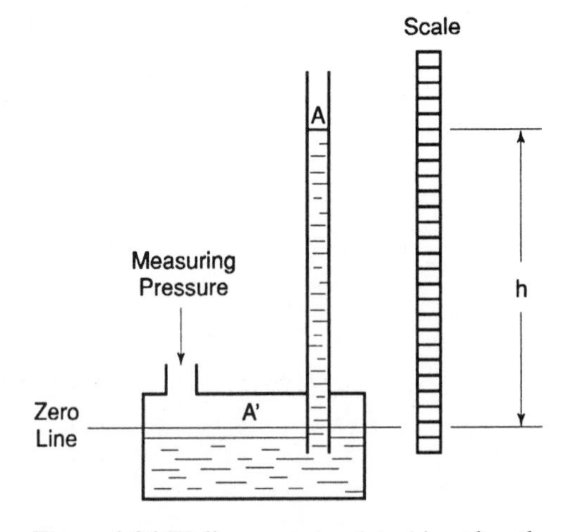

Figure 4-36 Well manometer. A to A' not less than a ratio of 1 : 10.

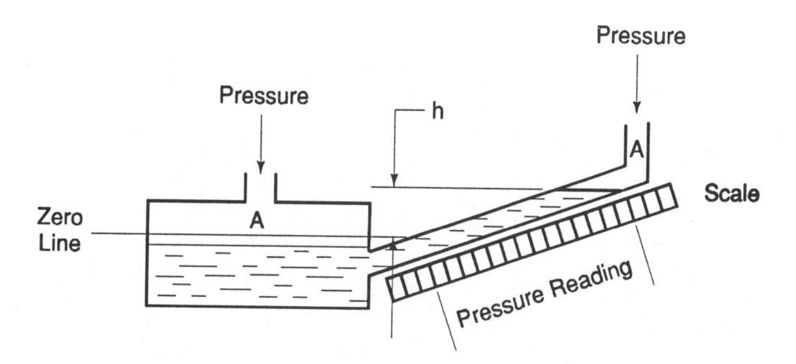

Figure 4-37 Inclined manometer. This manometer expands the scale for low differential pressures.

hypotenuse of a triangle. The angle reading expands the vertical scale, making a small difference in pressure easier to read, Figure 4-37. This instrument is also referred to as a draft gage because of its application to measure the differential pressure from the top and bottom of a chimney or flue.

Mechanical Pressure Devices

The Bourdon Tube - The bourdon tube has long been used by industry to measure pressures from 25 psi vacuum to 80,000 psi constant pressure. The C-type bourdon tube is a widely used pressure indicating instrument. The C-type tube is shaped by rolling a tube into an elliptical cross section while forming it into a letter C configuration. The C portion of the tube is curved to about 180°-270°, with one end bent back upon itself and soldered to a pipe fitting so that the gage can be mounted to its pressure source. The other end of the C is sealed and an attachment is added so that a linkage may transmit the tube's movement to the mechanical linkage system.

The tube, because of its material and work hardening, has become a spring element. When pressure is applied inside the tube, it will tend to become rounder and the shape of the C opens up the incomplete circle, providing a movement at the curved tip end. This movement can be correlated with the pressure within the tube. The attached linkage provides movement to a sector gear that mates a pinion gear; its shaft carries the pointer of the instrument. These gear and lever systems multiply the movement of the tube to the pointer, Figure 4-38.

The bourdon tube has variations of the C-type: spiral, helical, and twisted, Figure 4-39. All utilize Young's *modulus of elasticity* of the material to measure the pressure. Common materials used to make bourdon tubes are cartridge brass, trumpet brass, phosphor bronze, beryllium copper, monel metal, stainless steel, chrome molybdenum, and plastics. In addition to the selection of a material for the bourdon tube, the working environment must be considered. The instrument system designer must take into consideration the corrosion resistance of the equipment as well as the physical forces acting on the instruments.

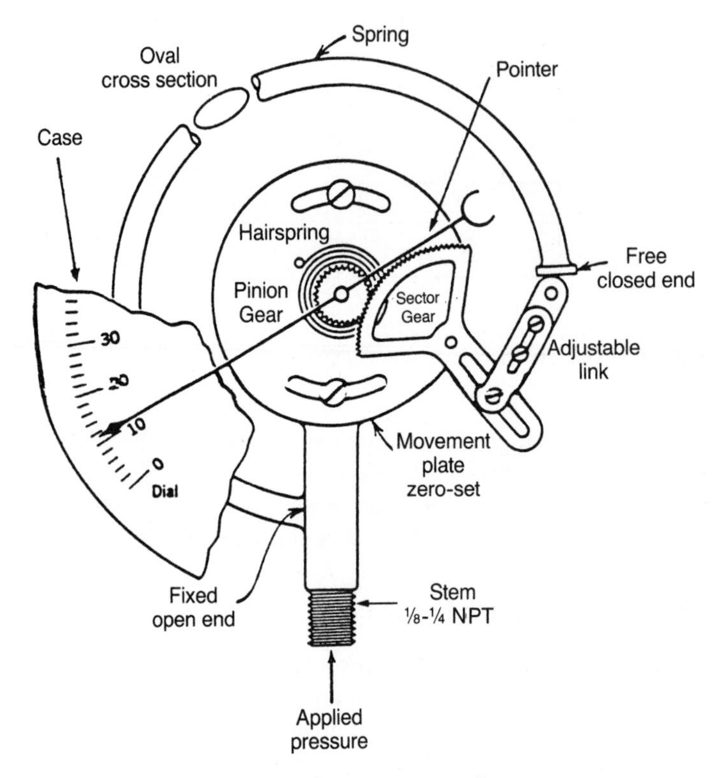

Figure 4-38 C-type bourdon tube pressure gage. Pressure gage and part names

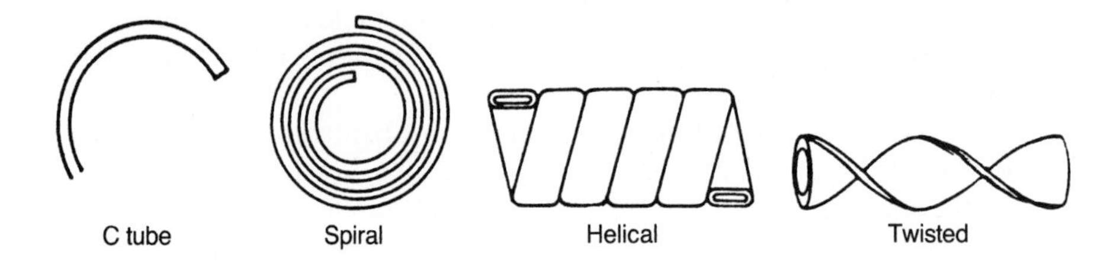

Figure 4-39 Bourdon tube sensing elements

The Bellows - Bellows are deeply convoluted cylindrical containers that have a number of applications in pressure instrumentation. The chief characteristic of the bellows is that they provide a greater mechanical displacement or movement when subjected to a given pressure. The physical response of the bellows is determined by the number of corrugations, the thickness of the moving metal, the diameter of the bellows, and the modulus of elasticity of the material used. The bellows can vary in size from 1/16 of an inch to six feet in diameter. The metals of the bellows are thin and flexible but ductile enough so that they will resist fatigue failure. These corrugated cylinders are manufactured by rolling or hydroforming. The ends and connectors are added by soldering. The metals frequently used are various brasses, bronzes, beryllium copper, stainless steel, monel, steels, and plastics.

Bellows are employed in a number of different ways. The simplest procedure of measuring a pressure is to apply the pressure or a vacuum to the interior of the bellows measuring device. Another method is to place the bellows in a container so that the pressure on the outside of the bellows and the pressure on the inside of the bellows is different. This is referred to as a nested bellows and will provide an algebraic sum of two different pressures by the positioning of the linkage or transmitter, Figure 4-40. Bellows can also be used to provide sealing for moveable mechanical members. Bellows may be used to isolate vibration between apparatuses and to provide temperature correction. They also can be used as efficient springs or be used for the alignment of shafts. Bellows may be employed as a small dashpot or *accumulator*. When in a pneumatic circuit, they can be used to transmit, divide, or multiply force and/or movement.

The bellow's primary function in instrumentation is to sense and transmit pressure signals. These signals frequently are transduced into electrical, electronic, or digital signals.

The Diaphragm - A diaphragm is a sensor made of thin flexible material that is

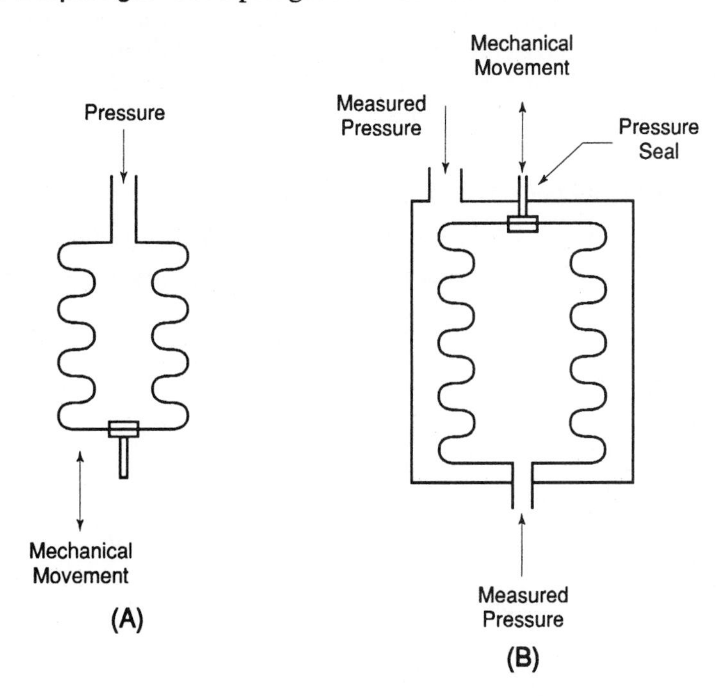

Figure 4-40 Mechanical bellows pressure transducers (A) Bellows; (B) Nested Bellows

supported by a circular rim. A mechanical fastener is attached to the center of the corrugated flexible diaphragm. The diaphragm is mounted so that the pressure to be measured is on one side and the atmospheric pressure is on the other. The difference in pressure causes the diaphragm to flex and provides a mechanical movement proportional to the pressure. The deflection of the diaphragm is a function of the material's modulus of elasticity; the diameter or area and thickness of the disc; the number, shape, and depth of the corrugations; and the pressure applied to the sensor, Figure 4-41. The output of the sensor causes the movement of a linkage. The linkage in turn can be attached to a mechanical, electrical, electronic, or digital device for the final signal output.

Like the other deforming sensors, diaphragms are made of brasses, beryllium copper, stainless steel, leather, neoprene, rubberized fabric, or plastic.

Diaphragms are used in instrumentation for mechanical applications other than sensors. They are used to seal one area from another, to provide a semimoveable linkage, and to provide a barrier to fluids or other contaminants.

The Capsule - Capsules are constructed of two or more diaphragms that are fastened at their edges and are frequently stacked in a series. The capsules are fastened as a unit through the center. They may also have a pressure connection between the units so that their expansion or contraction is additive. They sense the applied pressure and they transmit their new dimension to a mechanical position of the capsule, Figure 4-42. With the movement of an attached linkage, the information is relayed to any type of signal conditioning device: electrical, mechanical, pneumatic, or electronic. These devices are designed to cover low pressure ranges (0–5000 PSIG).

Electrical Pressure Transmission Devices

Many electrical pressure sensor transmission devices adapt mechanical movements to electrical devices, such as the potentiometer, rheostat, variable capacitor,

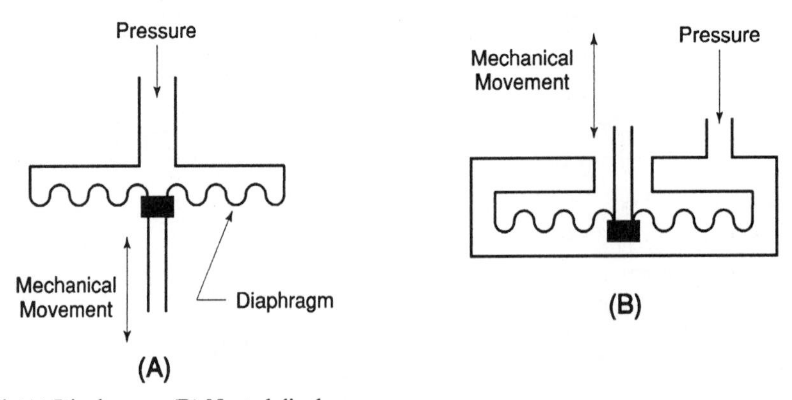

Figure 4-41 (A) Diaphragm; (B) Nested diaphragm

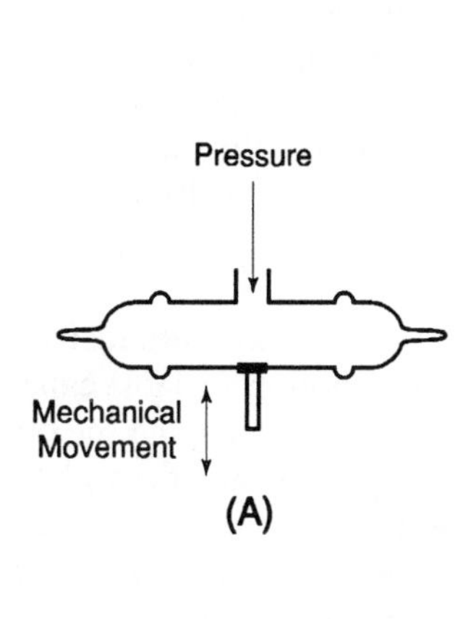

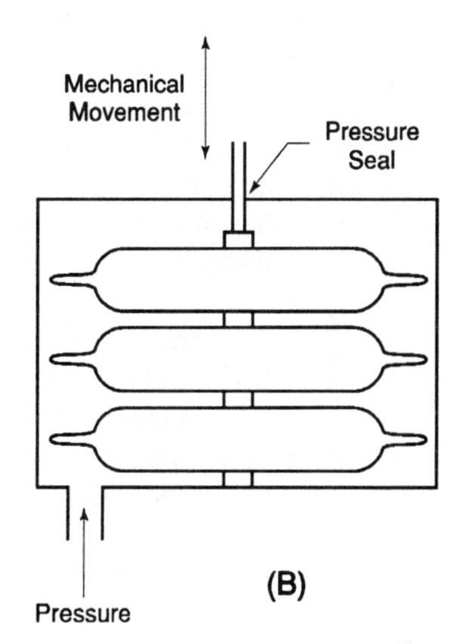

Figure 4-42 (A) Capsule; (B) Nested capsule

variable differential transformer, rotary variable differential transformer, variable reluctance transducer, variable inductor, strain gage, Wheatstone bridge, piezo-electric transducer, hot cathode ionization gage, and cold cathode ionization gage.

Potentiometer - If an output from a sensor is a linear motion, the movement can be converted to an electrical signal with the application of a linear motion potentiometer. The movement causes a slider to move over a resistor, changing the value of the resistance in the signal circuit. This change can be measured with a Wheatstone bridge, or it can be measured by a voltage drop caused by an increase in the resistance. This type of transducer is built with a variable resistance winding and a sliding contact that is moved by the linkage from the motion sensor such as a diaphragm or bellows. The same principle is applied to the angular motion of a semirotary potentiometer that may be attached to a device such as a helical bourdon tube, Figure 4-43.

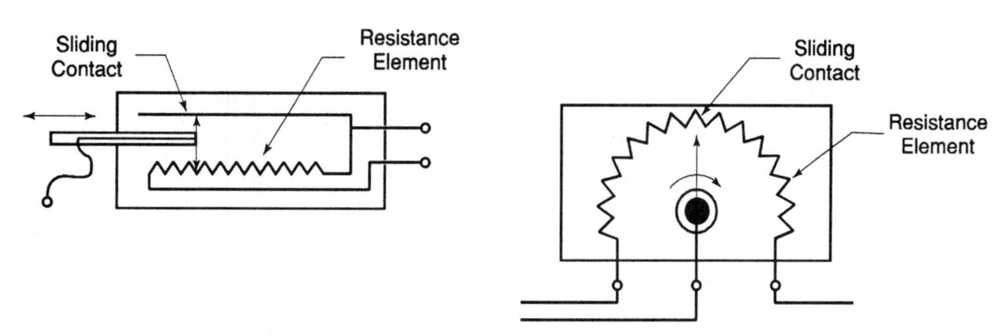

Figure 4-43 (A) Linear potentiometer; (B) Rotary potentiometer

Rheostat - The rheostat is a variable resistor connected in the circuit so that it controls the current, Figure 4-44. In the rheostat, only two connections are used. By moving the rheostat the slider on the resistance winding, the resistance can be increased to the maximum or reduced to the minimum by the amount of resistance available in the device. This change in resistance controls the current and also drops the voltage. Motion converters previously described can also be applied to the rheostat.

Variable Capacitor - Mechanical movements created by pressure instruments can be measured by electrical means utilizing a linear motion variable capacitor. This device is made up of two plates—one is fixed to its mount and the other is moveable by the variable. Two plates make up the capacitor, and the air between the plates provides the dielectric. When the plates are moved close together, the capacitance is increased; when the distance is increased between the plates, the capacitance is decreased. An AC signal is used to sense the change in capacitance that will alter the frequency of an oscillator, Figure 4-45.

Linear Variable Differential Transformer - The linear variable differential transformer is frequently used to convert a mechanical motion to a changing voltage signal. This is a linear position transducer. The transformer is designed with two identical secondary coils side by side with a third coil that provides the source of electromagnetic induction. The unit is constructed so that a soft iron core

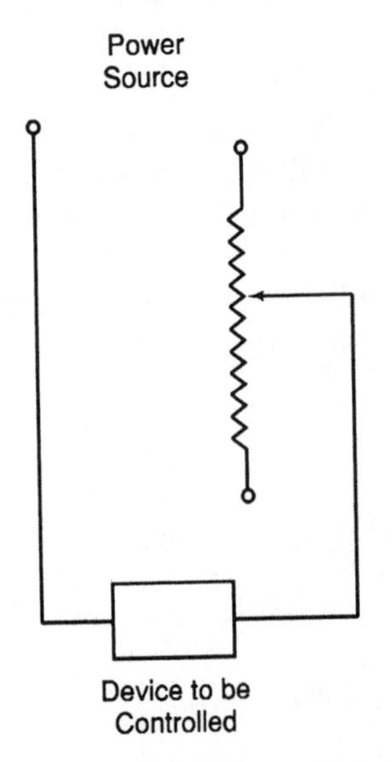

Figure 4-44 Rheostat. The rheostat is designed to adjust or control the current and may be rotary or linear in construction

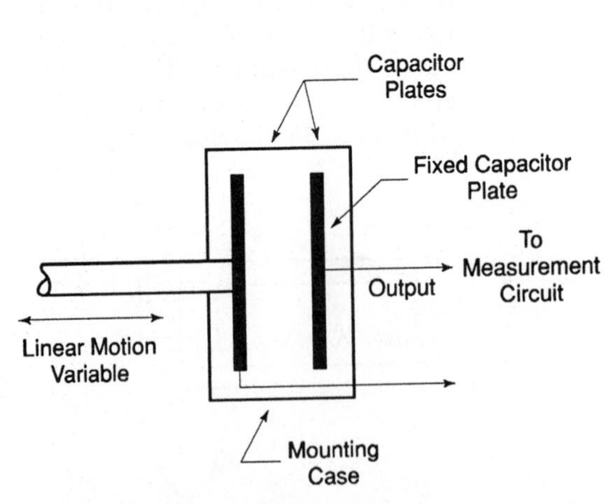

Figure 4-45 Linear motion variable capacitor

can be positioned inside the coils. When the core is in the center or null position, the output from the two secondary coils is the same. If the core is moved, the electromagnetic induction couples with one of the secondary coils more than the other. Thus, a higher output from one coil and less from the other produce a differential between the two coil outputs. If the core is moved in the opposite direction, the reverse action takes place in the secondary coils. The degree of voltage change in the secondary coils is proportional to the displacement of the core by the mechanical movement. The core is moved by the action of the mechanical sensor, and it provides an accurate electrical signal, Figure 4-46.

Rotary Variable Differential Transformer - The rotary variable differential transformer is a variation of the linear variable differential transformer except that the motion received by the unit is a mechanical rotary motion, Figure 4-47. This is a position or angular motion transducer. It again is designed with two coils wound side by side to make up the secondary windings and to produce the

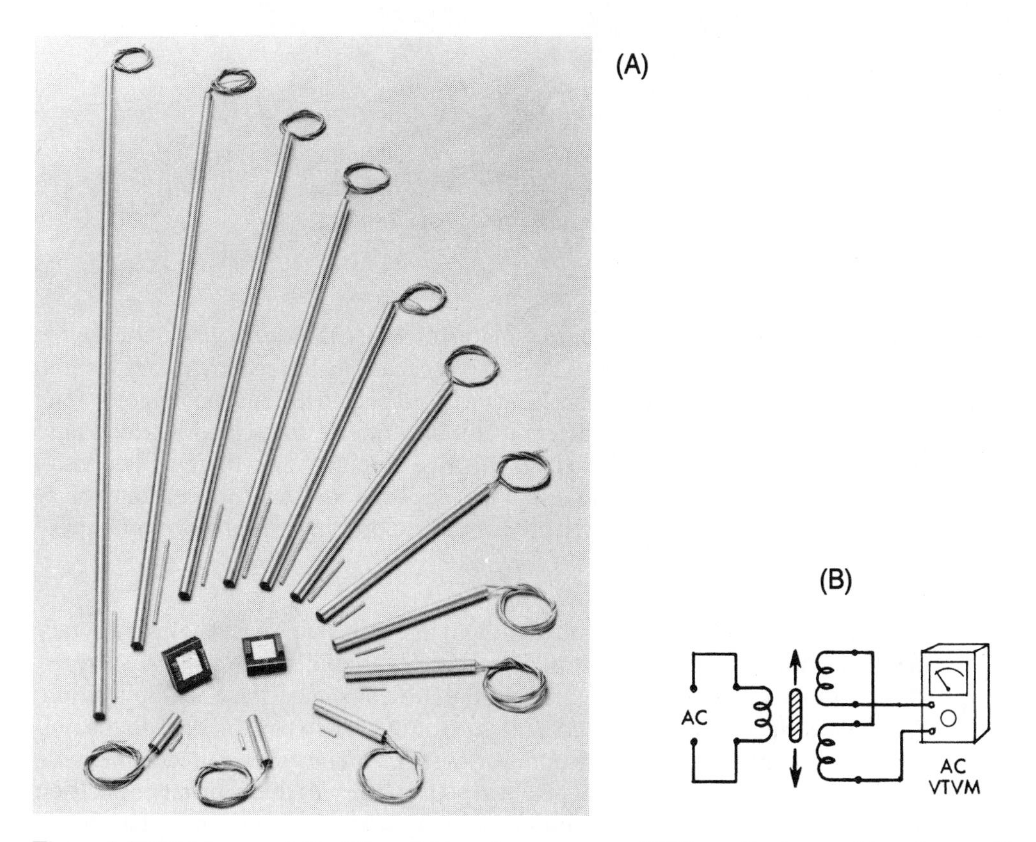

Figure 4-46 (A) Linear variable differential transformer sensors; (B) Linear displacement transducers with display *(Courtesy of Trans-Tek Inc.)*

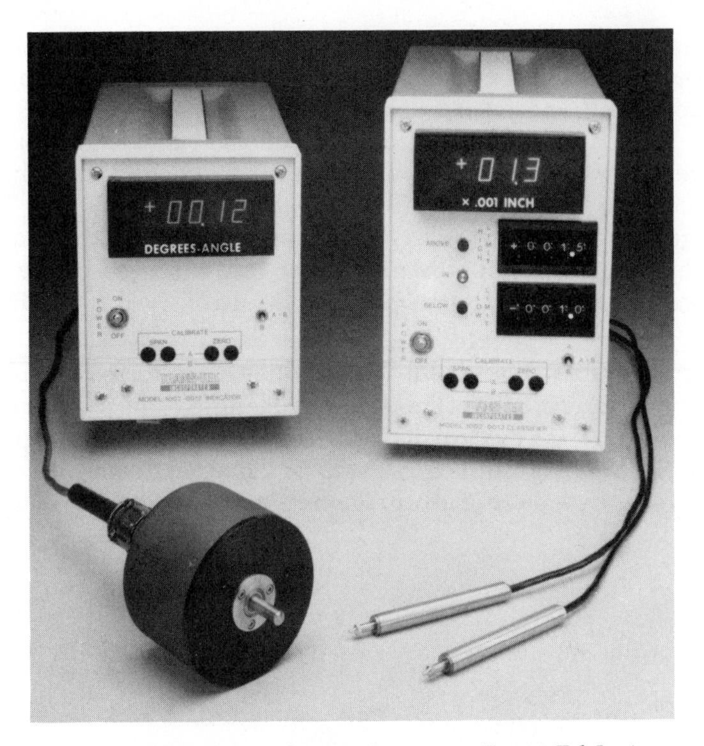

Figure 4-47 Rotary variable differential transformer *(Courtesy or Trans - Tek Inc.)*

differential voltage output. A third winding provides the electromagnetic source of energy. The core in this case is shaped and can be positioned in a rotary fashion to a number of different rotary positions with respect to the secondary coils. The position of the core provides different amounts of electromagnetic induction to secondary coils thus changing to coils' voltage output. Again there will be a null position where the outputs are equal but opposite and will cancel, but rotary movement in either direction will produce a proportional signal to the amount of movement.

Variable Reluctance Transducer - A variable reluctance transducer is a rotary or angular transducer designed similar to a rotary variable differential transformer. The two reluctance coils are identical but are connected in series-opposition. When the shaped iron core *(armature)* is positioned between the coils in the null position, the impedances of both coils are the same. The voltage drop is the same in each of the coils but they are opposite in *phase*. The variable reluctance position transducer's output varies with rotor position because the impedance of each coil leg is affected by the amount of magnetic iron in its AC field. A clockwise rotation will cause an increased voltage in one phase and a counterclockwise rotation will cause an increased voltage in the opposite phase, Figure 4-48. A rotation change can be transmitted equal to forty-five degrees on each side of the null position.

Variable Reluctance Induction Transducer - The variable reluctance induction transducer is formed by placing a coil around a permanent magnet. The unit is placed close to a rotating ferrous unit such as a gear or a vane mounted adjacent to the shaft element to be sensed. When the element, such as a gear tooth, passes the pole face of the magnet, the reluctance of the flux path between the magnet and coil is reduced and a voltage is induced into the coil. When the element moves away from the pole face, the reluctance increases and the induced voltage drops, thus producing a voltage pulse. These pulses can be counted against a time base to yield a rotational speed of the element, Figure 4-49.

Strain Gage - The strain gage is a device for measuring mechanical surface strain. By applying this resistance measuring element to a diaphragm, beam, or

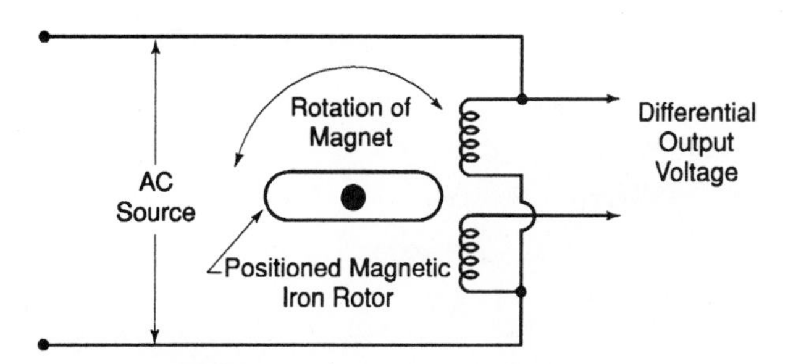

Figure 4-48 Variable reluctance position transducer. The magnetic rotor's position affects the impedance of each coil leg.

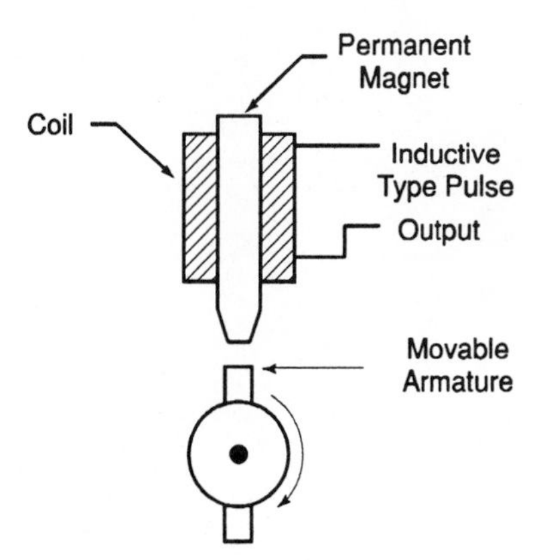

Figure 4-49 Variable reluctance transducer. The variable reluctance induction transducer supplies a voltage pulse with each passing of an armature vane

column, stress loads can be measured. When a bonded strain gage is fastened to the surface to be strained, an electrical resistance pressure transducer is produced that will correlate stress with resistance.

When a strain is applied to the device to be measured, the length of the gage increases while the cross section decreases. In both cases the resistance increases and thus provides a signal proportional to the load or strain.

Strain gages are made from metal wire, silicon, germanium, or a diffused silicon semiconductor or thin film of metal alloy. These materials are used in different physical forms or shapes. A frequently used form is a grid made of various foils. Foil strain gages are small and fragile and must be installed with care in order to receive accurate data, Figure 4-50.

Signal conditioning of strain gages is carried out by using the strain gages themselves as the active elements in a Wheatstone bridge configuration. The strain gage bridge will have added resistances included to control the excitation voltage, temperature, and zero compensation. In addition, metal foil and semiconductor transducers can have integral amplifiers included in the system.

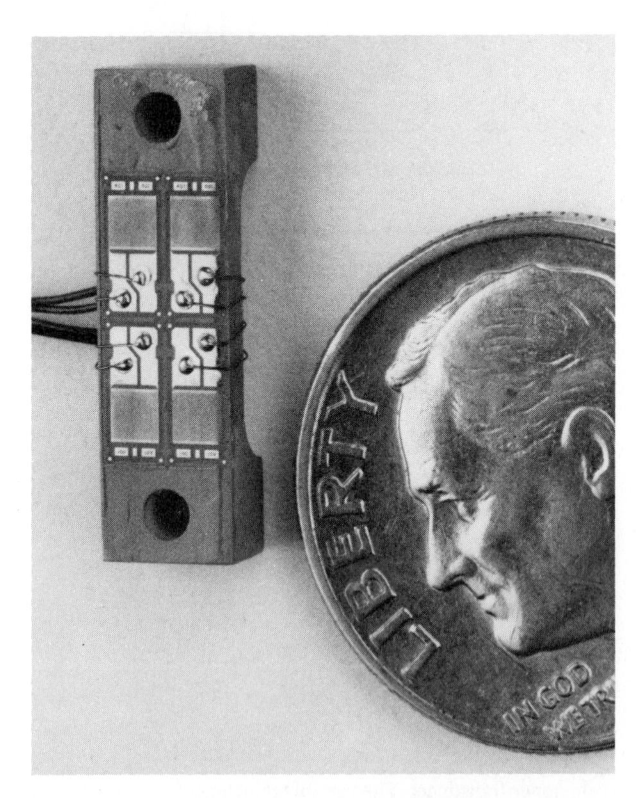

Figure 4-50 Typical bonded-foil strain gages *(Courtesy of Measurement Group Inc.)*

Piezoelectric Crystal Transducers - The piezoelectric crystal transducer applies the fact that certain crystals generate an electrostatic charge when they are stressed or placed under a pressure. The materials that perform in this fashion are quartz, tourmaline, rochelle salt, and ammonium dihydrogen phosphate crystals. Ceramic materials may have piezoelectric characteristics imparted to them by pressing, sintering, and polarizing the material. During manufacture, the ceramic material is exposed to an electrical field during a cooling phase that orients materials in the structure and provides the piezoelectric characteristics. These ceramic sensors should not be used in applications that exceed the temperatures recommended for the specific material. If the material is heated to the Curie point, the polarization (and thus its piezoelectric properties) will be lost and the element will not function.

Piezoelectric pressure transducers are used for the measurement of dynamic pressures that occur very, very rapidly such as ballistics measurements, rocket motors, pumps, pipelines, turbines, hydraulic equipment, fuel injection, and spacecraft applications, Figure 4-51.

Hot Cathode Ionization Gage - The hot cathode ionization gage is applied to vacuum measurement. An electric current is supplied to the heating filament in the gage envelope and the electrons are attracted to a spiral grid by a positive DC potential. As the electrons pass through the accelerating space, the gas molecules to be measured are struck and ionized. The gas ions are transported to the center of the gage to a collector with a negative potential. An ion current is the result that is proportional to the vacuum being measured, Figure 4-52. The torr is a vacuum measurement unit. It is the pressure needed to raise a column of mercury one millimeter at zero degrees centigrade and with standard gravity. The current of the ionization gage is near the amount of 100 milliamperes per torr, but is dependent upon the gas that is present. The pressure range is from 0.001 micron to 1 micron

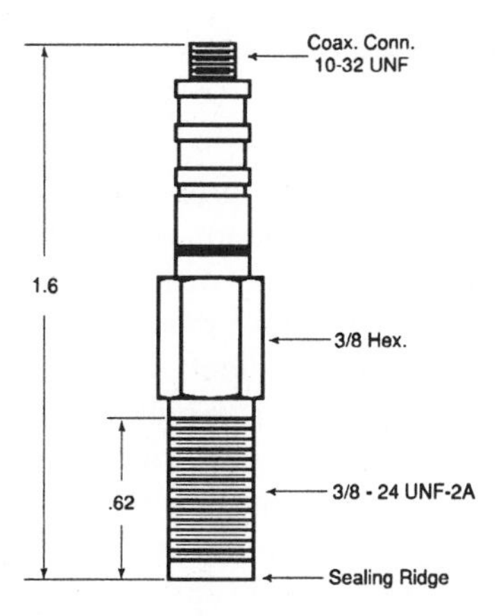

Figure 4-51 Quartz pressure sensor. A high pressure ballistic transducer *(Courtesy of Kistler Instrument Corporation)*

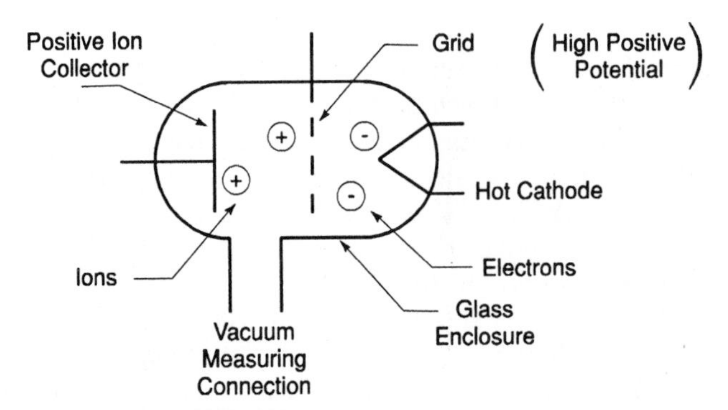

Figure 4-52 Schematic of a hot cathode ionization vacuum measuring gage

of mercury; if the pressure rises above 1 micron there is danger of a filament burnout. This sensor should not be used with gases that will deteriorate the hot filament. The range of this instrument is 10^{-2} to 10^{-11} torr.

Cold Cathode Ionization Vacuum Measuring Gage - The cold cathode ionization gage is similar to the above gage except that it does not use a heated filament to cause the electrons to be emitted from the filament, but rather applies a high potential field. The emission is less, but a magnetic field surrounding the tube causes the emitted electrons to travel in a spiral path. They collide and ionize more gas molecules and produce the ion current that is monitored. These sensors operate in the vacuum range of 10^{-4} torr.

The cold cathode has had its range increased by adding parts of the hot cathode to trigger the ion emission. This combination has increased its pressure measuring capability down to 10^{-14} torr.

Solid-state Electronic Pressure Devices

The integrated circuit pressure transducers provide an electrical signal that is linear with pressure.

Piezoresistance - Silicon has the property of changing resistance with the application of pressure; this phenomenon is known as "piezoresistance." It is this property that makes pressure sensing with silicon possible. Thus an integrated circuit pressure sensor chip converts a pressure signal to a change of resistance. Each sensor chip has an integral silicon diaphragm that deflects with applied pressure, causing different stresses in the bulk silicon. The pressure induced stress makes it easier or more difficult for electrons to move through the silicon. This stress increases or decreases the resistance based on the applied pressure. The sensor chip is electrically modeled after a Wheatstone bridge and has a varied voltage output, linear with the pressure applied, Figure 4-53. The devices are used in process control equipment. Sensing devices are applied to hydraulic and pneumatic equipment, energy management, medical diagnostic equipment, computer peripherals, engine and brake monitoring, robotics, automation, machine tools, and automotive diagnostics.

Integrated circuit pressure sensors are available in a number of types of pressure transducers. They are absolute pressure devices by which the pressures are measured relative to an absolute vacuum hermetically sealed within the sensor chip. These devices can be used for barometric pressure sensors. In this sensor the circuitry is located on the top side of the sensor die, and therefore hostile media such as aqueous or conductive materials should not be exposed to these surfaces. Backward gage devices measure pressures relative to ambient or atmospheric pressure. One port is always open to atmospheric pressure. The differential pressure gage provides an output that is the difference between two pressures. These are like differential amplifiers in which only differential and not common mode pressure changes create an output change. These are typically used in flow monitoring applications.

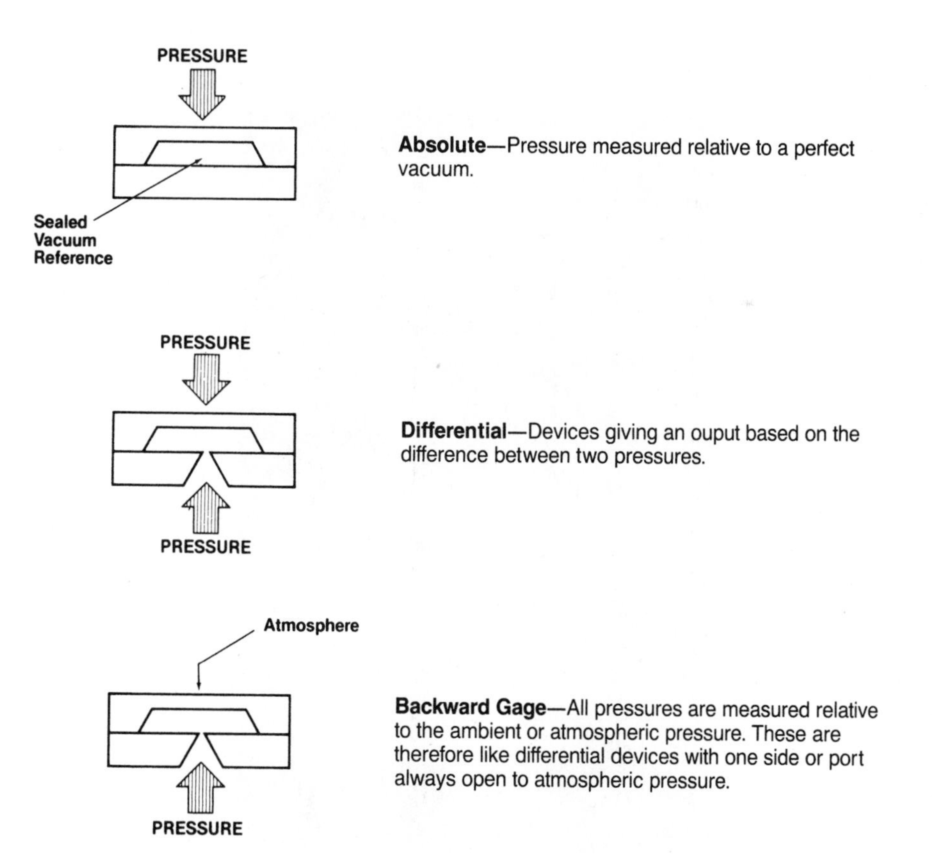

Figure 4-53 Diaphragm configurations for different transducers *(Courtesy of SenSym Inc.)*

This silicon technology is applied to a wide variety of pressure sensing applications by which corrosive liquids or gases are monitored. The sensors are laser trimmed for calibration with temperature compensation. These sensors are used in energy management, process control, robotics, sewage and water treatment, hydraulics, and agriculture vehicles, Figure 4-54A and Figure 4-54B.

Three levels of signal conditioning can be provided with integrated circuit pressure devices. A monolithic sensor consists only of the pressure-sensitive bridge. This sensor has design flexibility and low cost. A temperature-compensated monolithic sensor is provided by adding passive compensation networks on a hybrid ceramic to the sensor and laser trimming for proper performance. This compensates for temperature variations, while also calibrating the offset and span of the sensor. The signal conditioned transducers have calibrated offset and span and are compensated for temperature. In addition, the fully signal conditioned transducers provide temperature compensation, voltage regulation, and an amplified output signal.

Figure 4-54A Solid-state pressure transducer *(Courtesy of SenSym Inc.)*

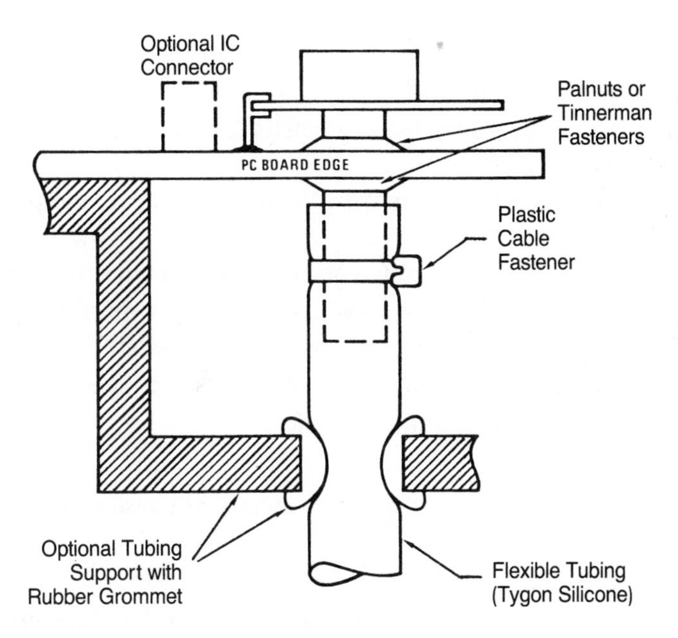

Figure 4-54B A typical installation on a PC board *(Courtesy of SenSym Inc.)*

SUMMARY/FACTS

- Temperature changes applied to unrestrained materials cause them to expand and contract equally in all directions.
- Temperature electrical instrumentation systems gather thermal data through the employment of resistance thermometry, diode thermometry, thermoelectrics, radiation measurements, optical pyrometry, thermistors, and solid-state integrated circuitry.
- Pressure is measured with four major systems: mechanical systems using liquid column devices; mechanical systems employing elastic element devices, such as bourdon tubes, bellows, diaphragms, and capsules; electrical systems employing the electrical elements of resistance, current, voltage, frequency, capacity, conductance, etc.; solid-state electronic systems including chip technology and integrated circuits.

REVIEW QUESTIONS

1. What is the function of a sensor?
2. What four systems, based on the characteristics of materials, are used in the measurement of temperature?
3. What physical characteristic of metals, liquids, or gases is the primary basis of mechanical measuring sensors?
4. How did the early liquid-filled thermometer provide a remote reading or recording?
5. What happens to a metal's resistance when the metal's temperature is increased as measured by the electrical resistance types of thermometers?
6. Name some metals used in the manufacture of resistance sensing elements.
7. Describe the voltage drop of a silicon diode.
8. Define a thermistor.
9. What is produced when two dissimilar metals are fastened together and the connection subjected to heat?
10. What would be the effect of applying intermediate metals in a thermocouple circuit?
11. What happens to radiant energy (electromagnetic waves) when it strikes an object?
12. What causes a change in the resonant frequency of crystals?
13. What is the name of the system that utilizes a solid-state silicon pressure sensor in an integrated circuit pressure transducer to sense a heated constant volume gas bulb which in turn produces an electrical temperature signal?
14. A fundamental liquid column instrument that measures absolute pressure is known by what name?
15. What instrument is used for measuring variable pressures and differential pressures?
16. What is a draft gage?
17. The C-type bourdon tube is a widely used industrial indicating instrument for measuring what variable?
18. Describe the function of the bellows.
19. What is the name given to a sensor made of thin flexible material supported by a circular rim?
20. What is an instrument that converts a linear sensor movement to an electrical signal?
21. What is the function of the rotary potentiometer (rheostat)?
22. Angular, rotary, or linear movements may also be converted into electrical signals by devices known as what?
23. What name is given to resistance changes elements that are bonded to the element to be strained (such as a diaphragm, beam, or column)?
24. Silicon has the property of changing resistance with the application of pressure. What is the name of this phenomenon?

25. What instruments are applied in the measurement of vacuums?
26. What gage employs a silicon die with a vacuum sealed on one side of the die and the pressure to be sensed on the other side of the die?
27. Describe a monolithic temperature-compensated solid-state pressure sensor?

5

SENSORS OF FLOW AND LEVEL

OBJECTIVES

Upon completing this chapter, you will have learned and be able to describe or explain the use and applications of these sensors:

- Sensors that function using the concepts of a differential pressure
- Sensors that measure flow in open channels
- Sensors that measure mass flow
- Sensors that measure by using positive displacement and metering pumps
- A group of sensors that measure level

INTRODUCTION

The role of fluids in today's technology is vital. Their characteristics of flow, level, and density must be closely monitored in order that industry may function in a rational manner. The devices that measure these phenomena are described in this chapter.

TRANSDUCERS USED TO SENSE FLOW DATA

Flow is fluid in motion and includes liquids, gases, and in some cases slurries. The measurement of flow is very important to industrial processing plants using water, steam, gases, oil, acids, basic solutions, materials in suspension, and other fluid-processing variables. Flow is measured as (1) a rate — the amount of fluid moving past a specified position at a given instant; or (2) total flow — the amount of fluid moving past a specified position during a designated period of time. These two precepts define the broad measuring parameters of flow.

Flow measurements made by sampling the flow in conduits are called inferred measurements. They infer or estimate information about the flow in the circuit. Total flow measurement measures the total volume of moving fluids past the measuring position.

Types of Flow Measurement Sensors

The measurement of flow applies to a number of types of instruments. Each type of device has its own characteristics. The classes of measurement instruments are: differential pressure, variable area, mass flow, oscillatory, electromagnetic flow, turbine, ultrasonic, and positive displacement.

Differential Pressure

Flow is frequently measured by using the principle of hydrostatic head. This principle uses the height or level of a fluid to infer flow. These instruments measure the difference in hydrostatic head from above and from below an obstacle placed in the stream. The pressure resulting from the differences of the hydrostatic head is in turn equated to the flow in the conduit.

The Orifice Plate - The *orifice* plate is a sensor that is used in a very large numbers of industrial processing plants. It is incorporated into many plumbing installations. To the untrained individual, it would seem to be a flange coupling in the plumbing. However, placed between the flanges is a plate with an accurately machined hole in it. The size and edge configuration is very closely specified so that data can be accurately drawn from the pressures differences that will appear on each side of the orifice plate.

If a constant flow of fluid enters into the system, it is expected that a like flow rate must exit the system. To do this, there must be a change in the movement of the fluid. The velocity changes as the fluid flows through the obstacle placed in the stream. In this case, the obstacle is the orifice plate causing the change. As the fluid's velocity increases, the pressure drops, causing a pressure differential between the upstream pressure and the pressure just below the orifice plate. The point of lowest pressure is called the "vena contracta." As the flow rate changes through the orifice plate, the pressure differential between the upstream tap and the vena contracta tap will change providing a signal that is proportional to the rate of flow, Figure 5-1.

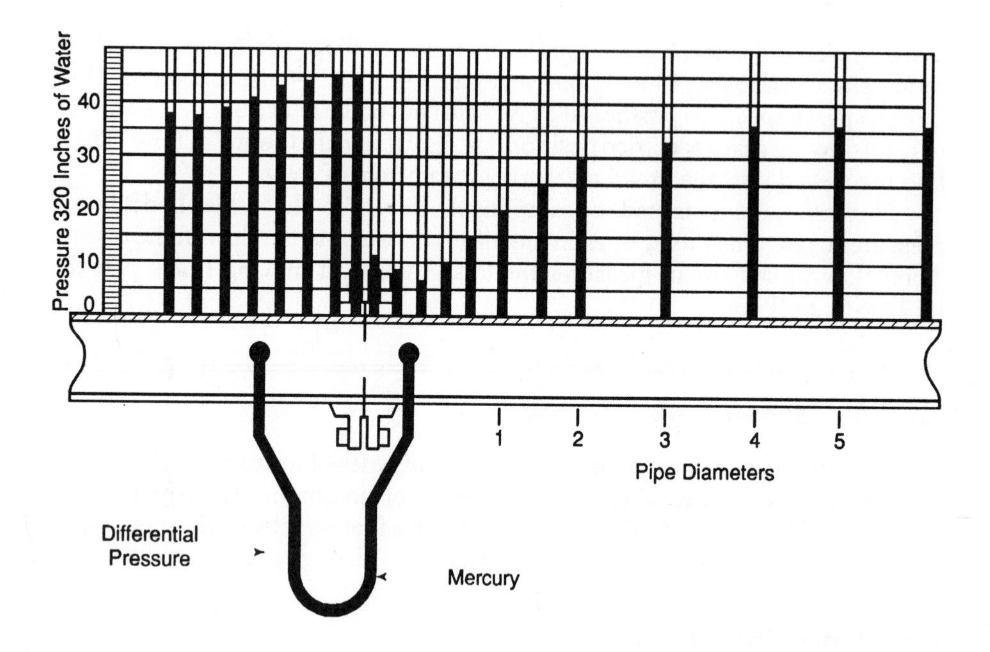

Figure 5-1 A principle of flow measurement. The differential pressure across an orifice

The rate of flow can be read directly by a number of differential pressure devices. They send signals to a readout or control equipment calibrated in barrels per day, cubic feet per hour, pounds per hour, gallons per minute, or other units of flow rate.

Square-edged orifice plates used in industrial flow measurements are of three types: concentric, eccentric, and segmental. The concentric orifices have a bore in the center of the plate, thus placing it in the center of the pipe's flow, Figure 5-2. This type of orifice plate is used where gases are dry, that is, where no condensates exist (such as in a dry steam). They are also used with clean liquids. However they are not used in measurements where the fluid is changing from gas to liquid or vice versa. The condition of the pipe interior, the orifice diameter, and

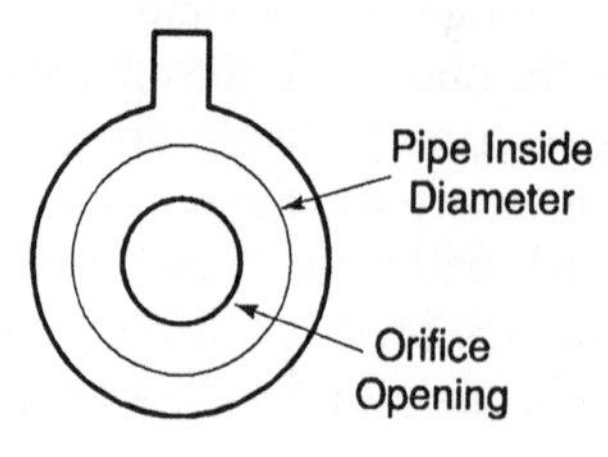

Figure 5-2 The concentric orifice plate

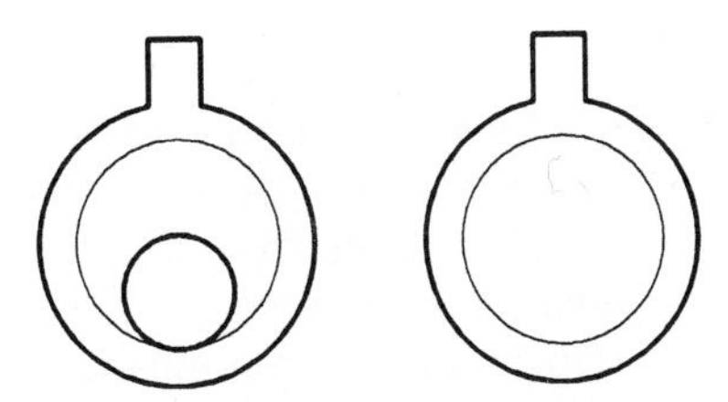

Figure 5-3 The eccentric orifice plate

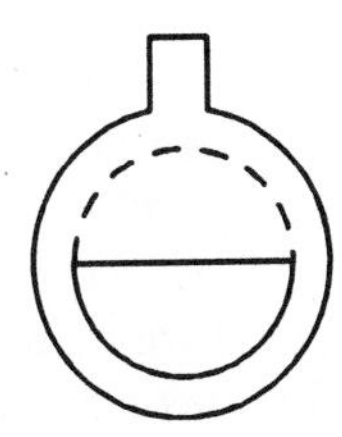

Figure 5-4 The segmental orifice plate

the fluid velocity are factors that influence the accuracy of the measurement. This is the most accurate plate of the basic three. When properly installed and maintained, it delivers data within one half of 1 percent of the actual value flowing in the system.

Eccentric orifice plates are recommended where condensed liquid may exist in gas flow measurements. In the eccentric orifice plates, the bore is usually at the base of the area of flow in the pipe. Thus the existence of liquids that have condensed out of the gases present in the flow system can be drained away from the measuring sensor, Figure 5-3. Accuracies of approximately 2 percent will be expected from this installation.

Segmental orifice plates are used in flow measurements with solids in suspension. Slurries of different materials can be measured. The lower portion of the plate is open for the passage of solids, Figure 5-4. The curvature of the lower portion of the opening in the plate is approximately 98 percent of the radius of the interior of the pipe's radius. This type of application will deliver accuracies of approximately 3 percent of the true flow in the piping system.

The Flow Nozzle - The flow nozzle is a device that is designed to reduce the pressure loss or to improve the recovery of the pressure quickly after the fluid has passed through a measuring element. The entrance of the nozzle has the contour of a near ellipse in cross section. Flow nozzles are applied where high velocity flow is measured. It is also used where condensible material such as wet steam is measured or where the velocity is beyond the range of orifice plates, Figure 5-5. Installations that employ high-pressure steam may cause erosion damage to the measuring system. Flow nozzles retain their calibration over long periods of time primarily because curvature of the throat minimizes the wear.

The Venturi Tube - Measuring flow with a venturi tube delivers the best accuracy. The venturi tube gains its efficiency from a formed entrance cone section that reduces the size of the flow to the size of the throat, where the measurement is taken, and then expands the size of the flow in the discharge cone gradually back to that of the original.

Bernoulli's theorem applies in the throat section of the venturi where the increase in velocity of the fluid creates a lower pressure area. On larger venturi tubes, a piezometer ring chamber encircles the area of the throat with a number of radial holes drilled into the throat for the measurement of the low pressure. The ring and a number of taps deliver an average of the pressure from the various sides of the throat to the differential measuring gage. Upstream from the throat and above the formed entrance cone, another annular piezometer ring is located that averages the pressure of the barrel. These two tap pressures are compared to yield the instrument's differential pressure, Figure 5-6. Tables have been constructed that take the differential pressure and the parameters of the venturi tube and provide the flow rate within the instrument. Readout devices can be calibrated and direct data can be recorded.

It should be noted that venturi tubes are fabricated or cast in brass or stainless steel and require a considerable amount of machine work. Large-sized venturi tubes are quite expensive. A lesser cost is one of the reasons for the use of the other differential pressure flow instruments.

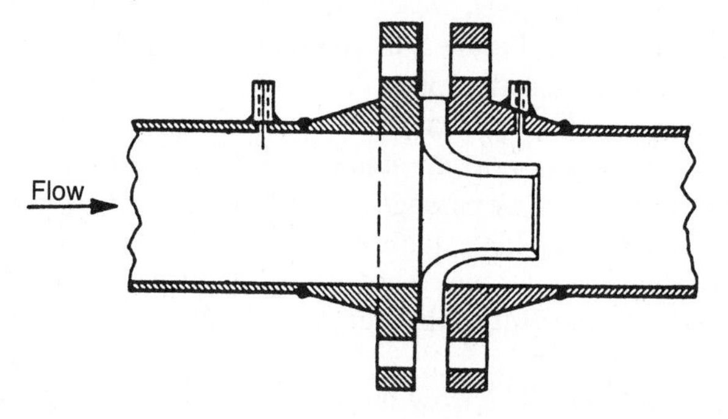

Figure 5-5 Flow nozzle

Figure 5-6 (A) A cross section of a "V" series venturi flow sensor; (B) Low-loss venturi flowmeters *(Courtesy of Preso Industries)*

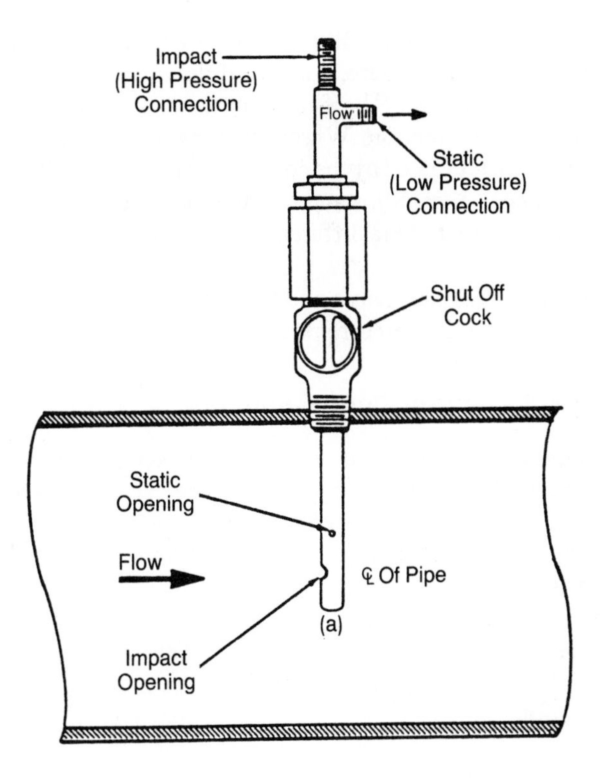

Figure 5-7 Pitot tube

The Pitot Tube - The pitot tube is a small cylindrical tube that is mounted in the stream of fluid (liquid or gas) to be measured. The upstream side of the tube has a small hole facing the fluid. The velocity of the fluid impacts on the area of the hole to produce an impact head pressure. The other pressure is obtained by placing another hole at ninety degrees from the axis of flow. The second pressure is referred to as the static head pressure. These pressures are compared in a pressure differential device that gives a signal that represents the rate of flow in the pipe, Figure 5-7.

The use of the pitot tube disturbs the flow very little, resulting in practically no pressure loss because of the measuring device. The velocity of fluid flowing in a conduit will often vary from the center to the edge of the flow caused by friction with the wall of the conduit. To gain an average flow reading, a number of readings can be taken across the stream and averaged together to obtain a more realistic value of the flow in the conduit. Instrument makers construct a pitot tube with a number of impact ports on the tube in an effort to obtain an average.

The pitot tube can be used in any clean fluid. It is employed in airplanes, air-conditioning and heating ducts, as well as in various industrial plumbing applications.

The Elbow Flow Sensor - Elbow flow elements utilize centrifugal force to develop a differential pressure. As a fluid flows around an elbow, it changes its direction ninety degrees. The fluid's natural tendency is to want to continue flowing in the same direction. When the pipe causes a directional change, an area of higher pressure is created by the centrifugal force. This increase in pressure can be tapped to sense a pressure and relate it to a flow rate. The output of these pressure taps is compared in a differential device to determine the flow rate, Figure 5-8.

Elbow taps are inexpensive and produce practically no flow disturbance or pressure loss. However their error is high, approximately plus or minus 5 percent.

Flow Measurement in Open Channels

Open channel flow is a flow that has a free surface open to atmospheric pressure and unrestricted as to the height to which it can rise in its channel. This type of measurement is applied to large volumes of liquids such as rivers, canals, flumes, ditches, bypasses, and channels that never completely fill such as tunnels, aqueducts, and sewers.

Large volumes of liquids are measured with weirs, flumes, and open nozzles that provide a change in hydrostatic head. This change in hydrostatic head is measured by applying a differential pressure gage or level gaging instrumentation.

The Weir - A weir is an engineered opening in an obstruction placed in an open channel that causes a backup in the liquid, Figure 5-9. This backup creates a change in the liquid's height behind the weir that can be measured and correlated with the flow rate over the weir.

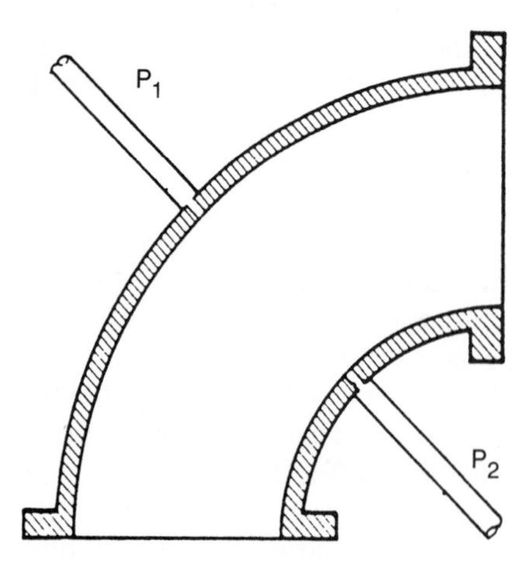

Figure 5-8 The elbow flow element

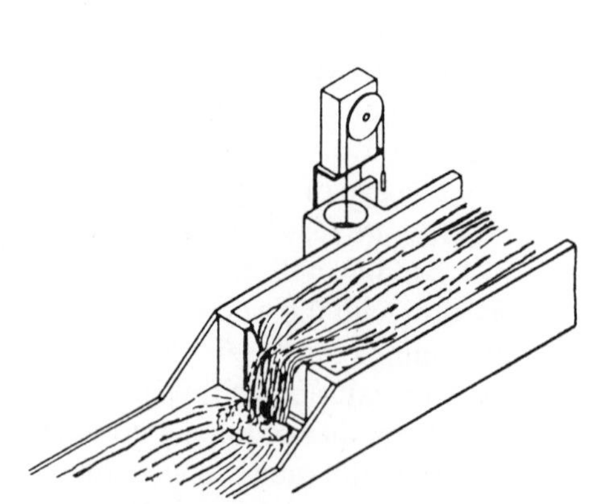

Figure 5-9 The weir

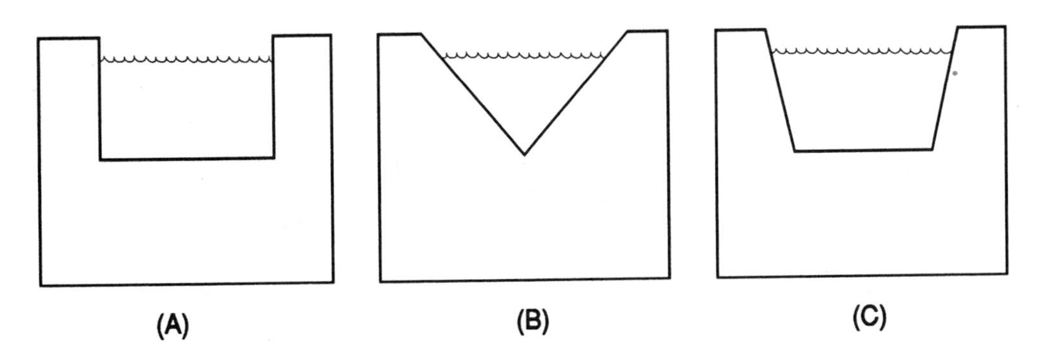

Figure 5-10 Common weirs: (A) Rectangular; (B) V-notch 60° and 90°; (C) Trapezoidal or Cipolletti

To accommodate different flow rates, different shapes of weirs have been designed. Their flow data is recorded in tables so that a specific size and height of liquid hydrostatic head results in a defined flow rate. The common weirs are the V-notch weir, the rectangular weir, and the trapezoidal or Cipolletti weir, Figure 5-10. The V-notch has either a 60° or 90° opening. This type of weir measures small flow rates up to as large as ten cubic feet per second. The rectangular weir is a large capacity weir, capable of measuring the flow of rivers. This is usually constructed in concrete with a flat bottom and vertical sides and may be part of a dam's spillway system, a low level obstruction across a river bed, or set in the channel of a canal. Data from very large flow rates are gained from this shape of weir. The trapezoidal or Cipolletti weir is similar to the rectangular weir, except that the sides are sloping.

When the liquid is flowing, the measurement is taken from the bottom or the sill of the weir to the top of the liquid upstream surface from where the liquid height has stabilized. This measurement is usually taken in a protected area (a "still well"), so that environmental conditions will not disturb the reading. Because this measurement is also a hydrostatic *head change*, the measurement can be read by a differential pressure device.

The Flume - A flume is a measuring section in a channel with restrictions in both its sides and channel bed. The Parshall flume, similar to a venturi tube, consists of an opening section, converging section, throat section, and diverging section. The bottom drops in elevation in the throat section and gradually rises again in the diverging section. The flume requires only one measuring point and that is a connection of the converging section to a still well where the liquid's height is measured. The measuring instruments usually measure the liquid's level, but because it is also a hydrostatic head, pressure measurements may be utilized.

The Parshall Flume - The Parshall flume is applied in irrigation and sewage water measurements. It is used where velocities are moderate and where solids

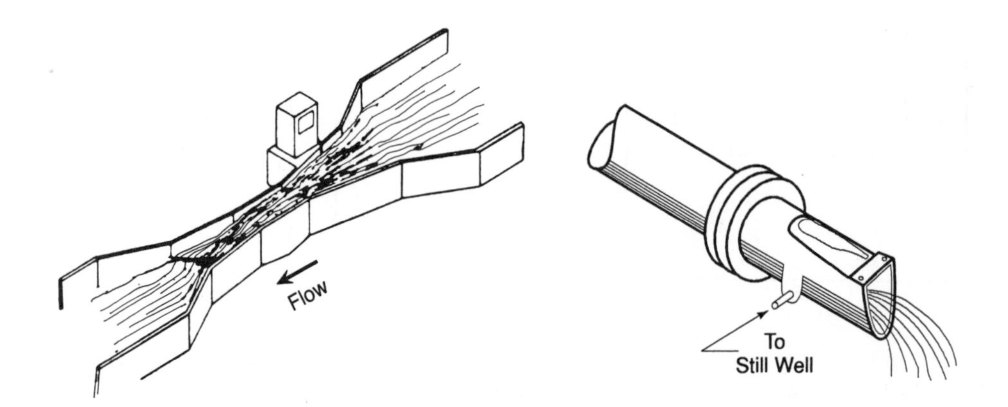

Figure 5-11 The Parshall flume

Figure 5-12 Open flow nozzle

may be included in the flow. The design is such that the flume is self-cleaning. It also provides a measurement with little energy or head loss. The Parshall flume is constructed from concrete, wood, fiberglass, or stainless steel. The latter are manufactured, shipped to the site, and set in place, Figure 5-11.

The data collected from the Parshall flume is a flow rate and indicated in cubic feet per second, gallons per minute, or millions of gallons per day. The measurement data is found in tables and is based on a function of the flume's width and the fluid's height.

Open Flow Nozzle - An open flow nozzle is used to measure the flow from a pipe or outlet that is not flowing at its total capacity. The open flow nozzle reduces the flow cross-sectional area and produces an increase in the hydrostatic head for the measurement. This device will measure the flow of slurries, sludge, sewage, and most forms of liquid industrial waste. The design is such that it is self-cleaning and requires very little maintenance. The measurement is taken from a single tap that transmits the fluid's level to a float-measuring instrument. Flow data is obtained from the manufacturer's tables correlating the flow nozzle's size and the height of the head measurement, Figure 5-12.

Mass Flow

In recent years, a number of devices have been presented to measure mass flow. Mass flow measures directly the mass per unit of time.

A number of instruments have been developed to make these flow measurements based on "nonintrusive techniques," meaning that no mechanical parts are introduced into the flowing stream to cause a pressure drop. These instruments utilize mass flow concepts—thermal mass, gyroscopic or Coriolis, and angular momentum.

The Thermal Mass Sensor - The thermal method of determining the flow rate has an intrinsic high reliability because it does not have moving parts and sensor elements that deteriorate with contact with chemically active flows. The thermal instruments can be applied to the measurements of liquids, gases, and slurries and are not affected by pressure, temperature, or radiation. One of these mass flow instruments is a thermal flowmeter that measures the rise in temperature of the

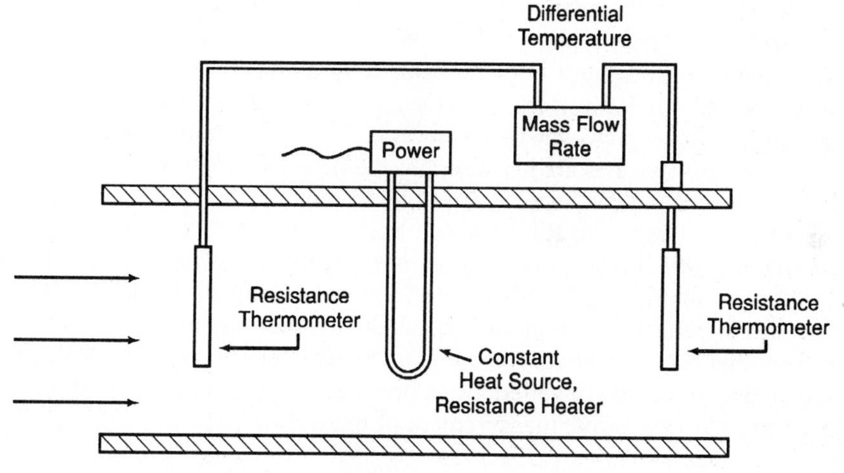

Figure 5-13A A mass flowmeter with an internal heat source

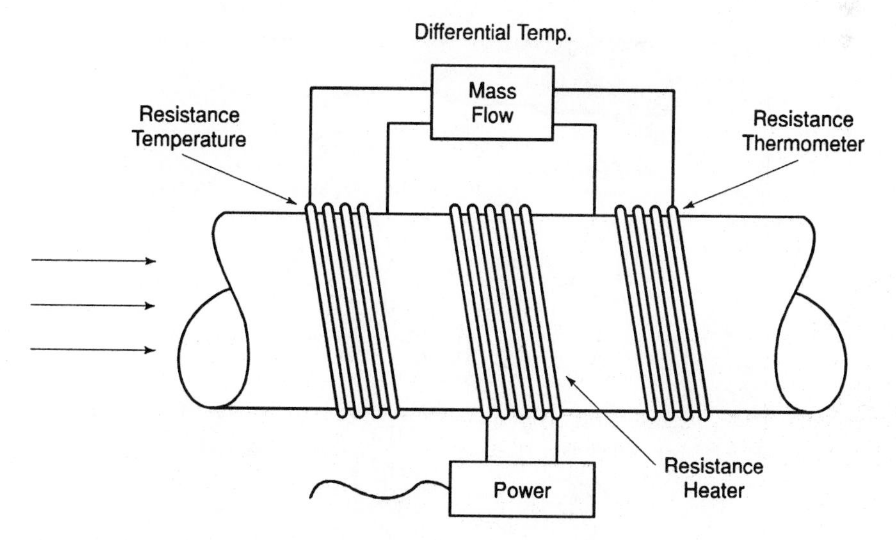

Figure 5-13B Mass flowmeter with the boundary layer externally heated

fluid stream as it passes through the heated section of the sensor, Figure 5-13. In these devices, a heated grid or a heated tube supplies heat to the fluid or the boundary layer of flow. There is a temperature sensing area above and below the heated area. A differential in temperature is measured and correlated with the mass flow and indicates directly in grams per seconds without additional corrections.

A Mass Thermal Dispersion Sensor - Another mass flow device applies thermal dispersion technology to mass flow metering. The device has two precisely matched platinum resistance temperature detectors and a component that assures that changes in media temperature affect the heater and the heated resistance thermal detectors at the same rate. This component maintains the instrument's accuracy even in the presence of transient temperature fluctuations.

The sensor probes are inserted into the flowing stream with a preferentially heated platinum resistance temperature detector. This creates a temperature difference between the RTDs which is greatest at zero flow and decreases in resistance as the flowing gas passes across the sensing element, cooling the heated RTD. The changes in flow directly and incrementally affect the extent to which heat is dissipated from the probe. Consequently, the magnitude of the temperature differential between the matched RTDs is decreased, Figure 5-14. This differential is electronically converted into a linearized signal output providing a highly accurate and *repeatable* measurement of mass flow rate.

Gyroscopic (Coriolis) Mass Flowmeter - Gyroscopic mass flowmeters of the Coriolis design are based on the fact that every moving body on earth is subject

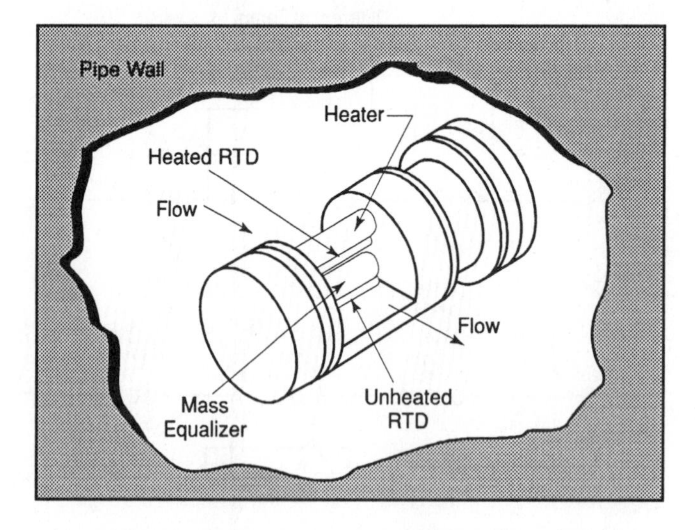

Figure 5-14 Thermal dispersion mass flowmeter *(Courtesy of Fluid Components, Inc.)*

to its angular velocity. To explain, an object dropped from a high tower will strike the ground a little east of vertical of its position from the top of the tower. This movement is the result of the earth's rotation during the fall. If a pipe were placed beside the tower and a sphere slightly smaller than the pipe were dropped into the pipe, it would rub the side of the pipe on the way down because of the earth's rotation and thus provide a force on the pipe's side. The result of this eastward force is called the Coriolis effect. If a fluid is substituted for the ball, it will respond as the ball and provide a force that will measure the mass of the material in the system. The measurement is linear without needing to adjust for variations in liquid properties (temperature, pressure, and viscosity). It is a true mass measuring device.

Coriolis meters are produced in various designs. One manufactured device consists of two oval, helically wound loops mounted side by side in such a way that the two conduits are vibrated in a periodic fashion. The direction of flow of fluid through the curved section of the conduit is opposite. A Coriolis force couple of equal but opposite forces results. The loop structure has sufficient mechanical elasticity to allow the mass-flow-induced Coriolis force couple to produce small, elastic deformations in the structure. A mass flow rate is possible from measuring the effects of these small elastic deformations, Figure 5-15.

The Angular Momentum Mass Flowmeter - The impeller-type of mass flowmeter has two elements that are involved in the measuring. An impeller provides the flowing fluid angular velocity, and a turbine measures the angular momentum of the flow that was imparted to the fluid from the impeller.

The impeller is oriented so that its shaft is parallel to the axis of the stream. The impeller is made up of a series of channels that are rotated by a constant-speed synchronous motor through a magnetic coupling. This rotation imparts the angular velocity to the moving fluid. Upon leaving the impeller, the fluid immediately enters the turbine channels that are not rotating. But the imparted

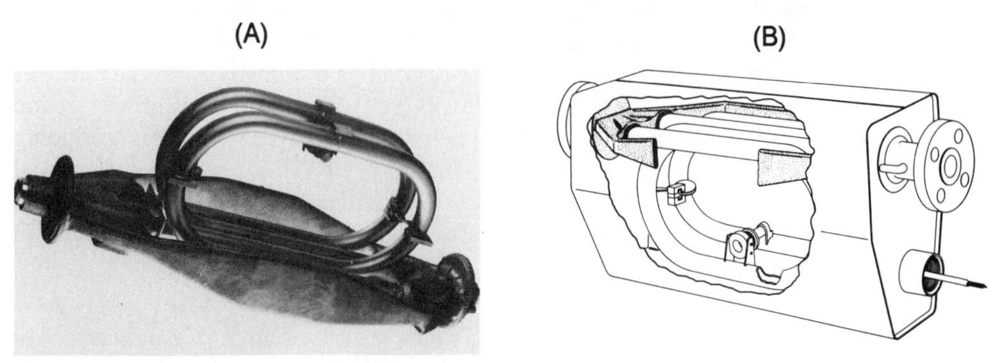

Figure 5-15 Coriolis flowmeter showing twin loops (A) The Coriolis effect is measured when an acceleration is applied to a rotating mass and is directly dependent on mass flow rate. (B) The Coriolis sensor will measure selectable variables — mass flow rate, density, and percent of solids. *(Courtesy of Exac Corporation)*

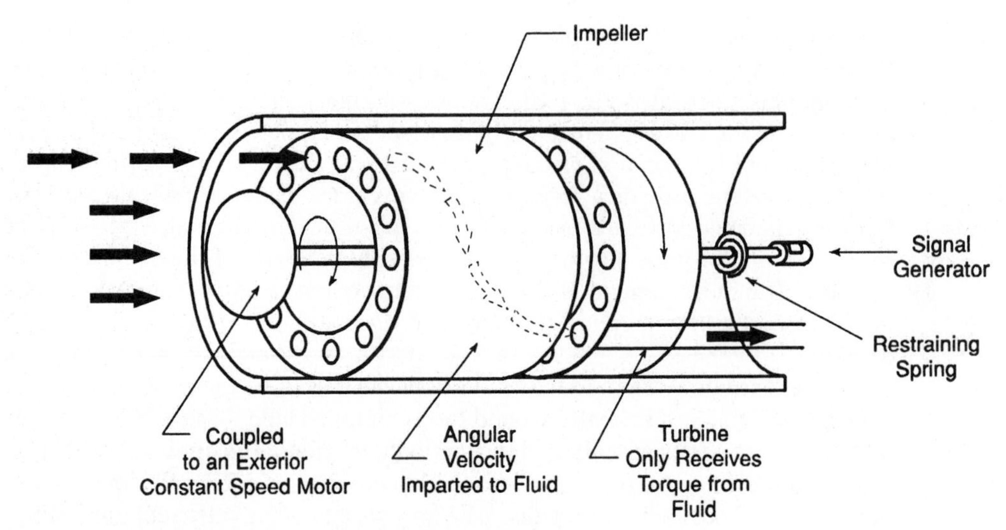

Figure 5-16 Angular momentum, impeller-turbine massflow sensor

angular velocity of the fluid tries to move the turbine. The turbine is restrained by a spring that in turn receives the torque from the fluid's angular momentum and positions a shaft whose orientation provides a signal that is proportional to mass flow through the sensor, Figure 5-16.

Oscillatory Flowmeters

The application of fluid oscillatory devices falls into two primary categories: vortex shedding flowmeters (bluff body) and fluidic (Coanda effect).

The Vortex Shedding Flowmeter - The vortex shedding principle is observed when water flows around a boulder in a stream or when a flag flutters in a light wind. A bluff or nonstreamlined obstruction is placed in a moving fluid. As the fluid moves around the obstacle, vortices or eddies trail away from the surfaces and flow downstream. The fluid breaks away from one side of the bluff object and then the other. This shedding occurs when the shear layer separates, producing a differential in pressure in the fluid, Figure 5-17. This pressure differential changing on either side of the obstruction is sensed frequently by a thermistor or by a piezoelectric device which generates a pulse signal that is correlated with volumetric flow.

The Fluidic Flowmeter - Fluidic flowmeters are based on the Coanda effect or wall effect. A stream of fluid when running parallel to a surface will attach itself to that surface. If the surface gradually curves away from the axis of the stream, the fluid will continue to follow the surface. Fluidic devices are designed to use this phenomenon and apply it to the measurement of fluid flow. A small amount

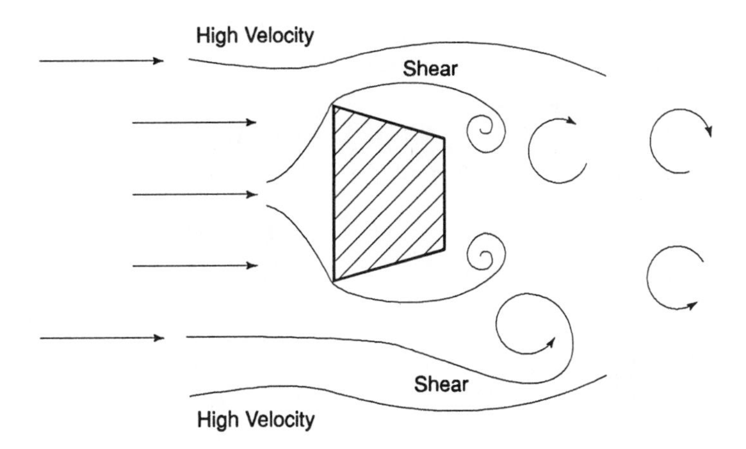

Figure 5-17 Vortex shedding phenomenon. Vortex shedding provides a turbulent wake. Karman vortices on opposite sides turn in opposite directions alternately downstream. The shedding frequency and flow velocity are related.

of the flow is diverted into a loop that is injected back into the original entering flow. This interrupts the attachment to the wall momentarily and produces a shift to an identical channel on the other side of the main flow stream thus producing an alternate oscillation.

Thermistors are located in the recirculation loops of the system. As each shift takes place, the return channel fluids cool the thermistor and produce an electrical signal, Figure 5-18. The changes in resistance represent a pulse signal that is counted and correlated to deliver a measure of volumetric flow rate.

The Vortex Precession Flowmeter - The application of the vortex precession principle is applied where the fluid is swirled by stationary prepositioned vanes creating a helical flow of a vortex around the central axis in the device. The velocity is higher in the center of the vortex than in the rest of the fluid. A heated thermistor senses a change in temperature because of cooling as the higher pressure vortex or swirl moves past. A constant current is supplied to the thermistor to supply the heat, and when it is cooled by the vortex, the thermistor's resistance is changed. The result is a pulse that is sensed, and the signal is conditioned to supply a frequency that will vary as the flow rate varies, Figure 5-19. To isolate the vortex measuring area from downstream plumbing disturbances, the instrument includes a flow straightening component to deswirl the fluid flow.

The Electromagnetic Flowmeter

Electromagnetic flowmeters use the laws of electromagnetic induction to produce a flow signal. The measuring tube is supplied with a strong magnetic field from the outside and induces a voltage in the flowing conductor. Electrodes are placed

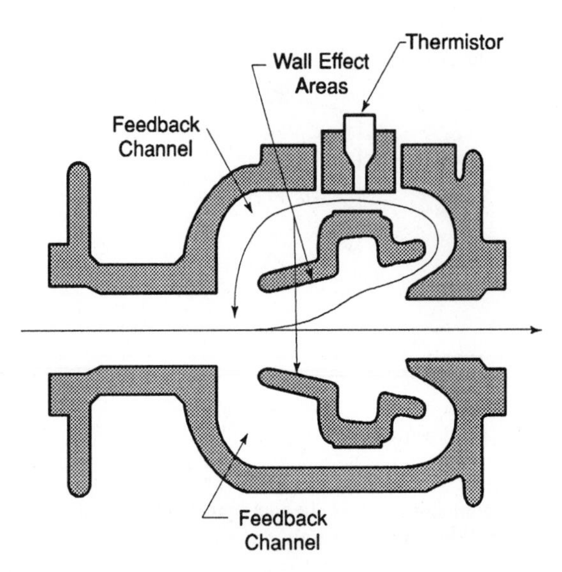

Figure 5-18 A cross section of a fluidic flowmeter. Part of the flow is diverted into the feedback passage which in turn breaks the wall effect of the flow and the flow switches. The oscillation is sensed by the cooling of a heated thermistor.

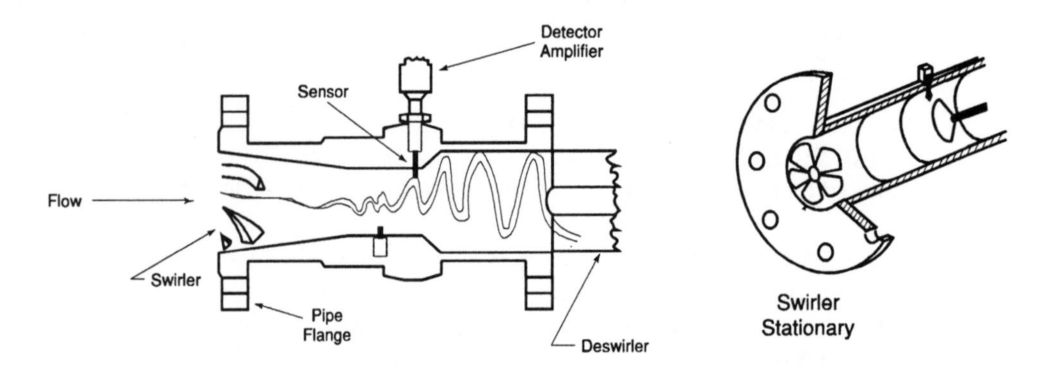

Figure 5-19 Vortex precision (swirlmeter)

ninety degrees from the induction coils and perpendicular to the flow. The electrodes transmit the induced measuring voltage to signal conditioning equipment with the output varying with the flow in the measured line.

The electromagnetic flowmeter is effective in measuring materials with difficult handling problems such as paper and other pulp stock, sewage, slurries, acids, and other chemically active materials. They are only applied to fluids that are electrolytic. These sensors are unaffected by changes in turbulence in the fluid, changes in viscosity or density, or variations in the plumbing system, Figure 5-20. The sensor requires a constant supply of electric power for the operation of its

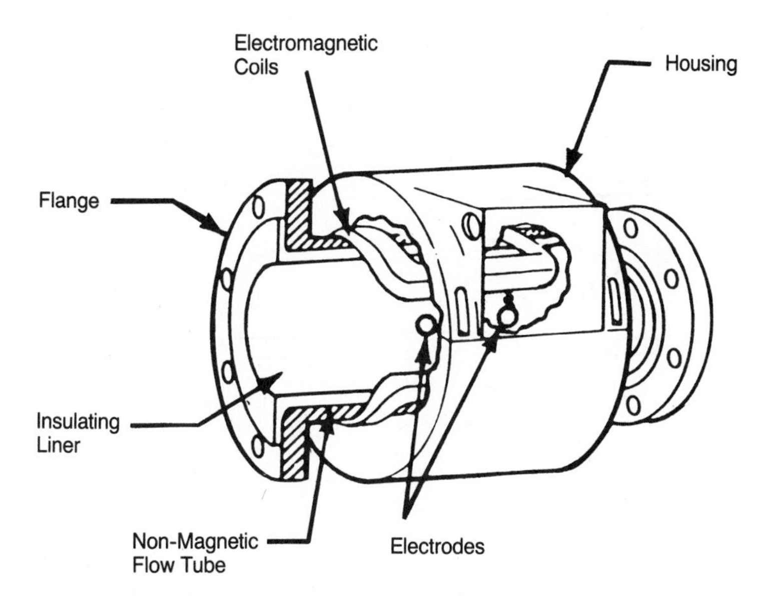

Figure 5-20 Magnetic flowmeter principles

magnetic field. The sensor can be installed in all common plumbing positions and will supply a signal with a very rapid response that is linear with flow through the sensor.

The Turbine Flowmeter

The turbine flowmeter is a flow tube with a turbine rotor mounted in the center. In front of the rotor is a flow straightener and bearing support and, in the rear, an outlet rotor bearing support. The meter housing is manufactured from a nonmagnetic stainless steel, while the rotor is made of a permeable material. The flow is through the tube onto the turbine rotor causing it to rotate at a rate consistent with the flow. On top of the housing, a magnetic pickup is mounted. A small, powerful, permanent magnet with a coil winding is placed where the permeable turbine blade passes through the magnetic field, causing a distortion of the field around the coil. This in turn induces an alternating current whose frequency is produced by the number of turbine blades per revolution passing the coil. The turbine rotor turns in relation with the flow rate; thus a specific frequency will represent a specific flow rate through the sensor, Figure 5-21.

These sensors require different types of bearings due to the various materials that pass through the sensor. With fluids that have minimum lubricating qualities, ball bearings are used; but with other materials, nonmetallic sleeve bearings provide trouble-free operation.

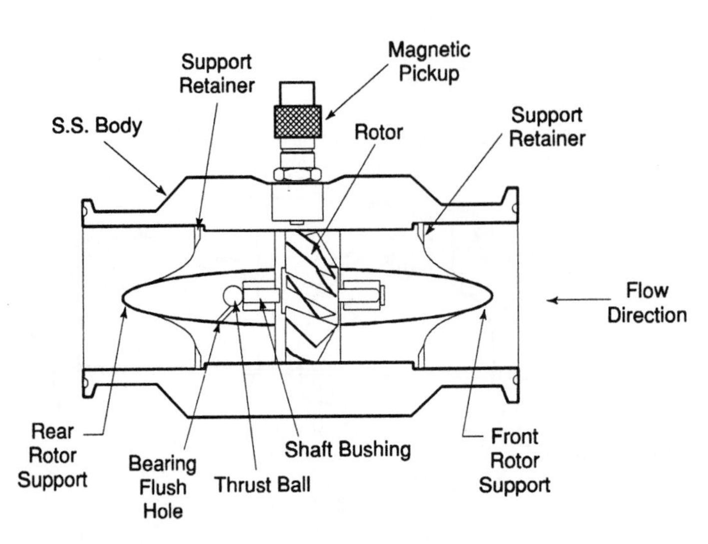

Figure 5-21 Turbine flowmeter

Because of slight variations in manufacture of turbine rotors, cases, and bearings, these flowmeters are calibrated at the factory. With wear, change of fluid viscosity, general dirty fluids, or corrosive material, the meter should be periodically calibrated in order to function at designated accuracy.

The Ultrasonic Flowmeter

Transit Time and Doppler Ultrasonic Flowmeters - Ultrasonic flowmeters are classified into two groups: transit time (pulsed type) and Doppler (frequency shift). Transit time ultrasonic flowmeters utilize sound waves traveling through a liquid. One or more pairs of piezoelectric transducers alternately send a pulse of sonic energy and then receive energy. The transducers are mounted at either forty-five degrees or ninety degrees to the flow tube axis. At the first pulse, the sonic pulse is directed downstream and its velocity is increased by that of the stream. The next ultrasonic pulse is directed upstream and the pulse is slowed by the velocity of the stream. The average pulse time is the difference between the time of these two pulses which in turn are proportional to the average velocity of fluid flowing through the instrument. A frequency of pulses is supplied to the transmitters, and as the flow changes, the transit time output pulses change, giving a signal that can be conditioned electronically, Figure 5-22.

Doppler reflection flow devices measure a shift in frequency caused by the movement of a fluid. This is like the change in the pitch of the sound from a whistle when the sound is coming from a speeding train that passes you, or the red shift in the light from a star that is moving away from you.

The flowmeter has an ultrasonic transmitting element and a receiving element. These piezoelectric devices detect a frequency change caused by the

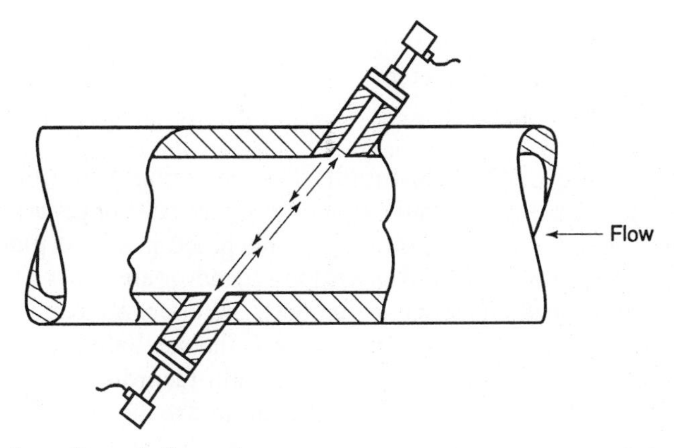

Figure 5-22 Transit time ultrasonic flowmeter

moving particles or bubbles flowing in the stream. If the flow is increasing, the frequency will be higher; or if the flow is decreasing, the frequency will be lower. The frequency shift is electronically conditioned, and the flow is correlated with the frequency change, Figure 5-23.

Positive Displacement Flowmeters and Metering Pumps

Positive displacement flowmeters are devices that utilize volumetric spaces and employ various mechanical *cavity* sweeping mechanisms to displace known volumes of fluids.

Measurement is achieved by applying hydraulic-type pumping elements but utilizing the energy in the fluid flow to move the volume measuring components to deliver a signal that is proportional to the flow rate. These mechanical devices

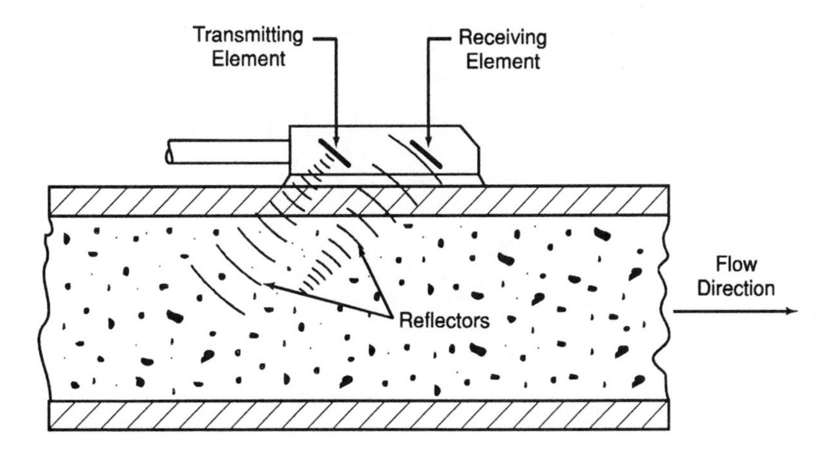

Figure 5-23 Doppler reflection flowmeter

are: oscillating-piston, nutating disk, vane, rotating impeller, peristaltic, diaphragm, and piston flowmeters.

The Oscillating-piston Flowmeter - The oscillating-piston measures the volume in the meter by using a slotted cylinder that oscillates or pivots around a division plate between the inlet and outlet port machined in a base plate. As the piston oscillates, the volume of fluid enters the expanding spaces and is swept around by the oscillating-piston until the outlet ports are uncovered and the fluid flows out of the meter. Spokes connect the center of the oscillating-piston to the hub that contains a crank pin that rotates a gear train recording each oscillation of the piston, Figure 5-24. The data can be transmitted to mechanical counter wheels or sent electrically to a remote recorder.

The Nutating Disk Flowmeter - The nutating disk flowmeter is designed with four chambers, two above the wobble disk and two below. The disk has a slot running from the edge to the disk's spherical bearing that divides the meter into two chambers and prevents the disk from rotating. A small shaft extends up from the axis of the spherical bearing that engages a second slotted disk and transmits the wobble of the nutating disk into a rotary motion for the gear train counting mechanism, Figure 5-25.

This flowmeter is widely used for measuring domestic and industrial water with service lines up to two inches nominal diameter.

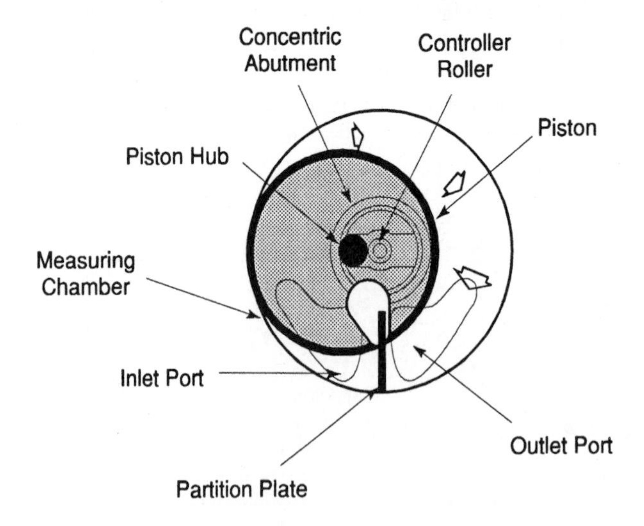

Figure 5-24 Principle of operation of an oscillating-piston meter

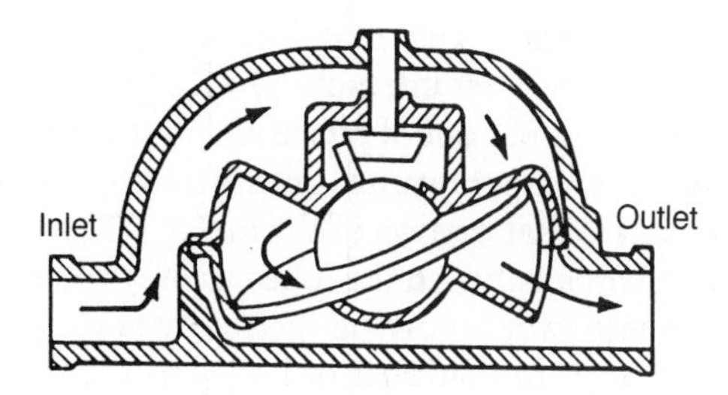

Figure 5-25 Nutating disk flowmeter

The Rotating Vane Flowmeter - The rotating vane flowmeter is constructed with a revolving cylinder with radiating slots cut to receive spring-loaded blades or vanes. This rotor system is mounted off-center in a housing so that the rotor nearly contacts the housing when the vanes are contracted. The vanes are extended at all times so that they contact the housing and provide a seal for the fluids. The fluid enters on the side where the vanes are extending to provide space between the case and hub and is carried to the opposite side of the meter where the vanes are retracting and the fluid leaves the meter, Figure 5-26.

The rotation of the shaft provides data that can be attached to a gear train and a counter; or the data can be transmitted electrically to a remote location.

The rotating vane flowmeter is used extensively in the petroleum industry for measuring crude oil, fuel oil, gasoline, and the many low viscosity products. These meters are very efficient and can be designed for very low flow rates to huge flow rates per hour.

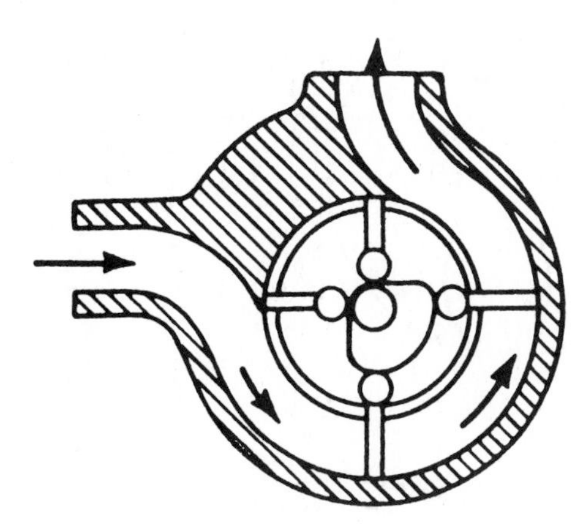

Figure 5-26 Rotating vane flowmeter

The Rotating Lobe Flowmeter - A number of positive displacement rotating impeller flowmeters have been designed. They include lobed-impellers and oval-shaped gear flowmeters. The lobe-impeller flowmeter is constructed with carefully machined lobes that interact. The impellers are geared together so that the lobes are ninety degrees out of phase and in rolling contact with each other to form a seal. The flowmeter housing is shaped so that the impeller forms a sliding seal as the lobe passes that portion of the meter. The fluid enters on one side and passes through the area between the lobes as they rotate and out the other half of the meter. The energy for revolving the rotors is supplied by the flowing fluid, Figure 5-27. The data information is provided by the rotation of the impeller shaft and can be applied to a gear train and counter or applied to an electronic pulse generator for remote totalization.

The Oval-shaped Gear Flowmeter -The oval-shaped gear flowmeter is a design utilizing the principles of a gear and a lobe pump. This meter applies precision-matched, oval-shaped gears as the rotating members of the flowmeter. The intermeshed gear teeth provide the seal between the rotating parts and also seal with the housing of the meter. The gears are assembled so that they are ninety degrees out of phase and thus provide a crescent-shaped space for fluid between the housing and the gear each time a gear makes a one-half revolution. The fluid flows from the inlet around the gears to the outlet with four volumes per revolution providing an accurate flow measurement, Figure 5-28. The data is collected from the rotation of the shaft by means of an electric impulse contactor or other selected accessories to interface with a data acquisition system.

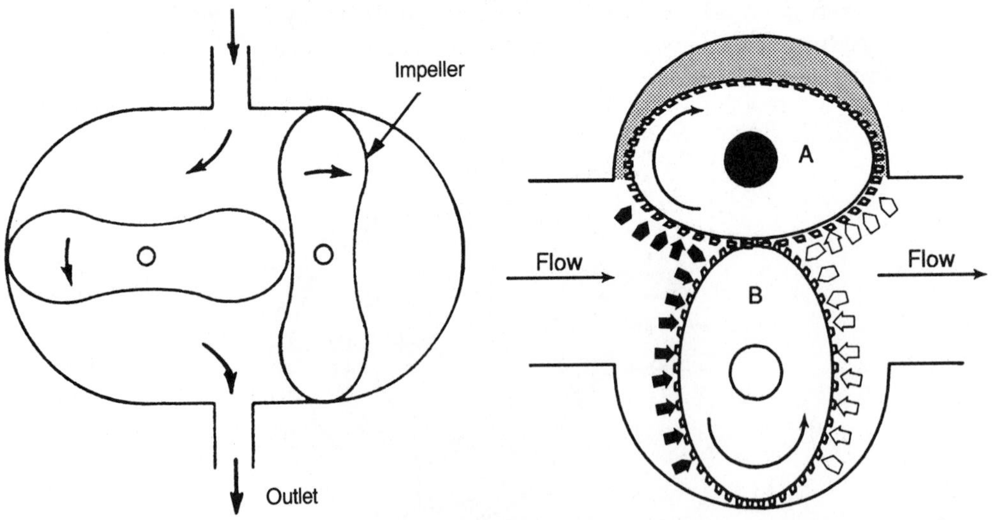

Figure 5-27 Rotating lobe flowmeter

Figure 5-28 Sectional schematic of an oval-gear flowmeter

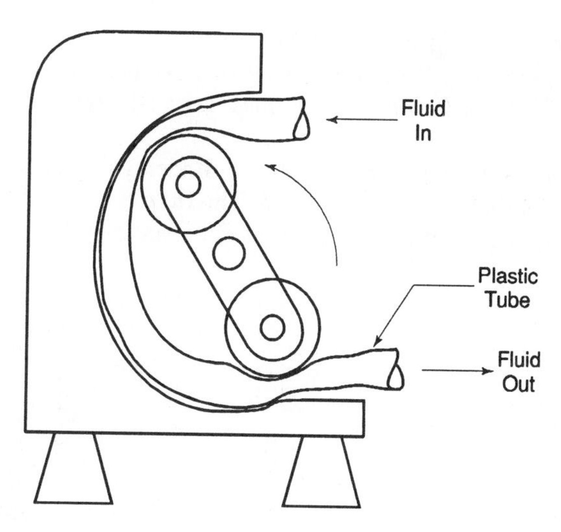

Figure 5-29 A cross section of a roller peristaltic pump flowmeter

The Peristaltic Pump Flowmeter - The peristaltic pump is not a flowmeter in the usual sense. It a metering pump that delivers specific rates of flow. The pump is designed with a circular housing that has a set of two rollers mounted on a shaft in the center of the housing. A pliable tube is inserted against the circular housing, and the rotor assembly contacts and squeezes the tubing flat as the roller passes making a moving leak-proof seal. Liquids or gases are introduced into the tube and the volume of tube between the rollers is the amount of fluid that will be measured per half revolution. The amount of fluid transported is equal to the volume of the tube purged per revolution and the number of revolutions per minute. Very low and accurate flow rates can be achieved by this device, Figure 5-29. Because fluids do not contact any materials other than the tubing, sterile conditions can be met, and medical application requiring accurate flow control can be achieved.

The Diaphragm Metering Pump Flowmeter - The diaphragm pump is another positive displacement metering pump that will supply a specific flow rate. It is constructed with a flexible membrane connected to a reciprocating push rod to provide movement and displacement of fluids above and below the diaphragm. The diaphragm is mounted between two circular shells to provide a fluid space between itself and the shell castings. The castings hold automatic valves so that when the push rod moves down, fluid enters the top half above the diaphragm while the fluid in the area below the diaphragm is expelled through another set of automatic valves. When the push rod reverses direction, the other side of the pump expels its fluid. The flow is controlled by the speed at which the push rod is activated, by the length of the stroke that the push rod has, and by the volume of the pumping cavities, Figure 5-30.

The Piston Metering Pump Flowmeter - The piston pump is a reciprocating device that provides a flow equal to the volume that is displaced with each stroke of the piston. This flowmeter supplies very accurate flow rates at very high pressures.

The meter is designed with a piston, cylinder, valves, and a push rod to provide the movement that displaces the fluid. The flow is controlled by the speed of the pump and by the volume of the cylinder, Figure 5-31.

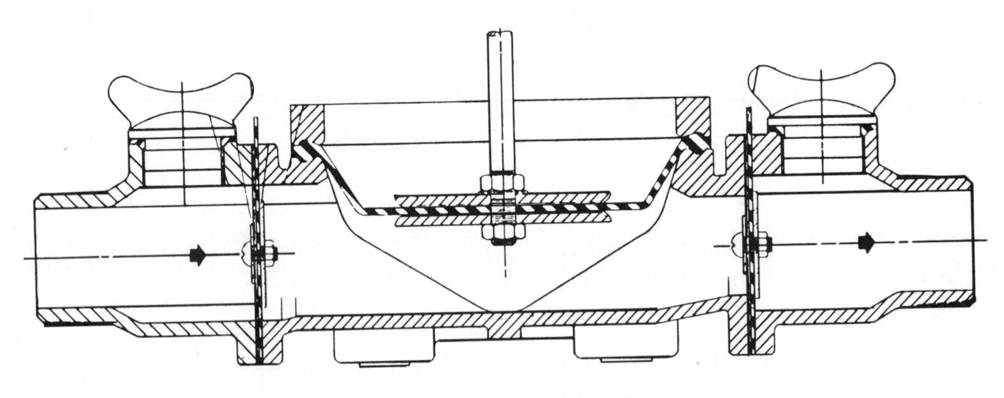

Figure 5-30 A cross section of a diaphragm metering pump flowmeter

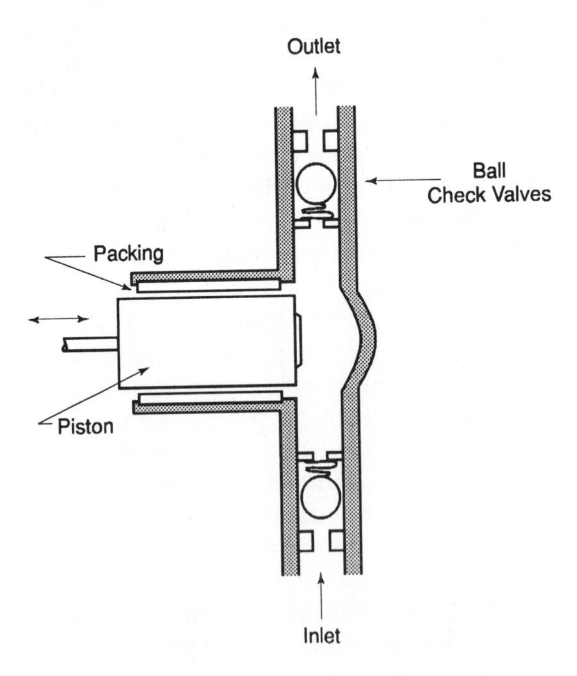

Figure 5-31 A cross section of a single piston flowmeter

SENSORS USED TO OBTAIN LEVEL DATA

In the production of numerous products, materials necessary for manufacturing are controlled by maintaining the proper level of liquids, powders, slurries, or granular solids in both open and closed containers, boilers, tanks, reservoirs, lakes, silos, bins, or hoppers. A factor that enters into level measurement is the shape of the container. If the sides of the container are parallel, the volume of the container at a specific level can easily be calculated. If, on the other hand, the shape of the container has an irregular configuration, the relationship between level and volume will require a more complex solution.

Level Sensors

Level measuring instruments may be classified into two large groups: direct data producing devices and indirect data producing devices. The first devices are designed so that an observation can be made directly, such as seeing a liquid level in a sight glass or tube. The indirect devices will infer a level where the data is transferred by a device to a readout system. An instrument such as a nuclear radiation level detector presents data that infers the level of a product in a storage tank.

Level Measurement Sensors

The measurement of level applies to a number of types of instruments that use different physical principles to make the measurements. The classes of instruments are based on the following: visual observation, hydrostatic head, buoyancy, electrical concepts, weight, nuclear radiation, ultrasonic, and solid-state devices.

Visual Level Observations

Visual observation includes measurements that have been used since early man to observe level. The dipstick is used to measure water level as well as most other liquids. Gasoline was measured at one time by placing a wooden rule in the tank to measure the level of gasoline in an auto's fuel tank. Today, the same method is used to gage the tank's fuel level in a service station, only the stick is longer. The oil level in the automobile is measured with a dipstick and provides scaled information within a quart of the proper oil level in the engine.

Plumb Bob and Tape - The level in a large storage tank is measured with a plumb bob and tape. The plumb bob is attached to a steel tape and lowered into a large tank of crude oil until the plumb bob either touches the surface of the oil or the bottom of the tank. Then it is slowly pulled out and the tape read from a reference point and the surface of the oil, or from the highest point that the oil reached on the tape if a bottom measurement was used. The tape is read in feet and inches of oil depth, indicating the level of the oil. The dimensions of the tank are measured and the volume of oil is calculated. On a tank that has previously been measured, the only data required is the level of the oil in the tank.

The Hook Gage - Accurate level measurements of fluids are also made with the use of a hook gage, applying the phenomenon of surface tension of fluids to make the measurement. The hook gage is a very sharp rod bent into the shape of the letter J, with the sharp area pointing up on the lower part of the letter. The measuring rod is vertically adjustable and passes through a rigid mount anchored to the side of a still well. The still well is connected to the level of the fluid body being measured. The mount and the rod are marked and scaled so the level measurement can be read. The scale is usually a vernier scale that can be read to the nearest thousandth of an inch. To measure, the hook is lowered into the fluid so that the sharp point is below the surface, and when the surface is still, the hook is slowly raised until the point breaks the surface tension of the fluid. At this point the instrument is read, and the level is indicated, Figure 5-32. This method is applied to large tanks, reservoirs and lakes.

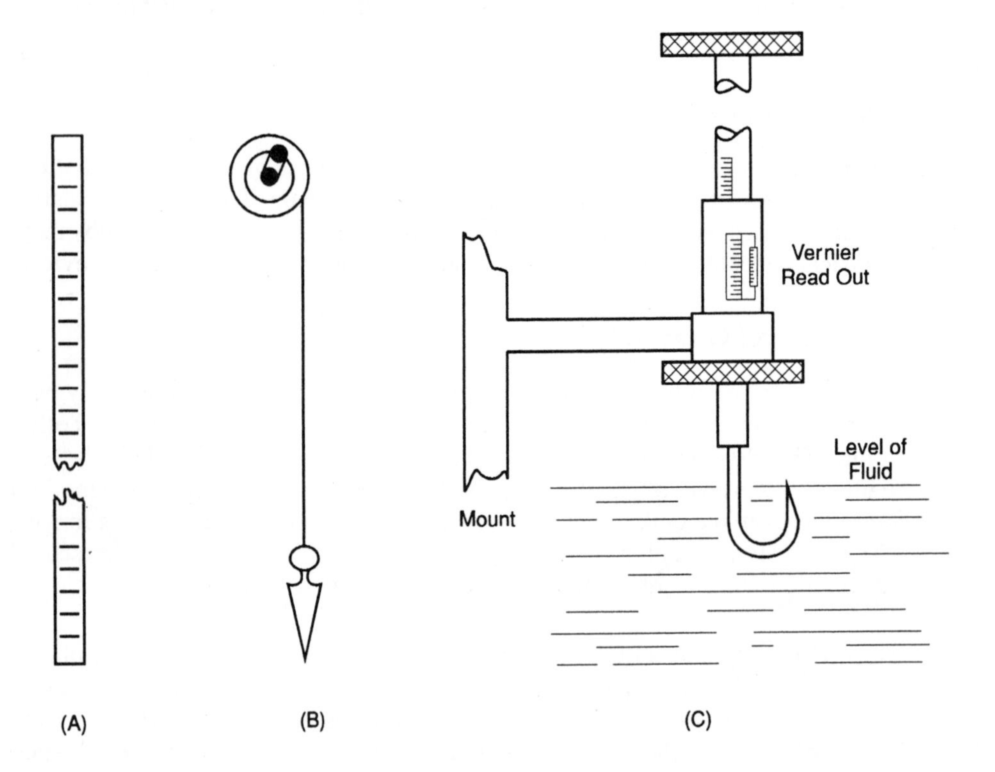

Figure 5-32 (A) Dipstick; (B) Plumb bob; (C) Hook gage

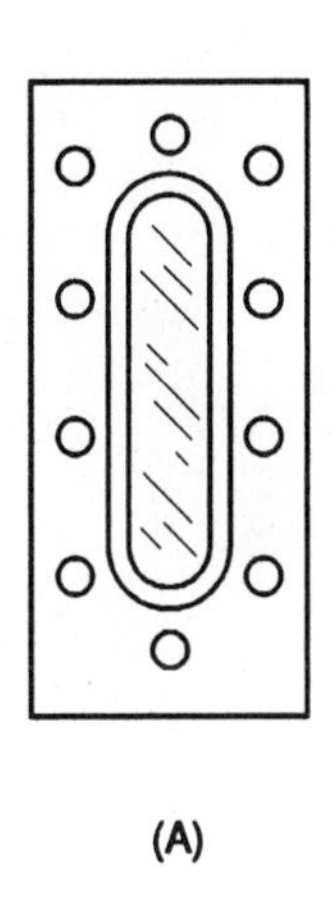

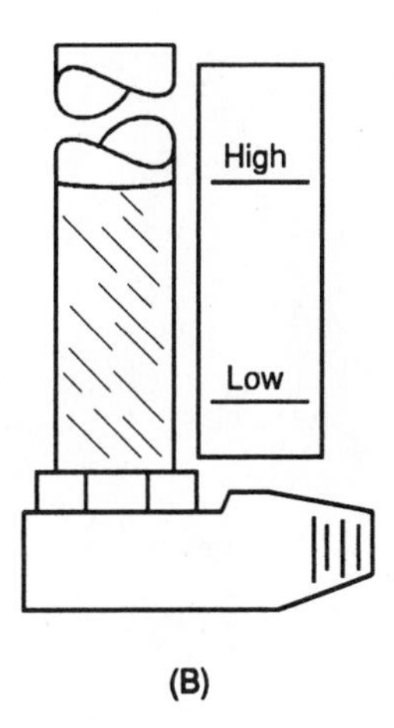

(A) (B)

Figure 5-33 (A) Sight glass; (B) Sight tube

The Sight Glass - A sight glass is a heavy tube or elongated window that makes it possible for an observer to see the level of the fluid directly. The level measured by this method is applied to clean liquids that are not viscous or sticky. It is used to measure the level of fluids such as in hydraulic oil reservoirs, food industry containers, and steam boilers. The transparent tubes and windows are made of a special tempered borosilicate glass, so that these tubes and flat glass windows can withstand the elevated temperatures and pressures. However, because these devices may be limited by their physical characteristics, this type of level measurement is not frequently used on a measurement greater than a three-foot span, Figure 5-33.

A Reflex Glass Level Sensor - In hot and high-pressure steam boiler level measurements, a reflex glass is installed in the gage body, sealing off the liquid or steam space against the atmosphere. On the side exposed to the medium, the glass has prismatic grooves. The rays of light penetrating from the outside are absorbed by the water (black reading) and reflected by steam to produce a silver-like reading. The black reading indicates the level of the water in the boiler.

A bicolor level gage is a transparent gage with a wedge-shaped centerpiece and an illuminator, equipped with one red and one green filter. The ray of red light is absorbed by water and, unhindered, penetrates the space containing steam, which then appears red. The ray of green light is absorbed by steam and penetrates water, producing a green reading of the liquid level, Figure 5-34A. For remote level reading, transmission units have been designed to make use of the same signals as the human eye—that is, the visual electronic sensors that will differentiate between red and green light. These sensor blocks are attached to the bicolor gage to provide remote indication and control, Figure 5-34B.

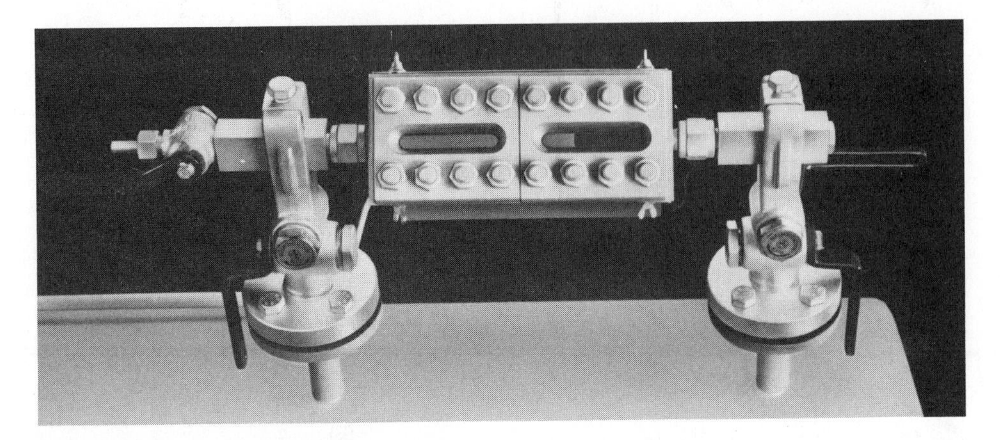

Figure 5-34A Bicolor level gage

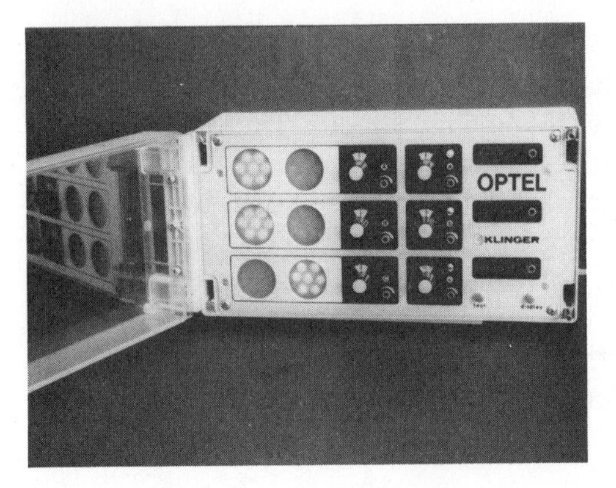

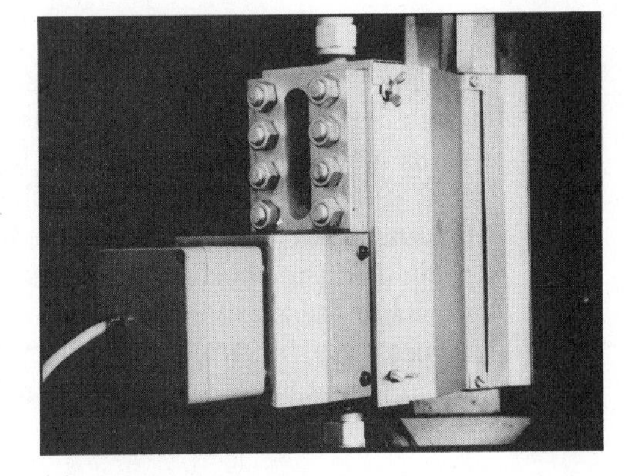

Figure 5-34B Remote level reading block *(Courtesy of Richard Klinger Inc.)*

The Slip Tube Level Sensor - Slip tubes are used to measure the liquid level of pressurized fluids such as are found in a butane tank. Slip tubes are of two types: a rotary tube, horizontal to the fluid's surface; a vertical type, perpendicular to the fluid's surface. A tube is built into the end of a horizontal round tank. The slip tube has a ninety degree bend in the end of the tube inside of the tank. The bent tube inside the tank is long enough to reach near the bottom of the tank when it is pointing straight down. To use the slip tube, the inside end is turned to the vertical position and a bleed valve on the outside end of the tube is slightly opened, and the pressurized gas is allowed to leak out. The tube is rotated toward the down position until liquid material begins to come out through the valve. At this point, the bent end of the tube inside the tank indicates the fluid level in the tank, and a pointer on the outside indicates the fluid's level on a dial, Figure 5-35.

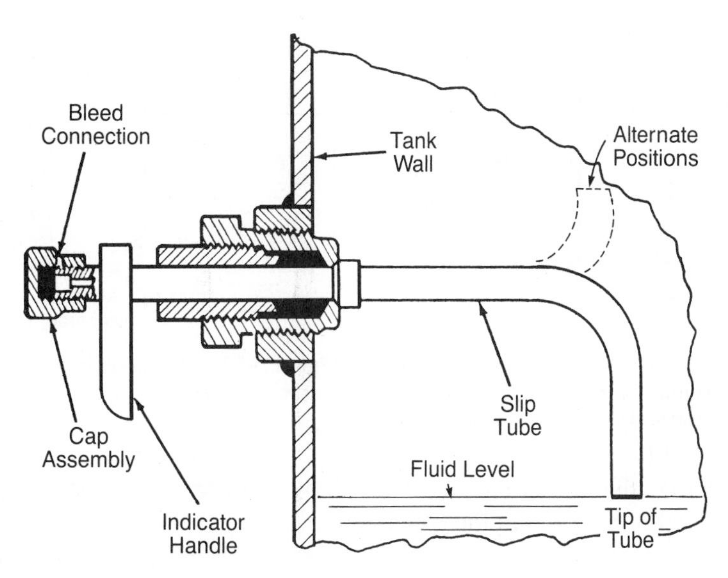

Figure 5-35 Horizonal slip tube level indicator

The vertical slip tube is installed on the top of the tank to be measured and, with the bleed valve open, is slowly pushed into the tank until the surface is reached and liquid starts to sputter out. This indicates the interface between the pressurized gas vapor and the level of the liquid in the tank. The level is read off of a scale that is inscribed on the slip tube itself. This system may be hazardous if the operator has not had the proper training—the escaping gas vapors may strike his face.

Hydrostatic Head Level Measurement

The principle of hydrostatic head (the generation of a pressure that is correlated with the height of a fluid above a sensor) is the basis for these instruments. The application of U-tube manometers, diaphragms, air bubblers, and other differential pressure detectors are many variations applied to the pressure measurement of level.

U-tube Level Measurements - U-tubes have been used to measure pressures for a long time. When one branch of a U-tube is attached to the bottom of a container, a hydrostatic pressure is sensed. The mercury or other manometer fluid will come to a height where the weight of the mercury will balance the weight of the hydrostatic head of the fluid under measurement. As the level of the fluid changes, so does the height of the mercury in the column. The mercury column is calibrated in terms of the fluid's level in the container, Figure 5-36. The higher the specific gravity of the fluid used in the manometer tube with respect to the fluid in the

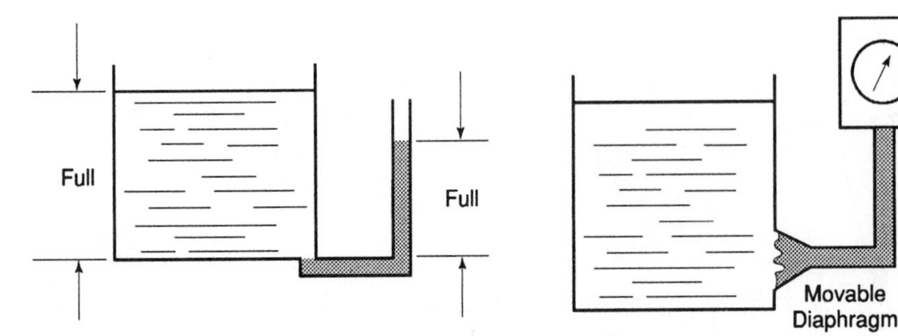

Figure 5-36 Single branch U-tube level gage. The height of the manometer leg will be a function of the difference of specific gravity of the two fluids.

Figure 5-37 Continuous hydrostatic head level sensor

container, the shorter the tube that will be necessary to indicate the total range of the level of the fluid in the container.

Hydrostatic Head Diaphragm Level Gage - Hydrostatic pressure liquid level measuring devices using diaphragms have many variations. One of the applications is the diaphragm box, which is an open bottom container with air or other gas in the device above the diaphragm. The diaphragm box is lowered into the fluid to be measured and is mounted at the low point in the fluid. The hydrostatic head acts on the diaphragm and pressurizes the gas above the diaphragm, sending a pressure signal indicating the level of the fluid in the container, Figure 5-37.

Gas-purge Level Sensor - The air-bubbler or gas-purge level measuring device is used in open containers and employs a flow of regulated gas to a tube that extends to the near bottom of the container. The gas pressure applied to the tube is slightly higher than the pressure generated by the hydrostatic head of the fluid under measurement. The excess gas will bubble off the end of the tube so that the pressure in the tube will equal the hydrostatic head and provide a pressure that is correlated with the level of the fluid in the container. This pressure may be transferred to a number of different varieties of pressure measuring units that can be calibrated to indicate the level of the fluid in the container, Figure 5-38.

Buoyancy Level Sensors

Buoyancy devices use floats or displacers to provide a measure of level. Floats are applied to mechanical linkage systems such as chains, tapes, levers, rods, and rotating shafts to transmit data on the level of fluids to an indicator or control system. Float devices are also integrated with other transmission schemes such as a movement of a float to convert a new float position into electrical values. A change can be made in voltage, amperage, resistance, inductance, reactance, or capacitance to produce a level signal.

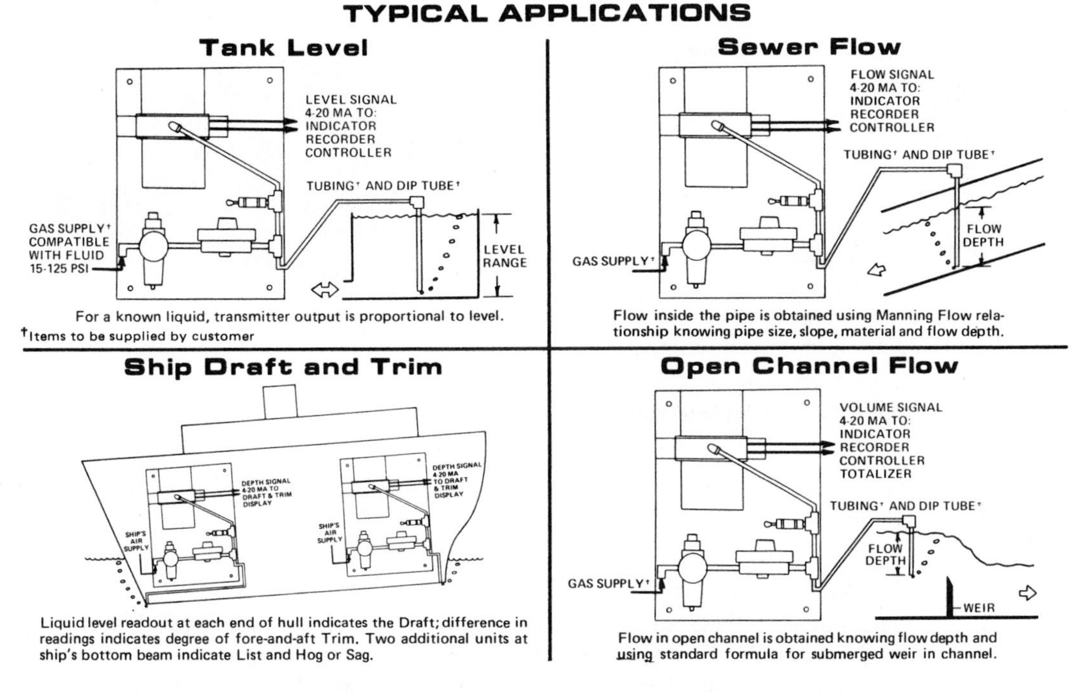

Figure 5-38 Level-bubbler or gas-purge system *(Courtesy of Computer Instrument Corp.)*

In other applications, the float position can be mechanically linked to produce a pneumatic signal where a level of fluid will be represented by a specific air pressure. The air pressure is used to activate flow valves, indicators, or controllers.

The majority of float-activated level measuring devices have a measuring range of zero to three feet. Some have a maximum of twelve feet, which for some industrial applications is limiting. Large tanks sometimes use a float and tape and can make up to one hundred feet in level measurement variation.

When the level mechanisms are based on the rotation of a torque arm, the data gathering is restricted to a movement of sixty degrees of rotation.

A magnetic coupling for level measurement is designed to pass the level measurement through the container wall. It is made by placing a magnet either in the float or connected by means of a linkage system that will operate switches or follow-on magnets. With these follow-on magnets, the container can be closed and no sealing of moving parts is necessary. These instruments can be designed with a very large level range.

Electrical Level Sensors

Electrical level measurements have been connected with float-activated mechanisms for some time, but recently the demand for electrical sensors has greatly

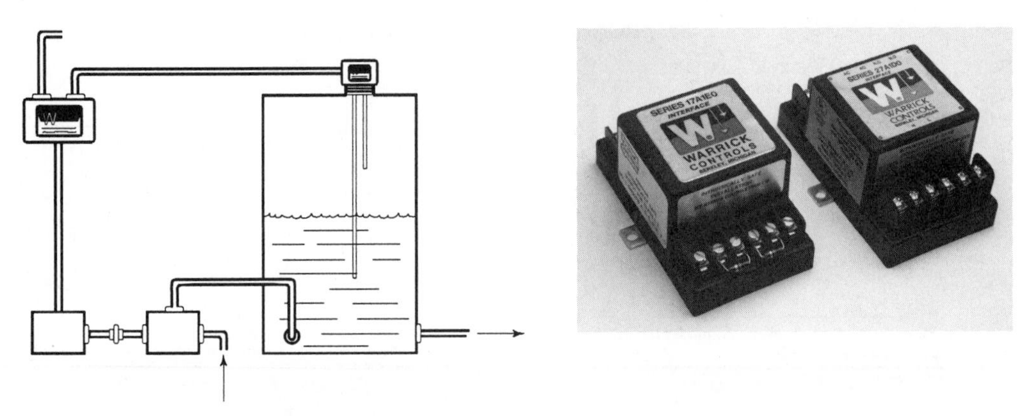

Figure 5-39 A rod conductive level measuring device *(Courtesy of Warrick Controls, Inc.)*

increased. Instruments have been designed that sense only with electrical phenomena.

Conductive Level Sensor - The conductance probe is one of the electrical mechanical methods of level measurement. It consists of a series of probes of different lengths. Fixed level sensing electrodes are installed vertically with their lower ends positioned at the levels at which the control is to be actuated. When a conductive fluid reaches the level of one of these conductors, an electrical circuit is completed and a pump or valve activated, Figure 5-39. This type of sensor has probes that signal it to start or stop. Alarms or other circuits can be added to this device.

Capacitance Level Sensors - Capacitance sensors measure the levels of liquids and granular materials because the electrical capacitance of the conducting probes employed vary with the level of the material being measured. They produce a continuous signal. The capacitance signal is produced by two conductors separated by an insulator or dielectric material. When an alternating current is applied to this device, the capacitance in turn causes a corresponding change in oscillator frequency that varies with the level of the material. In the measurement of dielectric materials, a probe is placed in the tank and the material itself becomes the dielectric; the tank wall becomes the other half of the capacitor. This technique is referred to as a "bare capacitance probe." In the measurement of conductive materials, a probe is designed with the capacity elements inside the probe coated with teflon so that the fluids will not physically contact the elements. As the fluid level changes around the probe, its capacitance changes. Its frequency change is measured by electronic detection, Figure 5-40.

Thermal Level Detectors - Thermal detectors for liquid level are designed by using a self-heated thermistor in a bridge circuit that changes its resistance when

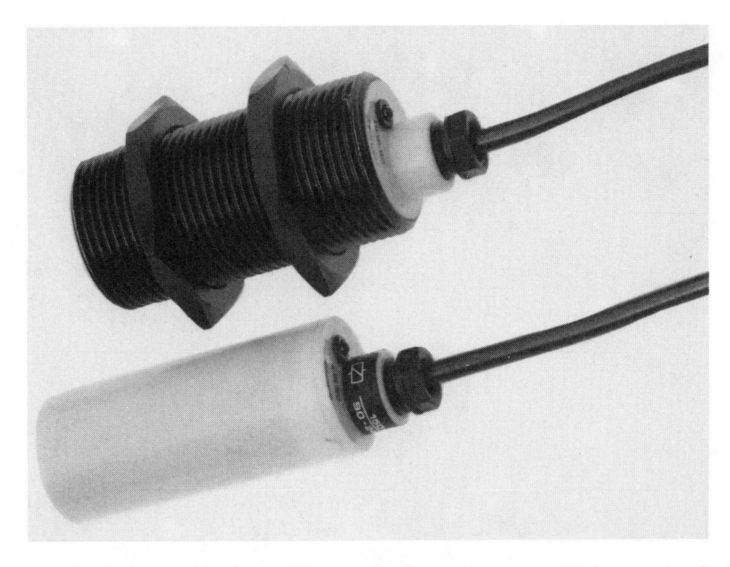

Figure 5-40 Capacitance level measurement. Capacitive sensor for solid, fluid, or granulated substances, either conductive or non conductive. *(Courtesy of Electronic Controls Corp.)*

a liquid reaches its level. The heat exchange unbalances the bridge to produce a signal that controls the system. The thermal dispersion principle of the device consists of two temperature sensitive probes connected in a Wheatstone bridge, and a low-power heating element that preferentially heats one of the probes. A temperature differential between the two sensors is greatest when it is in a dry condition and least when the one probe is submerged, since the liquid cools the heated probe. This difference in temperature results in a change of resistance that supplies a signal to the solid-state circuitry, thus controlling the interface.

These thermal detectors are mounted vertically in a series on the side of a tank. As the level of the fluid rises, the heat absorbed from the sensor correlates with the level of the fluid in the container. These devices can be exactly calibrated to detect the interface between these media: liquids, gases, slurries, and foam. They can discriminate between wet and dry conditions, Figure 5-41.

Photoelectric Level Control - In container-filling operations, the level is frequently controlled by a photoelectric cell. This system employs a light source, an optical system to make a beam, and a detector. When the light passes through the system and strikes the detector, the filling system is activated and the filling of the container takes place. As the level of the fluid reaches the beam, the beam is diffused, the circuit is disrupted, and the filling stops.

On automated conveyors in electronic assembly plants, the photoelectric cell has been replaced with a light-dependent resistor (LDR). The LDR is a solid-state device that lowers its resistance when a high intensity light strikes it. When the

Figure 5-41A Liquid level interface controller *(Courtesy of Fluid Components, Inc.)*

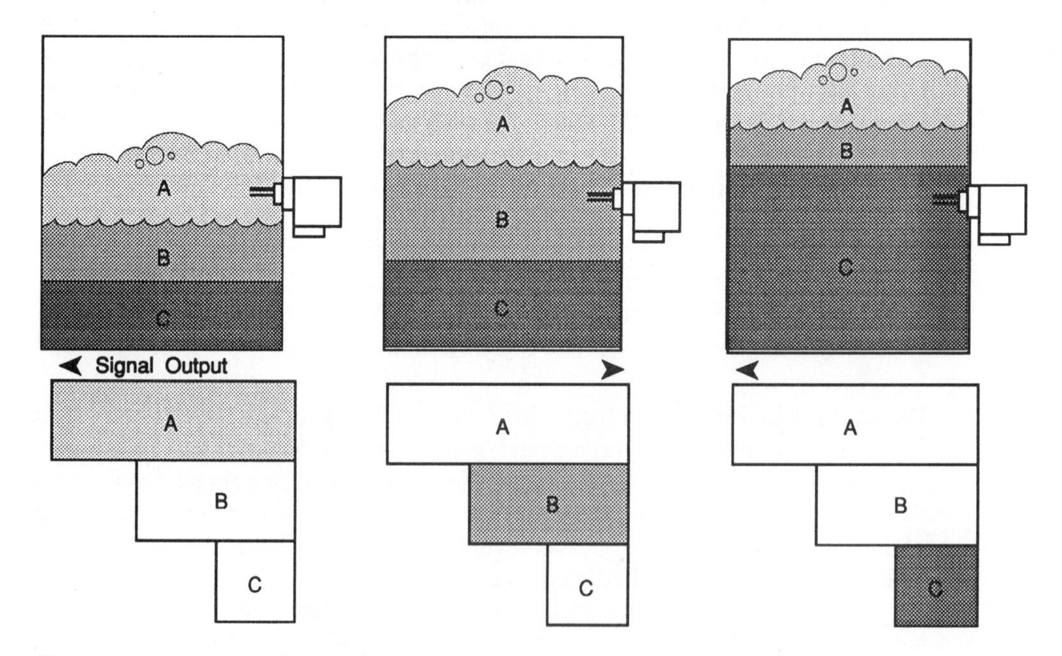

Figure 5-41B Interface signal output

light level is low, the resistance is high, Figure 5-42. This equipment is very reliable and can be used for the control of level as well as position.

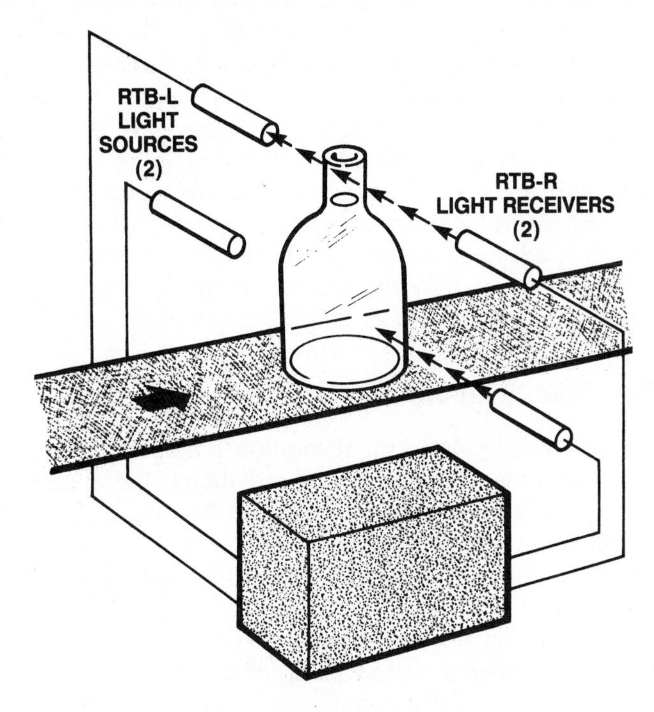

Figure 5-42 Light-dependent resistor control *(Courtesy of Photomation Inc.)*

Figure 5-43 Load cell—weighing a tank to determine level *(Courtesy of Daytronic Corp.)*

The Load Cell Employed as a Level Sensor

Weight can be used as a measure of level. With load cells constructed out of strain gages, total containers can be weighed and the output expressed as level of material in the container. The load cells are of the shear beam design. They are placed in a number of positions under the supports around the vessel being weighed. As the load increases, the strain gages in the load cells increase in resistance. The electrical signal is fed to a summation box where the signals are combined into one signal. The output is a change in resistance that is determined by a Wheatstone bridge and its data output is reported in pounds, Figure 5-43. This application is adaptable to almost any weighing problem, but is excellent for determining the level of solids, slurries, or liquids. The weight can be converted into volume, gallons, or level.

Nuclear Level Sensors

Nuclear level devices detect the absorption of radiation. The level of the product presents more or less material for the radiation to travel through in reaching the detector, which is most frequently a Geiger-Mueller counter. The more product that the radiation passes through, the less amount of radiation that reaches the counter. The strength of the signal represents a measure of the level of material in the container.

Another method of setting up a level sensor system is so that the intensity of a stable radiation source varies with the product level that is inverse to that of the above.

Level detectors are designed for on/off operation or continuous level measurement. The radiation source may be mounted at the desired level inside the container and the detector at that level outside the container. This plan provides a good signal for an on/off control. In another design, the source may be mounted on or near the bottom of the container, with the detector on top and outside the container. The radiation value is calibrated to cause the starting or stopping of the container filling equipment.

Continuous level radiation measurement is accomplished by placing a radiation source at the midpoint outside of the container and detectors down the side of the container. As the level changes in the container, uncovering the various detectors, the radiation causes a progressive strengthening of the signal. In the case of liquids, this can also be achieved by placing the source on a float within the tank and as the level changes the outside detectors will receive an increasingly strong signal indicating the lowering level of the fluid, Figure 5-44.

Radiation sources used are natural radium, cesium 137, or cobalt 60 isotopes. The half-life for radium is fourteen years, for cesium thirty-three years, and for cobalt five years. This should be considered in the selection of the instrumentation. Cobalt 60 is used only when high penetration is necessary, such as when the radiation must pass through thick-walled containers. Cesium is frequently chosen

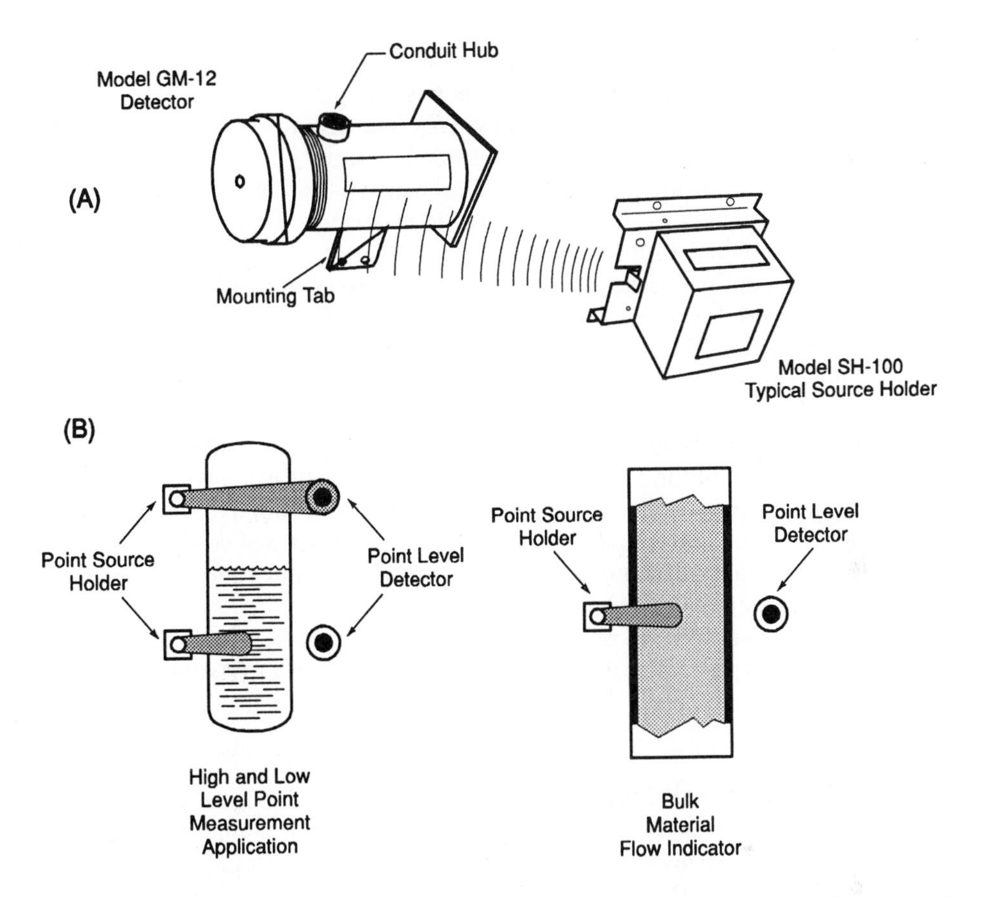

Figure 5-44 Non contact gamma ray level or flow measuring; (B) A gamma ray level or flow measuring system *(Courtesy of The Ohmart Corp.)*

because of its cost and low decay rate of 2–3 percent per year; it provides a good industrial choice. When artificial isotopes are employed, a license from the Atomic Energy Commission is required.

Ultrasonic Level Sensors

Ultrasonic level measurement can be applied to point measurements of level or be extended into a continuous measurement of level. This type of instrumentation may be used on liquid levels, interfaces such as foam above a liquid, or solid materials in a bin or tank.

The device develops a sonic beam with the energy from a piezoelectric crystal vibrating from twenty thousand to forty thousand hertz, depending on the manufacture and application. This sonic beam can be placed across the container with the transmitter unit on one side and the receiver on the other side. As the level of the fluid rises, it interrupts the beam and a signal is sent to the controller to take action, such as stopping a pump or conveyor system.

Another type of ultrasonic device is fastened along the vertical side of the container. As the fluid level reaches one in a series of detectors, that detector no longer transmits the sonic energy. The change in oscillation provides a change in frequency that is received. The level is determined by the final detector in the series in contact with the medium.

The continuous level measurement devices function in a time period. An ultrasonic pulse is required to strike a liquid surface and be reflected back. The time of travel of the pulse to the surface and back is correlated to the level of the fluid. These devices are frequently built in two sections -- one section is the transmitter and the other the receiver. The installations can take on different configurations. Some units are mounted above the fluid, while others are mounted at the bottom of the container inside the fluid. Units can also be mounted under the tank on the outside, transmitting and receiving through the bottom of the tank, Figure 5-45A. On this type of instrument the velocity of sound is changed with each chemical substance in the container. Temperature and pressure need to be considered in the calibration.

These sensors are applied where the materials to be sensed are constantly changing in physical properties, such as with solid waste, popcorn, puffed cereals, chipped fibers, potato chips, and ore.

Microwave Level Sensors - Microwave position and level control employ X-band microwaves. The sensors are mounted outside the container being measured. The microwave level control is used to monitor bins of sand, rock, asphalt, coal, etc., Figure 5-45B.

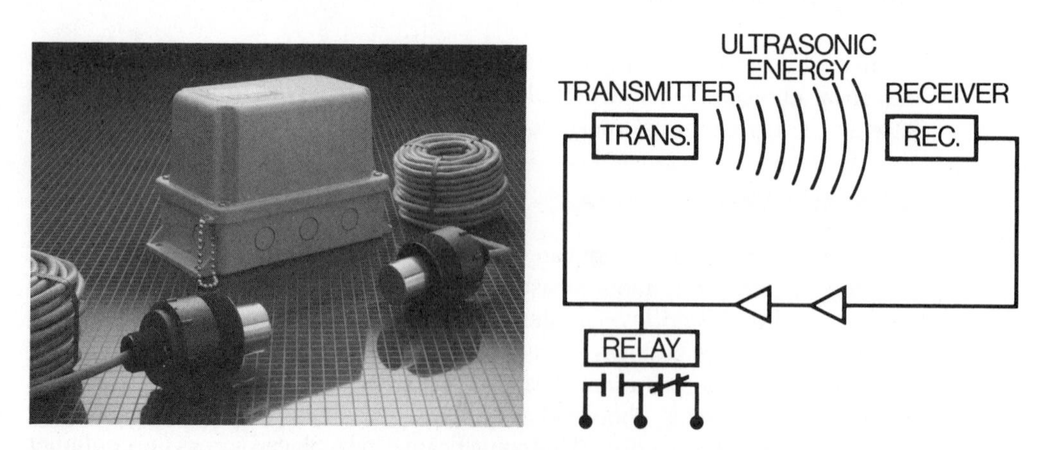

Figure 5-45A Ultrasonic motion, position, and bulk level sensor *(Courtesy of Delvan Inc.)*

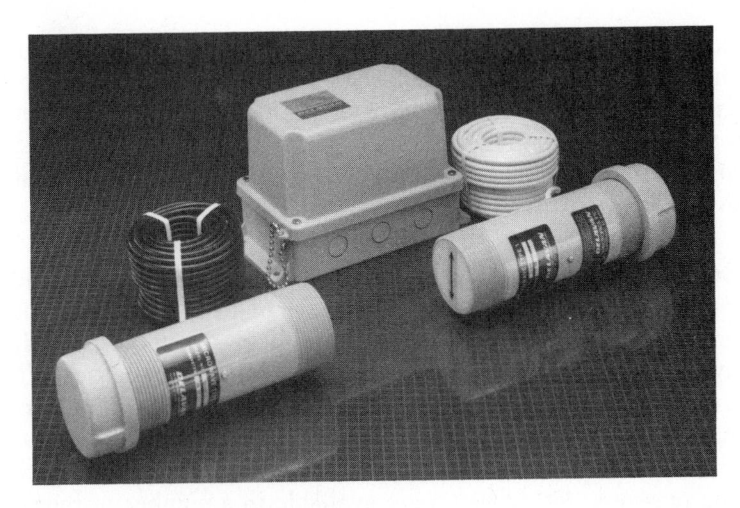

Figure 5-45B Microwave position and level sensor

Solid-state Level Control Sensors

Solid-state level control includes the application of digital electronics to instrumentation. These apply principles of the electrical phenomena of materials rather than mechanical principles to measure level. These instruments utilize pressure change, temperature change, resistance change, voltage change, capacitance change, frequency change, or other changes to provide signals. The signals are analyzed and amplified with the use of various operational amplifiers and circuits. These operational amplifiers can function as voltage comparators, voltage amplifiers, waveform generators, and multivibrators. Basic electronic elements such as resistors, capacitors, diodes, transistors, and other electronic devices are combined to produce networks. These networks are also combined into circuitry to produce voltage dividers and complex circuits. In the mid to late 1960s, a technology developed that resulted in microelectronics whereby thousands of electronic components could be combined on a small silicon chip. This revolution in electronic circuity was given the name of "integrated circuits."

Integrated circuits were designed that could perform a specific function and themselves be placed in a circuit. What resulted was a large number of integrated circuits that were standardized and could be plugged into a system to perform an electronic task.

The application of integrated circuits to circuitry containing discrete components and fabricated on an insulating material with wire or printed-circuit connections between components is referred to as "hybrid integrated circuitry." These hybrid circuits are capable of producing very small and powerful sensors and control devices that can be applied to most industrial variables that are measured.

Hybrid circuits are increasingly used for conditioning the sensor signal, for making the controller's decision, and for supplying the proper signal or voltage to the activator that changes the variable as it enters the system. To do this work, we see the application of potentiometers, inductors, variable capacitors, differential transformers, thermistors, strain gages, peizoelectric crystals, Wheatstone bridge, photoresistors, and phototransistors applied to solid-state devices to provide the signals for the numerous variables to be controlled.

A Solid-state Pressure Transducer for Hydrostatic Level Measurements - These instruments are based on the pressure of the medium on a silicon chip. The measured fluid contacts the backside of a silicon sensor chip, and the sensor chip converts the pressure to an electrical output. The pressure causes different stresses in the silicon, and these stresses make it easier or more difficult for electrons to move through the silicon. The change in the movement of electrons in turn increases or decreases the resistance of the circuit and supplies a signal proportional to the pressure. These gages are available in a number of ranges, the largest operating pressure range being 0 - 300 PSIG, Figure 5-46.

Figure 5-46A Solid-state pressure/level transducer *(Courtesy of SenSym, Inc.)*

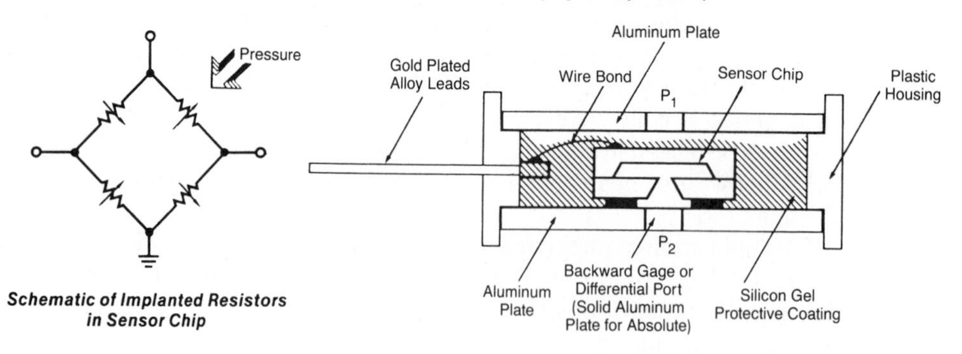

Figure 5-46B A backward gage pressure transducer

Laser Level Sensors

The laser level gage provides a concentrated light source at wavelengths selected for maximum transmission through combustion gases, water vapor, smoke, and dust. The beam dispersion is minimal, allowing for measurements at distances of many feet from the material to be measured. The level gage was developed initially for molten glass. But it can be used on any material that will reflect the laser beam and where the material level has to be controlled about some set-point, such as in continuous casting of metals.

The laser transmitter is set up at a fixed angle, typically fifteen to twenty degrees, perpendicular to the process material surface. The receiver is mounted at the same angle on the opposite side of the process. The laser beam is reflected from the process surface to the receiver. When the process level changes, the reflected beam is displaced in a vertical plane.

In the receiver, the reflected beam is optically scanned over a photoelectric detector. The time period during which the detector is activated is made proportional to the process level. Displacement of the reflected beam due to level change alters the detector activation time period, Figure 5-47A and Figure 5-47B.

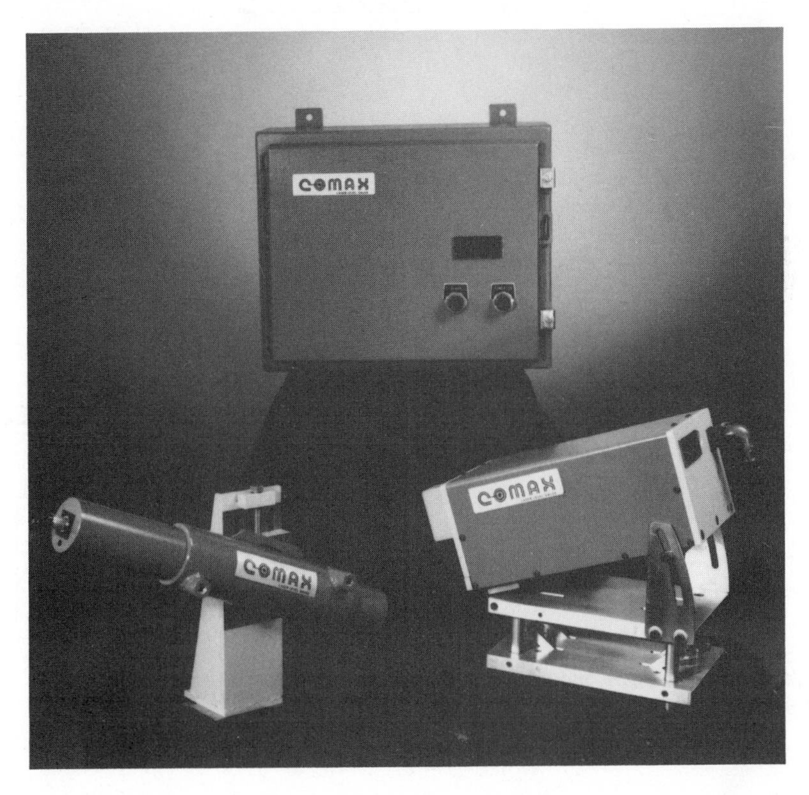

Figure 5-47A Laser transmitter, receiver, and wall mounting electronics *(Courtsey of Courser Inc.)*

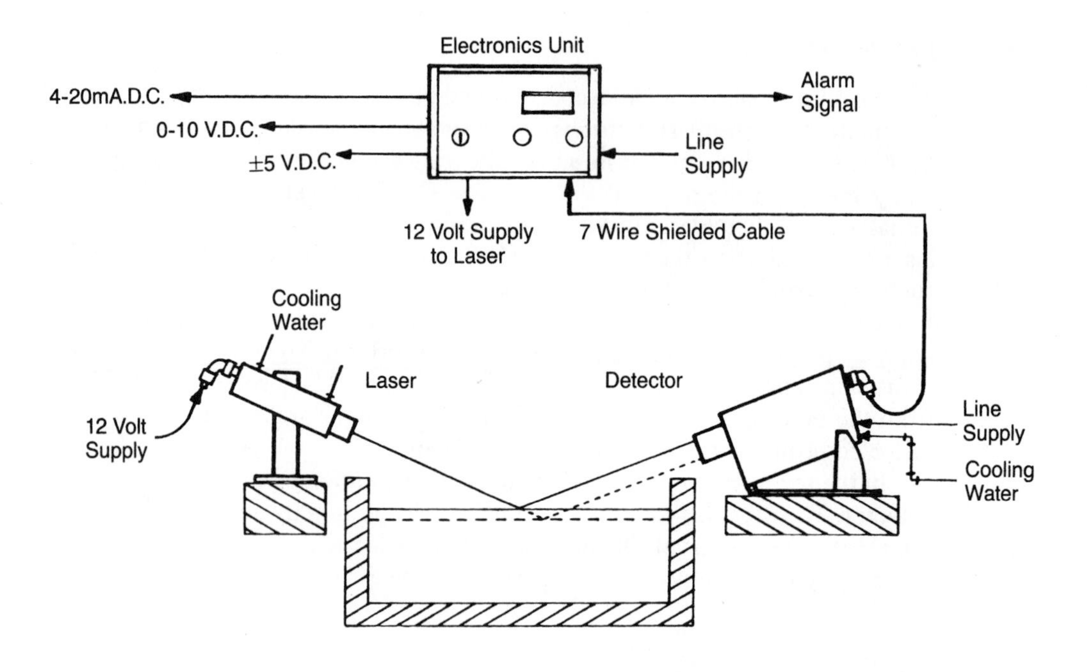

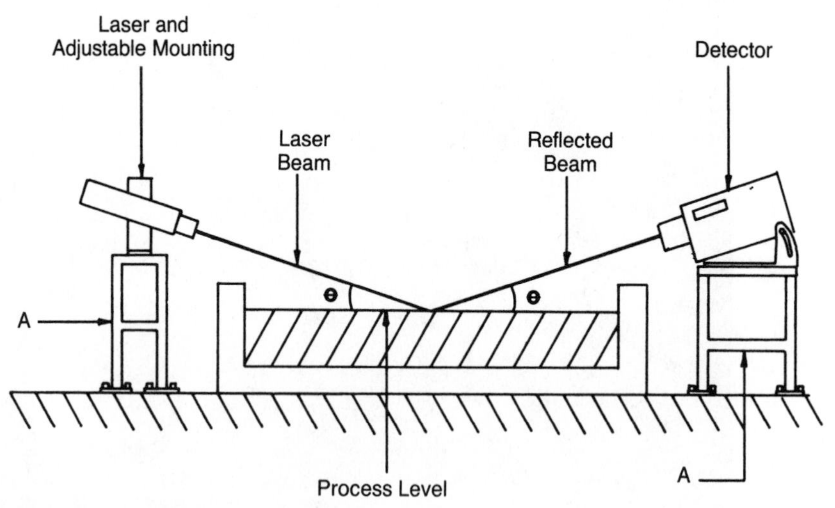

A - Rigid Mounting Structure Supplied by Customer
θ 15-60 Degrees

Figure 5-47B Laser level gage — general arrangement *(Courtesy of Courser Inc.)*

SUMMARY/FACTS

- Flow is fluid in motion and includes liquids, gases, and in some cases slurries.
- Flow is measured as a rate—the amount of fluid moving past a specified position at a given instant.
- Total flow is the amount of fluid moving past a specified position during a designated period of time.
- Flow instruments frequently use the difference of hydrostatic head above and below the sensor to infer a flow rate.
- A weir is an engineered obstruction in an open channel that causes a change in a flowing fluid's level above the weir. The level of the fluid above the weir is measured and correlated with the flow passing over the weir.
- Mass flow measures directly the mass per unit of time flowing through the measuring system.
- The positive displacement flowmeters are devices that utilize volumetric spaces and employ various mechanical cavity-sweeping mechanisms to displace known volumes of fluids.
- Level-measuring instruments may be classified into two large groups of instruments—direct data producing and indirect data producing devices. Direct devices are designed so that observations are made by seeing the fluid levels; indirect data is obtained from inferred instrument information.
- Frequently, level is obtained by weighing the total container. A load cell, constructed out of strain gages, weighs the system and the weight is correlated with the level of the substance within the container.
- Nuclear level detectors are often based upon the amount of radiation absorbed by the product. The higher the level, the more radiation that is absorbed. Thus the level is related to the amount of radiation received by the radiation detector.
- An ultrasonic beam is applied to level measurements where foam, liquids, or solids interface. The sonic beam is *attenuated* by the level of the substance.
- Solid-state devices are employed where hydrostatic pressures are applied on a silicon sensor chip. The output is an electronic signal for integrated circuitry that provides corrections of temperature and calibration resulting in a level measurement signal.

REVIEW QUESTIONS

1. What is a flow measurement that is a sample of a flow in a conduit?
2. What sensor is a disc with a hole in it that is placed within a flange of a fluid-carrying pipe?

3. Upon what principle does the above sensor depend?
4. Slurries of different material are frequently measured using a ______________ ?
5. What is one of the most efficient methods of measuring flow?
6. How is the flow of gases moving through large diameter pipes frequently measured?
7. How is flow in large open channels (e.g., an irrigation ditch) measured?
8. How are large volumes of water and sewage measured?
9. What is measurement of mass flow?
10. What are three types of instruments used to measure mass flow?
11. How does a vortex shedding flowmeter operate?
12. What is a turbine flowmeter?
13. How do positive displacement flowmeters and metering pumps measure?
14. Level-measuring instruments fall into two large groups: direct measuring and indirect measuring. How are these measurements accomplished?
15. Describe a sight glass.
16. What is the function of slip tubes?
17. The principle of measuring hydrostatic head as a measure of level of a fluid is the basis of many instruments. What is the relationship between hydrostatic head and level?
18. Name a measuring device that provides a hydrostatic pressure reading that correlates with the fluid's level.
19. Describe how a self-heated thermistor works.
20. In filling operations, level is frequently controlled by devices that sense by optical means. What common devices perform this function?
21. The entire container may be weighed to obtain level data within the container. What devices are used to achieve this measurement?
22. In what ways can nuclear radiation be used to detect the level of the material in a container?
23. In materials with a layer of foam, level measurement has been difficult to perform. What method has been developed to measure this?
24. Solid-state sensors employing silicon change resistance when pressure is applied. How can this principle be applied to sensing level?
25. How are lasers used to measure level?

6

DENSITY, SPECIFIC GRAVITY, VISCOSITY, HUMIDITY, ACIDITY, ALKALINITY, WEIGHT, FORCE, ROTATION, POSITION, MOTION, ACCELERATION, VIBRATION, AND DIMENSION SENSORS

OBJECTIVE

Upon completing this chapter, you should be able to describe or explain the applications of the sensors listed below:

- The sensors that control the variables of
  - Density
  - Specific Gravity
  - Vicsosity
  - Humidity
  - Acidity and Alkalinity
  - Weight and Force
  - Rotation
  - Position and Motion
  - Acceleration and vibration
  - Dimension

INTRODUCTION

The physical variables named in the chapter title are used in industry to control many manufacturing processes. Industrial manufacturing employs countless variations of these sensors in controlling processes during manufacturing.

DENSITY AND SPECIFIC GRAVITY

Density is the compactness of a material. It is measured as weight per unit volume or mass per unit volume. The units employed are grams per cubic centimeter, pounds per cubic foot, or pounds per gallon.

Specific gravity is a ratio of a material's density to another material's density. When dealing with liquids or solids, the standard reference material is water. When dealing with gases, the reference is hydrogen or air. The temperature and pressure, along with the characteristics of the material being measured, affect the specific gravity. For industrial measurements, the ambient pressure is used. Temperature compensation needs to be made if the material's temperature fluctuates.

Density and Specific Gravity Transducers

Density and specific gravity measuring transducers fall into a number of groups and are based upon buoyancy, displacement, differential pressure, radiation absorption, weight of a fixed volume, vibrating U-tube, ultrasonic attenuation, and thermal gas density.

The Hydrometer

The hydrometer is used to find the density or specific gravity of a fluid. It is based on Archimedes' principle of floatation—the weight of a floating body is equal to the weight of the fluid it displaces. The hydrometer is constructed of a glass float with a long, graduated stem for reading the specific gravity. A bulb weighted with lead shot or mercury below the float holds the instrument vertically in the fluid. The shot also calibrates the hydrometer so that it reads a specific gravity of 1 when placed in pure water.

Hydrometers are widely used where automated operations are not required. They supply data on antifreeze mixtures, salt solutions, acids, beer, wine, and dilution of milk. They can be mounted in standpipes with an overflow to provide a constant level. The hydrometer measures the specific gravity in this gently flowing fluid.

The hydrometer can be built into a remote sensor by placing the hydrometer in a tube similar to a rotameter housing. It incorporates a standpipe that controls the fluid's level and a guard to keep the hydrometer oriented. In this case, the stem of the hydrometer is opaque and a light source passes a beam through a slit to the stem of the hydrometer. When the stem of the hydrometer blocks the light traveling to a photocell opposite the slit, an electrical signal is transmitted.

A solid-state electronic device may be designed using the same procedure, except that an electronic photoresist and circuit is applied. It delivers great accuracy. In either case, the temperature of the fluid should be monitored.

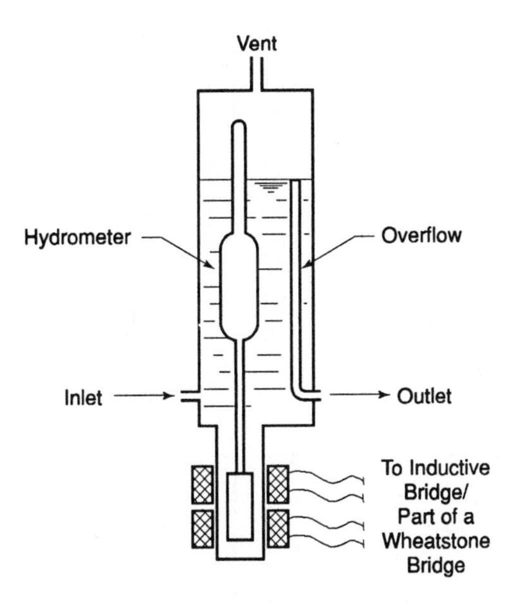

Figure 6-1 Continuous flow induction bridge hydrometer for specific gravity

The electrical application of an inductance bridge can be applied to a hydrometer by extending the length of a stem between the float and the weight. The weight is made of a soft iron or other magnetically permeable material. As the specific gravity changes, the armature's position changes inside the inductance coil, and an electrical signal is transmitted, Figure 6-1.

Air Purge or Bubbler

Density or specific gravity can be measured by the application of a bubbler. Differentials created by hydrostatic heads of fluid of different densities are compared. As was stated in the study of level, air bubblers purge air off the end of a tube in a fluid and create a back pressure equal to the fluid's hydrostatic head. This pressure varies with a change in the density or specific gravity of the fluid. The system can be employed to monitor the density or specific gravity of the fluid flowing through the system. For these measurements, two bubbler systems are placed in the fluid with the ends of the bubbler tubes at different depths in the tank. The difference between the pressures produced by the bubblers indicates the density or specific gravity, Figure 6-2.

Another method employed is to place one of the two bubbler tubes into a reference fluid. This compares the back pressure of a reference fluid with the back pressure of the process variable. Solid-state electronic differential pressure gages make this system of monitoring very efficient.

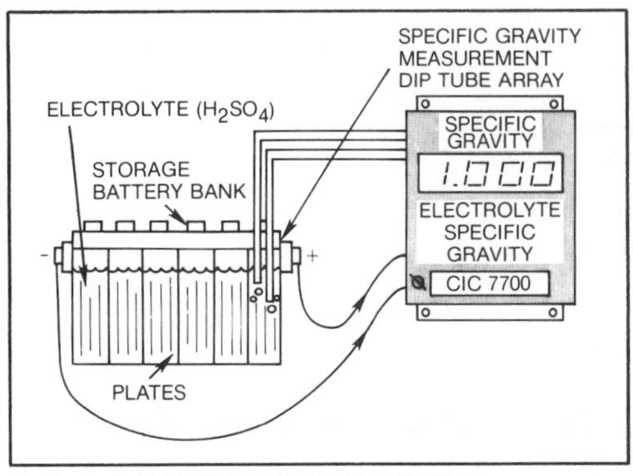

Continuous Sp Gr Monitoring of Battery Bank Electrolyte

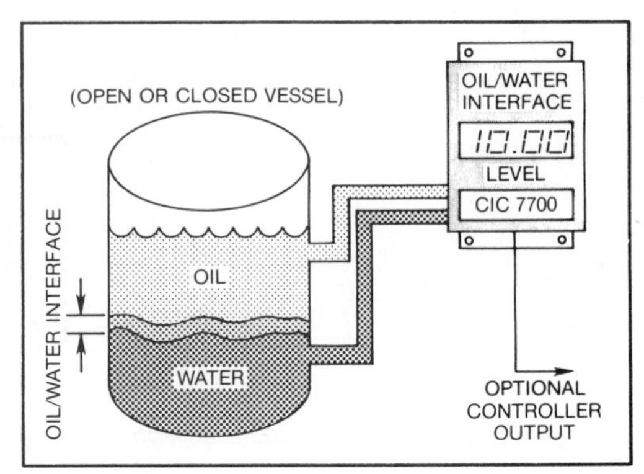

Oil/Water Interface in Oil Recovery System

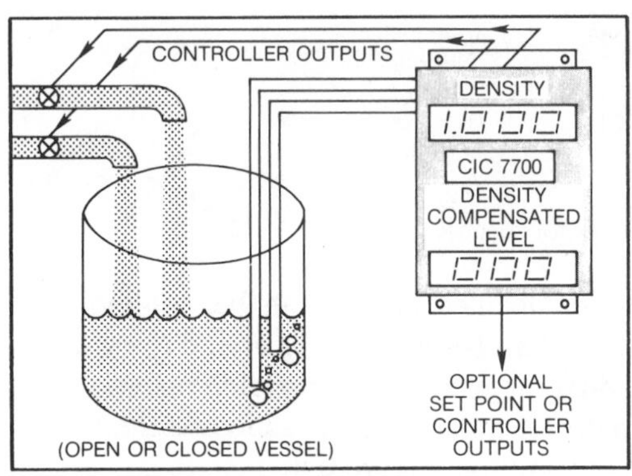

Density Compensated Level for Mixing Process Control

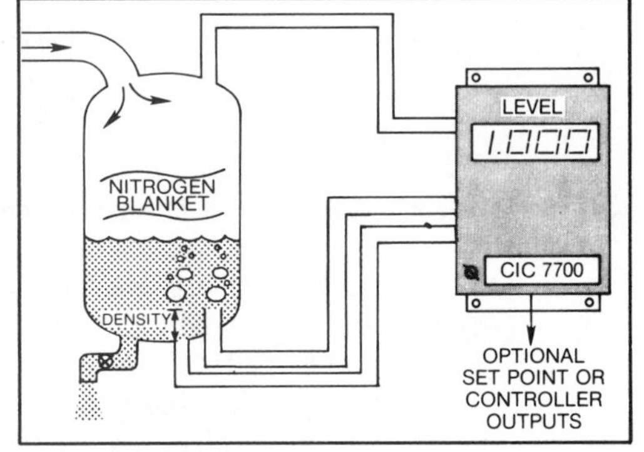

Pressurized Vessel Density/Level

Figure 6-2 Differential bubbler at two levels *(Courtesy of Computer Instruments Corp.)*

Radiation Density Sensor

Radiation absorption is applied to density measurement by passing a gamma ray beam down or through a pipe that contains the process flow being monitored. When the process product density increases, less radiation reaches the detector and therefore an inverted signal represents an increase in density of the process fluid. To get a satisfactory signal, the radiation beam is projected down an offset in the piping system. The vertical section in the pipe offset provides an additional depth of fluid for the radiation to pass through. This results in more radiation absorption taking place and an increase in the sensitivity of the signal. The sources of radiation are similar to the applications discussed in the section concerning level radiation sources. Cobalt 60 and cesium 137 are the materials used as radiation sources. The detector for the system most frequently used for density measurement is an ionization chamber. The chamber is made of two dissimilar

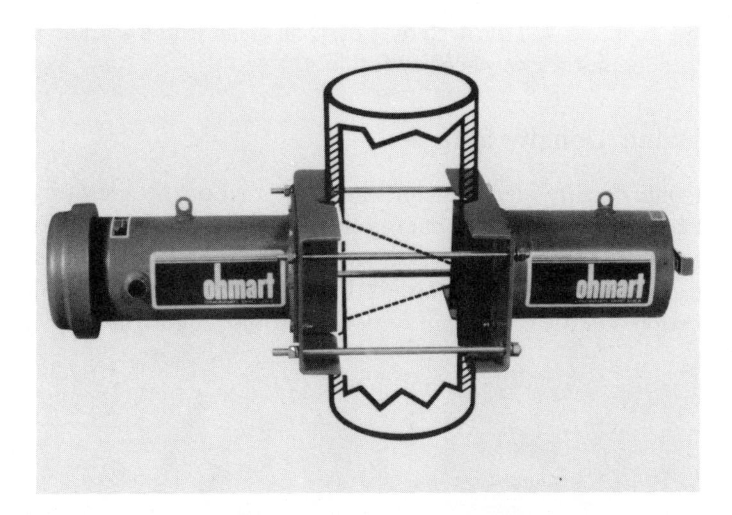

Figure 6-3 Radiation density sensor *(Courtesy of The Ohmart Corporation)*

metal plates subjected to pressurized gas. When the beam enters the chamber and strikes the gas, a very small current signal is produced, Figure 6-3. This signal is amplified to provide a signal that is correlated with the density in the system.

Density by Weight

Density measurement can be carried out by weighing the fluid as it passes through a fixed volume. The liquid is passed through a vessel, such as a small tank, tube, or bulb where the fluid is continuously flowing and the vessel is full. The vessel is fastened by flexible connections and the system is constantly weighed. If the weight of the fluid in the vessel changes, this is transmitted as data to indicate a density change in the fluid passing through the processing system. A controller will add materials to bring the fluid to the correct density requirements.

Vibration Density Sensor

The vibrating U-tube is a density measuring device based upon the natural vibrating frequency of a mass. The U-tube is vibrated mechanically along with the fluid to be measured that is running through the tube. If a change of density occurs in the fluid, the vibration frequency changes providing a signal that can be calibrated into a density value.

The instrument is designed as a U-tube welded into a frame at the natural nodal points of the vibrating tube. The tube itself becomes the elastic element in the system. At the bent end of the tube, an electrical transducer is mounted that provides vibration to the tube and receives a frequency change due to a change in density. The output of the coil is an AC voltage that is proportional to the density change, Figure 6-4. This system is used on clean fluids with no entrapped gases, with moderate-to-low viscosity fluids.

Ultrasonic Density Sensor

Ultrasonic density measurement can be carried out by measuring an attenuated sonic beam. The ultrasonic energy is transmitted across a fixed gap filled with a flowing slurry. The sonic instrument measures the percentage of suspended solids in the slurry. The suspended solids decrease the beam and provide a signal that is proportional to the specific gravity of the fluid, Figure 6-5.

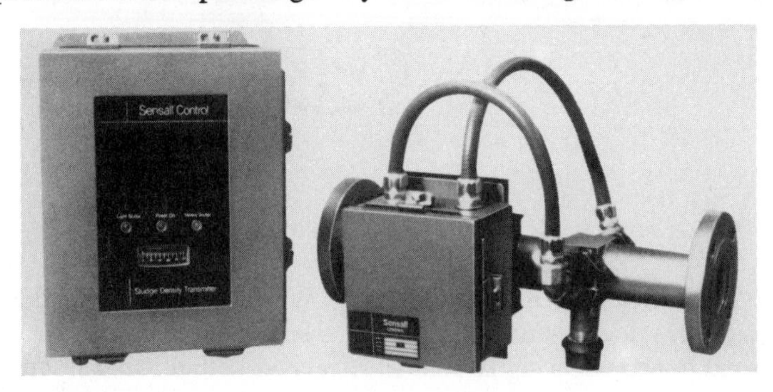

Figure 6-4 Vibrating U-tube density sensor *(Courtesy of Yokogawa Corporation of America)*

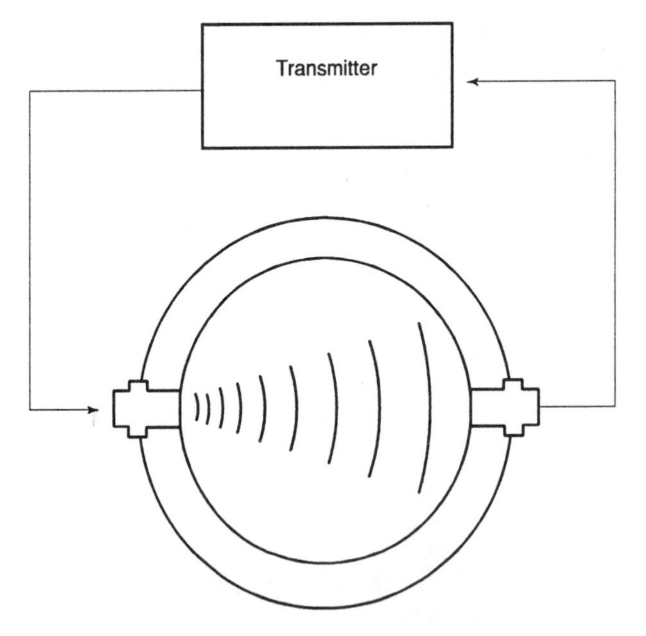

Figure 6-5 Ultrasonic density measuring *(Courtesy of National Sonics Div. of Xertex)*

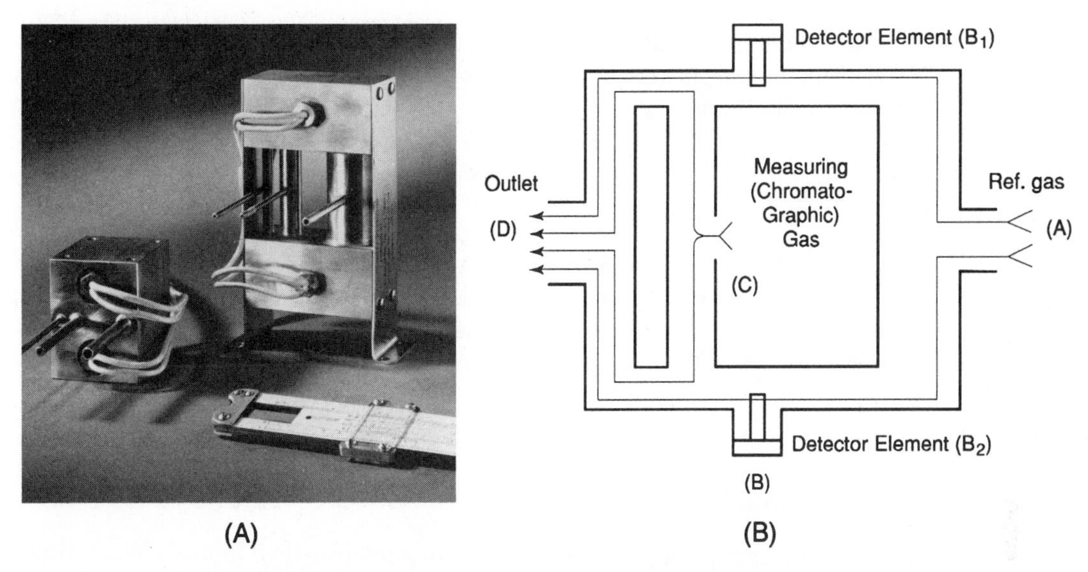

Figure 6-6 (A) Gas density thermal sensor; (B) A vertical pneumatic Wheatstone bridge. The internal geometry of the detector resembles a pneumatic Wheatstone bridge, the conduits of which must be in a vertical plane. *(Courtesy of Gow-Mac Instrument Co.)*

Gas Density Detector

Thermal density gas gages are based on the design of a pneumatic Wheatstone bridge and utilize a thermistor as a thermal conductor sensor. The Wheatstone bridge compares a reference gas, such as nitrogen or argon, with the gas under test by comparing their molecular weights. If the test gas has a higher molecular weight than the reference gas, the vertically mounted bridge will cause the gas to flow to the lower side of the bridge. This cools the upper thermistor. The heavier molecular weight gas flow in the lower part of the pneumatic bridge restricts the reference gas flow past the lower thermistor causing the lower themistor's temperature to rise. A differential in resistance between the thermistors yields a change in voltage that correlates with the density of the gas and provides a monitoring signal, Figure 6-6.

VISCOSITY SENSORS

Viscosity is a fluid's resistance to flow. Viscosity is the result of a fluid's shear stress and shear rate. In Newtonian fluids, there is a linear relationship between the amount of applied shear stress and the result of deformation or the shear rate. Gases and thin liquids tend to be of the Newtonian viscosity type.

Nonnewtonian fluids are not linear with respect to the applied shear stress and the deformation or shear rate. Synthetic oils and thermosetting plastics represent nonnewtonian fluids.

Viscosity Transducers

Instruments that provide a measure of viscosity in manufacturing are referred to as continuous sensors. A number of devices are available on the market, but only a few have been selected to represent the category of viscosity. The designs of sensors selected are these: continuous capillary tube viscometer, float viscometer, ultrasonic viscometer, and plastometer.

The Continuous Flow Capillary Tube Sensor

The continuous capillary tube is applied to thin laminar flow liquids. When flow in the capillary tube is held at a constant rate, the viscosity will be a linear function of the drop of the pressure between two pressure measuring points on the tube. Accurate readings can be maintained when the temperature is held at a standard reference point before entering the capillary tube. Calculations are made to calibrate the tube and deliver data on the liquid's viscosity. The inside diameter and the length of the tube can be adjusted to produce data that is most meaningful to the particular product that is being measured.

The pressure differential on the capillary tube can be measured by solid-state electronic differential pressure gages. They provide electrical signals that control the delivery of dilutent liquids and thereby control the viscosity of an ongoing process. The petroleum refining industry makes use of this sensor in a number of applications.

The Absolute Viscosity - The continuous viscometer for refinery applications employs a sample that is brought to constant temperature and pumped through a capillary at a constant rate of flow. A differential pressure transmitter is connected in parallel across the capillary and generates an output signal directly proportional to the absolute viscosity of the product sample.

The temperature of the sample is maintained constant by the immersion of the pump and capillary in an electrically heated oil bath with precise temperature control. A cooling water coil is provided to carry away excess heat. This coil is necessary especially for those applications where the temperature of the incoming sample and/or *ambient temperature* is close to the higher-than required bath measuring temperature, Figure 6-7.

These measurements are applied to continuous viscosity streams, such as lube oil, diesel oil, asphalts, and like products.

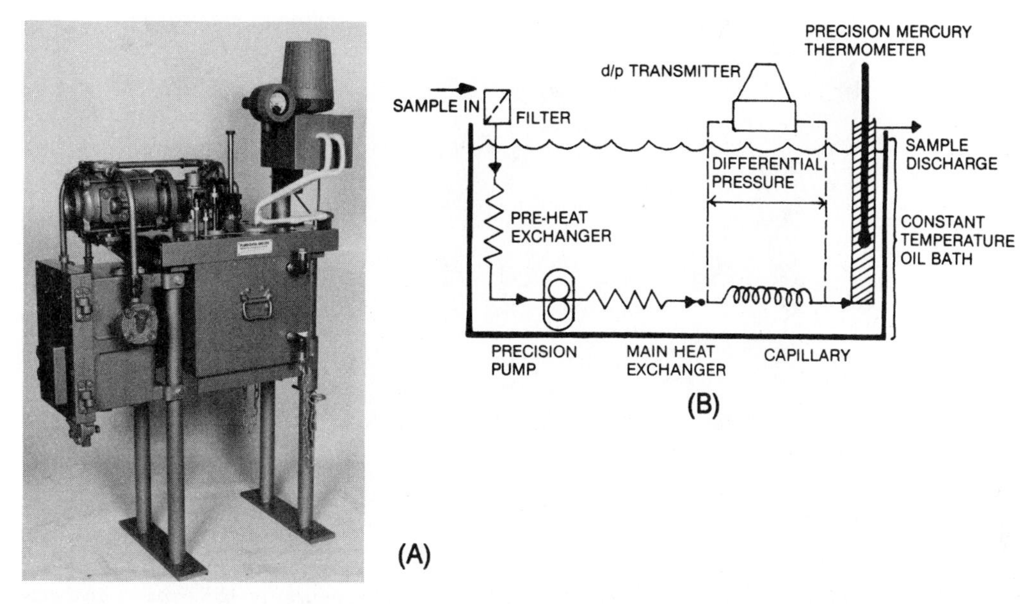

Figure 6-7 (A) Absolute viscosity; (B) Principle of operation *(Courtesy of Fluid Data Inc.)*

The Float Viscometer

The float viscometer is a variation of a rotameter. The liquid to be measured is held at a reference temperature and the fluid is pumped at a constant flow rate through the meter. The float size in the system is increased so that the viscous drag on the float will provide an accurate reading of the viscosity of the fluid moving through the production system, Figure 6-8.

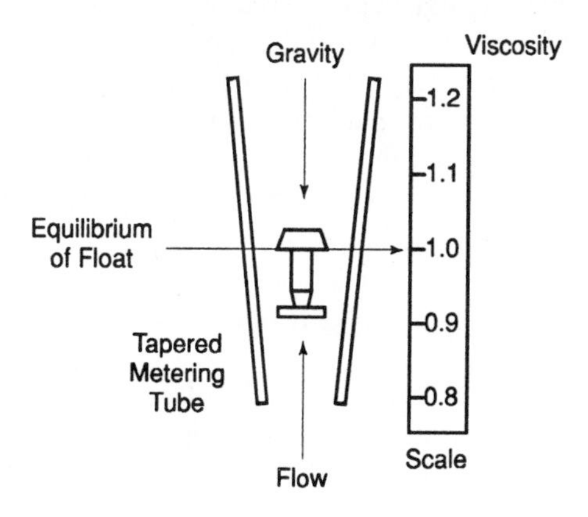

Figure 6-8 A variable area flow rotameter applied as a viscometer

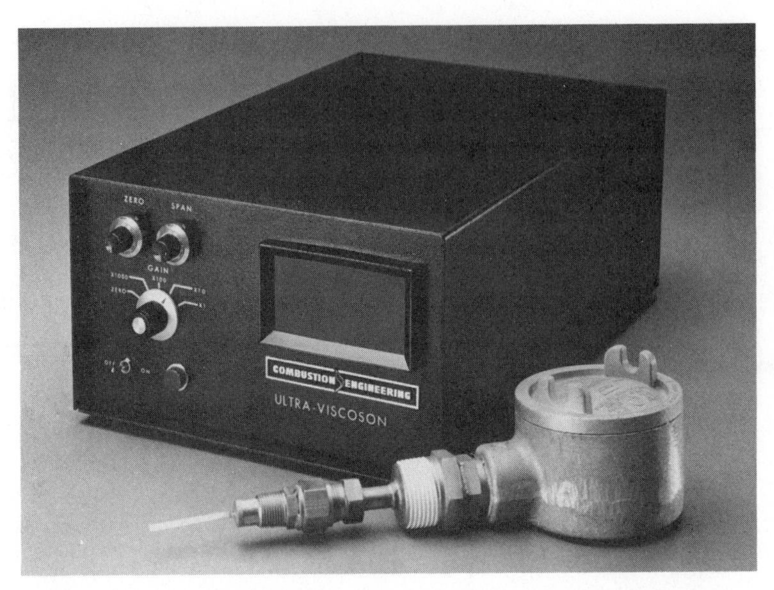

Figure 6-9 A continuous on-line explosion-proof viscosity detector *(Courtesy of ABB Process Analytics Inc.)*

The Ultrasonic Viscometer

The ultrasonic viscometer is designed using a probe made of a thin blade that vibrates due to pulses sent from a computer. Magnetic fields produced by the pulses of the computer result in sinusoidal waves along the blade. The blade is in contact with the liquid and is dampened by the shearing action of the liquid. The higher the viscosity of the liquid, the greater the dampening exerted on the vibrating blade. The viscous drag of the liquid dampens the ultrasonic vibration. The blade acts as a transducer and sends the dampened vibration signal back to the computer. This change in pulse frequency is proportional to the square root of viscosity times density in newtonian materials. The computer compared frequencies result in a scaled voltage that represents the viscosity of the liquid, Figure 6-9. The applications are for such newtonian liquids as lube oils and petroleum products, syrups, vegetable oils, etc; and for nonnewtonian liquids such as colloids, emulsions, and suspensions (starch, slurries, chocolate, paint, catsup, glues, inks, gelatin, and drilling mud).

The Cone and Plate Plastometer

Plastometers are widely employed by the plastic, rubber, and petroleum industries to monitor the behavior of plastic materials during the processes of manufacture. Plastic is formable at different stresses and temperatures. The shear stress and shear rate measurements of plasticity is correlated indirectly with measurements of a material's molecular weight. In the manufacture of plastic products, the control of the temperature and pressure provides the factors for a satisfactory product.

The cone and plate plastometer is a device that provides data on the viscosity of a solid material. It has a test chamber that is maintained at the temperature selected for the particular test sample. The sample is weighed and inserted between the rotor plate and the cone where the deformation takes place. The sample is heated by the test chamber until the test temperature has been achieved. The rotor surfaces have serrations machined into their surfaces to prevent the test material from slipping during the measurement. A constant-speed motor drives the upper rotor at a rate of usually two revolutions per minute. A shear results in the test sample between the cone and the plate. A shaft extends down from the cone and is connected to a calibrated U-shaped spring for measuring the torque delivered to the cone by the tested material. The spring is provided with strain gages that deliver a continuous signal to indicating and recording instrumentation. The result of the testing procedure is an average viscosity of the solid under test, Figure 6-10A and Figure 6-10B. A series of tests can be repeated at different temperatures to establish a recommended processing temperature.

HUMIDITY

Humidity is concerned with the amount of water vapor in an atmosphere. It is a very important factor in a number of industrial manufacturing processes. Mate-

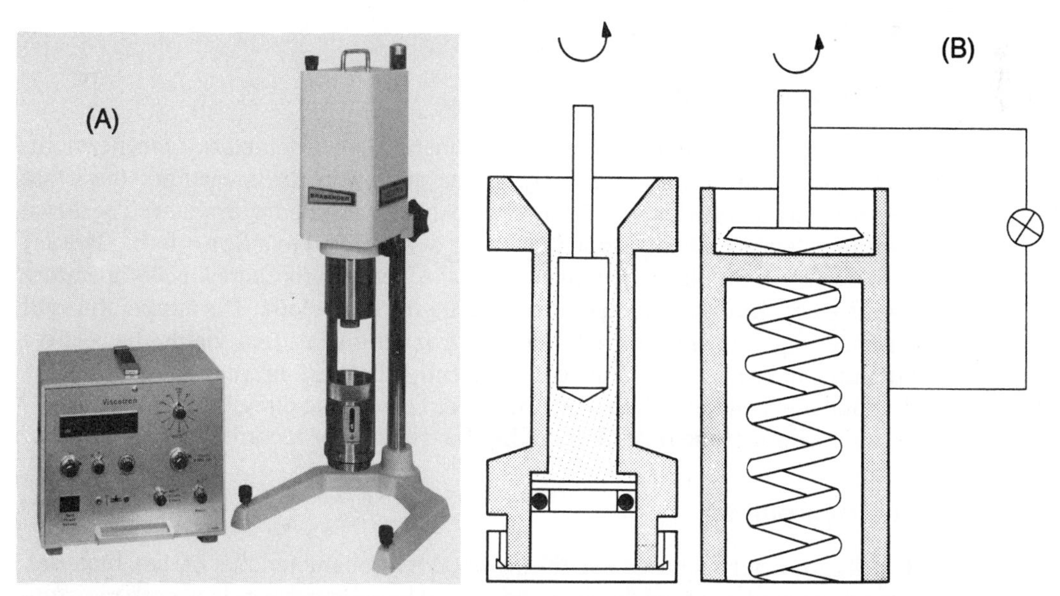

Figure 6-10 (A) Viscotron cone and plate lastometer; (B) Diagram of a cone and plate plastometer *(Courtesy of C. W. Brabender Instruments Inc.)*

rials that are extremely sensitive to humidity are ground and powered materials, textiles, woods, papers, plastics, and a number of chemical crystalline substances. Often we may think of humidity in atmosphere only in terms of air; but in industrial processes an atmosphere refers to other gases as well, e.g., nitrogen, hydrogen, chlorine, and special gases used in the semiconductor industry.

In the measurement of humidity, four common concepts are stated. The first is absolute humidity. This is the actual amount of water in a specific volume. It can be expressed in any unit of measure, as long as the chosen unit is consistent. This definition is not extensively used in industry, but is applied in highly technical and scientific studies. Second is specific humidity—the ratio of the mass of water vapor to the mass of dry air. This definition is applied in heating and air-conditioning calculations. The third definition refers to relative humidity and is concerned with the ratio of the actual amount of water vapor in the air compared to the maximum amount of water vapor the air would contain at the same temperature if saturated. The fourth concept related to humidity is dew point. The dew point is the temperature at which moisture begins to condense out of the water vapor that saturates the air. Dew point is measured at the temperature that a dew or frost begins to form on cool surfaces.

Humidity and Moisture Sensors

The measurement of humidity is based on a number of principles and devices. Among these are the following: psychrometers, hygrometers, dew point, electrical phenomena, microwave absorption, infrared absorption, and vibrating quartz crystal.

The Psychrometer

The psychrometer is constructed with two identical thermometers. One thermometer has its bulb exposed directly to the airflow to be measured; the other thermometer's bulb is covered with a cloth that is saturated with water. These two thermometers are swung around in the air or subjected to a flow of air. The dry bulb measures the air temperature, while the wet bulb measures the temperature produced by the evaporation of the moisture from the cloth. The temperatures of the thermometers are referred to a psychrometric chart that yields the relative humidity of the air flow, Figure 6-11. Recently, thermistors are substituted for the thermometers and integrate their values into electronic circuitry that produces a digital readout of the relative humidity for viewing or recording.

The Hygrometer

The hygrometer is based on the *hydroscopic* characteristic of the material. Ancient man understood that hair and animal membranes would absorb moisture and change their physical length. Early hygrometer designs utilized hair attached

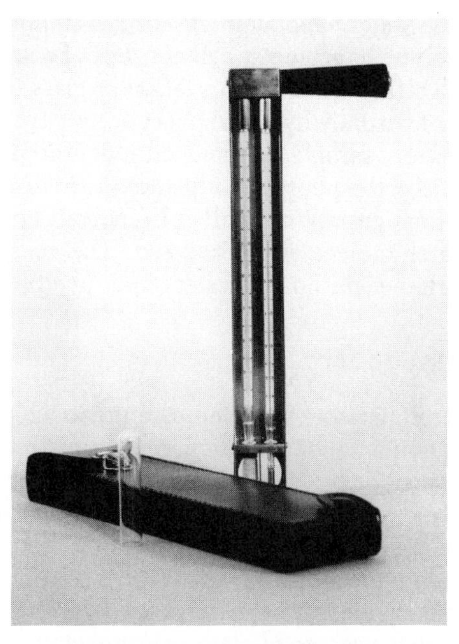

Figure 6-11 A sling psychrometer *(Courtesy of Weather Measurements/Weathertronics a Division of Qualimeters, Inc.)*

to a pointer. On damp or rainy days, the pointer was in one position; on dry or sunny days, the pointer was in another position. Thus an early weather gage was realized.

There are a number of hydroscopic materials that change their electrical characterstics when they absorb moisture. Materials such as aluminum oxide, lithium chloride, cobalt chloride, polymer films, and others are designed into instruments that function by absorbing moisture from the air and changing the material's electrical value. The sensor frequently used is a polymer base, a grid, or a thin-walled, hollow tube wrapped with tape that is saturated with a hydroscopic chemical. Around these forms two wires are wrapped that complete an electrical circuit as the moisture from the air condenses on the crystals and changes the electrical conductivity of the sensor. When the moisture content of the air increases, moisture is absorbed by the aluminum oxide layer and determines the conductivity of the unit. The moisture in the oxide layer allows an electric current to flow between the wires that correlates with the amount of moisture in the sampled environment.

The Electrolytic Hygrometer

The electrolytic hygrometer is based upon the change in the state of a material due to the electrolysis of water into oxygen and hydrogen. The electrolysis current is a measure of the water vapor content in a gas sample.

The electrolytic hygrometer cell consists of a small cell of two noble metal electrodes with bifilar winding on a teflon or glass tube. The wires are covered with a coating of partially hydrated phosphorus pentoxide that absorbs the water vapor in the gaseous sample. A direct current is applied to the electrodes and the water absorbed by the phosphorus pentoxide is dissociated into hydrogen and oxygen. The total current controlled in the cell is directly proportional to the number of water molecules hydrolyzed. The measure is reported to digital readouts as parts-per-million of water vapor by volume, Figure 6-12.

The Dew Point Sensor

The dew point temperature is the temperature to which the gas must be cooled to produce saturation and condensation. Dew point temperatures are readily converted to either vapor pressure units or to parts-per-million water vapor units. Dew point is a measure of absolute moisture content and can be determined with good precision over a wide range of humidities and temperatures. It is used for many industrial applications.

The optical condensation dew point hygrometer is a thermoelectrically cooled mirror that uses solid-state instrumentation. In the optical dew point hygrometer, a surface is cooled by a thermoelectric or Peltier cooler until the dew or frost begins to condense on the mirror. The condensate surface is maintained electronically in vapor pressure equilibrium with the surrounding gas. The surface condensation is detected by optical or, in some cases, electrical techniques. When maintained at the temperature at which the rate of condensate exactly equals the evaporation, the condensation surface is at the dew point temperature by defini-

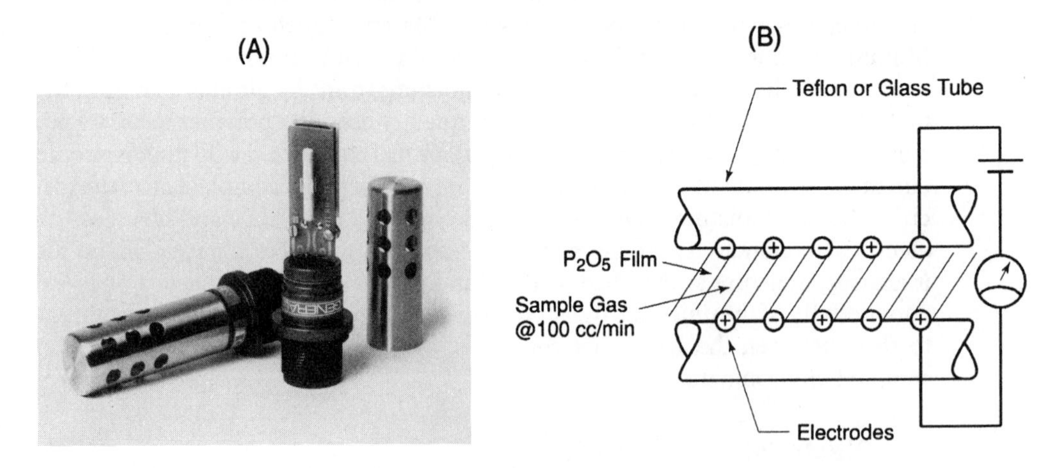

Figure 6-12 (A) Electrolytic hygrometer sensor; (B) Diagram of electrolytic sensor *(Courtesy of General Eastern Instruments Corporation)*

tion, Figure 6-13. The temperature of the surface when so controlled is typically measured with a platinum resistance thermometer, thermocouple, or thermistor imbedded in the mirror surface.

The Electrical Capacitance Hygrometer

Electrical phenomena are the basis of a group of hygrometers. One of these is the capacitance hygrometer in which a porous layer of aluminum oxide is deposited between two concentric cylinders. The two cylinders are electrically insulated from each other and become the capacitor plates. Two porous metal disc end pieces support the dielectric material. The cylinders hold the aluminum oxide in place and allow the gases to pass through the oxide. The moisture is either absorbed or desorbed by the oxide, but holds an equilibrium with the sample gas. The capacitance is amplified by the change of the dielectric constant of the material between the plates, which is a factor of moisture level in the gas under analysis. In the electronic circuitry, a fifteen kilocycle powered reference capacitor is switched alternately with the test capacitor. The difference in the capacitors' amplitudes is a function of the moisture content in the test gas.

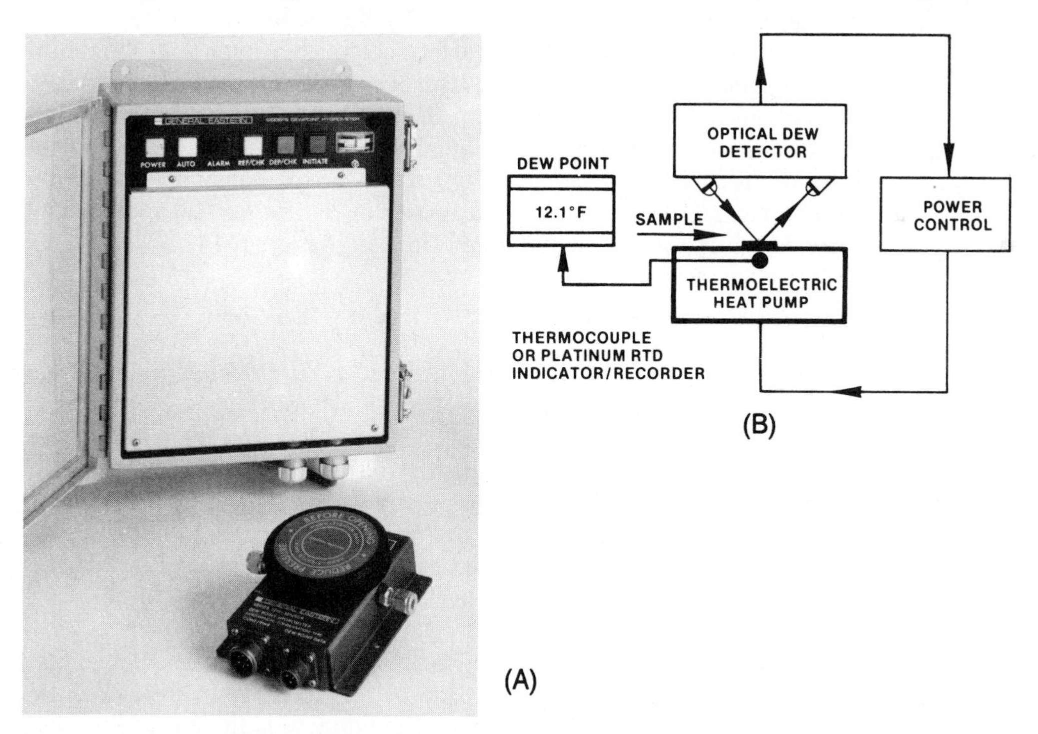

Figure 6-13 Dew point measured with a chilled mirror. (A) Dew/frost point industrial hygrometer; (B) Principle of operation condensation dew point model 1200 EPS *(Courtesy of General Eastern Instruments Corp.)*

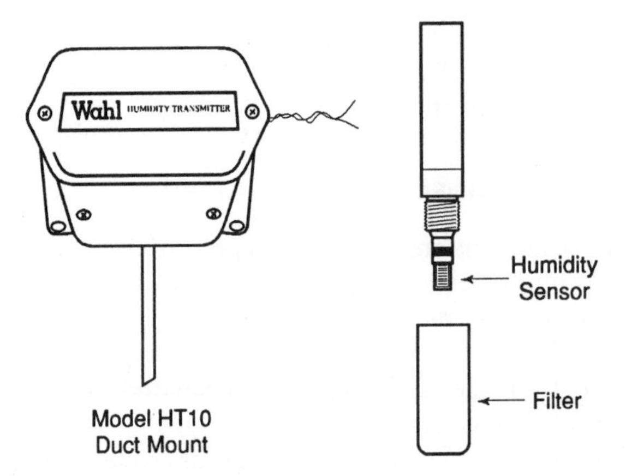

Figure 6-14 Gold and silver electrodes are plated on aluminum oxide for a single probe sensor *(Courtesy of Wahl Instruments Inc.)*

An Impedance Hygrometer

An impedance type hygrometer is formed by depositing a layer of porous aluminum oxide on an aluminum strip and then coating the oxide with a thin film of gold. The conductive base and the gold layer become the capacitor's electrodes. Water vapor penetrates the gold layer and is absorbed by the porous oxidation layer. The number of water molecules absorbed determines the electrical impedance of the capacitor which is, in turn, a measure of water vapor pressure.

Aluminum oxide sensors are frequently used in petroechemical applications where low dew points are to be monitored "in line," Figure 6-14.

A Microwave Hygrometer

The microwave absorption hygrometer functions on the principle that there is selective absorption of electromagnetic energy by moisture. A beam of microwave energy is passed through or reflected back from the moisture in the sample area where the beam is modified. The energy can be detected by loss or phase shift in the microwave beam. Water molecules greatly absorb the energy and in the K band (20.3–22.3 GHz) produces a resonant frequency no other molecules produce. This water molecule resonance energy absorption is detected and can be correlated with a percentage of moisture in the sample, Figure 6-15A, 6-15B, and 6-15C.

The instruments are frequently housed in a box or in a short pipe container that has a microwave window in the end made of teflon, polyethylene, or other microwave transparent window materials. The window is in contact with the fluid, slurry, or pastes, but provides no intrusion into the flow. As a receiver is mounted on the opposite side of the sampling area, it receives the modified microwave energy signal.

Figure 6-15A On-line noncontact sensor for solid flowable materials *(Courtesy of Kay Ray Inc.)*

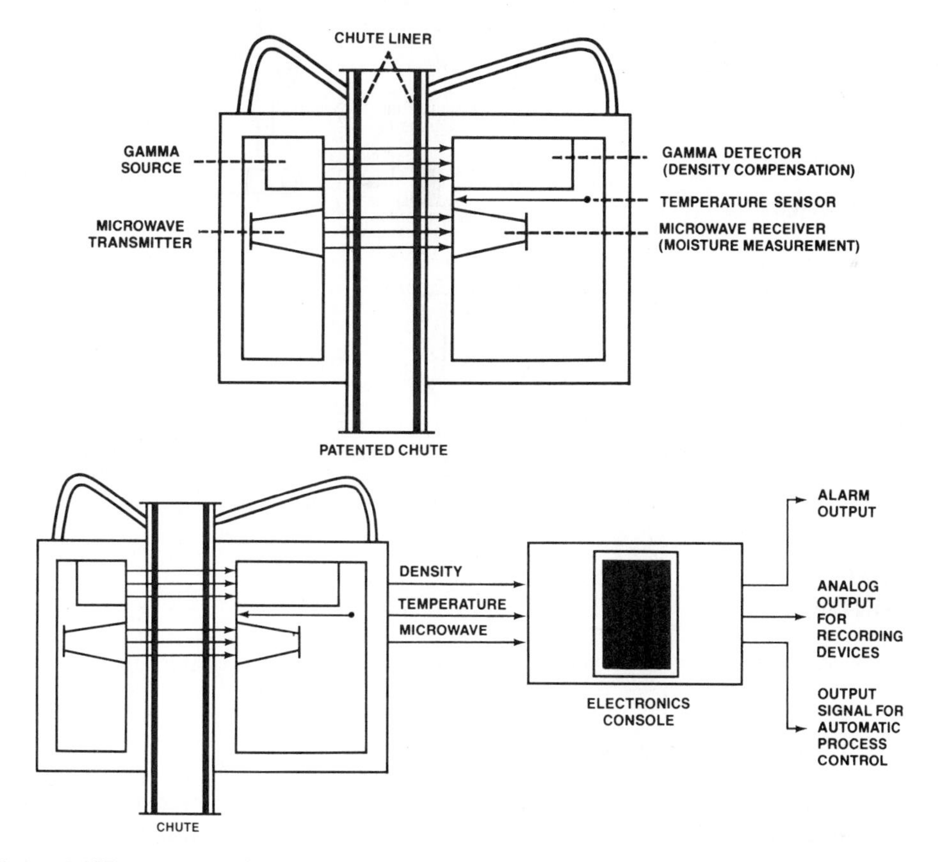

Figure 6-15B Gamma ray for density and microwave for moisture senors *(Courtesy of Kay Ray Inc.)*

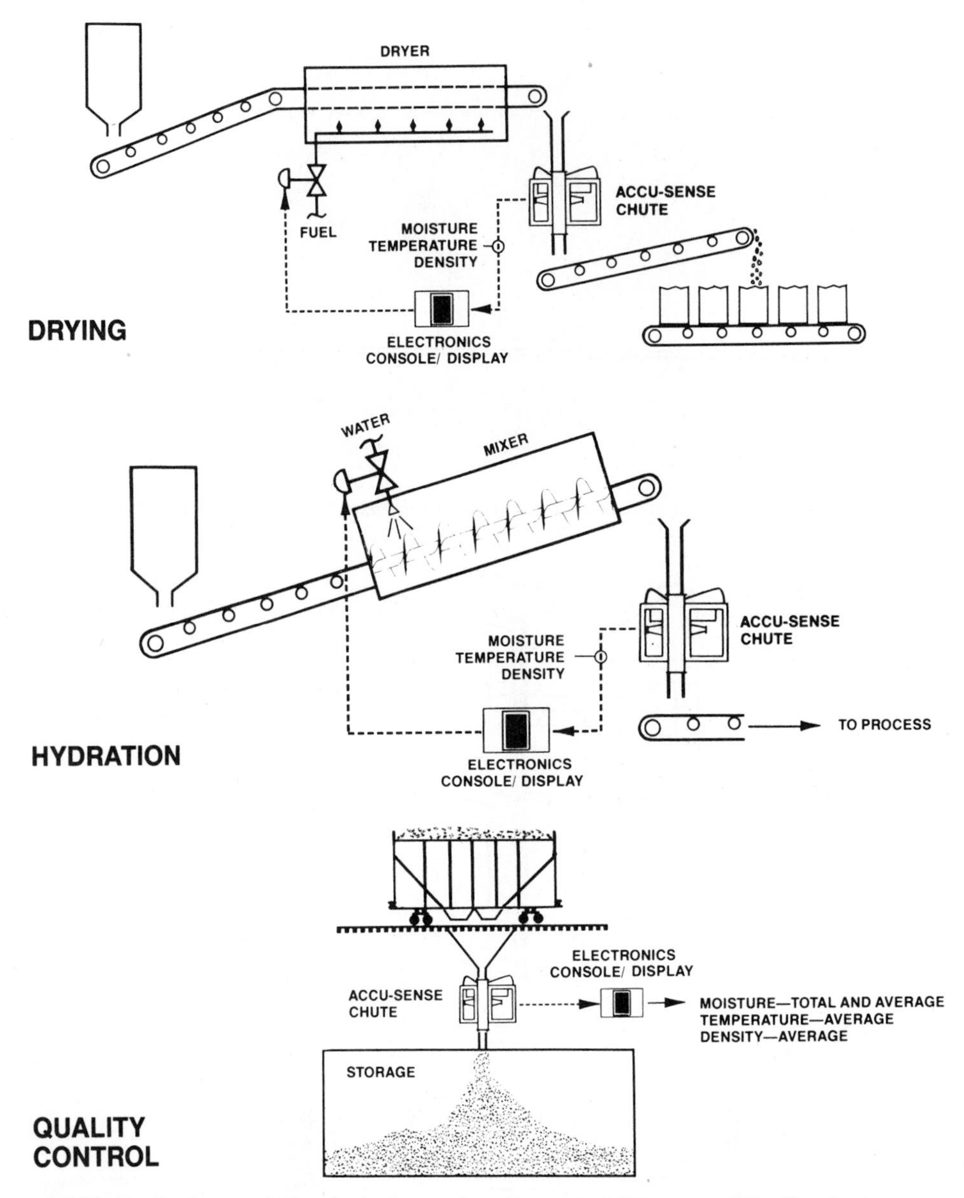

Figure 6-15C Applications to drying, hydration, and quality control *(Courtesy of Kay Ray Inc.)*

The Infrared Hygrometer

The infrared hygrometer obtains the water content by absorption of certain infrared wavelengths from an IR beam passed through the sample gas. A source beam is split into two equal beams—one passing through a reference gas and the other through the sample gas. The beams pass through filters to produce the appropriate wavelengths: one wavelength that is not affected by water vapor and the other that is affected by the water vapor. The intensity of the beams is alternately measured by a photoelectric cell, and the intensity of the wavelengths is compared. The dew point or the relative humidity of the gas is obtained from the measurement of the output voltage of the difference of the beam intensities measured by a photoelectric cell. Because of the type of construction of this instrument, it is often used to monitor other constituents (CO, CO_2, NO) by selecting the appropriate absorption wavelengths.

A Quartz Crystal Hygrometer

A quartz crystal can be sensitized to vary its vibrational frequency when subjected to a change of moisture in the gases that it is sampling. The sample gas is separated into two streams. One stream is passed directly over the sensor crystal. The other stream is passed through a dryer and then passed over the same crystal. These two streams are alternately measured and the frequency of the two streams monitored. These frequencies are then compared to a sealed reference crystal. A microprocessor carries out the computations that correlate the frequency change to the moisture level of the sample.

ACIDITY AND ALKALINITY

Hydrogen and Hydroxyl Ion Measurements

The acidity or alkalinity of an aqueous solution is inferred by its hydrogen ion concentration. This system was devised by Sorensen as the negative logarithm of the reciprocal of a given hydrogen ion concentration. If a solution contains 0.000001 (10^{-6}) grams of dissociated hydrogen ions per liter, the solution has a pH value of 6. The negative exponent need not be a whole number. The logarithm's reciprocal is given as the pH concentration number. Pure water dissociates into 10^{-7} Mols/liter of H+ and 10^{-7} Mols/liter of OH-; thus a pH of +7 or a neutral solution is assigned to pure water.

0 1 2 3 4 5 6 7 8 9 10 11 12 13 14
Strong Acid Neutral Strong Base

When the pH number changes by one, the hydrogen ion exponent level concentration increases or decreases by ten times.

pH — Acidity and Alkalinity Transducers

Three methods are used to determine pH concentration in solutions: the litmus paper test, the colorimetric technique, and the potentiometric measurements.

The Litmus Paper Test - The litmus paper strip is saturated with a chemical indicator. When it is inserted into the unknown solution, its color provides a reading: pink represents an acid solution; blue represents a basic or alkaline solution. The accuracy of this method is plus or minus one-half pH unit.

The Colorimetric Measurement - The colorimetric method is used on clear solutions with organic dyes and is used for laboratory titrations. The accuracy depends on the color discrimination of the human eye.

Potentiometric Sensors - The potentiometric method is the most accurate procedure and is most often used by industry. Instruments are designed so that they can be constantly monitored and corrected for the pH value of solutions. These measuring devices are based on electrolytic conductivity and are a measure of the ability of a solution to carry an electrical current. This ability can also be called "specific conductance" because it is defined as the reciprocal of the resistance in ohms of a one centimeter cube of the liquid at a specified temperature. Conductance is measured in mho which is ohm spelled backwards. In metals, the current is carried by free electrons, but in a solution, the current is carried by free ions.

The potentiometric measurement of pH requires two glass electrodes. The reference electrode is insensitive to the hydrogen ion concentration. They can be made of a silver chloride solution and a potassium chloride solution. If two solutions of different hydrogen ion concentrations are separated by a thin glass membrane, an electrical potential is set up as the result of a difference in the solutions' concentrations. This potential is measured. This potential becomes a reference against which other potentials can be measured. If one solution's potential and its pH are known, another solution's pH can be determined.

The glass indicator electrode is designed with a buffer solution filling a glass tube. The tube has a soft porous glass section. Inside the buffer solution area is a second area containing silver chloride that is connected to an external lead. The solution to be measured is in contact with the silver chloride chemical through the sensitive porous glass. In this process, an electrical potential is created. This potential is compared with the reference potential and the unknown pH is indicated, Figure 6-16.

Solid-state electronics have been added to these pH procedures. They provide the necessary amplifiers and voltage controllers so that these devices can be interfaced with microprocessors and yield an automated continuous pH control over industrial process solutions.

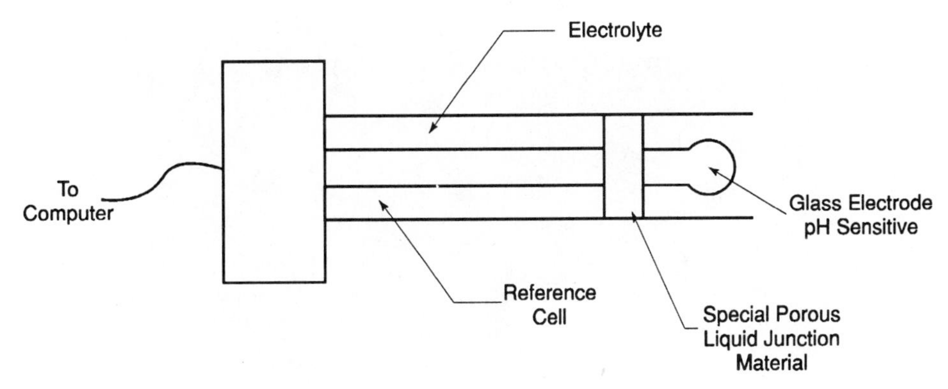

Figure 6-16 Combination pH glass electrodes

Measurement electrodes of material other than glass have been produced but are not widely used in manufacturing. These other pH sensors are the hydrogen electrode, the quinhydrone electrode, and the antimony electrode.

The pH reactions are sensitive to temperature changes, thus the temperature of the solution under evaluation will need to be stabilized by a thermo-compensation device.

WEIGHT AND FORCE SENSORS

Force is the push or pull applied on an object. Weight is the measurement of the force of gravity on the object. Industrial processes are usually concerned with weight to determine the size of a batch or the amount of ingredients needed to complete a formula. Weight measuring instruments are based on the following systems: lever, spring, hydraulic, pneumatic, and electronic.

Lever Systems

Mechanical lever systems have been is use since ancient times. The first class lever was used in the Far East and the Near East to weigh goods to be exchanged in commerce. Early lever systems were based on a steel bar or yard and a knife-edged fulcrum, with a pan on one end of the yard and a calibrated weight hanging from notches marked on the yard arm. Today's scales use the same concepts but have a series of lever combinations in a system. Platform scales, truck scales, railway scales, tank scales, hopper scales, bin scales, and batch scales are all examples of these lever systems. These systems of measuring weight are in wide use today because of their broad range of temperature correction, low period of natural frequency, and wide measuring load range. Mechanical lever systems of

Figure 6-17 Railroad track scales. A lever system railroad scale *(Courtesy of Fairbanks Weighing Division of Colt Industries, Inc.)*

scales are designed to economically cover a host of manufacturing process requirements, Figure 6-17.

Spring Systems

For lighter weighing systems, the spring scale is an economic method of obtaining a reasonably accurate measure of weight. The spring scale is based on Hook's law of elasticity, whereby a spring is calibrated as to its elongation versus the weight applied. The metal chosen for the spring is made of an alloy with a broad temperature modulus that reduces the effects of ambient temperature changes, Figure 6-18.

Pneumatic Load Cells

Weighing materials with a pneumatic load cell makes use of the force-balance concept. The pneumatic load cells have air chambers, diaphragms, and air nozzles that apply principles similar to a stacked pneumatic controller that uses a basic nozzle and flapper valve. A central shaft runs down through the instrument and rests above a nozzle seat that bleeds the air chamber. When a weight is placed on the cell, it bleeds the regulated air supply down through the bleed nozzle air outlet until the air pressure stabilizes. At this point, the pressure in the chamber generates a force equal to the weight of the load being measured. Extra chambers are added—one to provide a force equal to the weight of the container being filled; and other chambers that are interconnected with a small orifice to provide dampening of the system.

This type of load cell is employed in many types of check weighing, filling, and packaging operations. Because this system uses air, it is suitable for use around foods, pharmaceutical, medical, and explosive materials.

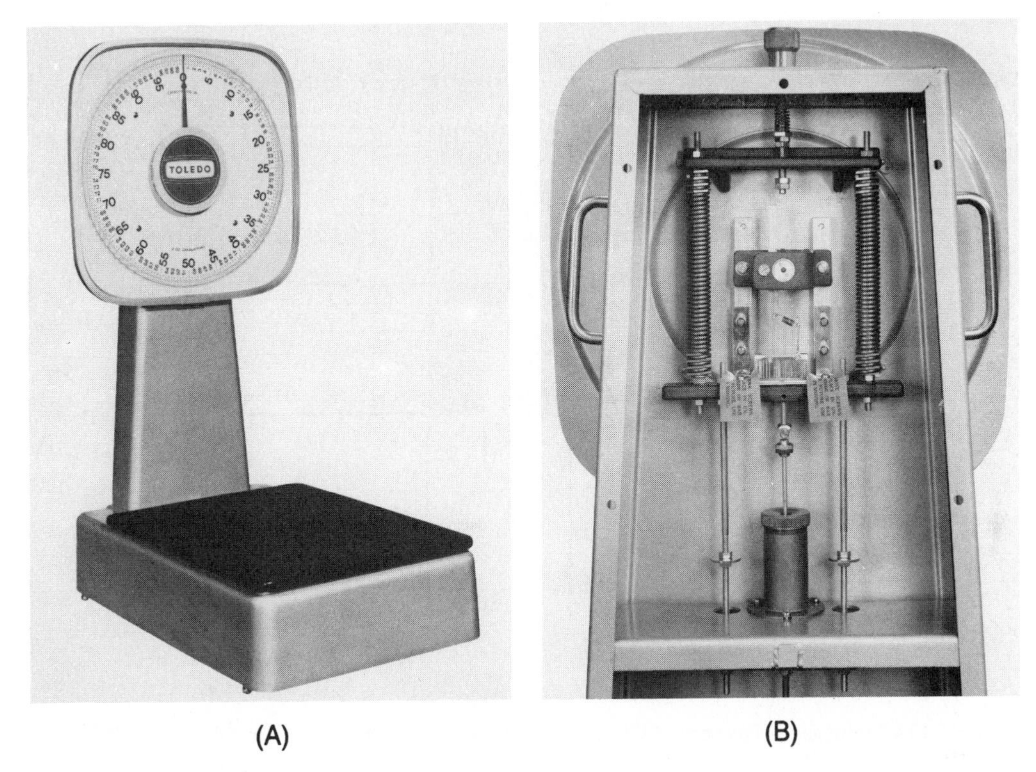

Figure 6-18 (A) A dial-type spring scale; (B) A spring system scale *(Courtesy of Toledo Scale Division, Reliance Electric)*

Electronic Systems

Strain gages are devices that change their electrical resistance with the application of a physical load. The load can be tension or compression and is correlated with a resistance change in the gage. Strain gages are made of a metal alloy that is insensitive to temperature change, but retains its resistance characteristics. They are made from small diameter wires or metal foils. The foils are frequently manufactured by applying chemical machining processes to obtain very intricate and delicate shapes, Figure 6-19A and Figure 6-19B.

The Strain Gage Load Cell

Strain gages are mounted on a column or a beam, the load-bearing member is mounted within the load cell. Whether a column or a beam is used depends upon the application selected. The strain gages are carefully bonded to the deflecting members. The gages are oriented in different positions so that one set of gages will gain in resistance when the column is receiving the load, while the other set

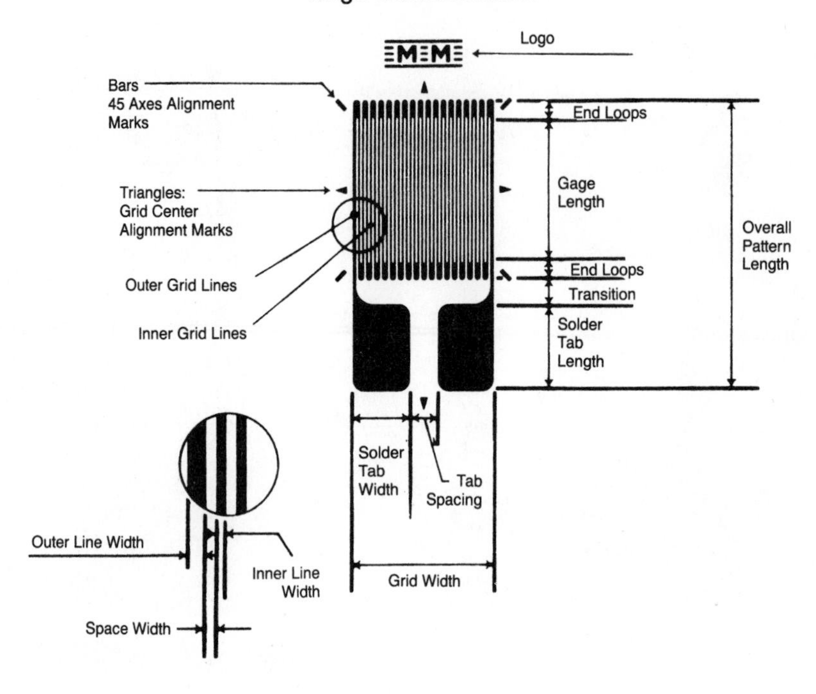

Figure 6-19A Strain gage *(Courtesy of Micro-Measurements Division of Measurements Group Inc.)*

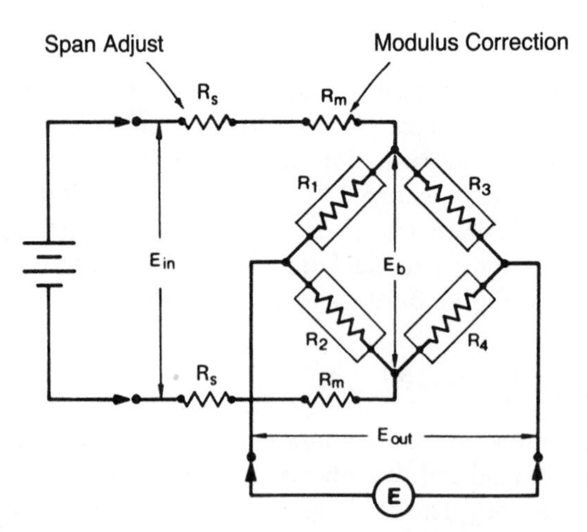

Figure 6-19B Typical strain gage transducer circuit *(Courtesy of Micro-Measurements Division of Measurements Group Inc.)*

(A)

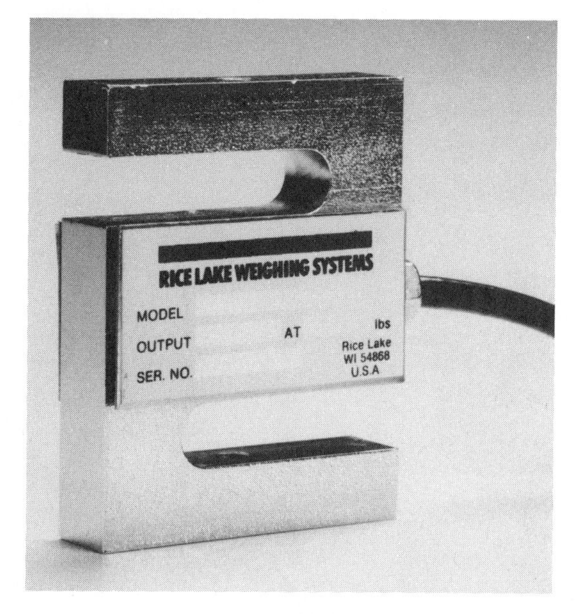

(C)

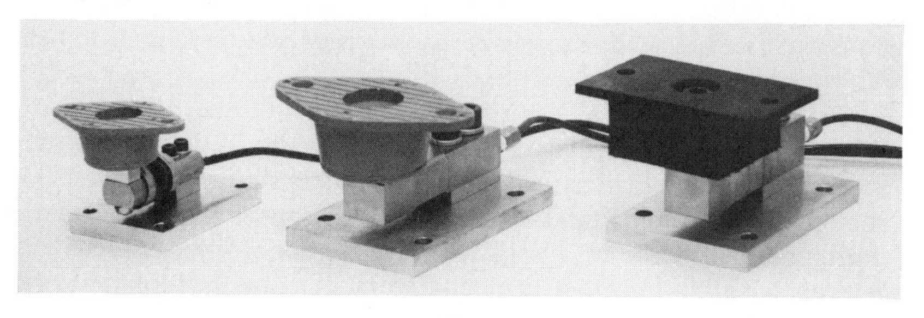

(B)

Figure 6-20 (A) Column-type load cell *(Courtesy of Transducers Inc.)* (B) Beam-type load cell; (C) S-type load cell *(Courtesy of Rice Lake Weighing Systems)*

of gages, usually oriented ninety degrees to the first set, will lose resistance. These gages are placed in a Wheatstone bridge; the gages take the place of the traditional resistances. The resistance change is calibrated into weight units and is transmitted to an indicator or recording instruments, Figure 6-20.

The Semiconductor Strain Gage Load Cell

Semiconductor strain gages have a gage factor ten to fifty times that of metal or foil strain gages and have become competitively successful in weight measuring. Semiconductor strain gages function on a peizoresistive principle defined as a

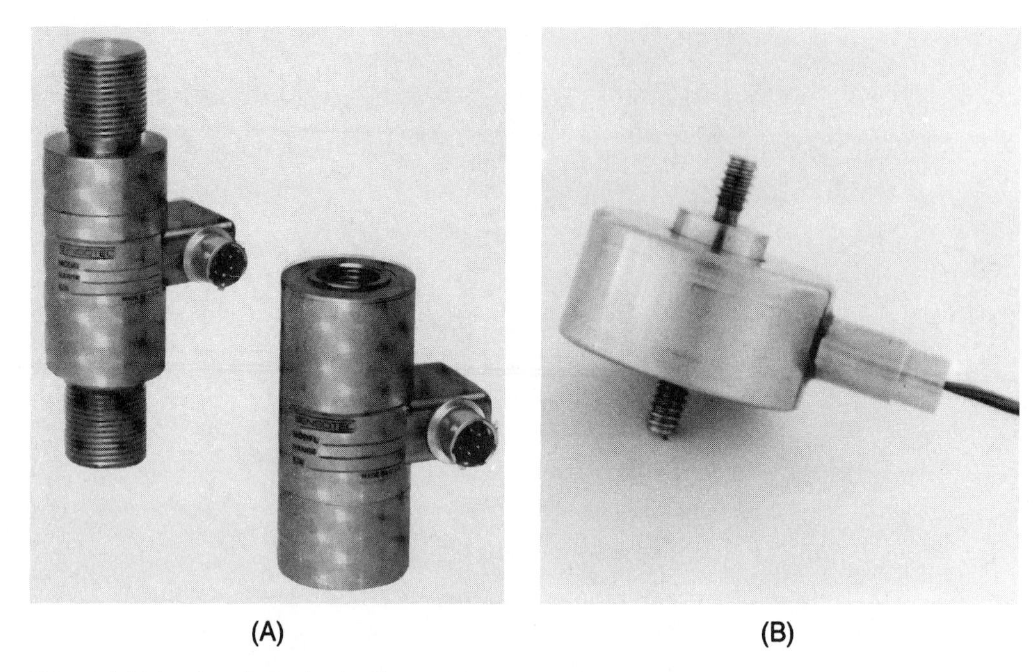

Figure 6-21 Semiconductor load cells (A) 5 VDC or 4-20 mA output range: 100-200,000 pounds; (B) Subminiature load cell for tension and compression: measurements range: 0-50 grams to 10,000 pounds *(Courtesy of Sensotec Inc.)*

change in electrical resistivity with an applied strain, Figure 6-21. This condition occurs whether the load has a dynamic or a static strain applied to the semiconductor. The deforming units to which the strain gages are bonded take on many configurations. Some are similar to bridge members; some are like yoke or ring devices; but all fall in the general configurations of either columns or beams.

ROTATIONAL SPEED SENSORS

Rotational velocity is called angular velocity. The most common unit applied to angular velocity is revolutions per minute (RPM). Industrial measurements commonly include speed data reported from rotating machinery that processes many different types of material. The data is taken from rotating shafts, wheels, gears, or pulleys.

Tachometers

Tachometers are used to measure rotational speed and are designed to utilize the phenomena of electromechanical generation, pulse induction, magnetic proximity pulses, photoelectric beams, stroboscopic lights, and solid-state electronics.

The Electromechanical Tachometer

Electromechanical tachometer generators are very reliable and have been widely used. They employ the concepts of the electric generator and motor. These devices use either a direct current voltage or an alternating voltage output. The voltage is proportional to the revolutions per minute of the equipment being measured.

The DC-type of tachometer generator contains a rotor made of stampings surrounded by a coil with lead wires coming out to a commutator. The rotor is mounted on ball bearings and turns inside a field coil. The field coil is also made of stampings and wound with wire to produce a magnetic field. The output of this small generator is proportional to the number of RPM of the drive shaft upon which it is mounted, Figure 6-22.

The AC-type of tachometer is designed like a two-phased induction electric motor. One stator winding is activated with an exterior AC source to *excite* the rotor. The rotor is short-circuited and induces a voltage in a second stator winding placed at ninety degrees to the first winding. The voltage induced in the second winding is proportional to the rotor's speed. The voltage output is scaled on the instrument as RPM.

Electro-magnetic Speed Sensor

Pulse frequency tachometers use a ferromagnetic toothed rotor mounted on the shaft whose speed is to be measured. A stationary mounted magnetic proximity sensor develops a pulse from each tooth as it passes by the sensor, Figure 6-23.

Figure 6-22 The tachometer generator *(Courtesy of Esterling Angus Instrument Corp.)*

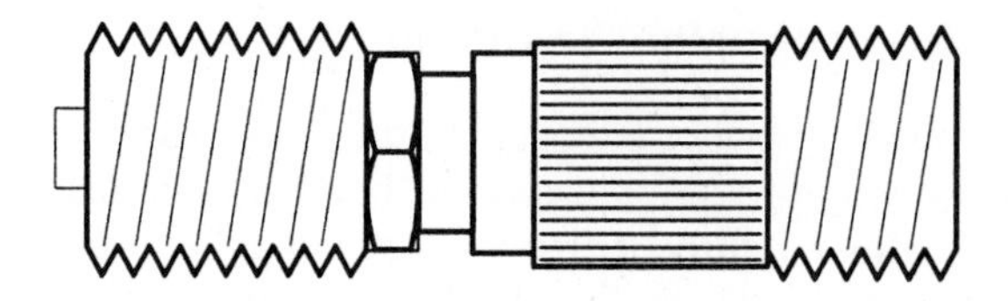

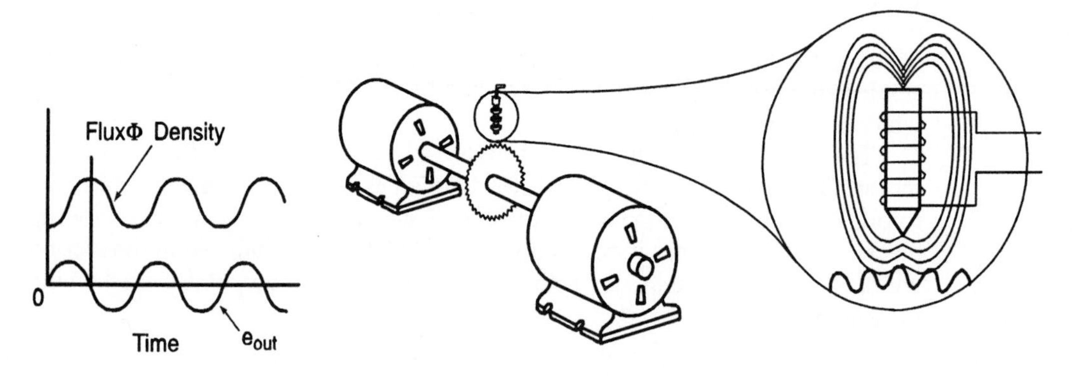

Figure 6-23 (A) Magnetic induction speed sensor; (B) Diagram of how it works *(Courtesy of Daytronics Corporation)*

Counting the pulses per unit of time results in a signal that can be conditioned electronically for an accurate reading. This method is popular because the problems involved with bearings, loads, shaft misalignments, or temperature changes do not enter into the measurements.

The above system may also function as a variable capacitance tachometer. The spindle has a reversing switch. A switch changes the polarity charging a capacitor that reverses twice with each revolution. The pulse from the capacitor is a function of the spindle's speed. The capacitor's output is displayed by a milliampere meter calibrated in RPM.

Magnetic Induction Transducers

The magnetic proximity speed switch can detect moving ferrous objects. These probes are induction pickup coils that provide a pulse when the ferrous object passes the probe. These devices are noncontacting and will provide an on/off type of output when the moving object is within a set distance from the probe. Turbines, conveyor belts, gear trains, and/or most rotating machinery can be monitored for speed or position with this device, Figure 6-24.

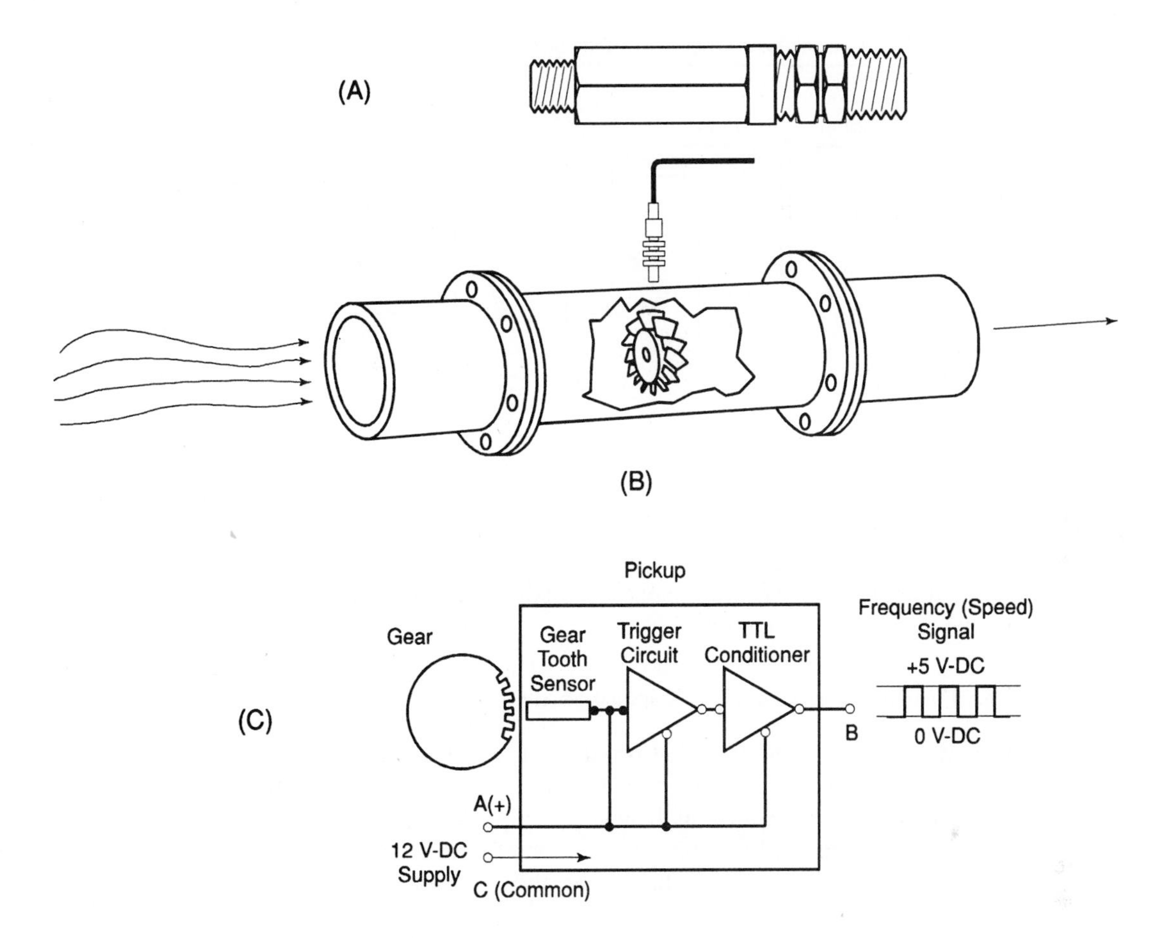

Figure 6-24 (A) Magnetic induction transducer; (B) Application; (C) Circuit for speed/position *(Courtesy of Daytronics Corporation)*

A Reflection, Holed, or Slotted Disc Photosensor

Speed can also be measured by a photosensor. These sensors can take a number of forms. One method is to focus a light source on a disc with alternate areas of light and dark segments. The light strikes the light areas and is reflected back to a photocell. The reflected light modulates a series of pulses from the photocell to a counting circuit that reports in RPM, Figure 6-25. The same principle can be used by passing the light beam through a disc with slots cut near the edge for the beam to pass through and strike the photo receiver cell. A count is the result again, but a stronger signal results because the beam passes directly to the photo-receiving cell. (This system is also used for position. Counting the number of openings a disc has rotated delivers the distance moved for a machine part fastened

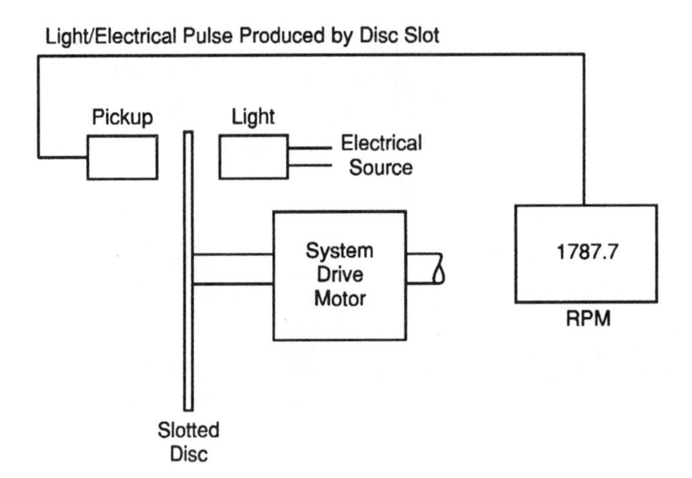

Figure 6-25 Speed or position applying a holed or slotted disc

to a lead screw. In this fashion, a position or distance a machine tool table is moved can be accurately measured.)

The Digital Stroboscope Tachometer

The stroboscope is a device that produces an electronically controlled flashing light with the aid of flash rate control circuits. The speed of flashing is adjusted until the flashes synchronize with the moving machinery and the RPM can be read off of the instrument, Figure 6-26. The operator must be aware of harmonic or subharmonic images that might appear and would result in false tachometer readings.

Solid-state Devices

Solid-state electronic devices apply light-emitting diodes (LED) to perform switching operations. With a small change in voltage (0.6 microvolts), these semiconductor technology devices allow a current to flow in one direction only. There is essentially no current flow when the direction of the current is reversed. In addition, this diode has the property of emitting light when it conducts. The light, in turn, can be used to activate a phototransistor that performs the switching function for an on/off counting operation, Figure 6-27.

The phototransistor is a light-sensing device that permits the flow of current when subjected to light and, in turn, activates solid-state switches or counting devices. It has a high electrical resistance for current in one direction and a low resistance in the opposite direction. The NPN or two-junction phototransistor is made in a single crystal. When the phototransistor is in a circuit and light falls on the NPN junction, the device conducts a current, Figure 6-28. The phototransistor

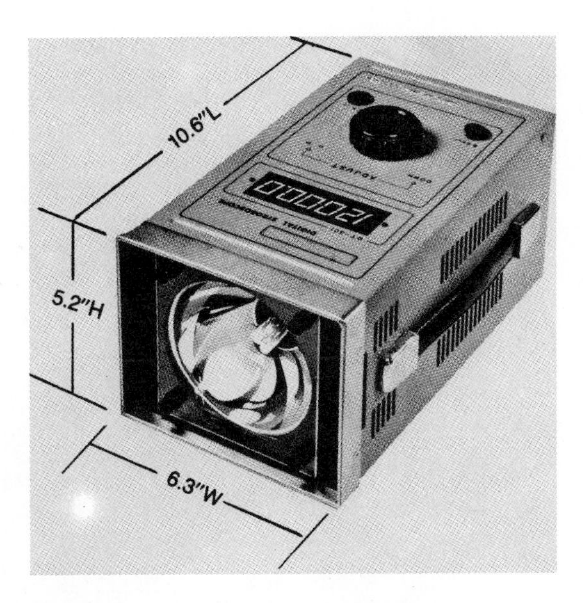

Figure 6-26 Digital stroboscope tachometer *(Courtesy of Electronic Equipment Co. Inc.)*

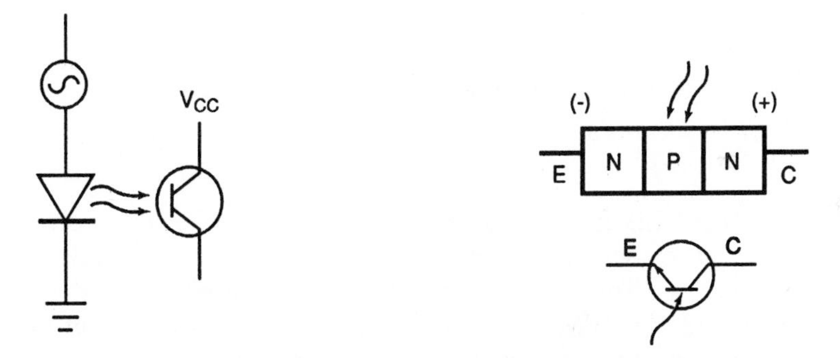

Figure 6-27 Light-emitting diode (LED) and phototransistor that performs counting and switching operations

Figure 6-28 Two-junction NPN phototransistor

is small, rugged, has a long life, and has greater sensitivity than the older phototube.

Microprocessor Technology

Microprocessor technology brings precision to the measurement of rotational speed by computing the time interval between successive pulses received from a rotating target. These pulses are compared against a megahertz internal crystal timebase. The resultant speed measurement is displayed directly in RPM on a

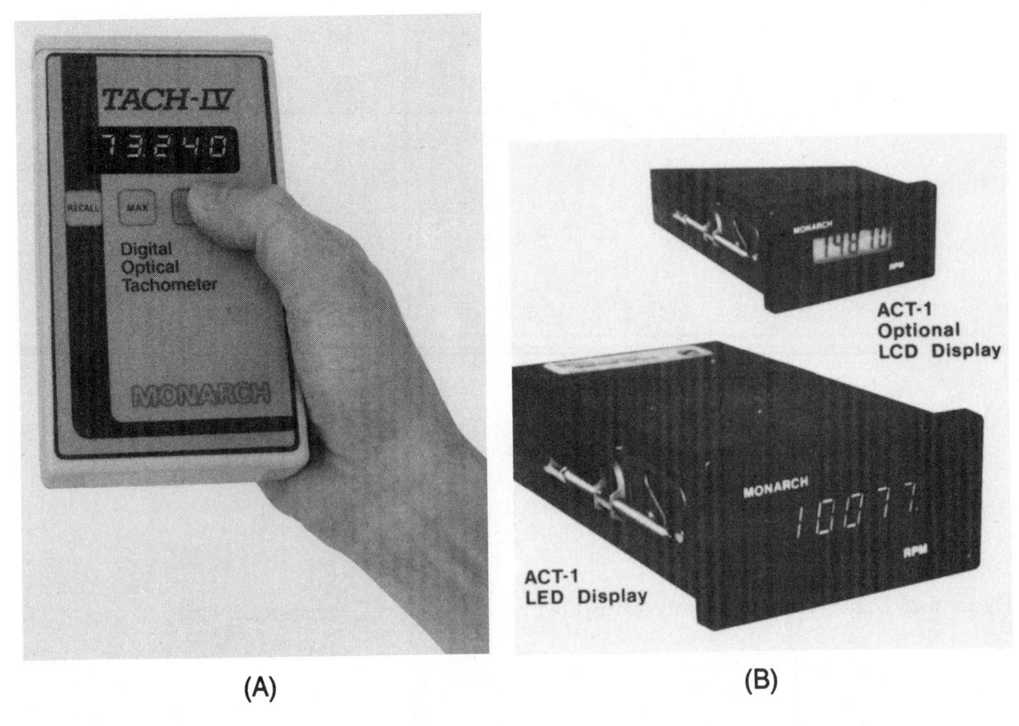

Figure 6-29 (A) Digital optical tachometer; (B) Panel-mounted digital tachometer *(Courtesy of Monarch Instrument)*

digital LED or LCD display. These tachometers are designed to receive different sensors — pulse signal from optical (reflective) sensors, proximity sensors, or magnetic sensors. The digital optical tachometer senses reflected light into the instrument and the RPM is electronically calculated. The output is displayed digitally on a panel or hand held readout, Figure 6-29.

POSITION DISPLACEMENT AND MOTION TRANSDUCERS

A position is a place, a location, or occupied site of an object. This may also refer to the orientation of a workpiece, robot arm, spot weld, or location of a hole to be bored in a casting.

Displacement is the difference between two locations. This can be a dimension, an offset, or a variance with what is required or needed. The difference may be a scaled measurement, a pressure, a temperature, a flow, or a voltage; but in any case it will be compared to some standard. The standard may be a required location or a set-point of any variable.

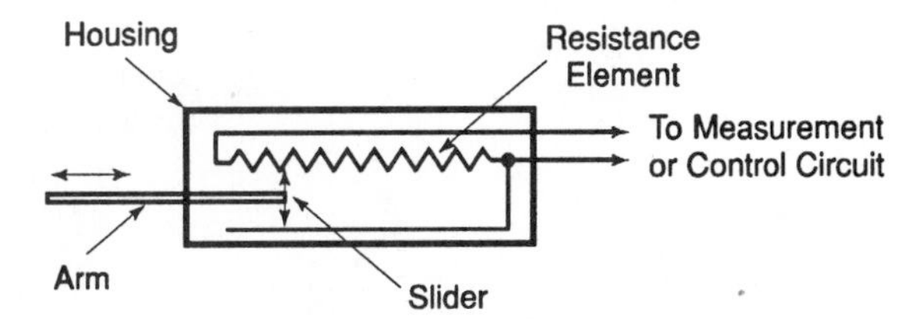

Figure 6-30 Potentiometric displacement transducer

Motion refers to the movement to a new location or position. Objects are detected as moving or have not moved, are present or not present.

The sensors used to make these distinctions are many, but the major groups are these: linear and rotary potentiometers, linear variable differential transformers, linear variable reluctance transducers, resolvers, synchros, pulse transducers, and *proximity* sensors.

Displacement Sensors

The linear and rotary potentiometers are displacement transducers that have been proven over the years and provide reliable data. The potentiometric devices convert linear or rotary displacement into resistance values or directly into voltage or amperage changes. The resistance values are very accurately transposed with the application of Wheatstone or other bridges. The voltage/amperage changes can be transmitted to electrical or electronic circuits to report changes to readout instrumentation, Figure 6-30. These sensors do suffer from slider and resistance wire wear as well as from environmental corrosion. The wire-wound resistance coils can cause fine resolution problems as well as electrical noise generated by the slider contact.

The Linear Variable Differential Transformer

A linear variable differential transformer (LVDT) is designed with three coils surrounding a moveable central core that provides a magnetic flux. The central coil is the primary coil and provides the excited magnetic flux produced by an AC source. The secondary coils receive the magnetic flux from the primary coil and an inductive voltage results. The secondary coils are usually connected in series-opposition so that when the core is in the central position, the output of the coils is zero. When the core is moved by the device to be measured, the voltage of one coil will increase, while the other coil's output will decrease. The voltage amplitude is linear with the displacement of the moveable central core, and also provides a phase shift that relates to the direction of movement of the core/sensor. The output of the secondary coils can be changed to a DC voltage by designing a circuit including diodes that will result in a bipolar output voltage, linearly proportional to the displacement, Figure 6-31.

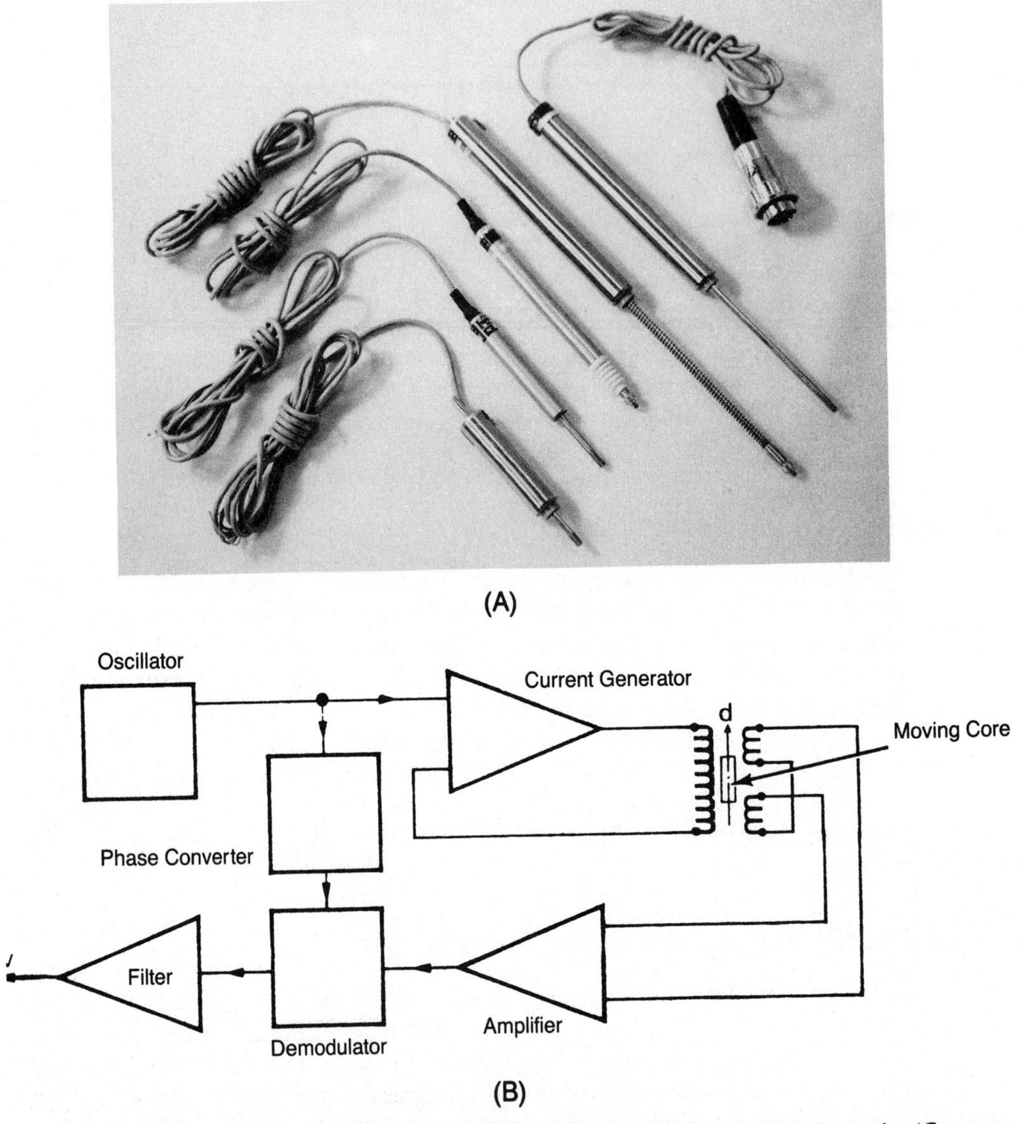

Figure 6-31 (A) Linear variable differential transformers; (B) Diagram of supporting electronics *(Courtesy of AEMC Corporation)*

The Linear Variable Reluctance Transducer

The linear variable reluctance transducer (LVRT) is a device that uses a variable reluctance to control the setting up of flux in a magnetic circuit. In an electrical inductive circuit, a ferromagnetic core is used to control magnetic permeability; and in a parallel way, in a magnetic circuit, reluctance is in opposition and may

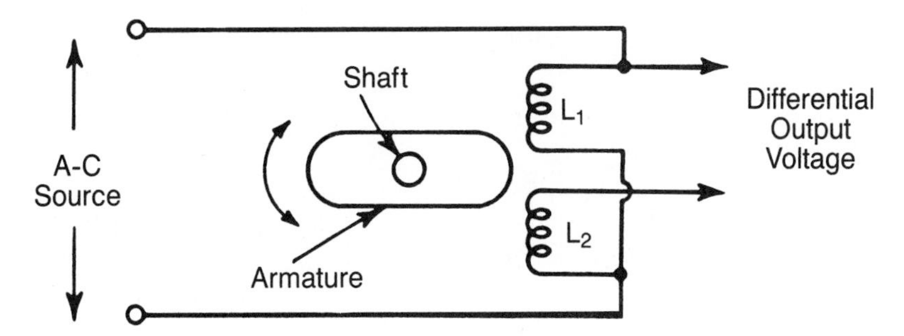

Figure 6-32 A rotary variable reluctive transducer

be used to control magnetic permeability. This alters the self-inductance of the secondary coils by a movement of the primary coil assembly. In this device, the moveable coil varies the flux density that changes the magnetic flux coupling with the secondary coils.

A coil movement of as little as three-thousandths of an inch will develop an AC signal output that can be measured by the circuitry. A reluctance transducer is successful because it is less affected by temperature changes than other devices, Figure 6-32.

A Resolver

A resolver is a motor-like device used to transmit the angular position of machinery data. It is a design of a rotary transformer with a primary winding on the rotor attached to the machine being controlled. The stator is wound like a motor and is the secondary in the frame of the resolver. The resolver functions with two stator windings ninety degrees apart. As the shaft is turned, it causes the voltage output of the stator to change with the changing angle of rotation. The ratio of these two outputs can be attached to another unit for an analog readout that will track the exact position of the resolver. Frequently, a digital converter is included in the assembly. A digital signal is sent directly to a microprocessor for position control of equipment.

A Synchro

A synchro is designed very much like a resolver, except that the stator has three stator windings each one hundred twenty degrees apart, Figure 6-33. As the rotor or primary coil is turned, the voltage induced in the secondary or stator is changed and these voltage changes are transmitted to another unit. These voltages are applied to a second unit's stator that will position the second unit's rotor to the same angular position as the original sending unit's position. The voltage output varies with the position of the rotor. The output can be converted into a digital value and applied to a microprocessor for robot or other positioning.

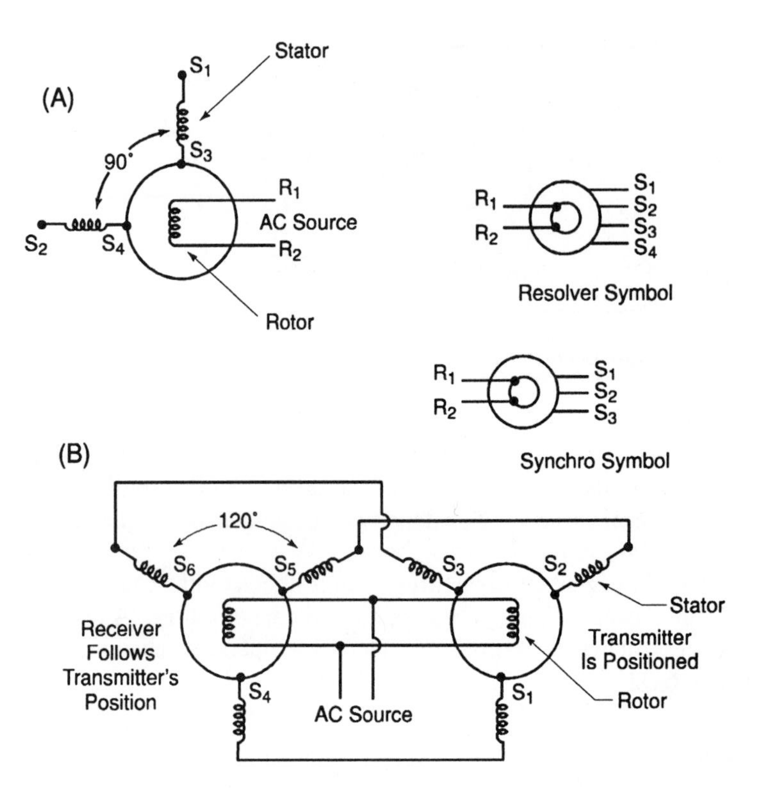

Figure 6-33 (A) A resolver diagram; (B) A synchro diagram

Three-member synchros consisting of a transmitter, differential, and transformer synchros are often employed to synchronize various parts of rotating machinery that can provide angular adjustment between rotating members such as drive shafts.

Synchros with solid-state digital converters provide servo loops with feedback and track all movement of the equipment. A second design samples the signals from the synchros, holds the data, and performs conversions to change or correct deviations in positions with the aid of a microprocessor.

Pulse Transducers

Encoders

Absolute encoders provide a unique binary address for each position. Incremental encoders provide a pulse for each incremental change of position; they start from a point and count to the present position.

The Absolute Encoders - An encoder will digitize linear or angular space positions. The spacing can be laid out on discs, plates, drums, or wheels that move with the part being measured. These spaces are digitized by inductive devices, optical devices, or mechanical devices. The output of the solid-state devices can be a series of pulses or counts, coded pulses, binary coded, or gray-to-binary codes. A gray code is a method used to eliminate position error by changing only one bit at a time of a binary or BCD code number.

The Incremental Encoder - The incremental encoder provides positional, directional, and velocity measurements to the controller. The incremental encoder has a light source that passes through a disc that contains alternate transparent and opaque divisions. As the disc rotates with the shaft, the sensor receives a series of light bursts on the optical sensor. These bursts of light are converted into electrical pulses by the sensor. The pulses represent a digital output, and the count of the pulses is compared with the count that was originally programed for that position. If the count is different, the controller will activate the feedback loop and the position will be corrected, Figure 6-34.

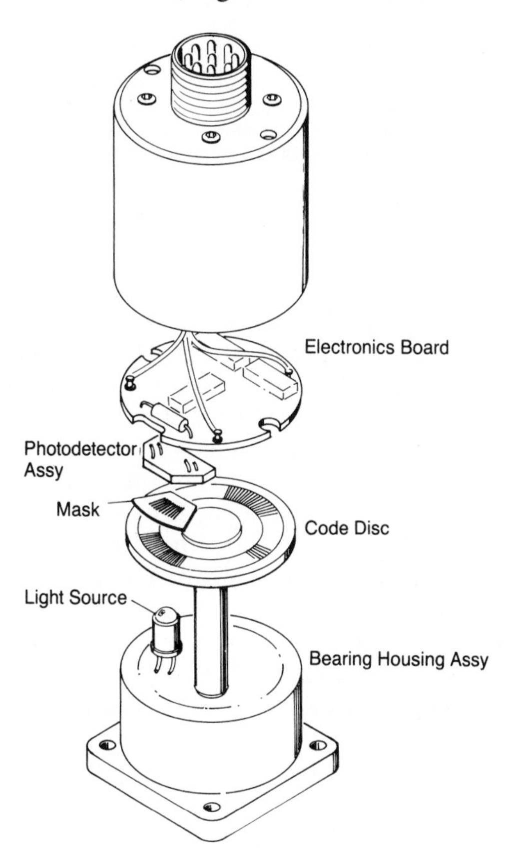

Figure 6-34 Incremental encoder *(Courtesy of BEI Electronics, Inc. Industrial Encoder Division Motion Systems Co.)*

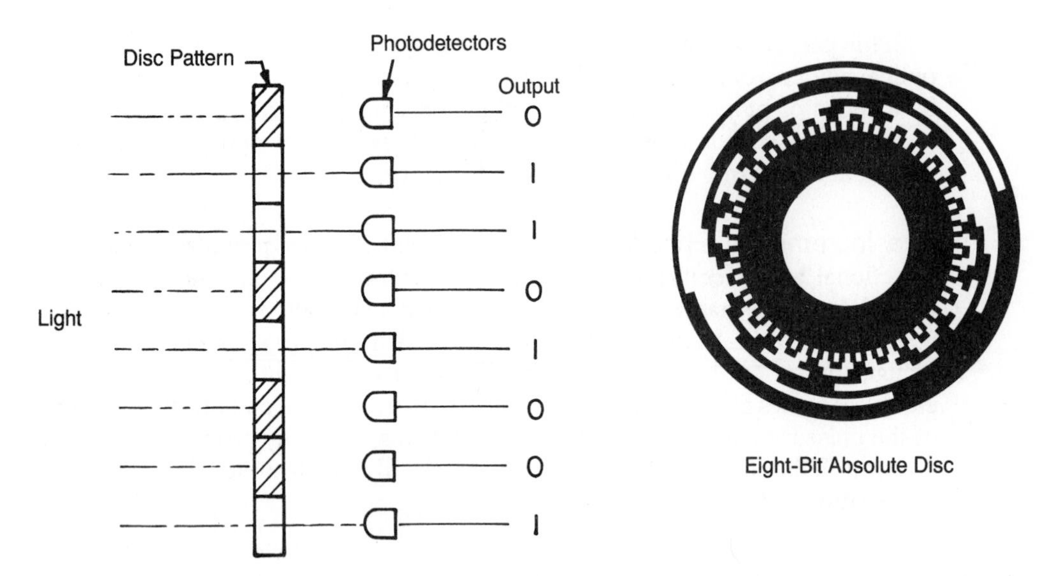

Absolute encoders are used in applications where a device is inactive for long periods of time or moves at slow rates, such as flood control, telescopes, cranes, etc.

Absolute encoders are capable of using many thousands of different codes but the most common are gray, natural binary and binary coded decimal (BCD). Gray and Natural Binary are available up to a total of 256 counts (8 bits).

Figure 6-35 Eight-bit output—absolute encoder *(Courtesy of BEI Electronics, Inc. Industrial Encoder Division Motion Systems Co.)*

The Construction of the Absolute Encoder - An absolute encoder employs a series of eight channels in concentric rings. These channels produce an eight-bit output signal of a position. The light passes through the disc as before, except that in each position, its channel represents a digital bit. All the channels produce a code by a pattern of light or no light falling on the sensors. The absolute encoder provides a "whole word" output with a unique code pattern representing each position, Figure 6-35. The codes employed by these systems are a binary, BCD code or a gray code. A gray code is a method used to eliminate position error by changing only one bit at a time of a binary or BCD code number.

The Incremental Modular Encoder - Pulse transducers employ digital electronic counters and measuring instruments that are designed for pulse generators. These generators frequently are two sets of unmodulated infrared photosensors that produce a signal for pulse-difference and position counters. If the shaft of the encoder is turned clockwise twenty-five pulses per turn, they will be added to the digital counter. If the shaft is turned counterclockwise twenty-five pulses per turn, these will be subtracted from the digital counter, Figure 6-36. These transducers are employed for positioning, tachometer metering, and temperature and frequency metering.

Figure 6-36 Incremental modular encoder mounted on a motor *(Courtesy of Disc Instruments, Inc.)*

The Proximity Sensor

Proximity sensors are designed to make use of the principles of inductive devices, capacitive devices, and optical devices. An inductive sensor is a sensor that gives a signal when a metallic object approaches the surface of the sensor. These sensors are supplied in a number of shapes: the I-shape senses objects approaching the front surface of the sensor; the U-shape gives a signal when a metallic object is inserted into the air gap; and the O-shape gives a signal when a metallic object is passed through the hole, Figure 6-37.

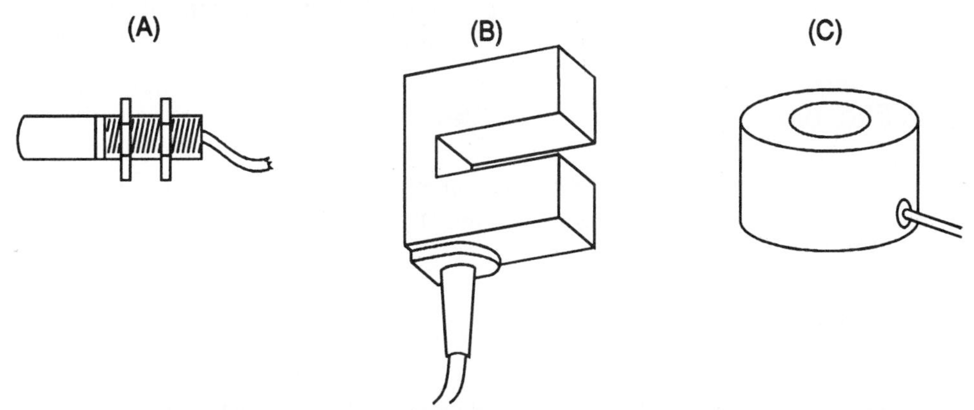

Figure 6-37 Inductive sensor shapes: (A) I-type sensor; (B) U-type sensor; (C) O-type sensor *(Courtesy of Electromatic Controls Corp.)*

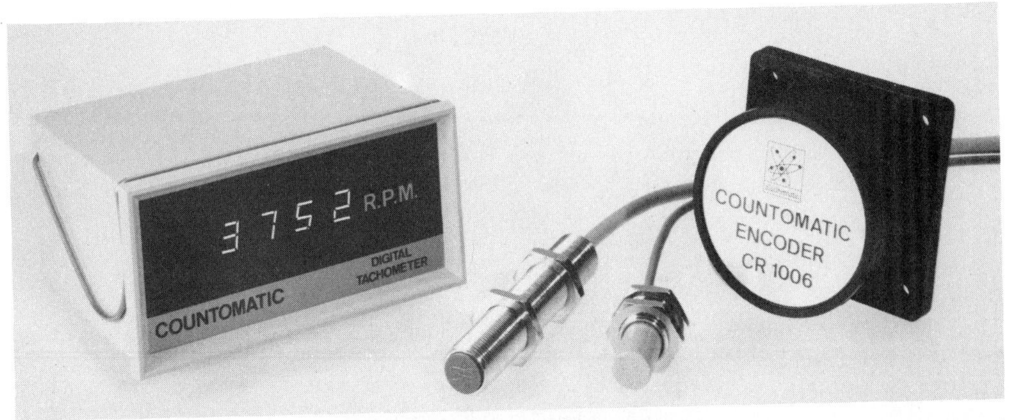

Figure 6-38 A tachometer with an encoder, inductive, or capacitive type input. A precision tachometer used in conjuction with an encoder, which gives 60 pulses per revolution and has a sampling time of one second. The RPM is displayed with a resolution of one RPM with a measuring range of 0-9999 RPM. *(Courtesy of Electromatic Controls Corp.)*

Solid-State Counting and Measuring Modules

Solid-state counting and measuring instruments are supplied in a number of functional modules depending on the application of the unit, e.g., totalizing counters, counting and measuring functions. Totalizing counters are designed for inputs by contact or transistors for measuring amperage or voltage or encoding for pulse counters, Figure 6-38. Instruments with these capabilities provide functional modules for ammeters, pulse counters, pulse difference/position counters, frequency meters, digital timer controls, pulse counter with preselections, stop watches, tachometers, synchros, clocks, and thermometers. The encoder pulse counters provide the great flexibility to quantify many industrial variables into an instrument that will totalize the input data. The sensors for the digital tachometer illustrated will receive data from an encoder, inductive, or capacitive sensors.

ACCELERATION AND VIBRATION TRANSDUCERS

Acceleration is the rate of change of velocity. Since velocity is the rate of change of displacement, then acceleration, velocity, and displacement are all related by time.

Acceleration Transducers

To measure acceleration, a group of sensors have been designed. They use a mass that is acted on by the rate of change of velocity of the object being measured. This resistance mass loads the sensing device. Five of these groups of sensors are described: *seismic* mass, linear variable differential transformer, capacitive, piezoelectric, and strain gage.

The Seismic Mass

The seismic mass was an early accelerometer. It uses a suspended mass and a stylus to record earth movements. This device, known as a seismograph, measures the intensity of earthquakes. Today, a seismic mass accelerometer may have a strain gage attached to the mass mount and a motion, vibration, or acceleration is sensed. The changes in the strain gage's resistance provide an accurate and reliable signal of the forces supplied to the instrument, Figure 6-39A and Figure 6-39B.

The Linear Variable Differential Transformer

The linear variable differential transformer (LVDT) and the linear variable differential reluctance transducer have been discussed earlier. These LVDTs are applied in acceleration sensing by using a seismic mass to move or change the position of a core or coil in relation to the secondary coils. The change in inductance produces a strong and reliable signal. Because alternating currents are utilized in this device, it can be part of an oscillating circuit for applications of telemetry.

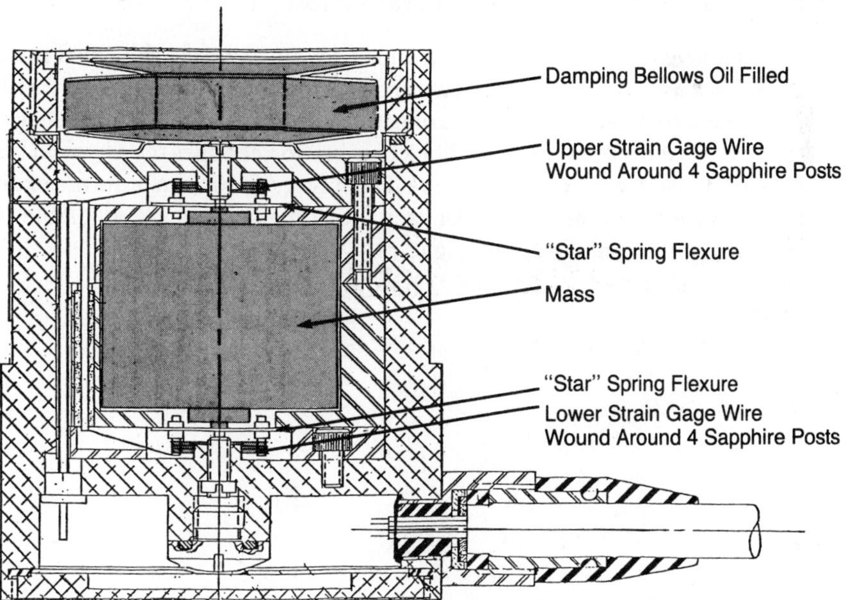

Figure 6-39A Schematic of a seismic mass applied to a strain gage to produce an accelerometer. Accelerometer cross section showing strain gage sensor. *(Courtesy of Transamerica Delaval Inc./CEC Instruments Division)*

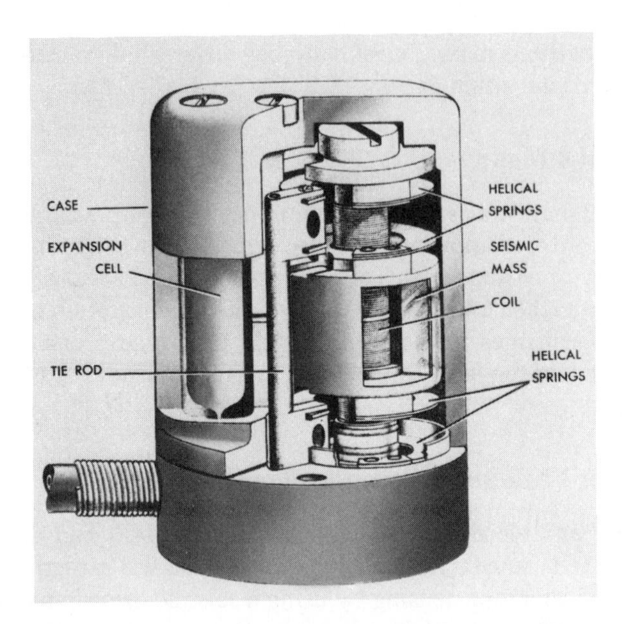

Figure 6-39B A cutaway view of an inductive vibration sensor. *(Courtesy of Transamerica Delaval Inc./ CEC Instruments Division)*

The Variable Capacitance Accelerometer

The variable capacitance accelerometer sensor consists of a thin, stiff metal disc and flexures assembled between two fixed insulated electrodes. The position of the seismic disc in relation to the two electrodes is proportional to the acceleration. The capacity of the unit varies as shock or vibration changes the physical distance between the electrodes and the flexure disc, Figure 6-40. The units are excited by external current supplies. These capacitance-type sensors are available in full-scale ranges from plus or minus 2g to plus or minus 600g.

Figure 6-40A High output accelerometer designed to measure shock, vibration and impact. A capacitance high-output linear accelerometer *(Courtesy of Setra Systems Inc.)*

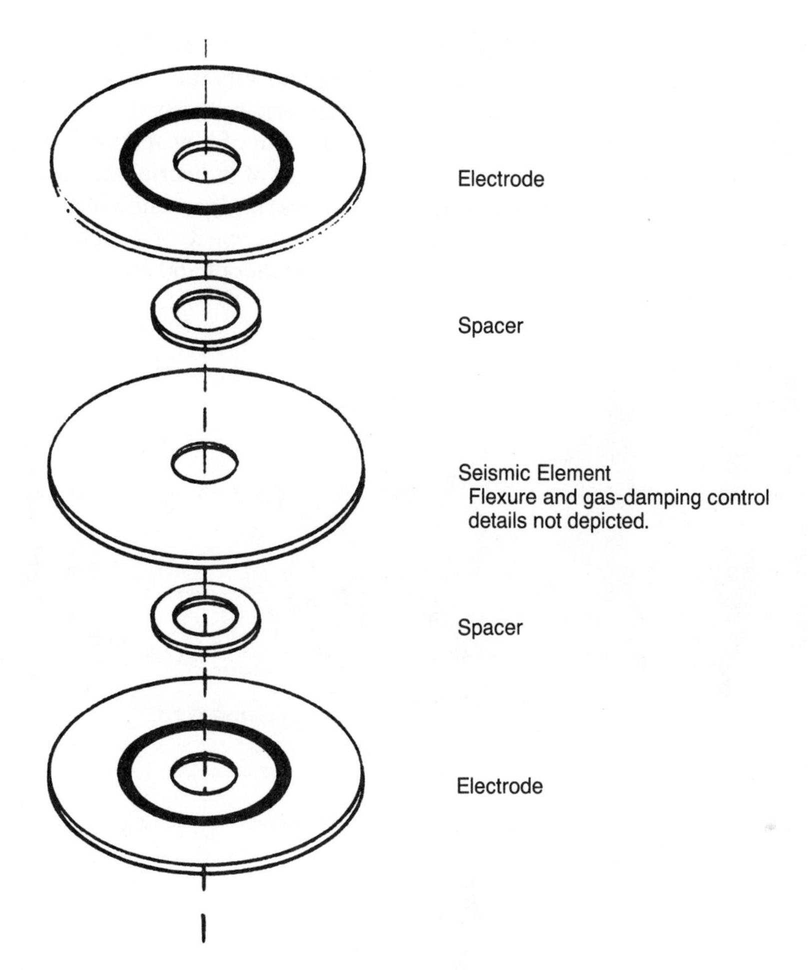

Figure 6-40B Diagram of a variable capacitance accelerometer *(Courtesy of Setra Systems Inc.)*

The Piezoelectric Accelerometer

Piezoelectric accelerometers generate an electromotive force (EMF) when quartz or synthetic crystalline materials of ceramics are subjected to load changes. A tungsten mass provides the load change when it receives a movement. This accelerometer is used only in a dynamic situation to record movements. The crystal generates an electrical charge proportional to the varying force that excites the crystal, Figure 6-41. Accelerometers of this type are designed in two configurations: the first in which the crystal functions under compression, and the other in which the crystal is subject to shear forces.

The Precision Quartz Accelerometer - Some accelerometers include an inverted quartz sensing element to minimize base strain sensitivity and built-in microelectronic signal conditioners, Figure 6-42A and Figure 6-42B. These devices measure the acceleration of shock and vibration motions over a wide range (from 1g to 500g) and under adverse environmental conditions. Quartz accelerometers are designed for the measurement of low and medium frequency vibration and shock motion on heavy structures of industrial machines, machine tools, vehicles, suspensions, engines, buildings, bridges, and vibration or impact testing machines. It transfers the acceleration aspect of shock and vibratory motion into a voltage signal compatible with readout and analyzing equipment.

Figure 6-41 A group of quartz accelerometers *(Courtesy of PCB Piezotronics, Inc.)*

Figure 6-42A A general purpose quartz accelerometer with built-in amplifier *(Courtesy of PCB Piezotronics, Inc.)*

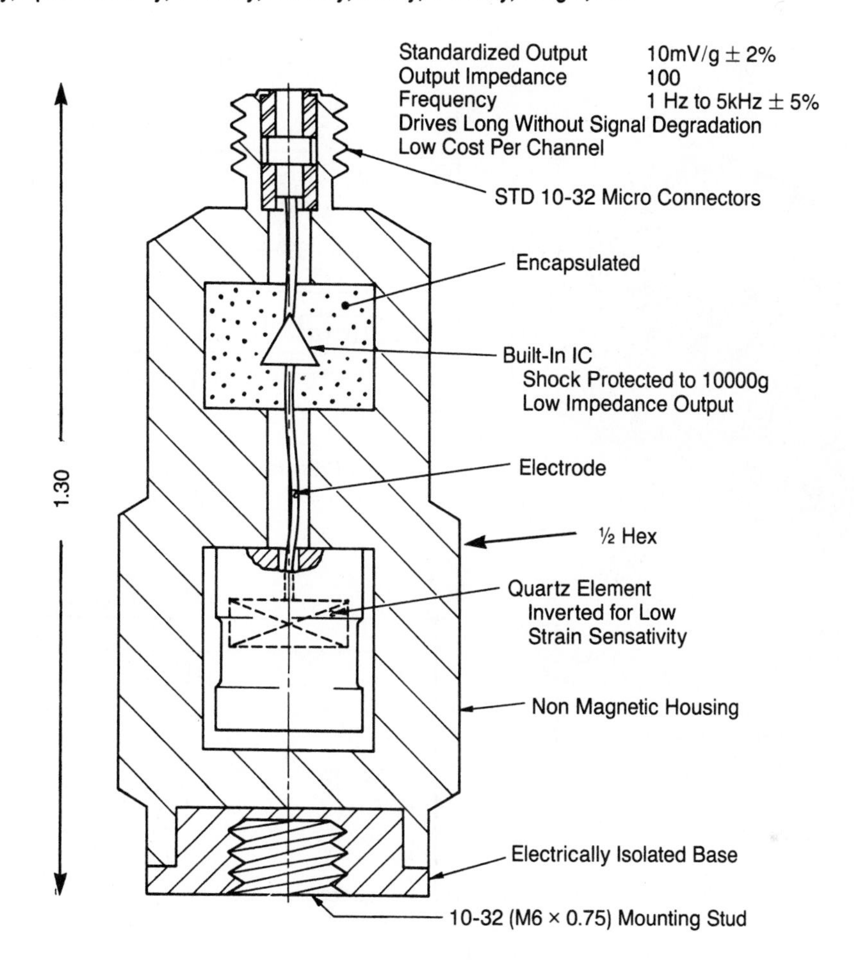

Figure 6-42B Diagram of a quartz accelerometer with built-in amplifier *(Courtesy of PCB Piezotronics, Inc.)*

The Strain Gage Accelerometer

Strain gage accelerometers function by a change of resistance as the shock distorts the strain gage, Figure 6-43. Acceleration measured by strain gages falls generally into two large groups of design: the unbonded wire strain gage and the bending type of strain gage. The wire types are designed with a suspended mass held in place by four strain gages. The mass is mounted with a container holding a dampener capable of moving with the shock. The strain gages are mounted externally and are wired so that the strain gages become part of a Wheatstone bridge. The change of resistance in the bridge provides the acceleration signal. This type of instrument responds to frequencies from zero to one thousand hertz and is stable over a wide range of operational temperatures. This type of

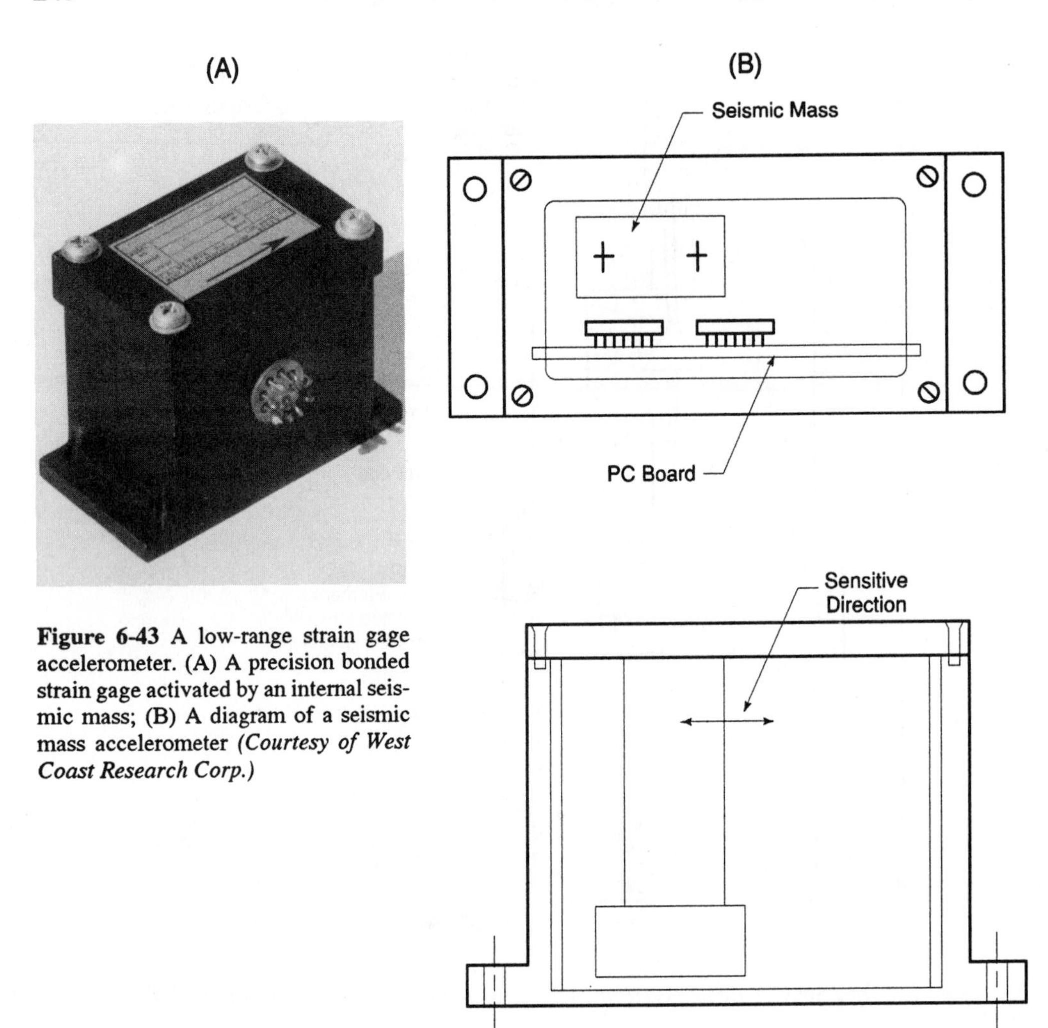

Figure 6-43 A low-range strain gage accelerometer. (A) A precision bonded strain gage activated by an internal seismic mass; (B) A diagram of a seismic mass accelerometer *(Courtesy of West Coast Research Corp.)*

accelerometer is useful in measurements on machining, vehicles, aerospace and marine structures, test stands, and seismic environmental measuring equipment.

The Bending-Type Accelerometer - The bending type of accelerometer is constructed so that the sensing member is a cantilevered mass fastened to two stationary, mounted strain gages. When a force is received, one of the gages is in compression and the other is in tension. These strain gages are connected so that they are one-half of the parts of a Wheatstone bridge that provides a variation of resistance with the application of acceleration, Figure 6-44.

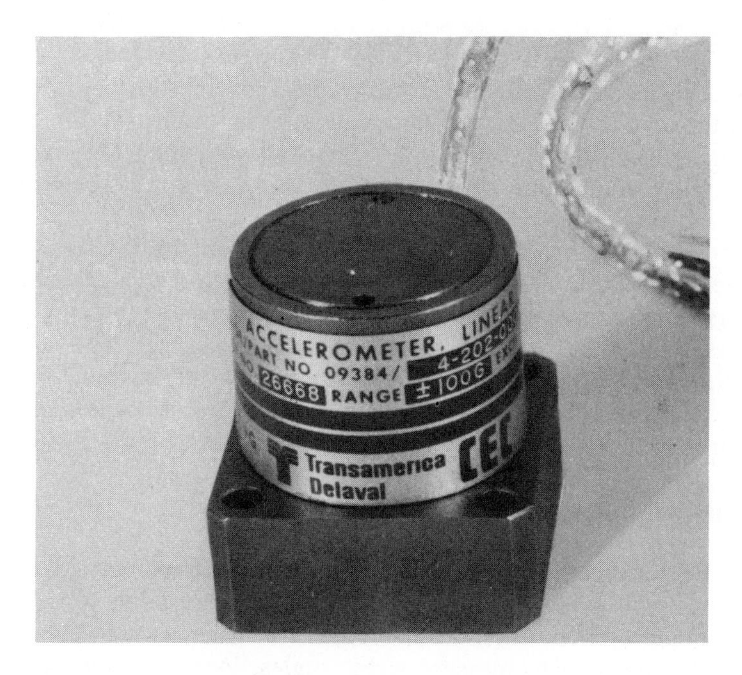

Figure 6-44 Bending-type strain gage accelerometer *(Courtesy of CEC Instrument Division)*

DIMENSIONAL TRANSDUCERS

The measurements fall into two large groups or classifications: those made by contact instruments and those made by noncontact instruments. The contact instruments include tools and standards of the direct reading instruments such as steel rules, tapes, square sets, rule depth gages, calipers, telescoping gages, small-hole gages, vernier calipers, vernier height gages, vernier bevel protractors, micrometer calipers, inside micrometers, micrometer depth gages, surface plates, gage blocks and sine bars, solid gages, microswitches, and others. These instruments make contact with the surfaces, distances, or positions that are to be measured.

Many of the contact measuring instruments are from the discipline of metrology and are concerned with laboratory and machine shop size measurements of tools, machines, and parts. These will not be discussed in this text. Measuring instruments that deal with the precise measurement of a distance moved (either for a short displacement or a very long distance), location, angular displacement, and flatness or straightness vital to robots and integrated manufacturing are the concern of this text.

Dimensional Transducers of Distance

Measuring instruments that do not make direct contact with the surfaces or spaces are referred to as noncontact instruments, such as: toolmaker's microscopes, air gaging, optical comparators, interferometers with monochromatic light and optical flats, autocollimator and optical tooling, laser interferometers, and proximity sensors.

In the high technology manufacturing industries, the trend is moving toward the noncontact distance and displacement transducers. Optical measuring technology can be combined with solid-state digital technology to provide readouts that interface with microprocessors. The proximity sensors have an advantage in that they can be smart sensors. They can provide information about the speed approaching or leaving a target or coming to a size or position and report this information in real time to a microprocessor and provide precise machine control.

The Precision Measuring Microscope

The precision measuring microscope is an optical instrument that does not contact the work or its location, Figure 6-45. It can measure in the X and Y planes to measure the width of lines, conductors, or the space between them. The microscope can be used to measure sizes, angles, locations, and positions. The instrument will include a digital readout and a table mount. In special cases the microscope can be mounted on a machine.

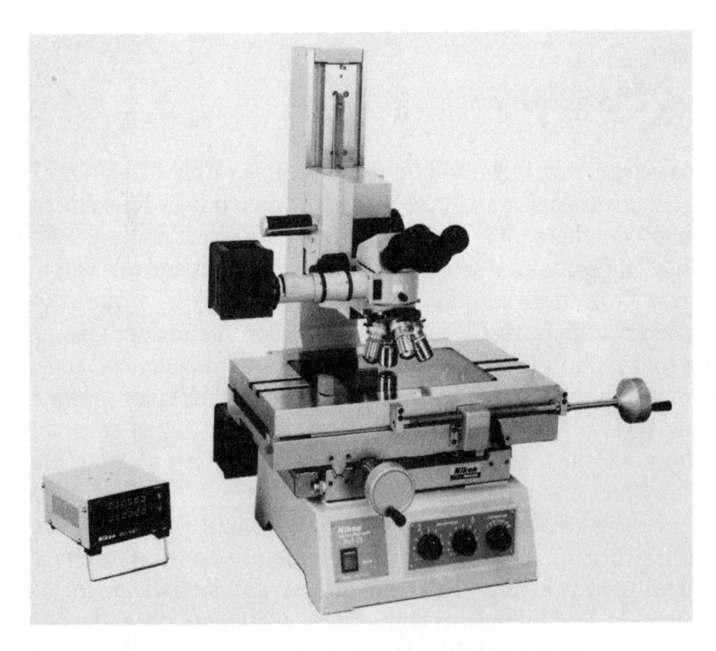

Figure 6-45 UM-3 universal measuring microscope *(Courtesy of Nikon Inc.)*

Air Gaging

Air gaging and comparators function on the same concepts as pneumatic control devices. Regulated air supply escaping from a small orifice will produce a back pressure that is linear with the distance from the orifice and the adjacent reference surface. There is no physical contact with the reference surface. Air gaging can be applied to measurements for checking roundness, size of deep holes, tapers, waviness, irregularity, concentricity, and size or distance from a surface.

Air devices are capable of measuring to very accurate dimensions and can be applied to the location of machine tool surfaces for location and positioning of parts in manufacturing. Pneumatic technology is capable of producing proximity functioning devices for high technology manufacturing systems, Figure 6-46A-C.

The Optical Comparator

The optical comparator projects either a reflection or shadow of an object placed before its system of lights, mirrors, and lenses. A master chart is placed on the screen and compared with a superimposed shadow projected on the screen. The images are magnified so that contours, dimensions, tolerances, threads, gears, and cutting tools can be examined and measured. The integration of light, shadow, fiber optics, and solid-state electronics provide a number of procedures for the automatic measurement of size, location, and position in manufacturing systems.

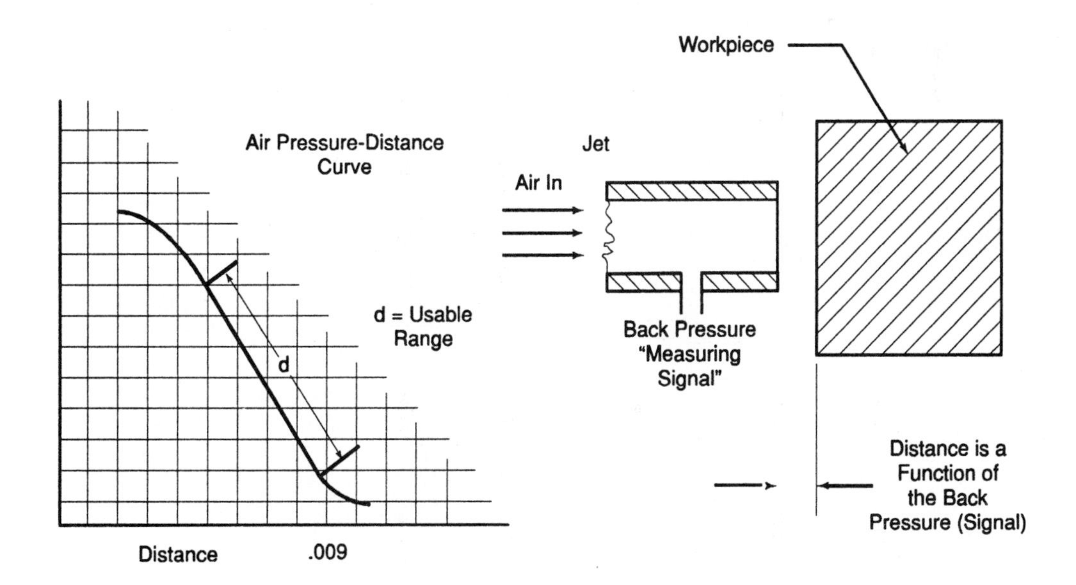

Figure 6-46A Principles of air gaging: A linear back pressure produced by orifice distance

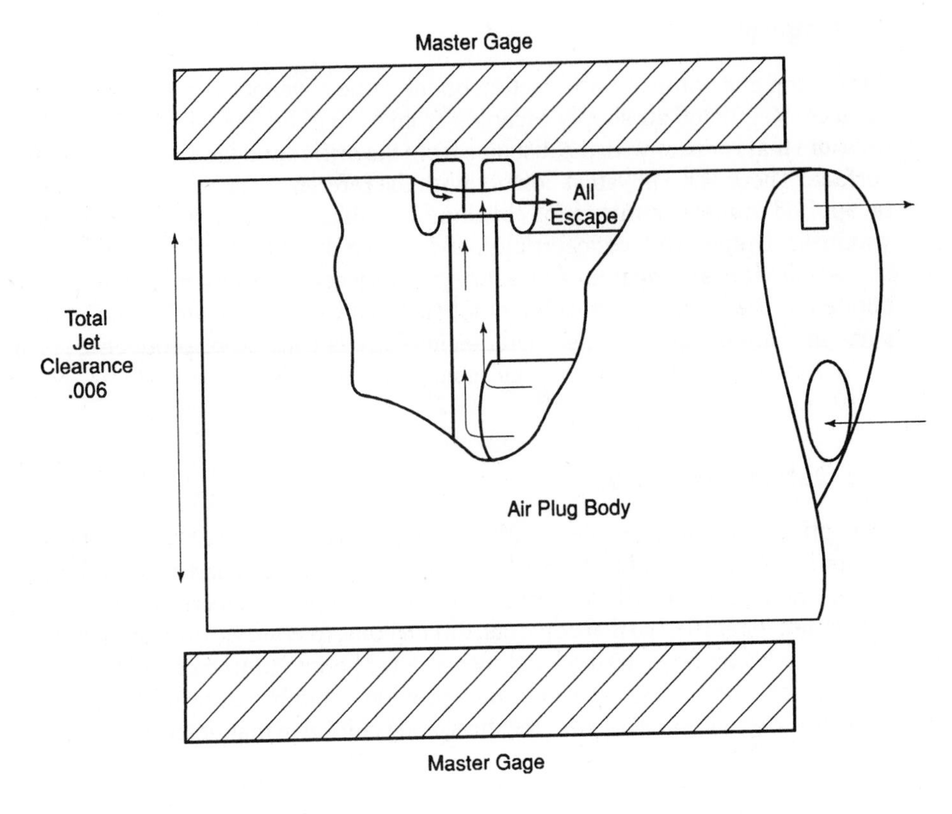

Figure 6-46B Principles of air gaging: Air gage plug back pressure calibrated to a master gage

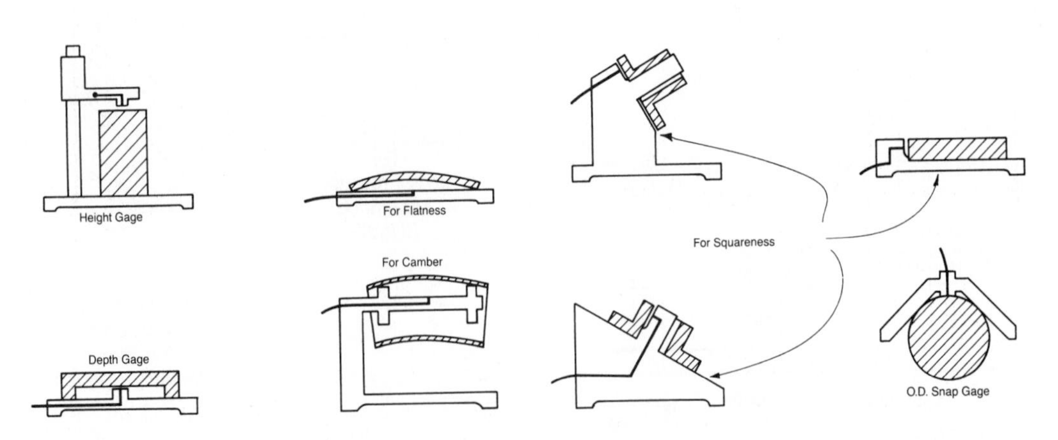

Figure 6-46C Principles of air gaging: Applications of air gaging

Interferometry Measurements

Interferometry measurements are made with the application of the physical phenomenon of light wave interference. These measurements are very accurate. They measure in the area of microinches. The measurement is actually based upon the primary International System of a wavelength, krypton 86, a red-orange line of colored light. A light source with a wavelength measured to this standard is used with a gage block to make a comparative measurement to the object under measurement. The item to be measured and a stack of gage blocks are wrung to a base called a "toolmaker's flat" made of steel, quartz, or a special glass disc. An optical flat made of quartz or special glass, that may vary depending on the size required (from two inches to ten inches in diameter, and from five-eighths of an inch to one and five-eights inches in thickness) is placed over the gage block and the item being measured. This will measure small deviations between the height of the gage block and the object being measured. The surfaces of the optical flats and toolmaker's flats used in the measurements are flat and parallel within one- to two-millionths of an inch.

A monochromatic light source is directed onto the optical flats. This light has a known wavelength. In the United States, it is usually from helium gas in a fluorescent tube and has a wavelength of 0.0000232 inches. This light passes through the optical flat to the surface of the gage block. There will be light reflected back to the eye from the measuring surface of the optical flat, as well as light reflected back from the surface of the gage block. If the total area of the gage block is in contact with the optical flat, certain of these two rays of light waves interfere or cancel each other, producing regular and parallel dark bands. In the case where the optical flat does not rest totally in contact with the gage block, there will be a small wedge of air between the two surfaces. The wedge of air increases the time of travel of the light rays between the reflection of the optical flats and the gage blocks. The same ray will be reflected back from the gage block out of phase with the original ray and will cause interference with the ray returning from the optical flat's measuring surface, and the rays will result in a dark band. If the work has a high area or a rolling surface, the dark bands will curve toward the line of contact indicating the high areas. The measurement is obtained by estimating the curvature of the band as a fraction of the band's total width and multiplying that fraction times the wavelength of the interference band. As an example, one-third a width of a band's curvature times the wavelength would be one-third of 11.6 millionths of an inch, which is equal to 4 microinches, Figure 6-47.

Optical Tooling

Optical tooling developed out of the need for the alignment of large fixtures and tooling such as the alignment of air frames, wings, ships, and large machine tools. It is an extension of surveying with its use of the telescope, theodolite, jig transit, mirrors, tooling tapes, and a collimated light source (the autocollimator). The concept of the optical lever was employed so that when a collimated light ray was sent down to the mirror target and back to the telescope, one image of the cross hairs would be seen if the position of the target was correct. If there was an error, two images would be seen and the alignment would be corrected. With the application of tooling slides and tooling tapes, the total fixture is in a theoretical box of a coordinate measuring system, Figure 6-48A and Figure 6-48B. Distance, height, angles, flatness, and alignment are controlled with optical tooling.

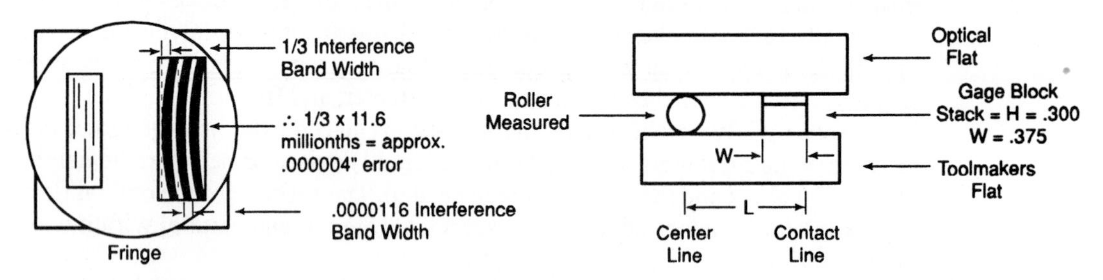

Figure 6-47 Interference bands reading to microinches

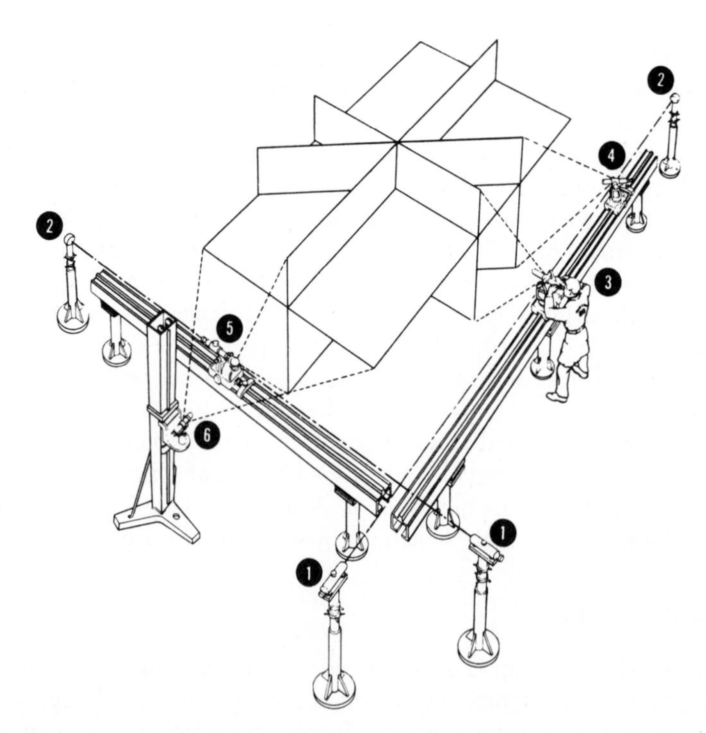

Figure 6-48A Alignment with metrological bars for electro-optical tooling *(Courtesy of Cubic Precision K&E Electro-Optical Products)*

Figure 6-48B A jig transit *(Courtesy of Cubic Precision K&E Electro-Optical Products)*

Computerised Autocollimation - Microcomputer control, provided with microcomputer control and data processing permit automatic measurement of straightness or flatness, with screen display and printout of the measurement, Figure 6-49A-C. Programs are available to check flatness and straightness. The rectangle and diagonal (Union Jack) method for flatness measurements offers the facility of displaying or printing out an initial diagram of the surface to be measured.

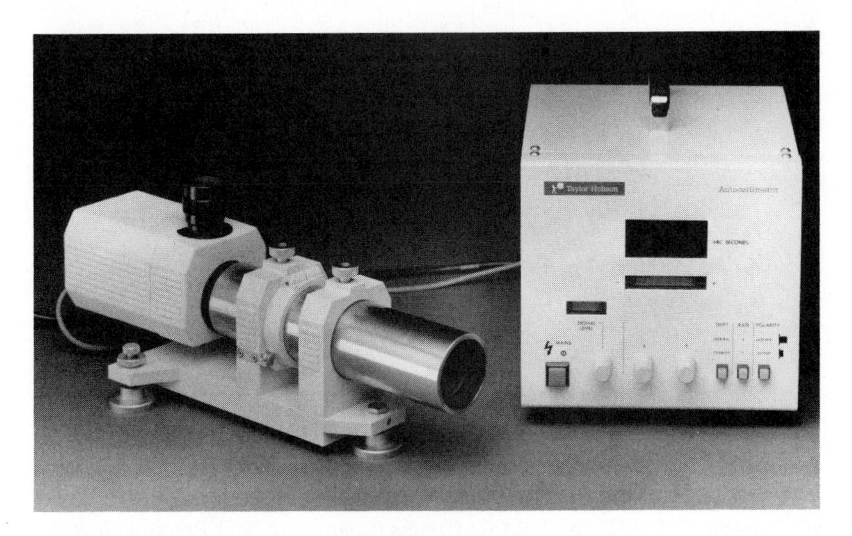

Figure 6-49A Autocollimator *(Courtesy of Rank Precision Industries Inc.)*

Figure 6-49B Checking a granite surface table *(Courtesy of Rank Precision Industries Inc.)*

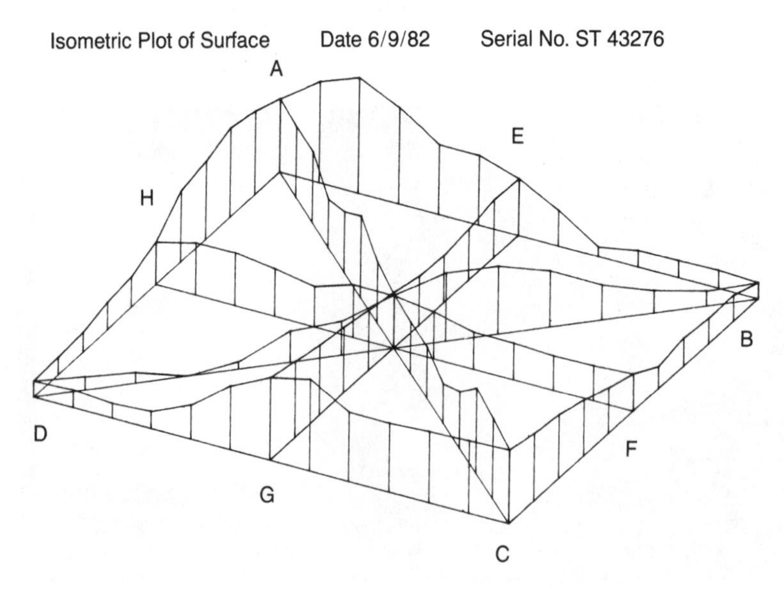

Figure 6-49C Isometric plot of "Union Jack" for flatness *(Courtesy of Rank Precision Industries Inc.)*

Analytical Industrial Measuring System - These measuring and alignment techniques are determined on the principle of three-dimensional triangulation. The theodolites measure both horizontal and vertical angles, providing the necessary data for an X, Y, Z (three-coordinate) solution. It is no longer necessary to move objects to a measuring room. Measurements are made by initially sighting in the theodolites at each other for angular orientation, then at two or three fixed reference targets for absolute orientation, and finally at an unlimited number of targets for measurement. The X, Y, Z coordinates of the points are sighted, computed, and displayed on a video display along with the distance from the preceding point. A printed copy is provided and results are recorded on a diskette, Figure 6-50. These measuring tools are employed in aircraft and aerospace manufacturing and rework; ship building and refitting; automotive, truck, and heavy equipment manufacturing; antenna manufacturing and Q.A., etc.

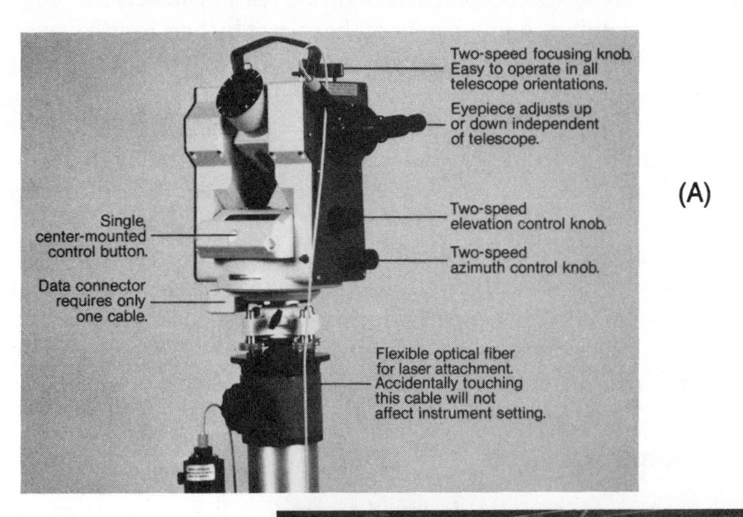

(A)

(B)

Figure 6-50 (A) Analytical measuring system with computer module and digital theodolite; (B) The alignment of an aircraft wing-section fixture with a theodolite and computer *(Courtesy of Cubic Precision K&E Electro-Optical Products)*

Laser Interferometry Measurement

Laser interferometry is the interlocking of interferometry and optical tooling concepts with laser and solid-state electronics to provide an outstanding measuring system. Laser interferometers measure linear, angular, flatness, straightness, and square displacements. Options can be added so that two measurements can be made simultaneously.

The laser provides a light source with a fixed phase relationship emitted from a focal point that results in a coherent light. The light source is directed to a half-silvered mirror that splits the laser beam — half of the beam to an internal mirror and half to a mirror mounted on the unit to be measured. The light from the mirror is reflected back to where the two halves are recombined. The distance to and back from the mirror has consumed travel time and the beams are no longer in precise phase. The interferometer combines the light waves and detects the "in phase" or "out of phase" condition. A movement of the mirror by one-fourth wavelength in distance results in a one-half wavelength change in the light's phase relationship and produces a change in the light's intensity. A photodetector is set to sense the change and count the pulses of light intensity. This output is multiplied by the wavelength of the light employed and is displayed as the distance in microinches. With the application of solid-state electronics, this data can be converted into a very accurate position, Figure 6-51A-C.

Laser-displacement measurement technology can be used to produce positioning accuracy in numerically controlled machine tools and coordinate measuring machines (CMM). This type of technology will deliver precision linear positioning to a submicron level for feedback for position control to numerous quality control and manufacturing applications.

Safety Proximity Devices

Proximity sensors are usually designed to employ the concepts of induction, capacitance, or photo-optics, Figure 6-52. A noncontact pulsed infrared sensor is one type of proximity sensor. It detects any surface or object entering its field of view regardless of material. It can respond to changes in colors, size, texture, and

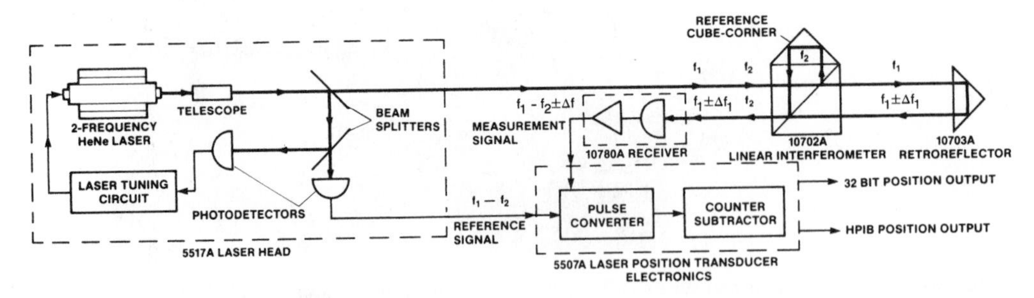

Figure 6-51A Theory of a laser interferometer system

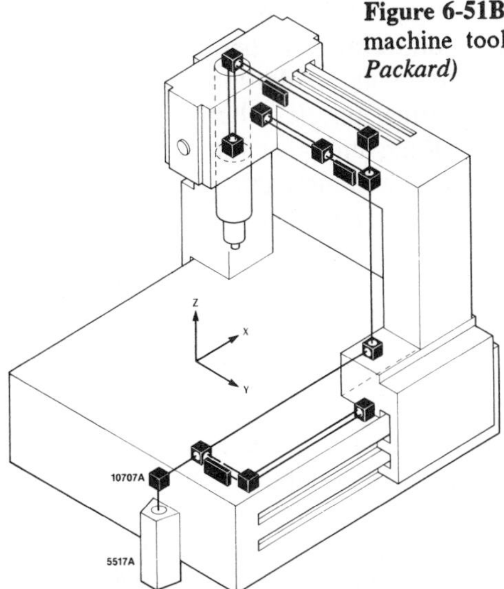

Figure 6-51B Three-axis linear interferometer for a machine tool application *(Courtesy of Hewlett-Packard)*

Figure 6-51C Laser position transducer system *(Courtesy of Hewlett-Packard)*

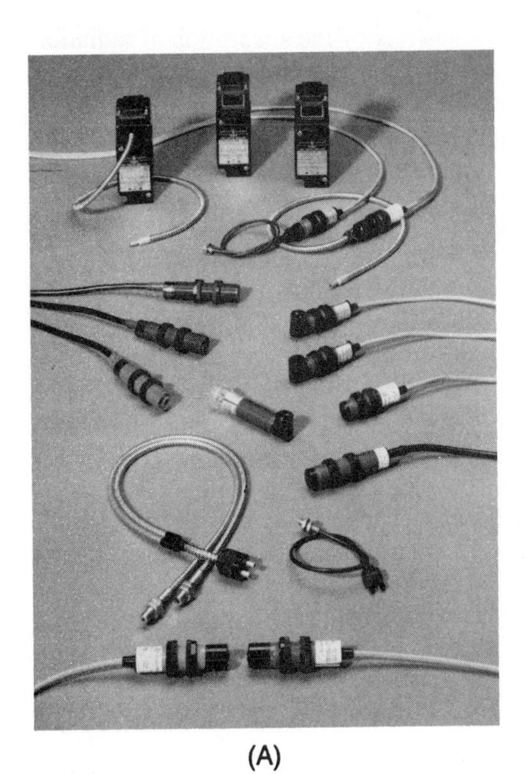

(A)

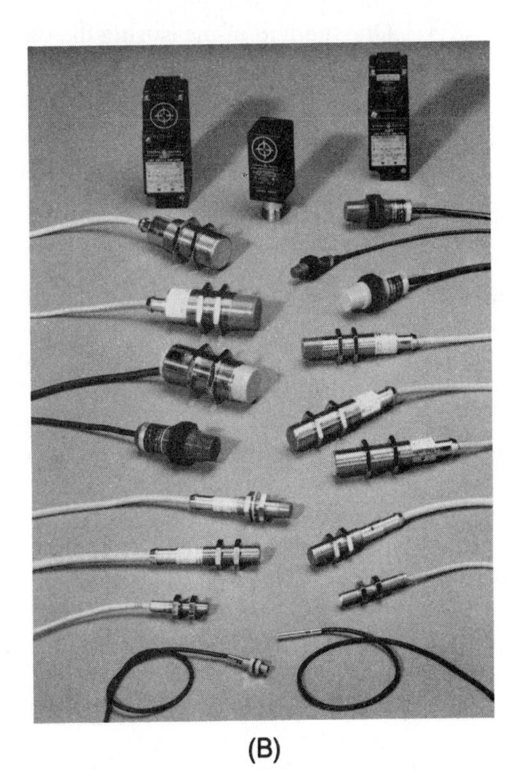

(B)

Figure 6-52 (A) Photoelectric sensors; (B) Proximity sensors *(Courtesy of GE Control Components, General Electric Co.)*

reflectivity of objects. Major uses include sensing, counting, routing, positioning, inspecting, measuring, code reading, web monitoring, and performing a wide variety of other automated process control functions. Additional applications include perimeter or intrusion protection alarms, burglar alarms, overheight barrier protection, liquid or solid level controlling, and other functions.

The sensor heads of these devices are sealed aluminum units with solid-state electronics throughout. They are designed to withstand washdowns and outdoor environments.

SUMMARY/FACTS

- Density is measured as weight per unit volume or mass per unit volume.
- Specific gravity is the ratio of a material's density to another material's density (usually water).
- Two bubbler tubes placed in a fluid with the tubes at different depths will deliver different pressures. This difference can be used to measure density or specific gravity of the fluid.
- One method of measuring the density of gases is by the absorption of radiation. When the radiation beam strikes a gas, between metal plates, a small current is produced that is proportional to the density of the gas.
- Viscosity measurements fall into two classes of substances: newtonian fluids, such as gases and thin liquids; and nonnewtonian substances, such as synthetic oils and thermosetting plastics.
- A continuous flow viscometer applied to the petroleum industry is of the capillary tube type.
- The dew point is a very important variable to many industries. It is a measure of absolute moisture content at a specific temperature and can be measured over a wide range of humidities and temperatures.
- The impedance hygrometer is formed by depositing a layer of porous aluminum oxide on an aluminum strip and then coating the oxide with a thin film of gold. The conductive base and the gold layer become the capacitor's electrodes. The number of water molecules absorbed by the oxide layer determines the electrical impedance of the capacitor and provides a measure of the water vapor pressure.
- The measurement of acids and bases frequently employs the potentiometric method. These devices are based on electrolytic conductivity and are a measure of the ability of a solution to carry an electrical current. The solution's conductivity indicates the solution's pH value, which is its hydrogen ion concentration.
- Strain gages are devices that change their electrical resistance with the application of a physical load. The load can be tension or compression and is correlated with a resistance change in the gage.
- The microprocessor digital tachometer receives successive pulses from a rotating target and compares these pulses against an internal crystal timebase that results

in a very accurate RPM output on a display.

- Position can be measured by an encoders. The absolute encoder provides a unique binary address for each position measured. The incremental encoders provide a pulse for each incremental change of position. They start from a point and count to the present position.
- The proximity sensors are based on inductive devices, capacitive devices, and optical devices. Inductive devices used in industry give a signal when a metallic object approaches the surface of the sensor. These sensors are used in many counting and positioning applications.
- The precision quartz accelerometer measures low and medium frequency vibration and shock motion on heavy structures, machines tools, bridges, and engines. It transfers the acceleration aspect of shock and vibratory motion into a voltage signal compatible with analyzing and readout equipment.
- Measurements can be made with the application of interferometry. This is an application of the physical phenomenon of light wave interference. Laser interferometery is used to measure linear, angular, flatness, straightness, and square displacements.
- Safety proximity devices are necessary to protect workers from dangerous areas. The noncontact pulsed infrared sensor is a sensor that will provide this security.

REVIEW QUESTIONS

1. Describe an early lever system for measuring weight.
2. In what types of scales for weighing trucks, railway cars, etc. are a series of those early lever systems used today?
3. What type of scales are based upon the elasticity of one of its parts and calibrated to read according to its elongation versus the weight applied?
4. Which device applies Pascal's laws of fluid pressure to a gage that is calibrated to read weight?
5. Electronic load cells are designed with ________________ mounted on a column or beam, the load bearing member.
6. Which type of load cells applies the force-balance concept using air chambers, diaphragms, and air nozzles?
7. What is the instrument that measures the speed of rotating machinery?
8. Sensors that measure the speed of rotating machinery use different means: 1) magnetic, 2) capacitance, and 3) induction. Name a sensor that employs each respective phenomenon.
9. How are photosensors used to provide a tachometer signal?

10. What is the name of the sensor that is comprised of a flashing light synchronized with moving machinery, and whose instrument dial reads out the RPM?
11. How can a light-emitting diode and a phototransistor be designed to function as a tachometer.
12. What is the word that refers to the orientation of the workpiece, robot arm, spot weld, etc?
13. What is displacement?
14. What is the relationship of variables and a standard in measurement of displacement?
15. What is the rate of change of velocity called?
16. What is a function of a synchro with a digital converter?
17. Name the two types of encoders described in this text.
18. Linear variable differential transformers change ______ in some accelerometers to provide a strong signal.
19. Strain gage accelerometers function by a change in ________.
20. Dimensional transducers are classified into two large groups. Name them.
21. What measurements are based upon the phenomenon of light wave interference?
22. The measurement, alignment, and positioning within a "theoretical" box of a the coordinate measuring system is accomplished by using the principles of ________________.
23. A technology that unites interferometry and optical tooling concepts with a specific light source and solid-state electronics provides very precise measurements over a broad range of distances. What is that measurement system called?
24. How can a noncontact pulsed infrared sensor serve as a safety device?

7

DECISION AND CONTROL OF THE SYSTEM

OBJECTIVES

Upon completing this chapter, you will be able to describe or explain how these control concepts work and are applied:

- The concepts of a control loop
- Process response characteristics
- Modes of process control
- Controller technologies
- Distributive control systems

INTRODUCTION

The understanding of control in the various manufacturing processes is a major responsibility of the manufacturing supervisor. His knowledge will include information concerning the application of process sensing, process requirements, and process modes of control. These include the application of microprocessors and microcomputers as they apply to process control, the application of direct digital computer technology, and the utilization of distributive control systems.

THE PROCESS CONTROL LOOP AND RESPONSE CHARACTERISTICS

A control loop is a group of instruments designed to regulate a system and maintain that system at a desired value. The loop is made up of a process to be controlled, a sensor that measures the amount of the variation in the process's condition, the set-point (the desired value), a controller to make a decision (whether the process is at, above, or below the set-point), and a final control element that adjusts the system to the desired value of the set-point and to change the variables' values as needed, Figure 7-1.

Response characteristics of a system are the result of resistance, capacity, mass, and dead time within the system. All of these characteristics interplay to produce transfer lag, i.e., the time it takes to make an adjustment to the total system.

The control loop sensor obtains information on the existing condition. This value is sent to its controller that compares it with what is desired. The controller then produces an output for a change as required. This change is the error signal. The error signal in turn is amplified and transmitted to a final element, e.g., a control valve. The final element causes the change or correction in the system. The total system — the control loop — functions as a unit.

Types of Control Loops

There are two types of control loops: one is referred to as an open-loop and the other as a closed-loop system, Figure 7-2. The open loop system does not have a self-correcting or feedback portion in the control loop. It would be analogous to

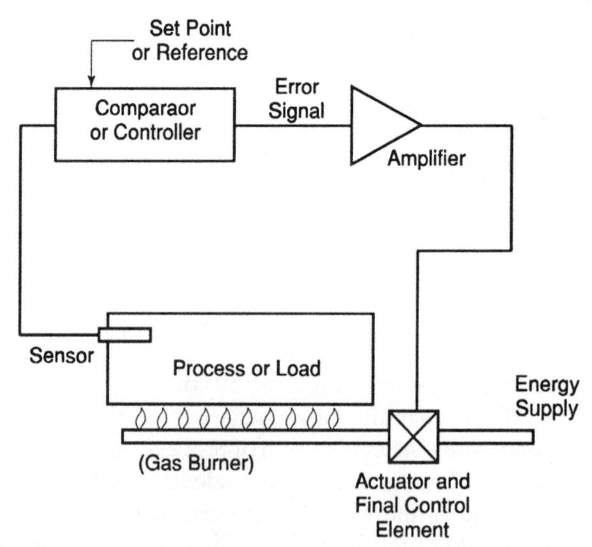

Figure 7-1 Block diagram of a control loop system

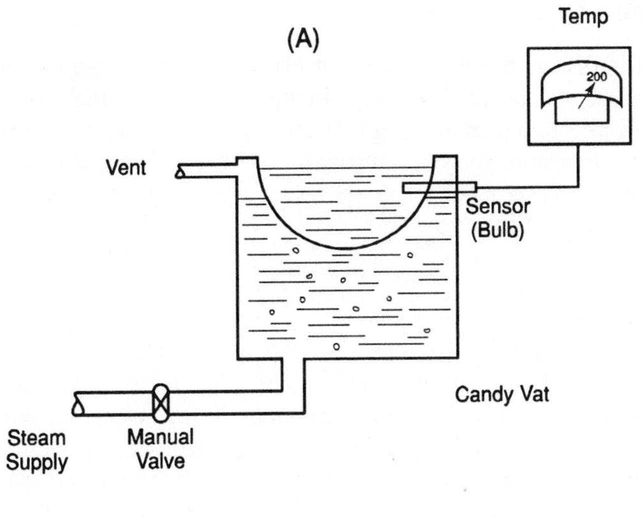

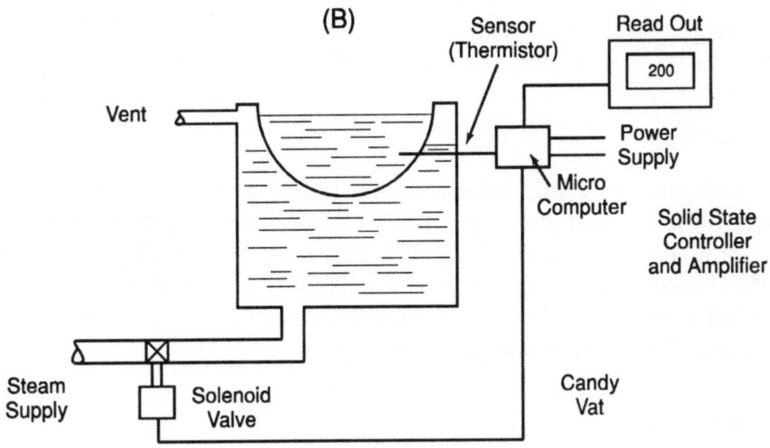

Figure 7-2 (A) Open loop diagram; (B) Closed loop diagram

a man measuring the temperature of a process and writing it into a record book. The open loop indicates the condition of a system and records what is happening in the system, but it does not take corrective action by itself to restore the system to the desired value.

The closed loop system provides feedback to the system under control. The system is measured and recorded as above but the data is also compared to a set-point for what is desired. If the desired value was not achieved, corrective action is taken by the controller and actuator. The closed loop has a number of technologies and modes of control that may be selected to accomplish the method of providing a feedback loop.

Process Response Characteristics

Process load is one of the dynamics that brings about change in a system and makes it necessary to control the change in the system. A change in loading (such as adding cold steel to a heat treating furnace) or a change such as the addition of the rate of flow to a sewage system provide a need for an adjustment in order to keep the system under control. In an operating system, it is a change of load that requires an adjustment in the system.

Resistance

Resistance is the response characteristic that retards a change in the variable brought about by a change in set-point or load. This retardation may be caused by various insulation materials, such as layers of steam, water, ceramic material, container walls, or other materials that reduce the speed at which energy may be transferred from one medium to another. These phenomena reduce the rate at which energy is transferred to the process. Resistance is one characteristic that adds a time factor to the response of the system, resulting in a response curve rather than a straight-line change in the system, Figure 7-3.

Capacity

Capacity may be thought of as the opposite of resistance. Some parts of the system have the ability to store energy and to release that energy to keep the system at the present energy level, Figure 7-4. If in a large furnace the brick work is at the operational temperature and a cool load of steel workpieces is added to the system, the hot areas of the furnace will quickly supply energy back in the furnace to retard a drop in the primary sensor's response.

Dead Time

Dead time may occur in a complicated industrial process. With events such as that of a change in the set-point of a system or a change in load, it is possible to have

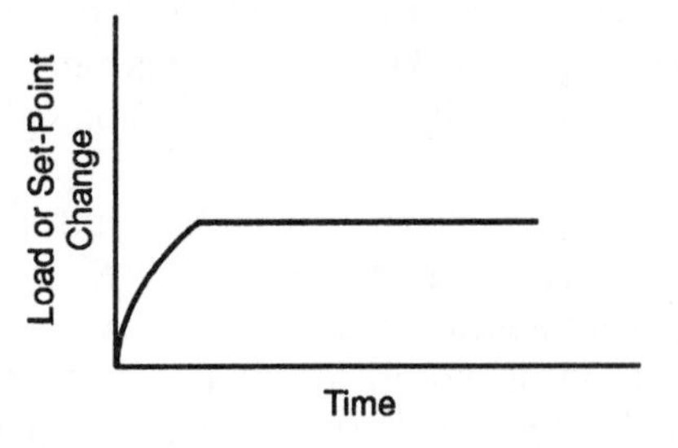

Figure 7-3 A low resistance system response curve

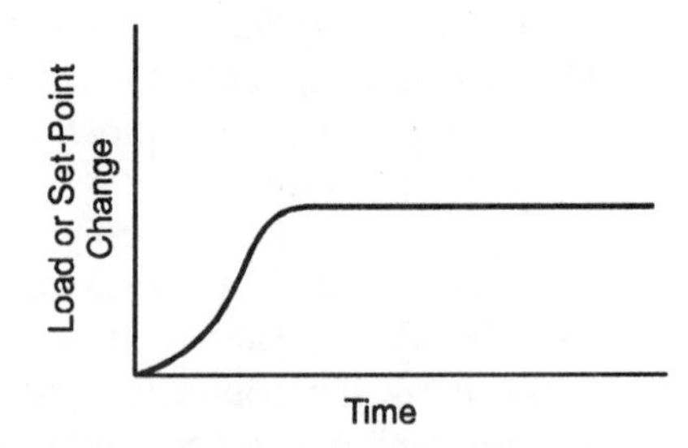

Figure 7-4 A capacity response curve

a period of time when no change is evident in the system. With the system working, the process *reaction curve* may remain flat. This is characteristic of dead time, Figure 7-5. This is the result of there being more resistance than capacity in the system.

Transfer Lag

Transfer lag is the time required for the process to move from one set-point to a new set-point or to recover from a change in the load. In a complicated system with many interfaces, this time for the energy change to pass through may be considerable. The result of this change is a process reaction curve and the total time involved is transfer lag, Figure 7-6.

Self-regulation

In the case where the mass of a product entering a stable system does not exceed the capacity of the system, *self-regulation* is possible. The process tends to adjust itself toward a balanced condition. An example of this concept is a tank filling with water from a spring. If the water flows into the tank at a rate larger than the flow rate leaving the tank, the level of water in the tank rises. The water level increases and results in a greater hydrostatic head, increasing the pressure on the bottom of the tank. Because of the increased hydrostatic head, and thus the pressure, the flow rate increases through the lower outlet. At some hydrostatic level, the out-flow rate will equal the in-flow rate producing self-regulation.

Modes of Process Control

Controllers employ different techniques of process control depending on the degree of accuracy required and the level of technology required to produce the desired accuracy. Controllers are designed using mechanical, pneumatic, electrical, hydraulic, or electronic technologies. In each case, the controller's function is to respond to and correct any deviation of the variable from the set-point.

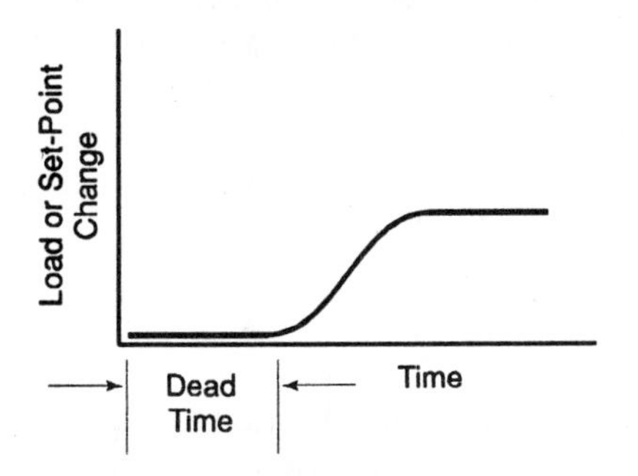

Figure 7-5 Dead time response

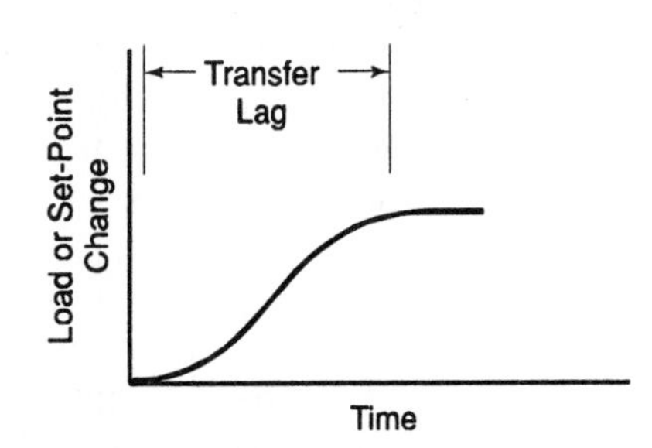

Figure 7-6 Transfer lag

Controllers employ one of three basic techniques to provide control: analog control, which has continuous data collection; digital control, which has discrete data collection; and hybrid control, which uses a combination of continuous and discrete data collection.

On/off Control

A basic device is the on/off or two-position control. This mode of control has two states: on, where the energy is added necessary for the system to raise it to the required set-point; and off, where the energy supply is removed because the system is at or above the required set-point. These devices are frequently used in heat control, such as a bimetallic sensing element controlling a set of electrical contact points or a mercury switch. These devices are referred to as thermostats and are widely used in heating systems, refrigerators, and water heating tanks in homes and in industry. In addition, this mode of control is applied to devices that employ various float switches, pressure switches, and position devices employing microswitches in many configurations.

Frequently, heating units have a thermal time constant and thus a slow reaction rate. The actual temperature will fluctuate above and below the set-point. This error around the set-point results in a band or "differential gap" through which the sensor must pass before a change in the control occurs. This band is helpful because it dampens the system's sinusoidal input, and cycling is minimized. This mode of control works well in systems in which the process reaction rate is slow, the transfer lag and dead time is small, and the cycle time is long. The on/off mode of control is available in all types of controllers (mechanical, pneumatic, fluidic, electrical, hydraulic, electronic, or digital).

A digital on/off control is obtained with an operational amplifier used in a voltage comparator configuration and two voltage dividers. With this application, a voltage representing the upper set-point of the process or condition being controlled is received by the (-) terminal of the op amp. The voltage from the second voltage divider (sensing device) becomes greater than that of the reference voltage, the op amp inverts, and the output of the op amp is a low that turns off an optical relay or other solid-state device and the circuit energy supply is off.

When the voltage received by the (+) terminal of the op amp is greater than the set-point or reference voltage, the op amp inverts again and its output is high, starting the units that supply energy to the process or condition. This is a solid-state on/off control system, using an op amp to provide a fast and accurate control, Figure 7-7 and Figure 7-8.

Proportional Control

Proportional control is applied where a smoother control action is required than an on/off control. The proportional control is a throttling or modulated action of a continuous measurement provided by the controlled variable. The actuator

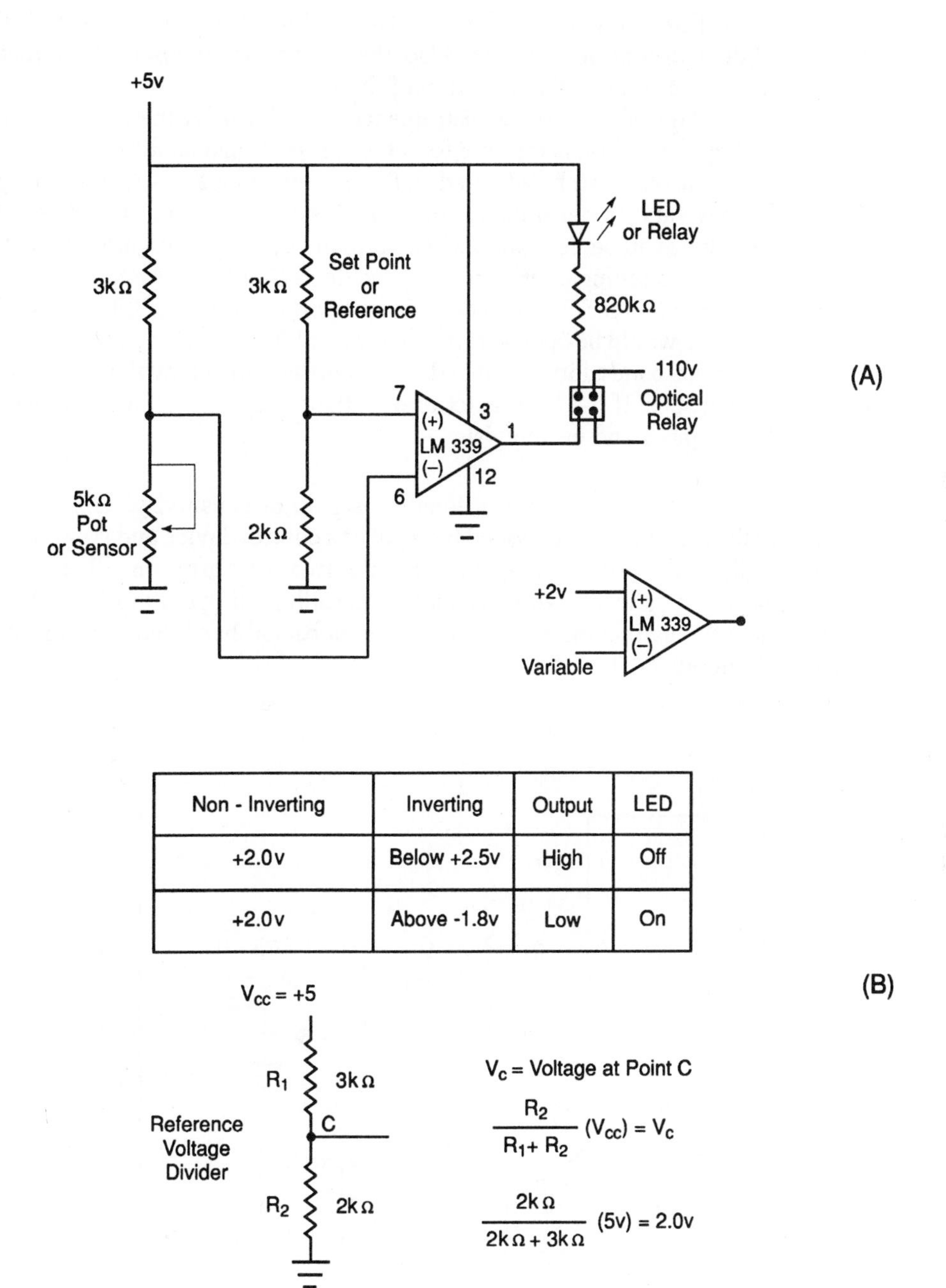

Non - Inverting	Inverting	Output	LED
+2.0v	Below +2.5v	High	Off
+2.0v	Above -1.8v	Low	On

$$\frac{R_2}{R_1 + R_2}(V_{cc}) = V_c$$

$$\frac{2k\Omega}{2k\Omega + 3k\Omega}(5v) = 2.0v$$

Figure 7-7 (A) A schematic of an on/off digital control circuit; (B) A voltage divider for op amp control

movement or final element, such as a valve, is directly proportional with the amount of deviation of the variable. Also, there is no movement of the actuator unless there is a change in the deviation of the variable.

The percentage of range of the instrument required to drive the actuator of the valve to full open position is referred to as the "proportional band" or "throttling range." The controller can be adjusted so that its percentage of the proportional *bandwidth* may vary. For example, a functioning system is set up so that its final element or valve is 50 percent open when normally operating, Figure 7-9. When a controller is functioning over a one hundred degree temperature range and the instrument's set-point is fifty degrees with a proportional band of 20 percent, the actuator or valve would be open at forty degrees and fully closed at sixty degrees. If the proportional bandwidth is reduced too much, the controller will function like an on/off controller. If the band is too broad, the controller will not respond to system disturbances and be sluggish.

Gain - Another concept of proportionality is gain or sensitivity. This term is defined as the ratio of output divided by input of a control device and is expressed mechanically as the ratio between the length of two lever arms on either side of a pivot, or as the ratio of two resistances controlling an operational amplifier. Mathematically, gain is the reciprocal of a proportional band that is reported as a decimal, Figure 7-10.

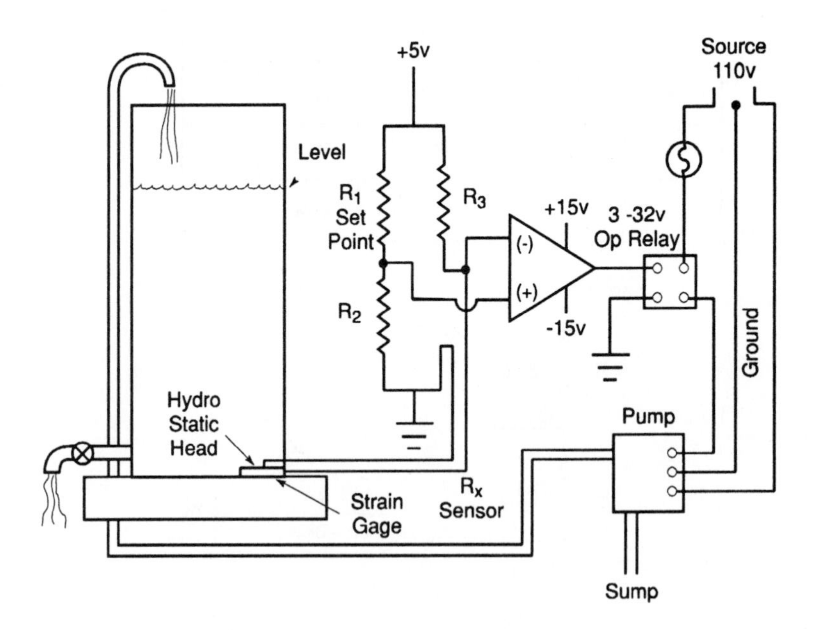

Figure 7-8 On/off or two-position liquid level solid-state control circuit

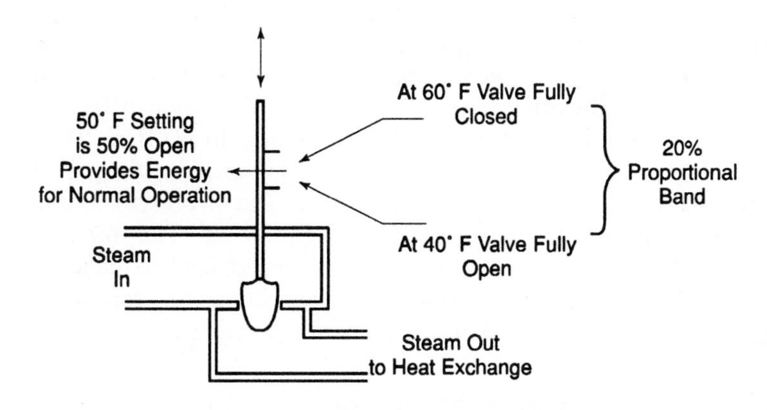

Figure 7-9 Valve travel and proportional band in proportional control

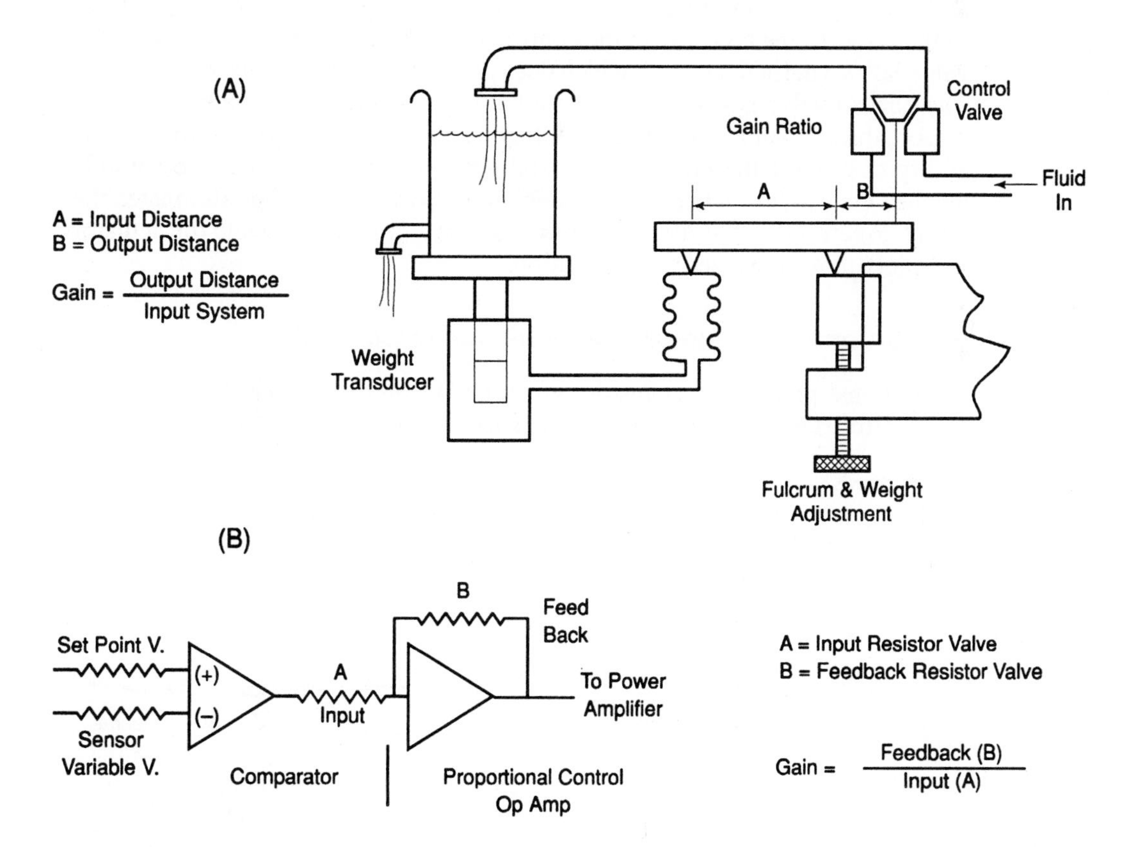

Figure 7-10 (A) Schematic of mechanical gain (mechanical); (B) Schematic of electronic gain (op amp)

Offset - A disturbance of a load change in a proportional control system requires a new valve position to bring the system back to a steady-state condition. The *proportional band* presents a characteristic of only one valve setting within the proportional bandwidth. This results in a controlled value different from the set-point. This variation from the set-point is referred to as offset. The offset can be reduced by narrowing the proportional band or increasing the gain. If the band is narrowed too much, the system will oscillate. The offset will continue as long as the load change value is maintained, Figure 7-11. This offset error can be eliminated with a manual reset bringing the system back to the desired set-point. Automatic control is not present when offset occurs with proportional control. In this system, recalibration to the set-point requires an additional mode of control.

Proportional-plus-integral Control

Proportional-plus-*integral control action* automatically adjusts the offset error due to load changes in the proportional control system. This addition has the ability to adjust the valve when there is an error in the system, at a rate proportional to the error. In the case where the control valve stem velocity goes to zero, the valve holds a definite position which does not change as long as the error is zero. The control valve position is proportional to the integral of the error. For this reason, this system is referred to as integral control. The integral action has the ability of moving the proportional band above the set-point until the variable reaches the set-point. The integral element is slow moving, but eliminates the offset, Figure 7-12. The electronic systems move more quickly and the actuator is rapidly moved to a zero error position.

Proportional-plus-integral-plus-derivative Control

Proportional-plus-integral-plus-derivative (PID) control action is the most effective controller mode because it responds successfully to the widest range of error.

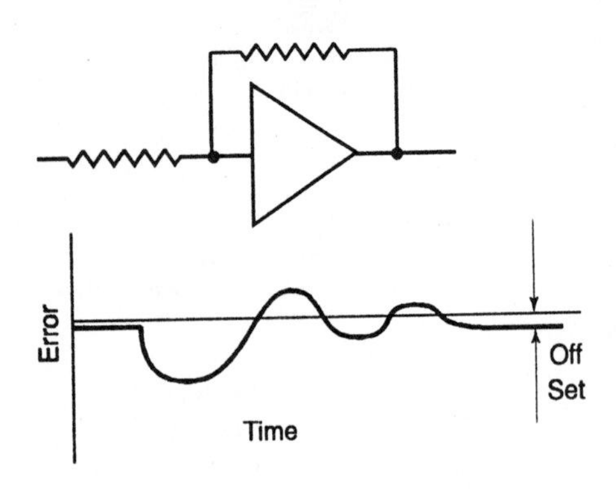

Figure 7-11 Proportional only operational amplifier control illustrating control offset

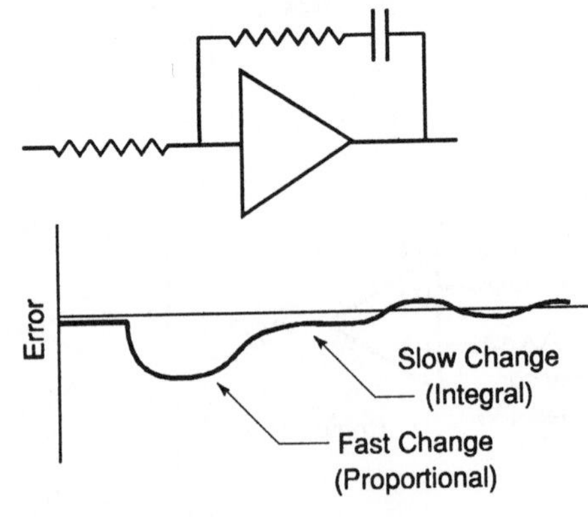

Figure 7-12 Proportional-plus-integral control action (single op amp)

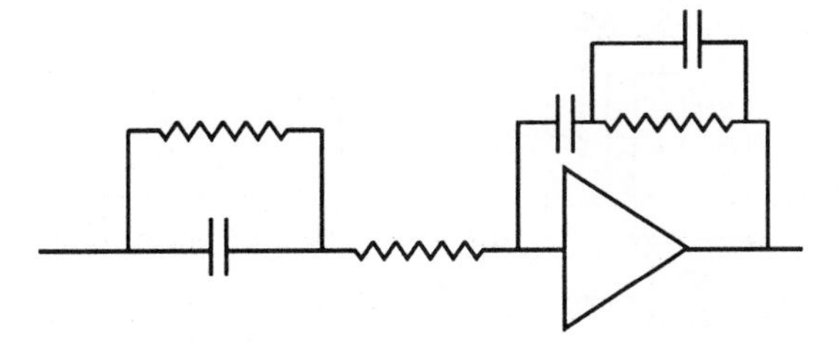

Figure 7-13 The three-mode controller—proportional-plus-integral-plus-derivative control action (PID) control

It induces reactions to load disturbances, transfer lag, dead times, and various amplitude and frequency responses. This control combines proportional action, integral action, and derivative action into one controller. The derivative element produces an output proportional to the rate of change of the error. This action corrects the position of a control valve by opposing change by an amount that is proportional to the rate of change of the controlled variable. *Derivative action* either adds or subtracts energy to the system. The derivative action dampens the response to load changes and provides a stabilizing effect on the control system. Because derivative control reduces system oscillation, it allows the proportional gain to be set at a higher value, thus increasing the speed of response of the controller to system disturbances. The more rapidly the variable error increases the rate of change or error, the greater is the initial overshoot error. The derivative action aids in this correction. Derivative time or rate action does not change the system's offset, but the integral action in the three-mode system performs this function, Figure 7-13.

Hybrid Control Systems

A hybrid control system is the application of a computer to a system in which parts of the system utilize analog data collection and control devices, while other parts have direct digital computer control, Figure 7-14. These systems are interfaced so that communication between the systems is complete. The application of the computer to the control system provides problem-solving flexibility because the computer can be programmed to optimize the control condition. The gain, drift, or other parameters of the system based on the experience learned from the system are used to adapt the system to its most efficient operation.

Controller Technologies

The *modes of control* (proportional, integral, derivative, and hybrid) are implemented by various technologies: pneumatic, hydraulic, electrical, electronic, and digital.

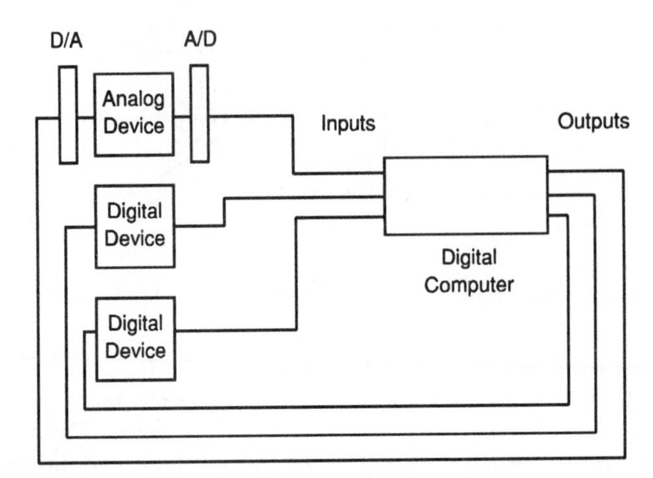

Figure 7-14 A block diagram of a hybrid system. A hybrid system requires analog-to-digital converters to interface the signal with a digital computer.

Pneumatic Controllers

Pneumatic controllers have been in use by industry for a very long time. They had been widespread in the control industry until the application of solid-state electronics. Pneumatic controllers are still widely used in industry and are especially valuable in various types of explosive prone industries. The various modes of control stated previously were developed with the emphasis of pneumatic control and this theory has been applied to other technologies of control.

The pneumatic control concepts are based on the physics of the behavior of gases. If a regulated supply of air is applied to a tube with a restriction in the tube, the air volume passing through the tube will stabilize with a flow of air leaving the tube. If a nozzle is added to the end of this tube, an air jet is formed, passing the air out into the atmosphere. When a plate or baffle (flapper), is placed in front of the jet of air, a back pressure will result behind the nozzle. The back pressure behind the nozzle in the tube has a linear relationship with the distance of the flapper from the nozzle, Figure 7-15. This linear relationship is confined to the area of movement that is characteristic of the particular design of the instrument, but it is very valuable in providing an accurate sensing pneumatic relay.

The flapper and nozzle are used in a number of pneumatic control devices. The position of the flapper is moved by various possible sensing units of the phenomenon being measured. This sensed data is converted into a pneumatic signal (pressure) that is processed through the controller, Figure 7-16 (page 276). The controller may be designed for any or all of the modes of control mentioned previously. The error signal from the controller (pressure) is transmitted to a final control element. This is received as a pressure variation and can be transmitted by any of the modes of control.

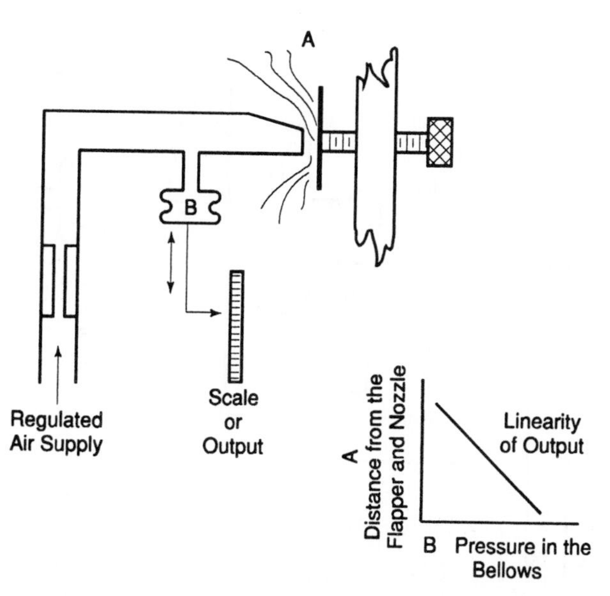

Figure 7-15 *Linearity* of a pneumatic relay

The set-point of the system is provided by a set screw that can increase or decrease the pressure on a spring. This spring pressure opposes the force exerted on the flapper by the sensor. This adjustable pressure delivers the means of setting the controller to the required value.

The gain is adjusted by changing the difference (ratio) between the length of the lever arm from a pivot point to the set-point screw and the length of the arm from the nozzle and the set-point screw. This is to establish the ratio of input to output of the controller.

With the addition of bellows and an adjustable leak, a reset (integral) action is supplied. The set-point thus is adjusted to provide a band that involves correction for the offset of the proportional action controller. In this case, the pressure leaving the relay valve and going to the control valve is fed back into a nested bellows. The energy from the pressure causes movement in the bellows and pressurizes the throttling bellows which results in a change in the flapper/nozzle distance. This movement makes a correction in the system, but as the system approaches a new set-point, the adjustable leak allows the energy reaching the throttling bellows to be reduced. The result is a reduction in the amount of overshoot as the new set-point is reached. The adjustable leak is open to the atmosphere so that if the pressure falls below ambient pressure, air will leak into the system and provide a correction on the opposite cycle, Figure 7-17 (page 277).

A pneumatic derivative control is designed by adding an adjustable restriction between the pneumatic relay output and the integral or reset bellows. This restriction provides a rate (derivative) at which the change in pressure will reach the reset bellows and the throttling bellows. This restriction performs a damping

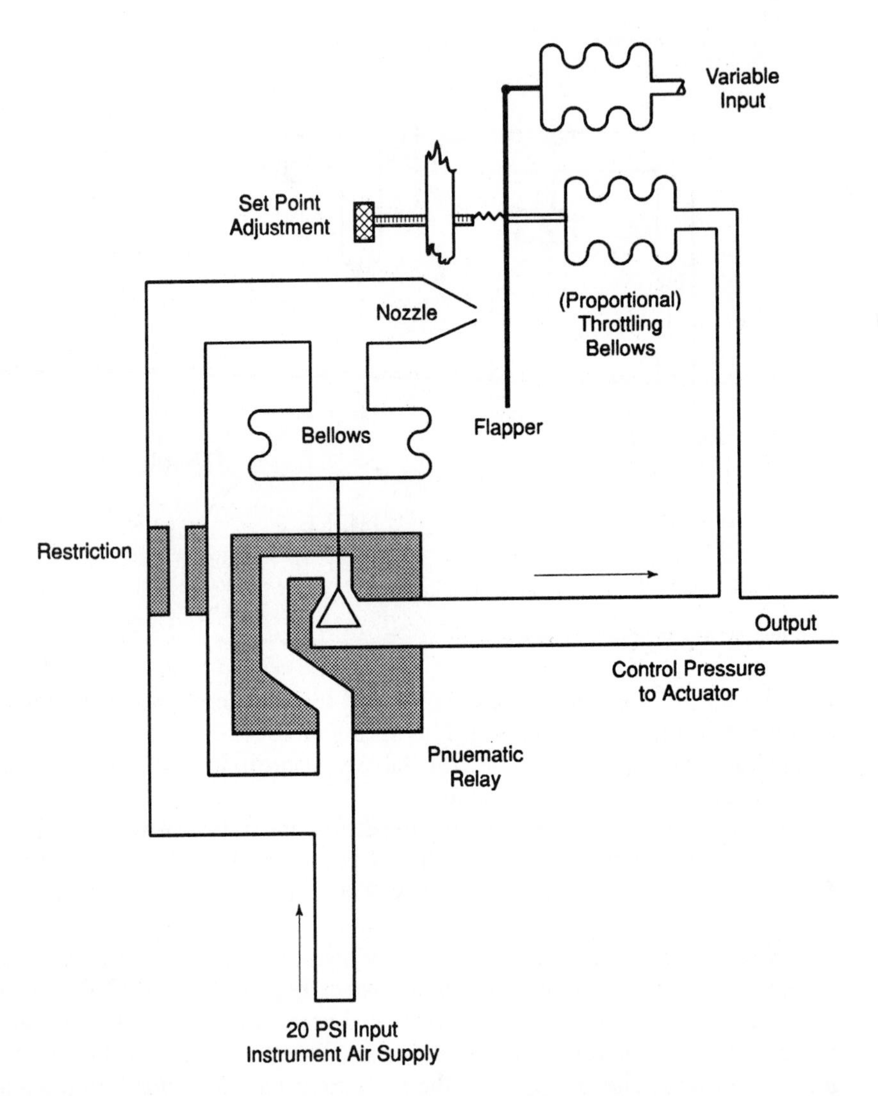

Figure 7-16 Schematic of a proportional-action pneumatic controller illustrating the flapper nozzle and pneumatic relay

function so that the reset and throttling bellows will have a different time constant and therefore not overcorrect the system above or below the new set-point, Figure 7-18 (page 278).

Fluidic Controllers

Devices that employ the dynamics of fluids flowing through their conduits are a unique group of controllers. These devices are constructed with flip-flop or

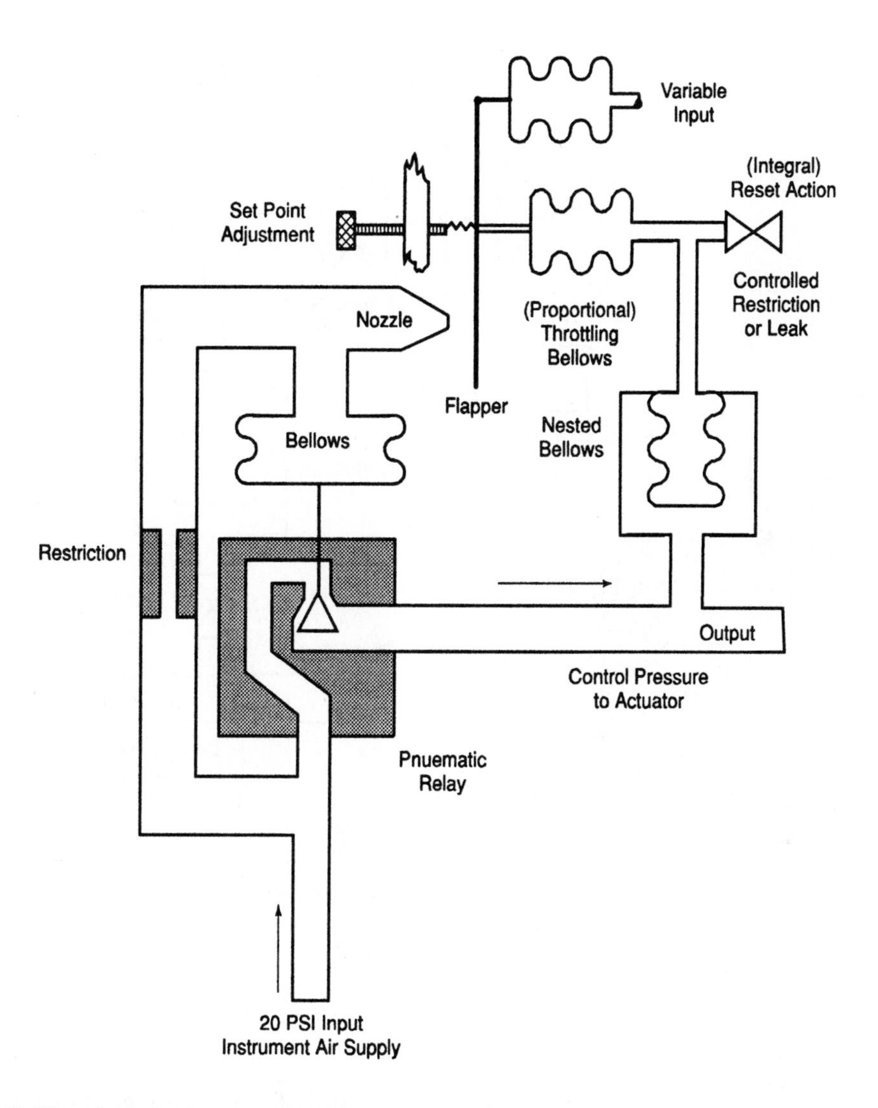

Figure 7-17 A schematic of a proportional-plus-integral action pneumatic controller with additional bellows and adjustable leak

bistable amplifiers, proportional amplifiers, oscillators, or pulse counters. They are based upon the principle referred to as the Coanda effect. Other names applied to this principle are the "wall effect" and the "boundary-layer effect."

The bistable amplifier is a basic unit for digital control and uses the same digital logic as is utilized in binary logic for electronic devices. The wall attachment device is a two-state logic device. This fluidic amplifier diverts a power jet from one channel to another with the application of a small pulse or leak to the fluid being controlled. When the small pulse or leak enters the left control

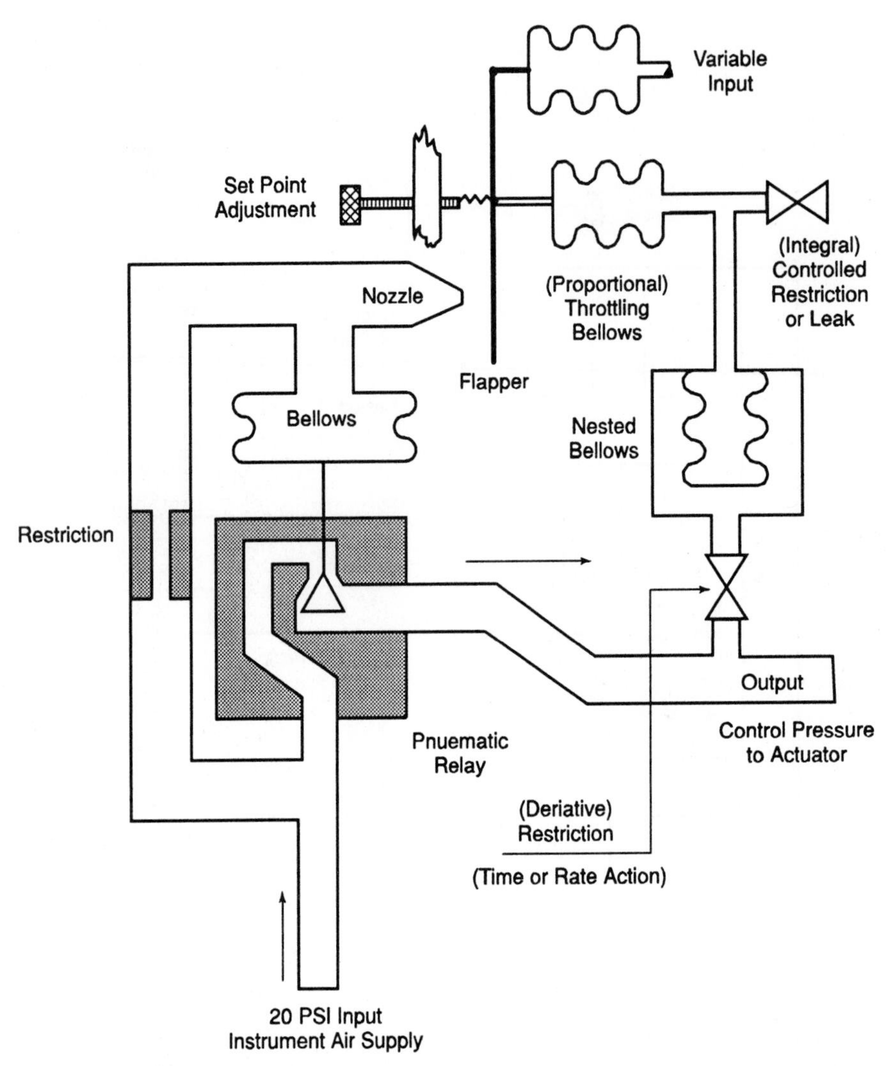

Figure 7-18 A schematic of a proportional-plus-integral-plus-derivative action pneumatic controller with the addition of an adjustable restriction

port, the power jet is switched from the left channel to the right channel and remains there because of the Coanda or wall effect. This essentially causes the memory of the device. When a pulse or leak enters from the right control port, the power jet will switch back to the left channel. The switching is caused by the pulse or leak disturbing a small vortex by the port and causing a separation bubble to grow and release the power jet from the wall attachment.

One application of a flip-flop controller is a diversion valve that switches the flow to maintain a fluid level. Its action is very rapid. Since there are no moving parts, little maintenance is required. And since they function with various media, such as gases, fluids, and slurries, they are very adaptable, Figure 7-19.

Hydraulic Controllers

Hydraulic technology is employed where large power requirements are necessary. Some of the controlling units in this technology are directional control valves, *sequence valves*, and servovalves.

Hydraulic *directional control valves* are of a number of types. The spool valve is very reliable and efficient and delivers trouble-free service. It can control extremely high levels of energy. The spool valve slides axially inside a bore and is classified according to the flow conditions created. There are internal flow paths for each valve position. The valve can have two, three, or four positions and also a series of valve connections. As an example, a two-position valve with a rest or neutral position does not allow fluid to enter a hydraulic cylinder, but it does allow fluid to exhaust from the cylinder.

A three-position, four-connection valve has three working positions. The central position is the neutral or locked position. At one extreme position, the valve allows fluid to enter one end of the cylinder and exhaust from the other end simultaneously. In the other extreme position, the valve allows a cross-over flow path—the fluid enters the previously exhausted cylinder and exhausts from the

COANDA EFFECT OR WALL ATTACHMENT

Water Tap

Water Will Attach Itself to Finger Deflecting the Stream

Water

Water In

Low Pressure

High Pressure

Friction of Water on the Wall Slows the Water, Causing a Lower Pressure Than the Rapidly Moving Water. The Higher Pressured Water Moves Toward The Low Pressure Area, Bending the Stream to the Wall.

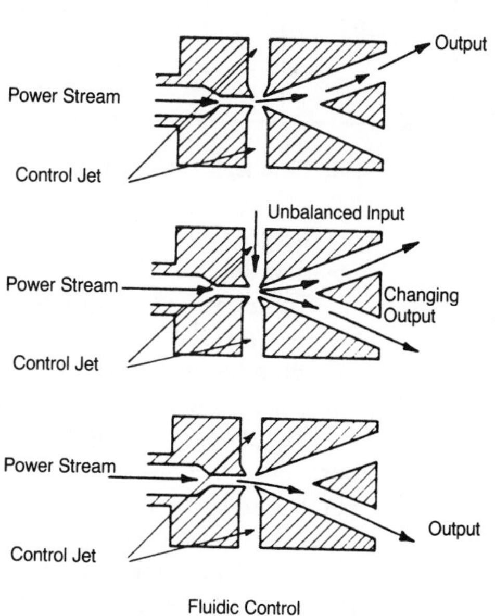

Fluidic Control
Main Stream Locks in Position Until Disrupted By an Input Stream.

Figure 7-19A Fluid control

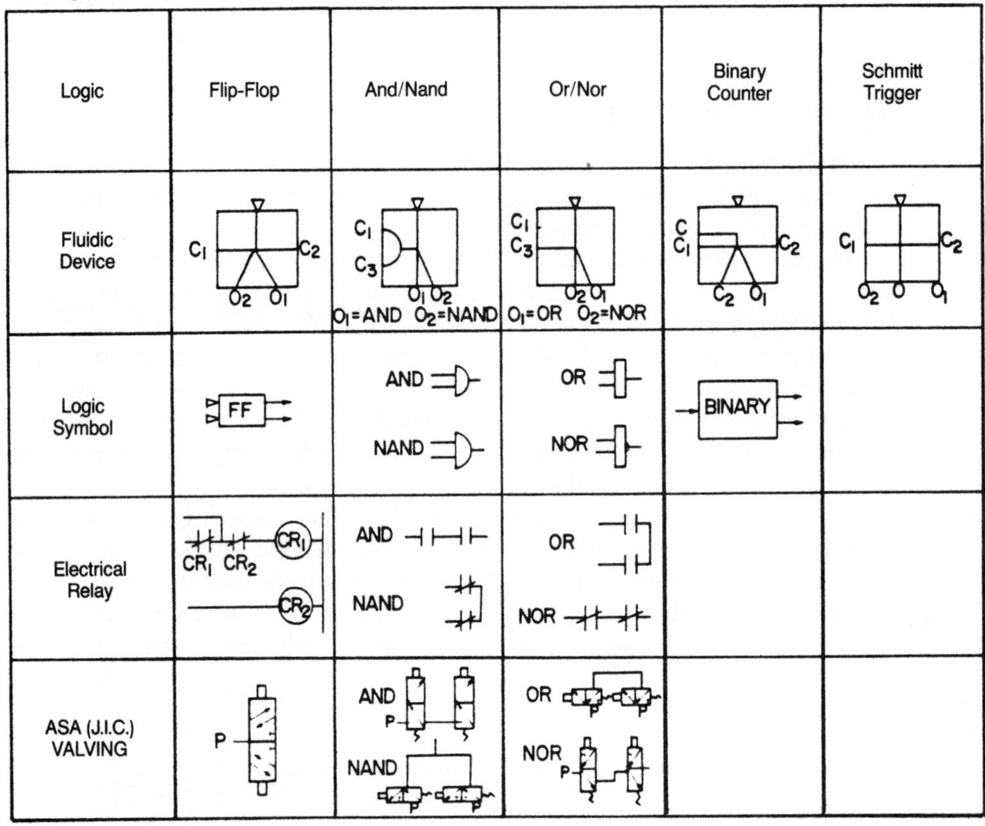

Figure 7-19B Fluid logic equivalants

previously filled cylinder. This is a four-way control valve with locked center position, Figure 7-20.

The above valve is referred to as a "closed center valve." By selecting a different spool configuration, the same valve housing could have valve ports open to each other when the valve is in the normal centered position. In this case the valve is referred to as "open center."

The primary function within the valve is the movement of the spool to various positions, thus shifting the flow to other paths within the valve and providing control to the final element actuators, Figure 7-21 (page 282).

A common control arrangement combines hydraulic valves with electrical controls. The spool valve may be positioned by a solenoid or a *pilot pressure*. Or it can be activated in a series by both a solenoid and the pilot pressure. The signal to the solenoid may be an electrical signal from a computer.

A hydraulic sequence valve controls the order of operation between two branches of a hydraulic circuit. This device is commonly used to regulate the

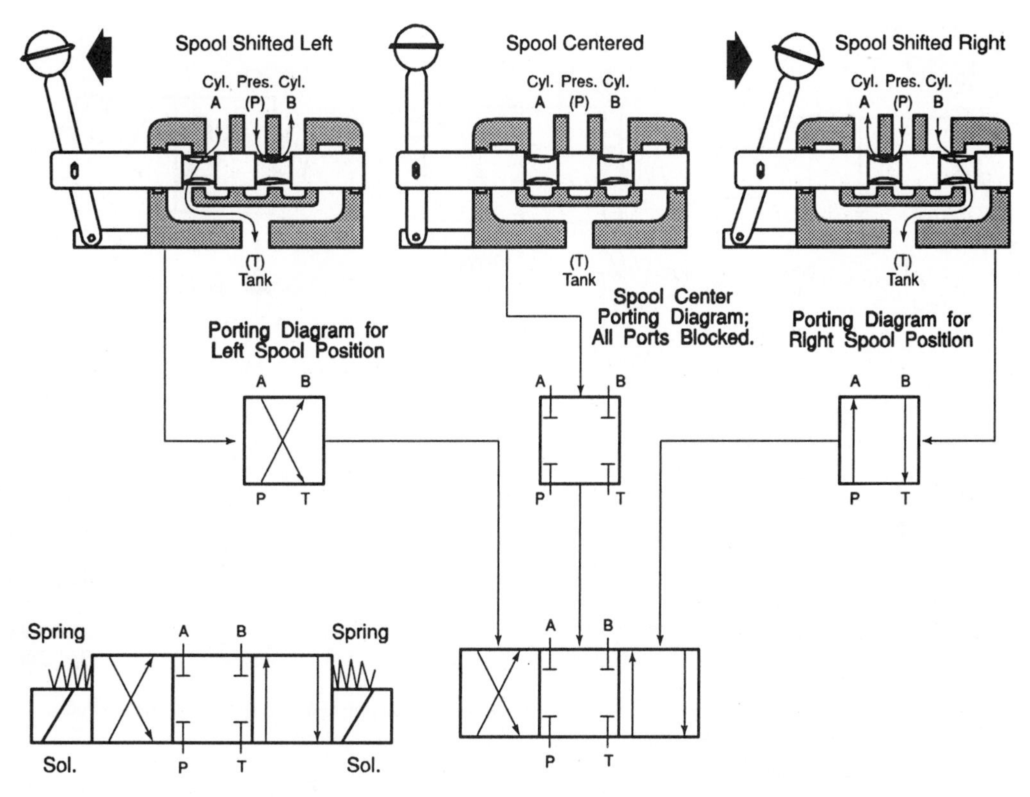

Figure 7-20 A three-position, four-connection, four-way directional control valve with a locked central position

operating sequence of two separate work cylinders, so that one cylinder begins its stroke when the other completes its stroke, Figure 7-22. This type of control is very simple and reliable and is used in many machine and control operations.

A hydraulic servovalve controls the direction and quantity of fluid flow in proportion to an input signal. This allows a low energy signal to control a very large amount of hydraulic energy with precise control of position and speed. It is a very precise directional control valve that usually contains feedback capability for closed loop operations. The servovalve can be activated with mechanical, hydraulic, pneumatic, electrical, or electronic devices.

Position control of automatic pilots, missiles, military equipment, robots, computer integrated manufacturing equipment, and many applications of production machinery employ a low-displacement, electrical torque motor to activate a two-stage, three-way hydraulic control valve. The valve supplies hydraulic fluid to the pilot ends of the main spool, controlling the source supply of hydraulic pressure going to the working hydraulic cylinder or other actuator (hydraulic motor). A feedback link is supplied by attaching a linear transducer to the main hydraulic piston rod driving the working member. The data transmitted from the

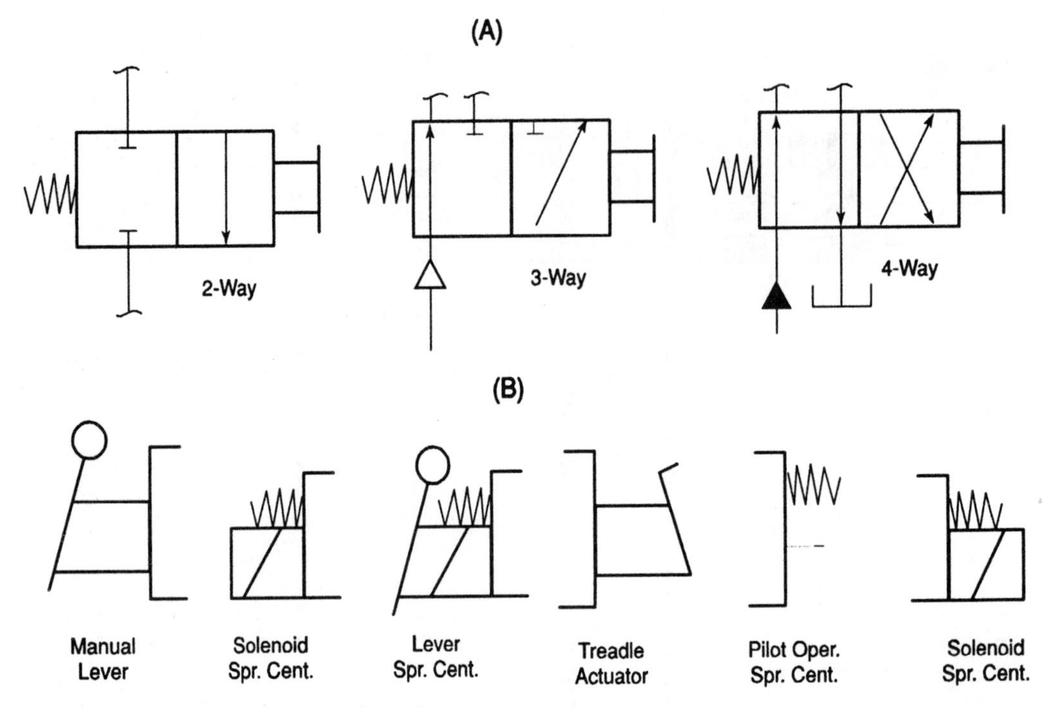

Figure 7-21 (A) Solenoid-operated, four-way directional control valve; (B) The spool may be shifted by a number of different operating devices.

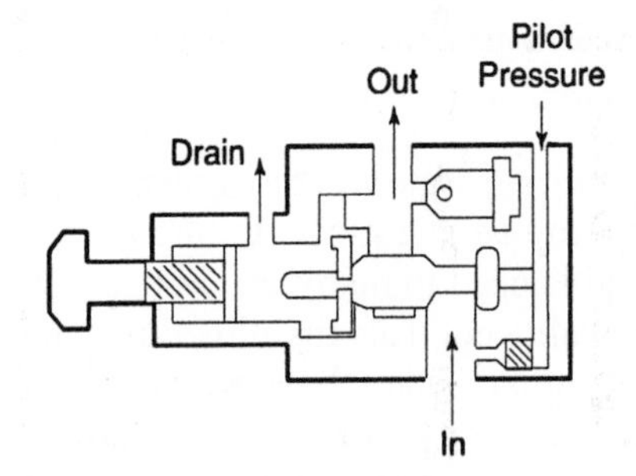

Figure 7-22 A schematic of a hydraulic sequence valve

transducer is compared in a controller to ascertain if the correct setting has been reached. If the desired position has not been achieved, a new signal is sent to the torque motor and the cylinder is repositioned. Highly sensitive servosystems can have a high-frequency pulse (a *dither*) superimposed upon the hydraulic system. The dither prevents the hydraulic seals from taking a set that would lower the unit's sensitivity to position change. The electrical hydraulic servovalve is employed in high-technology areas where large amounts of energy and control are required, such as in industrial robots, Figure 7-23.

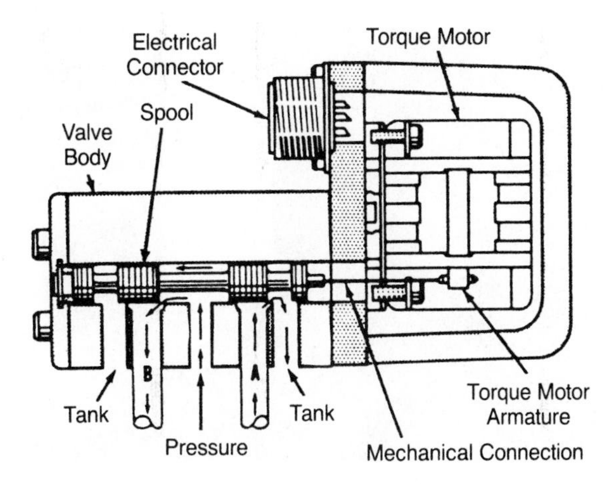

Figure 7-23 A schematic of an electrical hydraulic servocontrol

Programmable Controllers

The programmable logic controller (PLC) was designed to improve on the relay technology that was employed in large existing automotive manufacturing machinery. The relay manufacturing controls functioned on relay bank ladder logic, that is, one operation was completed before another function was started. All of these switching and sequencing operations were performed on a time base. The programmable logic controllers are applied to control many types of industrial operations, such as manufacturing lines, process control, measurement and gaging, material handling, and environmental pollution control.

The programmable logic controller is a digital computer that provides the additional functions of sequencing, timing, counting, logic, and arithmetic operations. The programmable controller has a dedicated computer designed for an industrial environment that is rugged. Because of its ladder logic, it is understood and programmed widely by engineers and technicians.

These controllers are maintained by electricians or technicians, because the repair is done by replacing modules rather than troubleshooting parts to find the operational problem. The controller often has its own diagnostic program that aids in the location of the troublesome module.

The programmable controller is usually programmed in contact symbology that is very similar to symbols used in programming relay ladder logic, however they may be reprogrammed in other languages.

The programmable controller, PC as it is also called, was developed because of the loss of production and the cost of rewiring relay banks whenever model or product changes were introduced into the manufacturing line. The PLC provided the opportunity to reprogram these machines and production lines at much less cost, with great reliability of operation. The programmable controller is small and

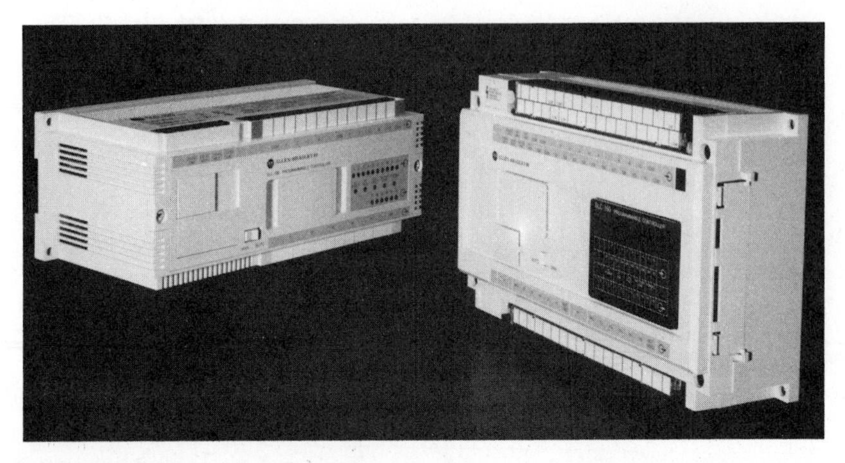

Figure 7-24 Programmable controller—shop-floor-control over manufacturing machines *(Courtesy of Allen-Bradley Company)*

frequently located near the manufacturing operation. It contains a programming device, a central processing unit, memory, I/O interfaces, and a power supply.

The output of the programmable controller interfaces closely with the processes controlled by such equipment as relays, switches, starters, motors, and various light displays. In addition, the PLCs will communicate with supervisory PLCs, i.e., other PLCs interacting with local area network (LAN) communication buses, Figure 7-24.

Digital Controllers

Digital controllers are a natural outgrowth of the application of large-frame computers that supervise a total system's functions down to individual processes. The development of the microprocessor brought about economic, realistic, and discrete process control. To perform this function, samples of the data are taken from an analog sensor and converted to a digital signal with the aid of an analog-to-digital converter. The digital converter delivers a signal that is compatible with other digital devices and represents the original signal. The advantage of a digital system is that it is economical, reliable, and the speed of response is improved.

The digital controller exploits the accuracy and error reduction inherent in digital applications. Digital electronics provides communication between different instruments and the computer, each having a discrete time of interaction. With the application of the computer to the control process, an increase in the complexity and potential of control functions is provided. The digital computer is capable of being programmed so that it functions under PID control. The final element in the control system will contain a digital-to-analog converter that takes the computer's output and delivers it in an analog form for the movement of the final element, Figure 7-25.

Figure 7-25 An industrial process control room *(Courtesy of Leeds & Northrup Co., A Unit of General Signal)*

Because of hazardous conditions, such as flammable materials or other factors, an industrial process may employ both an analog system and a digital system to control different parts of the system. With the application of various converters, they can interact and perform their required functions. This type of application is referred to as a hybrid system and is under the control of the digital system.

Analog control systems require as many controllers as there are processes to control. If a production plant requires ten processes to be under control, ten controllers and loops would be needed. With the application of a digital system, one computer would scan all the sensors on a time-sharing basis and be free to perform other functions.

A digital system can be expanded by rewriting the program to accommodate the new functions to be added. Functions such as new control loops, mathematical operations, alarms, quality control, inventory, scheduling, and a number of different forms of management data required for the processes to be controlled can be added.

With the computational ability of the computer, many control parameters based upon past performances of the system can be programmed and analyzed and stored within the computer for an optimum performance. At a later time, this data is used to calculate and select the best control values available for the process operation. This use of the data and the computer is called an "adaptive control system."

In critical process application, a complete system of analog control as well as a complete system of digital control are installed to provide an automatic backup control capability. The digital system usually controls the process while the analog system monitors the process so that control is available in the event of a failure in the primary digital controller system.

MICROPROCESSOR

The sensor microprocessor is a large-scale integrated chip or a group of chips that contains arithmetic, logic, and control logic that instructs and sequences. These functions on a single chip are referred to as a microprocessor unit (MPU). The importance of the MPU is that it has great versatility in the possibilities of being able to combine various chips for I/O interfacing into memory chips with some programming ability. The microprocessor is usually a dedicated device, used to control the operation of equipment, such as instruments that control level, temperature, flow, or the like. The microprocessor can have a permanent form of programming and will perform its function reliably over long periods of time. This is different from the microcomputer in that the microcomputer can be reprogrammed and continually modified by its user. The microprocessor is designed to perform specialized tasks.

The single chip microprocessor/microcomputer has provided a new class of instrumentation referred to as "smart" instruments. These instruments, because of their internal programs and arithmetic logic unit (ALU), can perform the surveillance of a vast number of sensors, sample their outputs and average their results, and perform statistical operations on the data, such as standard deviations. The output data is compared with the set-point to provide a process corrective error signal. In addition, these instruments can be programmed to periodically recalibrate themselves so that operational errors, such as drift, are eliminated, Figure 7-26.

The microprocessor-based instrumentation has advantages in the reporting of the condition of ongoing real time manufacturing processes. Because of their electronic interface properties, they can communicate decision data to a host of different electronic display instruments. The microprocessor can also be the input to a microcomputer, minicomputer, or the corporate mainframe computer. Instrument control rooms are usually made up of a series of CRT terminals that display the status of a whole manufacturing facility. This information is used to control the facility and is the raw data for the generation of management reports and production estimates and trends.

MICROCOMPUTER

The microcomputer is a small, complete computer with the essential components mounted on a printed circuit board. It is composed of large-scale integrated circuits, a microprocessor, that contains an arithmetic logic unit, instruction set, registers, and computer control unit. The microcomputer circuit board contains a

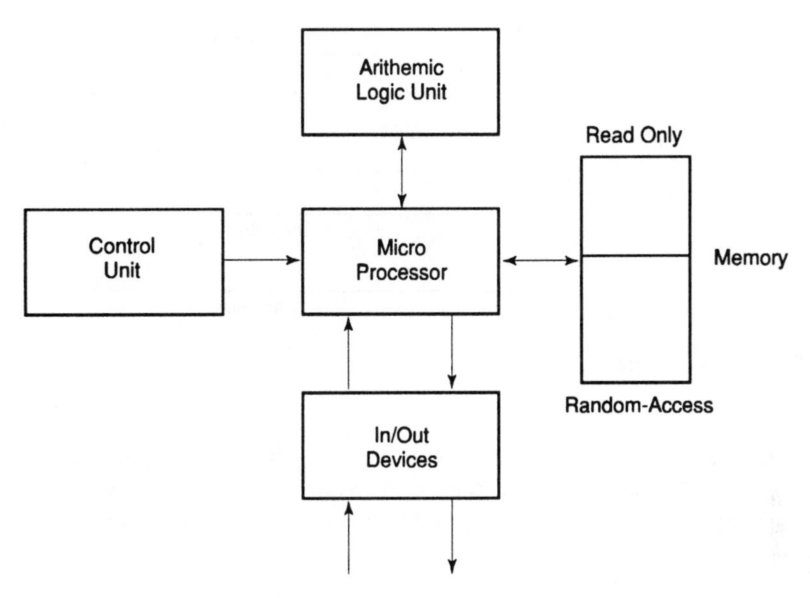

Figure 7-26 Microprocessor/microcomputer controller diagram

microprocessor chip as a central processing unit. The microcomputer board is interfaced with additional chips containing a solid-state memory called a "random access memory" (RAM). The RAM is referred to as a "volatile" memory because if power to it is lost, the memory is erased. When the memory is active, it can be written to by means of a keyboard or read from with a CRT or printer. The microcomputer also has a "read only memory" (ROM) that permanently retains its information even when the power is removed. This comes into play when starting the microcomputer because the ROM provides permanent information and procedures preprogrammed into the chip that are necessary to start the microcomputer. The output of the CPU is sent to circuits that convert logic-level signals to actuating power for devices such as relays and motors for control.

Single-chip computers contain a CPU, read/write memory, RAM, programmable read only memory, input/output circuits, a clock, and frequently other specialized circuits, all within one integrated circuit. These devices are primarily dedicated controllers for such devices as printers or small appliances. These are the essential units to make up a microcomputer, Figure 7-27.

DIRECT DIGITAL CONTROL TECHNOLOGY

Direct digital technology is described as one in which the machine or process is interacting with and being controlled by a digital computer. In the industrial setting, sensors measure the variables and transmit directly to the mainframe

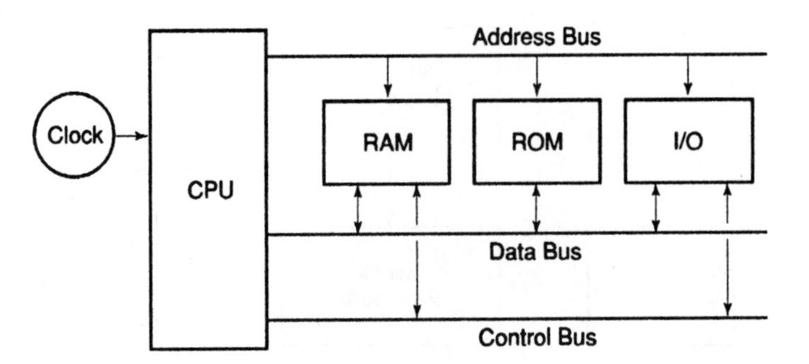

Figure 7-27 A block diagram of microcomputer essentials

computer. The sensing may be done on a chemical process, on a computer numerically controlled machine, or on any number of machines or processes. The chief responsibility is to get reliable information to the computer in a form to which it can respond.

Analog Data Input

Information or data leaving a sensor is transmitted in either analog or digital form. Until the last two decades, instrumentation data was transmitted largely in analog form, meaning continuous information coming from a sensor as a voltage change, amperage change, resistance change, frequency change, pressure change, temperature change, position change, force change, and so forth. In order for a digital computer to respond to this continuous data, it requires a change into a coded digital form. The A/D convertor (ADC) is made with a comparator differential amplifier whose output is a function of the difference in voltage between two signals. With the addition of gates and a flip-flop, an output is created that is a sharp transition between zero and five volts representing a binary 0 or 1 that can be received by the computer. If the input is an AC signal, it is converted into DC voltage levels and handled as above. In AC signals that transmit information by their frequency changes, the signal is converted to a squarewave and presented to the computer as a train of different DC voltage levels, Figure 7-28.

Digital Data Input

Digital signals can come directly from angular and linear digital encoders. These encoders provide a binary code that is received in digital series information, pulse rate information, or in a parallel binary word transfer from other devices to go directly into the computer.

Another type of signal only needs to know what state the sensor is in, such as a 1 condition or a 0 condition. This type of data frequently comes from switch positions, relays, limit switches, optical sensors, or safety devices. When the

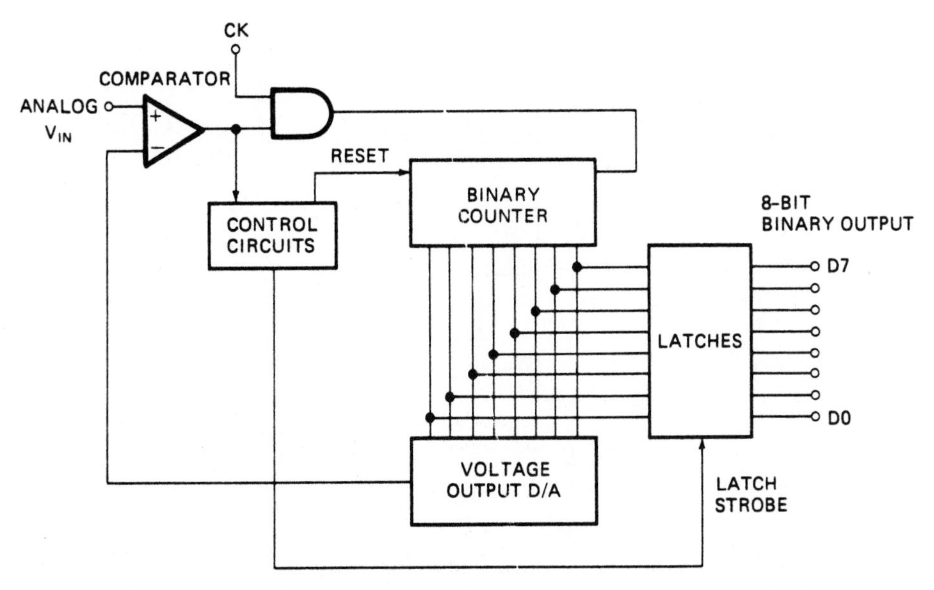

Figure 7-28 A schematic of analog-to-digital conversion applying the ramp voltage technique

sensor reports a change in important information, as in the case of safety devices, this change may require an action by the operator. In this case, an interrupt causes a temporary pause in the computer's routine so that the corrective action can be carried out.

Digital signals that are received from external memory devices such as tapes, discs, and keyboard inputs require additional control and timing circuitry to communicate with the computer. The responsibility of the interface electronics is to prepare these various types of signals to be received by the computer.

Computer Output

The computer output is the command that goes out from the computer to the unit being controlled. As with inputs, these commands will be in either analog output or digital output commands.

Analog Outputs

The signal leaves the computer as a digital signal and must be changed to control the actuator devices that are analog machines. These actuators are often switches, servovalves, electrical motors, graphic displays, or strip-chart recorders that provide the change in the variables or information. The analog devices require a changing voltage signal rather than a digital signal. To bring about this conversion of signal, a digital-to-analog converter is employed, Figure 7-29.

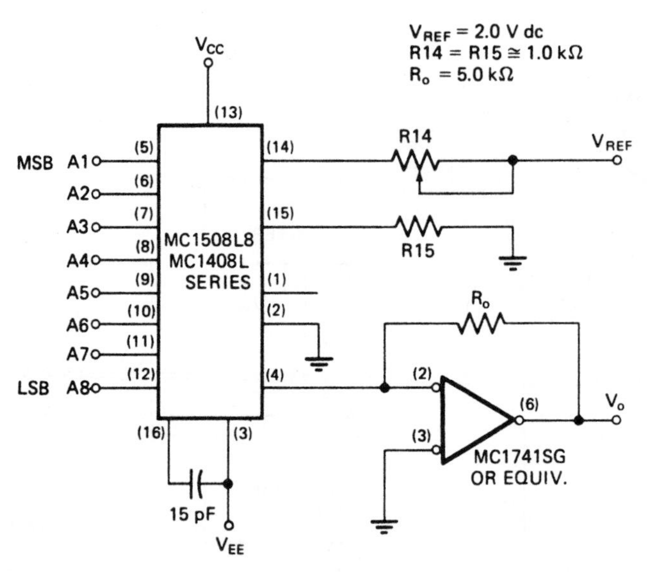

Figure 7-29 A schematic of an operational amplifier digital-to-analog converter

DIGITAL COMPUTER CONTROL

One of the advantages of digital control is the flexibility that can be built into the computer functions by the application of programming. Programming is a series of intelligent instructions designed to control the machine and cause it to perform the operations required by the programmer. The program is written by the programmer in a language that the machine can understand, a binary machine language. It takes human ideas and translates them into demands. These requirements are programmed into a logic of 0s and 1s that can command the machine to react and produce accurate responses or control as the programmer directed.

Software is defined as all the activities associated in the development and operation of a system except for the hardware or physical equipment. The software is anything that contributes to the control of the computer. The physical elements of the computer and the production equipment are considered as hardware.

In developing software for computer instrumentation and the automation of production, a program development plan is designed:

1. Define the problem. Carefully state, in whatever terms necessary, the scope, requirements, and specifications that are to be performed. State what the solution of the problem will be in measurable terms. This will make it possible for the problem to be evaluated easily upon completion and help in the correction process. It will also contribute to the decision processes as to what hardware will be required.

2. Determine a method of solution. Most of the time, there will be more than one procedure for solving a problem. Design a program that is simplest and most cost-effective. Some techniques are the flow chart, top-down design, modular program, or others.

3. Code the program. The program is started in a symbolic language such as BASIC, FORTRAN, or other languages. In turn, the program is translated into the language of a microprocessor, the machine language of 0s and 1s.

4. Run the program and verify the program's operation. New programs usually have some errors that need to be located and corrected. Isolating this fault and providing the correction is called "debugging."

5. Validation of a program is the testing of the program to see if the requirements and specifications stated in the definition have been achieved. The program is run with different inputs to check if the program and the microcomputer outputs delivered are equal to the designed outputs.

6. Documentation of the above processes is required so that if it becomes necessary for the user or maintenance personnel to modify the program at a later date, they can follow the developmental procedure. These people need to understand the original requirements, specifications, flow charts, program listings, and memory maps in order to correct for the user's needs, Figure 7-30.

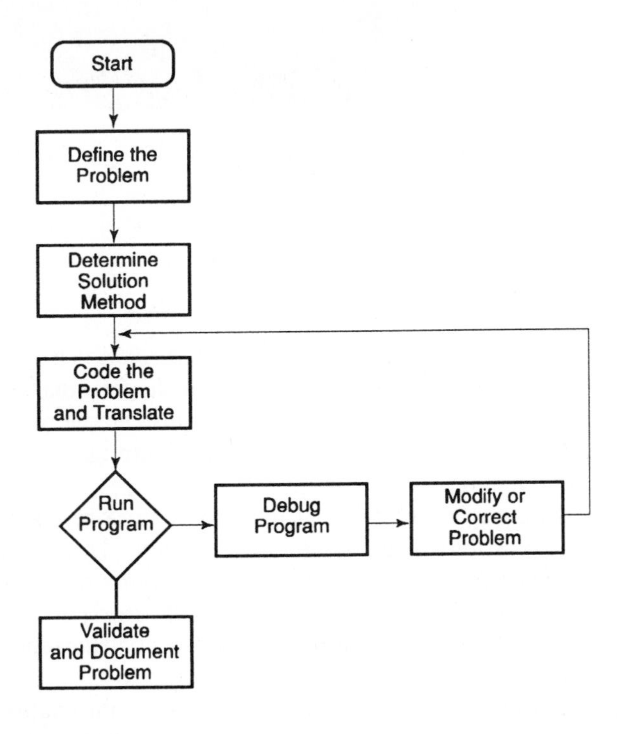

Figure 7-30 Program development

Figure 7-31 A control-based operations workcenter *(Courtesy of Siemens Capital Corp.)*

DISTRIBUTIVE CONTROL SYSTEMS

A *distributive control system* allows an operator station in the control room to use a series of CRT displays and a keyboard of the terminal to provide additional instruction and flexibility of action to the many remote microcontrollers. These microcontrollers or field controllers are placed throughout the production facility near the production process being controlled.

"Smart microprocessors" that interface with a series of dispersed sensors provide the basis for a distributive control system. These enlarged microprocessors or single chip microcomputers have a set-point memory, power, and speed that provide calculations, logic functions, and storage and retrieval functions. They have the ability to follow complex instructions to reprogram the microcontroller.

The microcontrollers perform direct control of automation processes through a long-distance communication network. The control task is divided between a central supervising computer and the microcontroller. This microcontroller is a smart microprocessor based controller and provides direct control over the various manufacturing actuators, such as relays, control valves, and motors. These microcontrollers are all integrated and interacting. A sixteen-bit, single-chip microcontroller is able to communicate with other processors over a serial link with peripherals with a parallel bus. These microcontroller chips provide high-performance control because they can handle large data words for sophisticated control systems.

The advantage of the distributed control system is that the control room or operations workcenter keeps the various personnel of the organization up-to-date about operation conditions occurring any place in the production system, Figure 7-31. It allows the operator to understand and react to process problems. It also provides current raw data so that patterns and trends are revealed quickly and decisions or actions may be carried out using accurate and reliable information. Control strategies can be changed with the application of software rather than changing hardware. Complex control modes changes are possible through the application of additional software.

The data and information from a distributed control system are very important to a shift operator in the recovery of a process upset. This accurate and current information will aid in bringing the systems back to maximum production, while keeping other processes within their parameters.

In addition, plant management also uses the data and trends provided with the data for planning and balancing of production rates, for controlling inventories of raw materials and products, and for the compilation of managerial reports for corporate decisions.

SUMMARY/FACTS

- A control loop is a group of instruments designed to regulate a system and maintain that system at a desired value.
- There are two types of control loops: one is referred to as an open loop and the other as a closed loop.
- The closed loop provides feedback for correction of the system under control.
- A change in process load is one of the dynamics that brings about the necessity of a correction in a system.
- Basic controllers employ different modes of process control to meet the processes needs. They are on/off control, *proportional control*, proportional-plus-integral control, proportional-plus-integral-plus-derivative control, hybrid control, direct digital control, and distributive control.
- The programmable controller is a rugged dedicated computer that is widely used in manufacturing and process control where the processes controlled primarily require timing, sequencing, and switching.

- A distributive control system allows an operator station in a control room to use a series of CRT displays and a keyboard of the terminal to provide additional instructions and flexibility of action to many remote microcontrollers. These smart microcontrollers or field controllers are placed throughout the production facility near the production process being controlled.

REVIEW QUESTIONS

1. What is the name given to a group of instruments designed to regulate a system and maintain that system at a desired value?
2. What is the difference between an open loop and a closed loop control system?
3. What does a change in a process load indicate?
4. What is the reduction in the rate of energy transfer from one medium to another called?
5. In measurement and control, that reduction is measured in terms of __________.
6. In process control, what is dead time?
7. What is transfer lag?
8. What is a two-position control that is either fully open or completely closed called?
9. Control that has a throttling or modulating action that can be positioned at different energy rates is referred to as ________________.
10. What is the ratio between output and input of a control device known as?
11. What does the proportional-plus-integral control action automatically adjust?
12. The derivative element in the proportional-plus-integral-plus-derivative control produces an output proportional to what?
13. What methods of control are combined to produce a hybrid control system?
14. In explosive prone industries, which controllers have come to play a valuable role?
15. When a flapper in a pneumatic controller is placed in front of a jet of air, what is the result?
16. By adding a bellows and an adjustable leak to a proportional-plus-integral action pneumatic controller, what is supplied?
17. To design a pneumatic derivative control, what is added between the pneumatic relay output and the integral or reset bellows?
18. What is a bistable amplifier?
19. In hydraulic directional control valves, spool valves are classified according to what criteria?
20. What is the primary function within the valve?

21. A device that allows a low energy signal to control a very large amount of hydraulic energy with precise control of position and speed is called a ____________________.
22. Describe how an op amp can be used as a voltage comparator.
23. Why is it an advantage for sensors to have a digital readout?

8

CONTROL ACTUATORS AND CONTROL VALVES

OBJECTIVES

Upon completing this chapter, you will be able to describe or explain how these actuators and control valves work and where they are applied:

- Electric actuators
- Pneumatic actuators
- Hydraulic actuator controls
- Hydraulic actuator power supplies
- Final control elements (control valves)

INTRODUCTION: ACTUATORS

The control actuator is the unit in the system that receives a signal from the controller and uses that signal to produce a change. This results in a new condition in a valve, switch, solenoid, motor, etc., that changes the controlled variable in the process.

Signals may arrive in different forms: electrical voltage signals ranging from 3 to 115 volts; electrical current signals that vary from 4 to 20 milliamps; pneumatic pressure signals that vary from 3 to 15 pounds per square inch; and digital signals with transistor-transistor logic (TTL) voltages in serial or parallel format.

The end unit of the control system is the actuator. This is a *servomechanism* which accepts the error signal and converts it into a mechanical motion (e.g., rotary, eccentric, linear, or reciprocating motion) to make a change in variable.

Electric Actuators

Electric actuators appear in different forms. An electric motor is commonly applied for the delivery of mechanical energy for movement of a final control element. The electric motor may be used to power a gear train that opens or closes large valves in hydroelectric power producing installations. In low technology manufacturing, the electrical power supply used to provide the energy for work is the AC induction motor. It is used to drive hydraulic pumps, transfer devices, and rotate machinery. Medium technology employs stepper motor technology for rate of change and positioning of tools and machinery. The stepper motor receives a pulse that rotates its shaft a specific amount with each pulse. In so doing, the stepper motor can control the movement of tools, machinery, and manipulators with data from a punched tape or microcomputer. High-technology manipulator equipment utilizes DC brushless motors or AC motors. These motors are of both rotary and linear types and are used because they do not contain brushes to energize the rotor for their magnetic fields. Because a commutator is not used, these electric motors do not generate electrical sparks that can result in dangerous environmental conditions.

Linear Motion Actuators

A linear actuator is any actuator that develops a forward or return motion along a straight line.

Solenoid - An electromechanical activator very commonly employed is the solenoid. It is designed with a magnetic moveable core surrounded by a coil producing a magnetic field. The strength of the field causes the core to move. The core is attached to the device that is to be positioned. Solenoids can be digital in operation by having the applied voltage move the core to its maximum position. When the voltage is off, a spring returns the core to its minimum position. This action produces an on/off device and thus a digital actuator, Figure 8-1. A vast number of applications utilize this on/off mode of operation.

Relay - A relay is a special application of the solenoid. These solenoids are designed so that they operate sets of electrical contacts. These contacts can be connected so that they are in the normally open or normally closed contact position. In addition, the contact leads can be crossed-over so the polarity of devices can be reversed. The relay has a common application of a remote switch

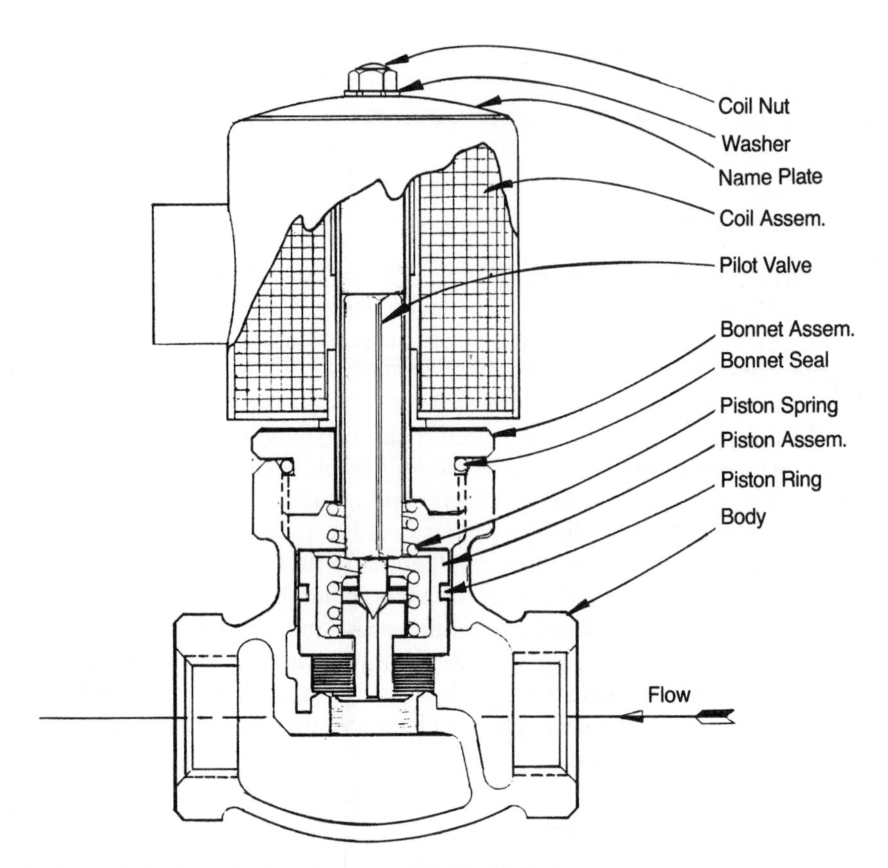

Figure 8-1 A typical solenoid valve *(Courtesy of D. Gould Co. Inc.)*

Figure 8-2A Solenoid-operated state optical relay

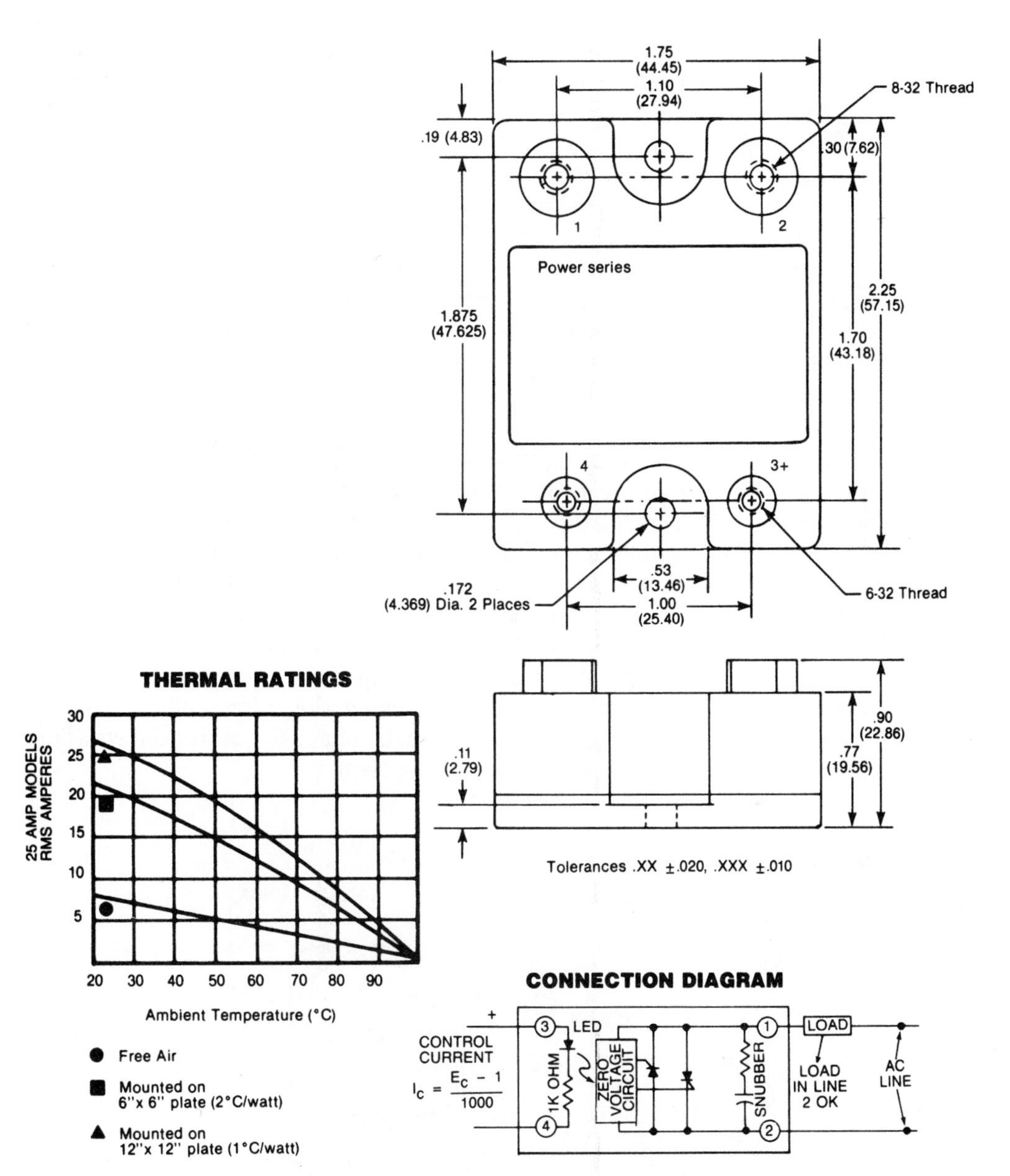

Figure 8-2B DC voltage control solid relay *(Courtesy of Opto 22)*

for the control of large dangerous voltages by a less hazardous low-voltage energizing circuit, Figure 8-2A.

Solid-state relays are employed because of the lack of mechanical moving parts and the high reliability. They use a light-emitting diode to trigger a light-sensitive transistor to complete the required circuit. These devices can be activated by a TTL voltage and control very large voltages without any physical contact of the conductor in the relay. Common voltages are three to thirty-two volts from the electronic circuitry that control up to four-hundred fifty volts AC for the actuator, Figure 8-2B.

An electrical analog device in the form of an amplifier receives a low-energy signal and increases the signal to a high energy level. These units may be used to power other analog devices with their increased proportional signal.

A power amplifier delivers power to the solenoid. As the power is increased, the movement of the core will vary and provide a series of continuous positions. This analog function is used in electrical braking mechanisms and other similar applications in which the solenoid provides increasing linear motion with an increase of electrical energy.

Rotary Motion and Linear Torque Actuators - Rotary motion solenoids and torque motors provide precision control movement. Normally a rotary motion solenoid does not have a large range of movement. An example of this type of analog device is the voltmeter with an D'Arsonval movement. In this case, an increase of voltage interacts with the magnetic field of the permanent magnet and produces a torque in a moving coil. This causes the coil and pointer of the instrument to move to a new equilibrium position indicating the voltage.

In another application, the analog linear torque motor positions a pilot spool valve to control the amount of hydraulic fluid entering a hydraulic servovalve, Figure 8-3. This analog actuator receives a small voltage signal but delivers a huge amount of controlled energy to a hydraulic system.

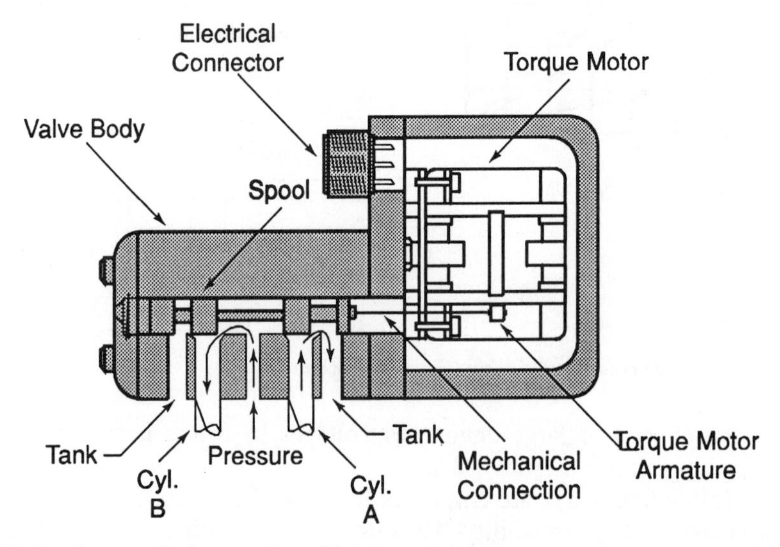

Figure 8-3 Hydraulic controlled servovalve with torque motor

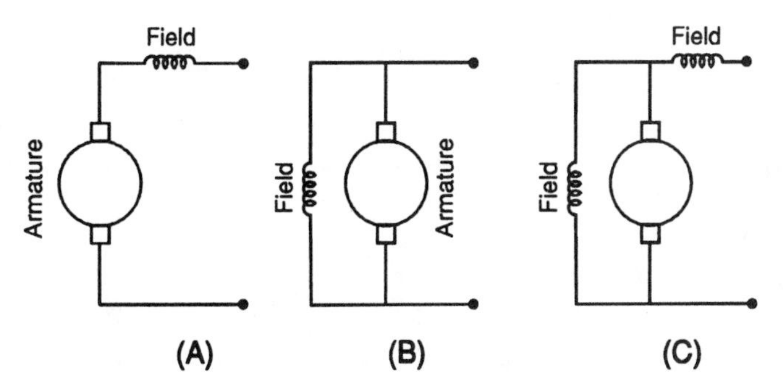

Figure 8-4 Three connections of direct current motors: (A) Series-wound motor; (B) Shunt-wound motor; (C) Compound-wound motor

Direct Current Electric Motors

Direct current electric motors are actuators designed to apply the principles of electromagnetic attraction and repulsion of magnetic poles to supply torque to the armature and shaft. Direct current motors are reversible, and their speed can be accurately controlled. These characteristics make them very valuable for the control of many types of manufacturing process variables.

Direct current motors are manufactured with three internal connections: a series-field-connected motor; a shunt-connected motor, in which the armature and field are connected in parallel; and a compound-connected motor, in which two field windings are provided with one field being a shunt winding, Figure 8-4.

Universal Motors - The series-connected motor has the armature and the field windings connected in series with the source of power. The field and armature coils receive the same current flow and therefore both have windings with relative few turns of heavy wire. In motors with this construction, the torque increases as the speed decreases. These motors deliver extremely high torque at low speeds. When the armature speed is low, the counter voltage is low and the amperage flow is high. As the armature's speed is increased, the counter voltage increases and reduces the amperage flow. This high starting torque is an advantage of this type of motor. The motor should be attached to some type of load or a runaway motor condition will result with damage to the motor.

In a series-excited motor, the speed of the motor varies with the load that is applied. This type of motor is also called a universal motor, because it will run on direct current and alternating current. If the motor is running on direct current and the leads are reversed, the motor continues turning in the same direction. This is because in the series connection, the current was reversed in both the armature and field windings and the direction of rotation did not change. The direction of rotation is a function of the right-hand rule and, with both fields changing polarity together, the motor will accept an alternating current, Figure 8-5A.

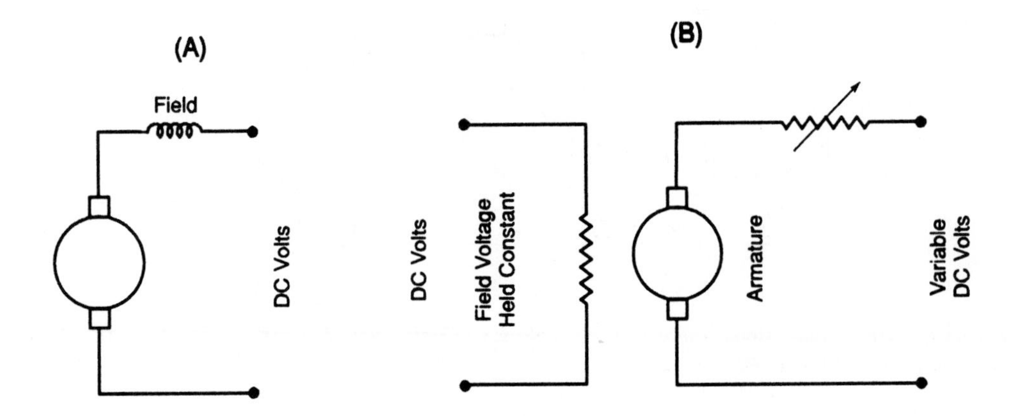

Figure 8-5 (A) Schematic of a series-wound, self-excited DC motor; (B) Schematic of a separately-excited DC motor

In a separately-excited motor, the armature voltage is variable and the field voltage is stabilized. This results in a motor in which the output torque remains constant and the horsepower output directly proportional to the speed. In the other case where the field voltage is variable and the armature voltage is stabilized, the horsepower remains constant and the output torque is proportional to the speed, Figure 8-5B. These relationships provide for additional motor control.

The Shunt Field Electric Motor - The shunt field motor is wound so that the field winding is connected in parallel with the armature winding, Figure 8-6A. The field winding is wound with a higher resistance using a finer wire than that of the armature. The speed of this motor is almost independent of the armature current up to the motor's full load. The result of this shunt-wound motor is a constant speed.

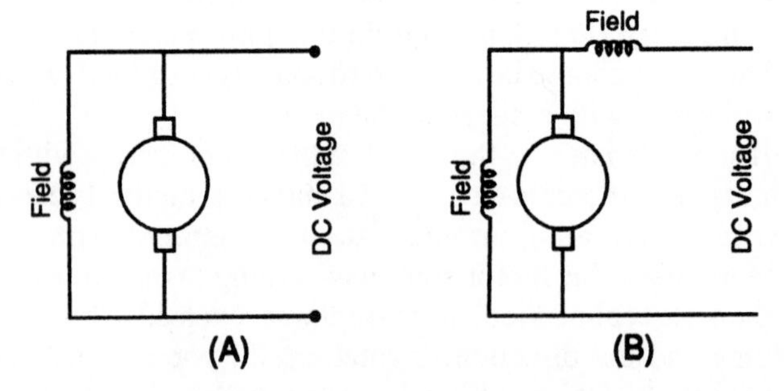

Figure 8-6 (A) Schematic of a shunt-wound DC motor; (B) Schematic of a compound-wound DC motor

The Compound Motor - The compound motor is wound to take advantage of the high starting torque of the series-wound motor and the constant speed of the shunt-wound motor. Two field windings are used: one with a few windings with a heavy wire, in series with the armature; the second winding is of many turns, with a finer wire, connected in parallel with the armature, Figure 8-6B. This motor delivers a very high starting torque and will maintain a constant speed under a changing load. Its speed will not run away when the load is removed.

Speed Control - The control of the speed in a series DC motor can be realized by placing a rheostat in series with the motor, Figure 8-7A. When the rheostat decreases the current in the field and armature, the rotational speed is reduced. As long as the load on the motor does not change, the motor speed will remain stable. When the resistance supplied by the rheostat is reduced, the motor will increase in speed.

The shunt-wound motor has within its design a speed control discussed above. But in addition, a rheostat can be placed in series with the armature to control the current received and thus vary the motor's speed. Also the rheostat can be placed in series with the field only and thus reduce the motor's speed, Figure 8-7B.

The compound motor's speed is controlled by placing rheostats in series with the field or armature or both. This provides a range of control with various combinations of resistances.

DC Motor Rotation Reversal - To reverse the direction of rotation of DC motors, the polarity of only one part of the motor is reversed with respect to the other. When the motor is running in one direction, it can be reversed by interchanging the leads to the armature or to the field, but not both. With DC motors, generally the leads to the armature are reversed, Figure 8-8.

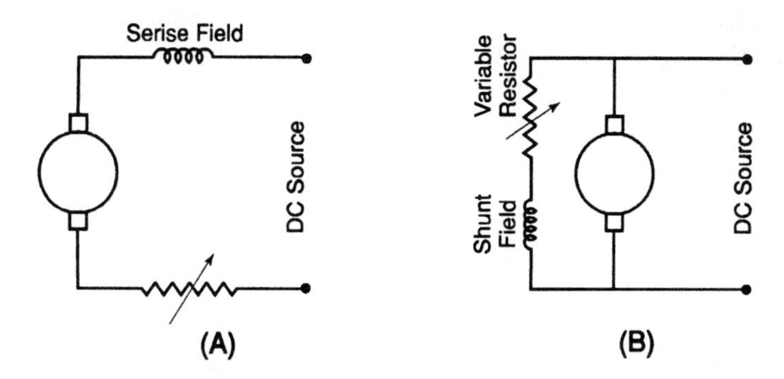

Figure 8-7 (A) Schematic circuit for the speed control of a series DC motor; (B) Schematic circuit for the speed control of a shunt DC motor

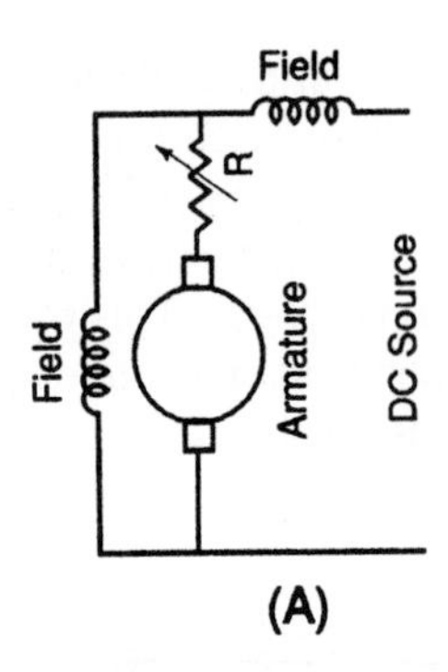

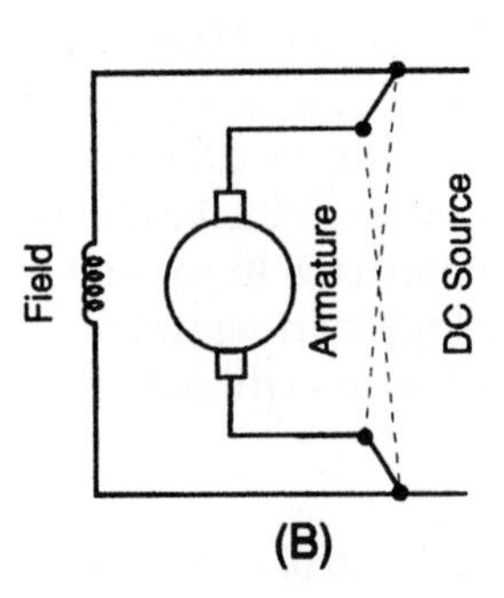

Figure 8-8 (A) Schematic circuit for the armature speed control of a compound DC motor; (B) Schematic circuit for the reversing of a DC motor (—— represent the alternate position)

Solid-state Motor Controls - The technology of control of DC motors has been rapidly transferring to solid-state control systems. These phase-control converters are provided through the application of a thyristor referred to as a chopper or SCR. The SCR is a device made up of a solid-state fused unit of a diode and a transistor. The SCR or thyristor is an electrically controlled silicon rectifier. The thyristor is switched on and off at a very rapid rate ending in a width of pulse modulation. The longer the thyristor is on, the wider the pulse and the more energy available to the motor.

This type of control is applied to heavy equipment that requires accurate control as well as large amounts of power, such as steel rolling mills, mine hoists, light rail systems, rotary machinery, and large control valves.

Until recently, AC motors have been considered as constant-speed sources of power. However, with solid-state control systems technology, speed control has been realized with a variable change of frequency. With the application of the thyristor, power transistor, and the gate turn-off thyristor, solid-state polyphased AC motors are provided with speed control. The types of motors that utilize polyphase variable speed control are squirrel cage induction motors, wound rotor induction motors, synchronous, and synchronous-reluctance motors.

These circuitry additions to the power supplies of electric motors are available when constant speed is essential for all manner of rotating manufacturing machinery.

Alternating Current Electric Motors

In alternating current motors (like the DC motors), the motor action is caused by conductors in a magnetic field. They are attracted or repelled depending upon their magnetic polarity, resulting in a rotary motion. The different characteristic is that the polarity of the device is constantly reversing. Because of this action, the AC motor may be visualized as a special transformer with the secondary winding coils free to rotate (rotor) and the primary winding coils placed around stationary pole pieces (stator).

There are different types of AC motors. These vary primarily because of the means of producing a magnetic field by the armature or rotor. The main groups of AC motors are these: the induction type, the synchronous type, and the universal type.

The Induction Motor - The induction motor is designed with copper or aluminum bars in place of the soft iron material and coils in the armature or rotor. The squirrel cage armature is an arrangement of conductors shorted so that each conductor provides a complete circuit, Figure 8-9B. The stator is connected to an AC source that induces very large current in the bars of the armature. This heavy current generates a strong electromagnetic field around each conductor. The outcome results in an interaction between the magnetic fields. This interaction provides the torque and rotation.

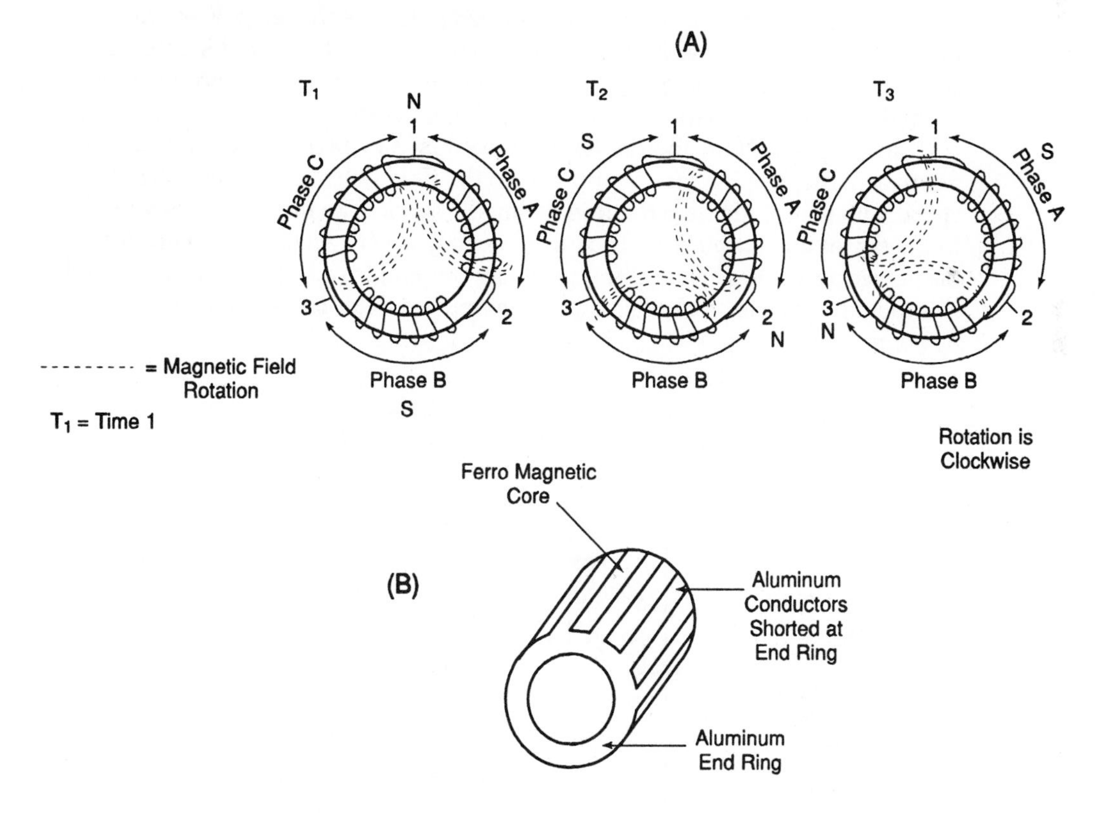

Figure 8-9 (A) A three-phase motor schematic demonstrating the rotating magnetic field phenomenon; (B) Squirrel cage rotor of an induction motor

The magnetic field in the stator is rotating around the stator. As it rotates, because the power coming in is a phased alternating current, the magnetic flux is constantly building up and collapsing. This action induces a current into the rotor. The rotor's polarity then is attracted to or repelled by the stator's magnetic field causing the rotor to rotate and deliver the power, Figure 8-9A. The speed of the rotation of the rotor is almost equal to that of the magnetic rotation within the stator. However, because of inertia, friction, or load, there is a difference between the two. This difference is referred to as "slip." It provides the cutting of the magnetic lines of flux and induces the current into the rotor.

The polyphased electric motor is the common motor used in manufacturing to power grinding mills, conveyor belts, hoists, mixers, pumps, as well as a host of power sources for rotating machinery.

The Synchronous Motor - A single-phase shaded-pole synchronous motor has a stator that is wound as that of an induction type of motor. However, the armature or rotor does not depend upon the induction of current for its magnetic field, except for the starting of some motors using a shaded pole. The shaded pole is produced by placing a solid copper loop around a small portion of the field's poles. This shaded coil causes a reaction to provide its starting torque. The coil produces only a small torque when starting and running, Figure 8-10.

Small synchronous motors have a rotor made of permanent magnets. The permanent magnets' poles are attracted to the field poles and synchronize their speed with that of the field rotation within the stator. The motor rotates with the frequency of the line current. The line frequency is very accurately controlled by the power companies. Thus, this type of motor is employed where constant speed is important, such as in clocks, fans, counters, and timer cams where RPM must be controlled.

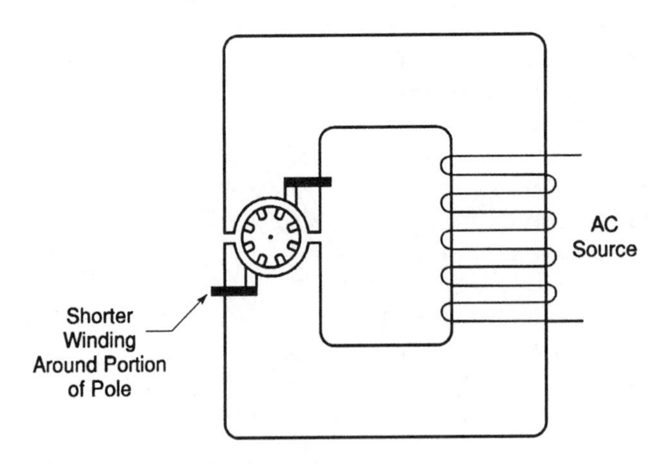

Figure 8-10 A schematic of a shaded synchronous motor

Large-size AC synchronous motors, those above one horsepower, are provided with direct-current excitation supplied to the rotor through slip rings. The direct current is supplied to the rotor from a generator attached to the shaft of the synchronous motor. These motors will supply large horsepower to industrial loads and are powered by either single or polyphased power. The large motors convert electrical power into mechanical power very efficiently at a constant speed. They also improve the power factor of the system. These large motors are used in applications that run for extended periods of time, as in power stations, pumping stations, and mines.

The Universal Motor With AC Power - The universal motor is a series DC motor with an AC power supply. In this application, both the rotor and stator core are made of laminated material to minimize eddy currents. Care is taken to control the sparking at the commutator. This type of motor is inherently a constant-speed motor. Therefore control will be gained by the use of reduction gears and clutches. This type of motor is used on vacuum cleaner motors, fans, and numerous industrial timers.

There are variations on the general types of motors made of combinations. With the addition of different components, such as silicon-controlled rectifiers, capacitors, centrifugal switches, brush configurations, rotor and stator windings, they become motors with specific characteristics required for various functions.

Digital Motor Actuators

Solid-state technology is capable of delivering a very accurate number of electronic pulses to actuator circuits. This pulsing accuracy is transformed into precise movements.

The Stepper Motor - The stepper motor is a direct digital actuator. Rotation is achieved by a number of discrete pulses that cause the rotor to move by one position per pulse. Pulses are delivered one at a time or in a train of pulses to cause a measured movement or continuous rotation. These systems are very accurate and will hold their last position until a new signal is received by the stepper.

Permanent-magnet Stepper Motor - The permanent-magnet (PM) stepper motor is designed with a rotor of a permanent magnet with two or more magnetic poles. The stator is designed with a series of pairs of magnetic pole pieces with two sets of windings common on each common stator with opposite magnetic generation coils. When one coil is energized, the pole pieces reverse their magnetic poles so that the permanent magnetic rotor positions itself between two opposite but like pole pieces. The coils that energize the pole pieces are controlled by solid-state switches that reverse the polarity of the pole pieces. With pairs of pole pieces at different positions around the rotor, this switching and changing the

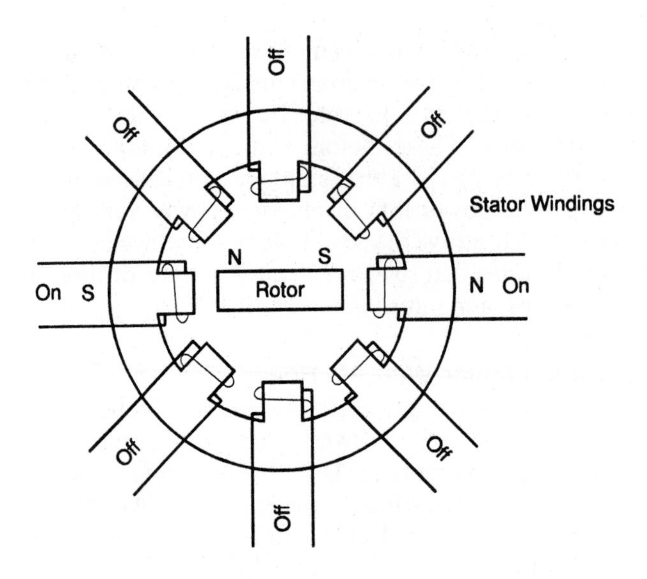

Figure 8-11 Schematic of a permanent-magnet stepper motor

polarity of the pole piece faces, causes the rotor to rotate one position. A series of pulses reaching these switches advances the rotor one pole for each pulse, Figure 8-11.

This stepper motor has an advantage because it responds directly to a digital code that can be interfaced with solid-state electronics. Thereby all the flexibility of computer control is available for industrial processes and manufacturing automation with absolute position accuracy. The stepper motor is applied to positioning machinery such as numerical controlled machines, robots, automated punching machines, and many other positioning and measuring machines.

Variable-reluctance Stepper - The variable-reluctance stepper is designed with a rotor of a magnetic material rather than a permanent magnet. This provides less torque but more positions are usually available. The rotor and stator poles or teeth are aligned by electromagnetic fields and orient the rotor position with the nearest set of energized poles, the position of least magnetic reluctanace between the rotor and the stator pole face, Figure 8-12. When the power is off, the rotor does not hold its position or detent torque as would a rotor designed with a permanent magnet. This is not a serious problem because the power is required to remain on during step movements. The variable-reluctance stepper motor is the positioning

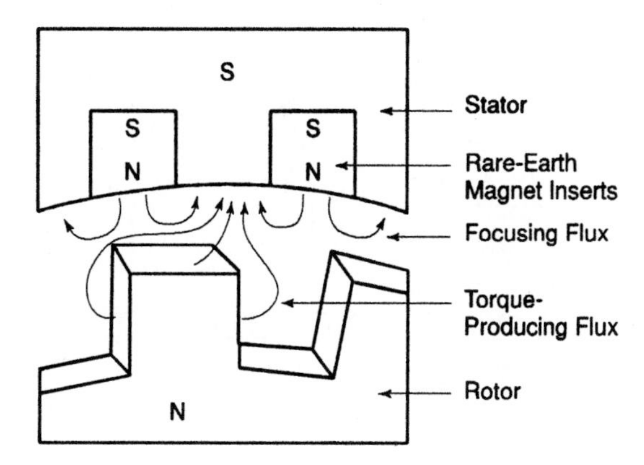

Figure 8-12 Rare-earth and alnico magnets create a high-performance stepper motor

motor for many computer peripherals, valve actuators, positioning tables, and indexers.

The Brushless DC Motor - Brushes in motors wear and produce arcing under load, causing commutator wear and damage that limits the operation of the motor. To eliminate these problems the motor was redesigned, reversing the motor's structure and electronically commutating these motors. In a brushless DC motor, the armature is stationary and the field rotates. The design is opposite to the traditional DC motor in which the armature rotates. The armature of the DC brushless motor contains from two to six coils that are pulsed with the aid of two or four transistors per coil. The transistors commutate each motor coil by providing pulses that are switched by sensors mounted on the rotor. These sensors deliver a transistor-switching sequence so that a rotating magnetic flux results within the coils. This rolling magnetic flux is at a fixed angle to the flux produced by the permanent magnets in the rotor. The interaction of the flux fields produces a torque that is directly proportional to the current supplied. This type of motor has a low starting torque and is applied to equipment such as blowers or speed reducers to increase the torque available for work. Because of the lack of brushes, sparking, carbon dust, and RF noise, these motors are ideal for clean rooms, vacuum chambers, and volatile environments.

The DC Brushless Linear Motor - The brushless DC linear motor supplies smooth, simple, straight-line movements. This motor is designed using highly efficient rare-earth magnets responding to electronic noncontact commutation. The motor can provide accelerations of 4gs and velocities to 60 inches per section without backlash or stepping action.

(A)

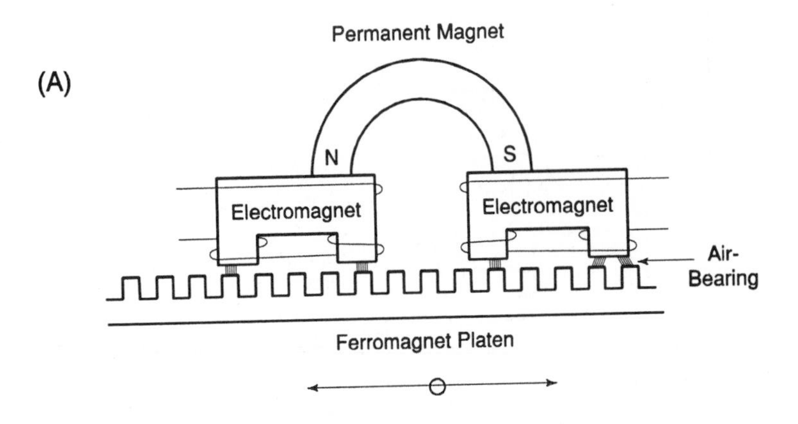

(B)

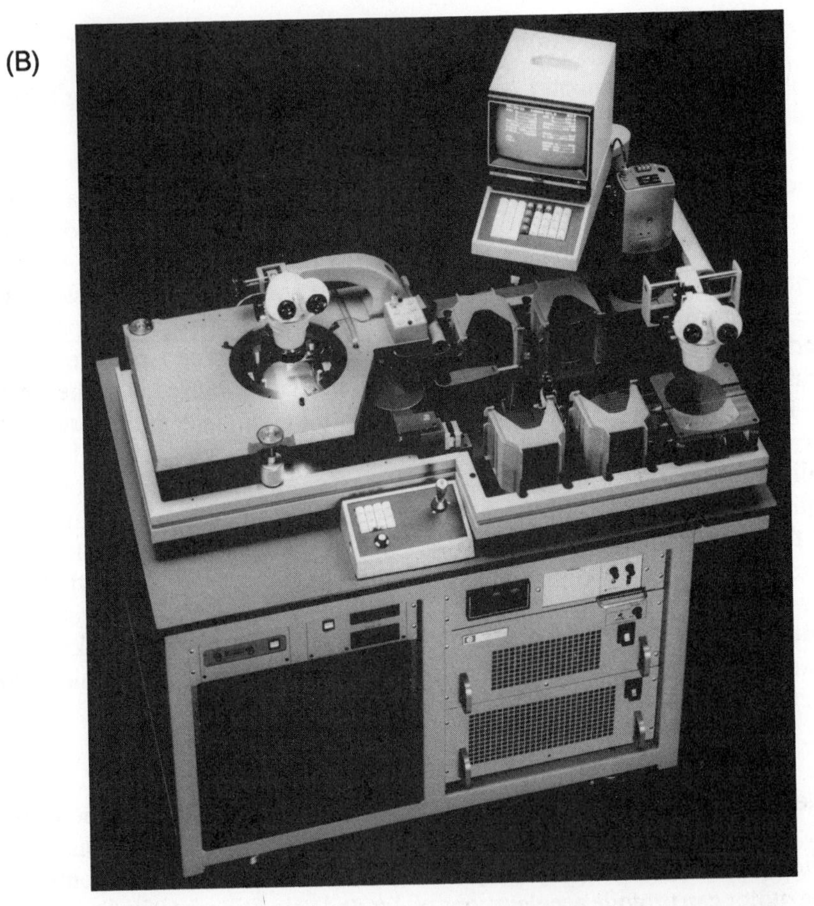

Figure 8-13 (A) Simplified linear motor; (B) A linear motor application *(Courtesy of Xynetics/Electroglas)*

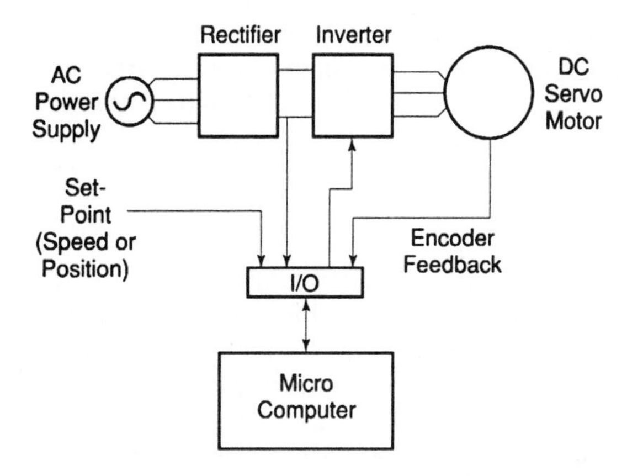

Figure 8-14 Schematic of a DC servomotor drive

A brushless linear DC servomotor application with a linear encoder or laser interferometer transducer can provide position control from .001 inch to .000001 inch accuracy depending on the feedback device employed. These motors are clean and spark-free, making them adaptable to clean rooms, vacuum chambers, and volatile environments. These linear motors have provided a revolution in motor-positioning table design, Figure 8-13.

DC Servomotors - DC servomotors are motors that are used as prime movers with instrumentation for control application, such as computers, numerically controlled machinery, and robots. The applications are where machinery is started, relocated, and stopped quickly. These movements are performed with great accuracy. Direct current motors allow precise control over position and speed of these actuators. The flux field in the servomotor's stator is always on when the system is active, either by using a permanent magnet for the motor's field or by supplying the field constantly with DC power. The direction and amount of servo movement needed to produce a speed or position is controlled by power to the rotor, Figure 8-14. The voltage to the rotor is controlled by feedback from a speed or position encoder supplying a microcomputer with an error signal. The DC servomotor's power can be supplied by transistor amplifiers that in turn can deliver up to ten horsepower under the control of a microcomputer.

AC Servomotors - The AC servomotor is designed to provide rapid starts, stops, and reversals in equipment. It is employed in instrumentation and computer applications. The AC servomotor is small in diameter with a low-inertia rotor for rapid movements. These motors are modified two-phased reversible induction motors. The windings are with two phases at right angles to each other. When in operation, these windings are supplied a fixed voltage from an AC source. The control winding is excited by the variable control voltage from a servoamplifier. The amplifier is under the control of the error signal resulting from the comparison of the equipment's desired condition and the actual measured feedback information.

Pneumatic Actuators

Pneumatic actuators are operated by employing either pressure or vacuum to units to provide an energy conversion. This energy is applied to a number of devices to produce a mechanical movement to the final element in the control loop. The common devices used are applications of diaphragms, bellows, cylinders, and pneumatic motors. Most of these devices supply linear motions. The pneumatic motor can be designed to provide either linear or rotary motion. With this capability, materials entering the manufacturing system are controlled whether they are solids, liquids, or gases. Examples of such materials are coal, fuel oil, or natural gas. In other processes, acids, salts, sugars, slurries, fillers, or chemicals may be metered into processes and controlled for the various manufacturing processes.

Pneumatic Diaphragm Motors

Pneumatic diaphragm motors have long been in use by industry. This actuator is frequently an air diaphragm motor valve. The actuator receives air pressure signals varying from three to fifteen pounds per square inch. The signal is analog and is sent from the controller through a pipe to the pneumatic actuator, Figure 8-15.

The final control element in many industrial control systems is a valve. It controls a fluid or gaseous medium supplying energy or product to the system.

Bellows

Bellows are used as linear motion converters when a longer movement is required by an actuator, Figure 8-16A. A flexible sealed joint is required to pass a moveable mechanical movement through a solid barrier, Figure 8-16B. Stems, levers, or angular deflections are examples of this flexibility.

Pneumatic Mechanical Actuators

Pneumatic cylinders are applied to control valve designs where a one-quarter turn of the valve provides full range from opened to closed. These actuators are usually of the Scotch yoke design, Figure 8-17A. The rack-and-pinion gear design is also used to provide the rotation for a valve, Figure 8-17B. The energy is supplied from the pressure source to a pneumatic cylinder.

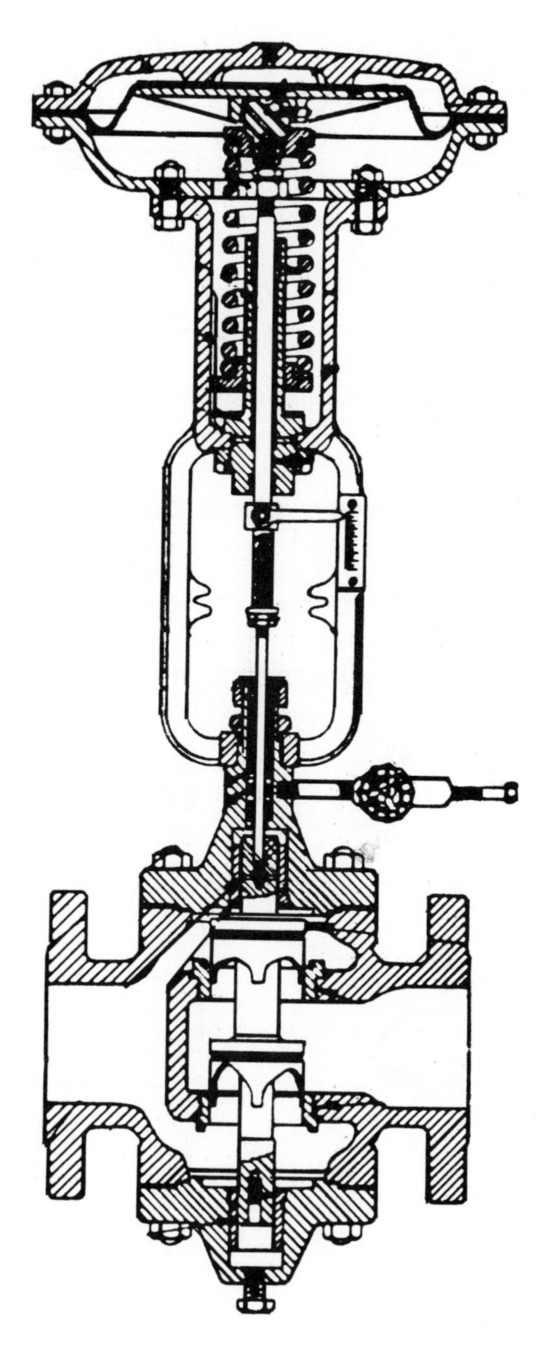

Figure 8-15 Schematic of diaphragm motor valve

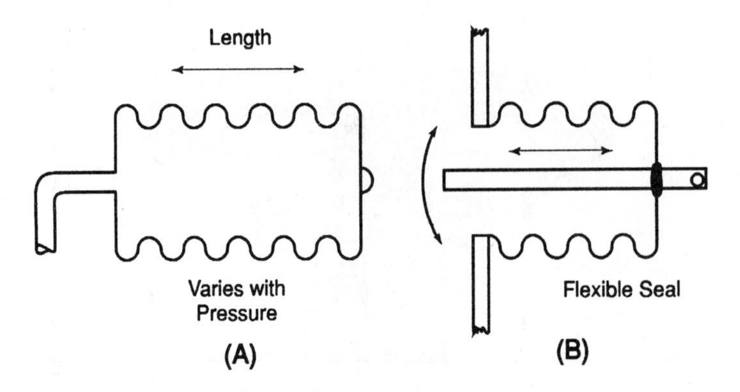

Figure 8-16 (A) Bellows linear actuator; (B) Bellows flexible seal

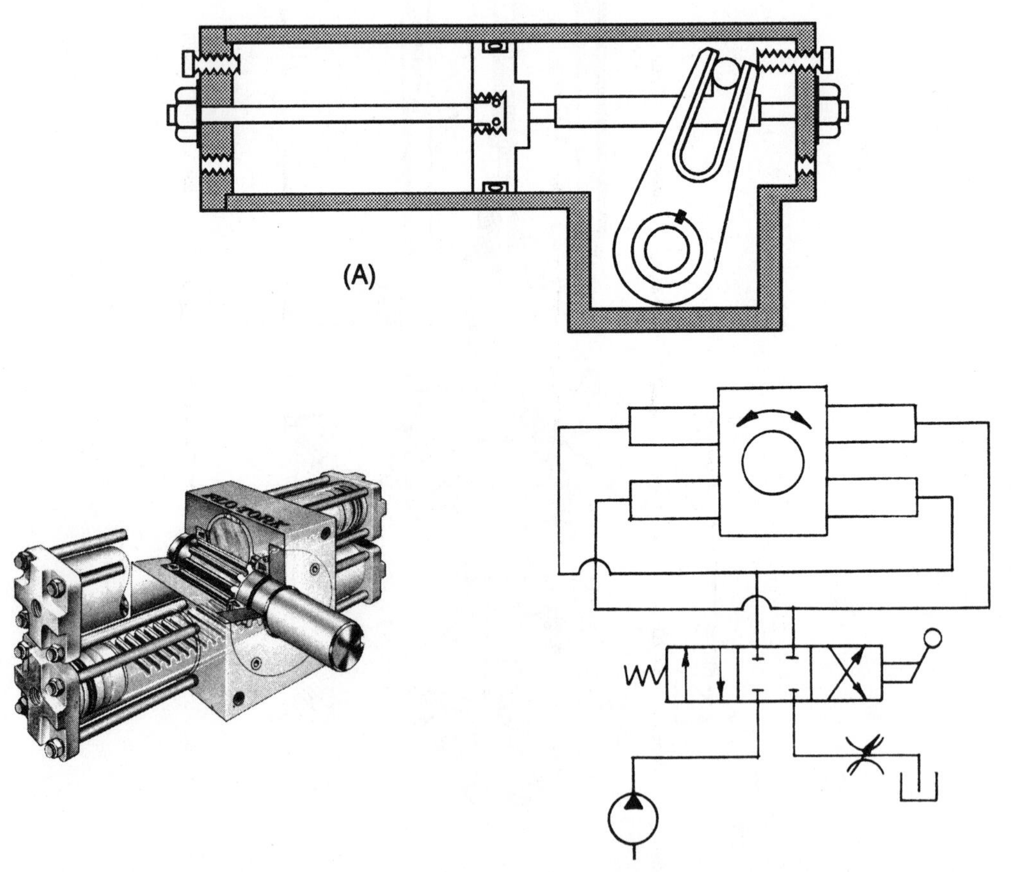

Figure 8-17 (A) Pneumatic cylinder Scotch yoke actuator; (B) Pneumatic rack-and-pinion gear actuator *(Courtesy of Flo-Tork, Inc.)*

Pneumatic Rotary Motor Actuators

Pneumatic rotary motors are employed when large gear driven actuators require a considerable amount of energy to move the active parts of the control valve or gate. This application utilizes a pneumatic *four-way valve* to control the movement of the motor. The four-way valve is controlled with a pilot air system or an electrical torque motor for remote closed loop control, Figure 8-18.

Hydraulic Actuator Controls

A number of hydraulic relay actuators to control a valve or other final element are available on the commercial market. The designs fall into three groups: the jet pipe valve, the flapper valve, and the spool valve.

Hydraulic Jet Pipe Valve Relay

A mechanical input signal positions a pivot jet pipe that transfers an impact force of fluid to develop a pressure in one end or the other of a working hydraulic cylinder. A small displacement of the pivot jet pipe causes the cylinder to deliver the necessary energy to control a valve or other final element, Figure 8-19.

Hydraulic Flapper Valve

Hydraulic flapper valves and nozzles are designed similar to the pneumatic ones. In the hydraulic design, there are two nozzles with a single flapper between them. A small mechanical displacement of the flapper moves the flapper away from one nozzle and closer to the other. When this condition exists, a differential output pressure flows either to one or the other end of a double acting cylinder, thus moving the control valve, Figure 8-20.

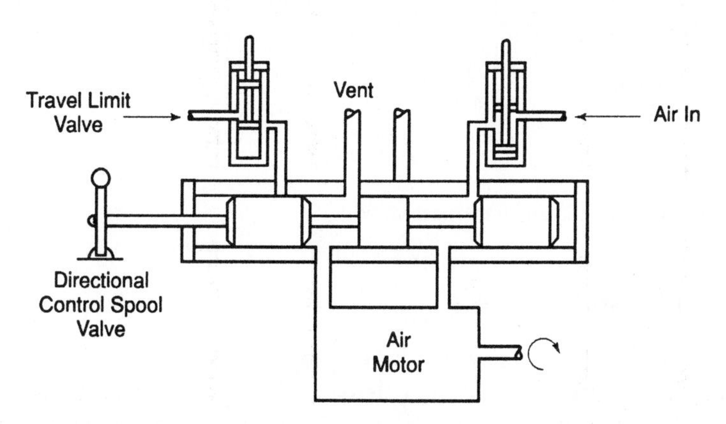

Figure 8-18 Air motor rotary actuator

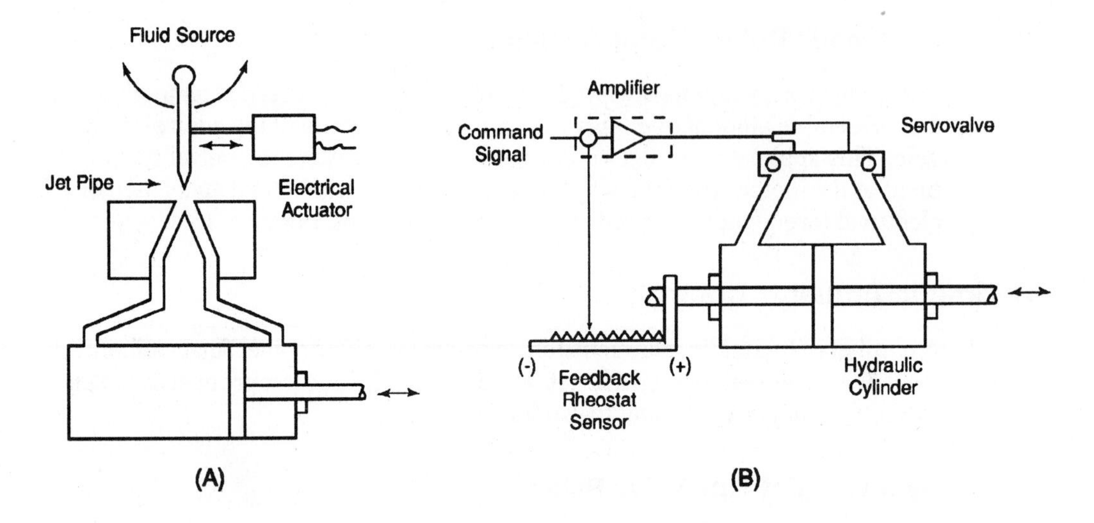

Figure 8-19 (A) A schematic of a jet pipe valve relay; (B) A schematic of a servovalve

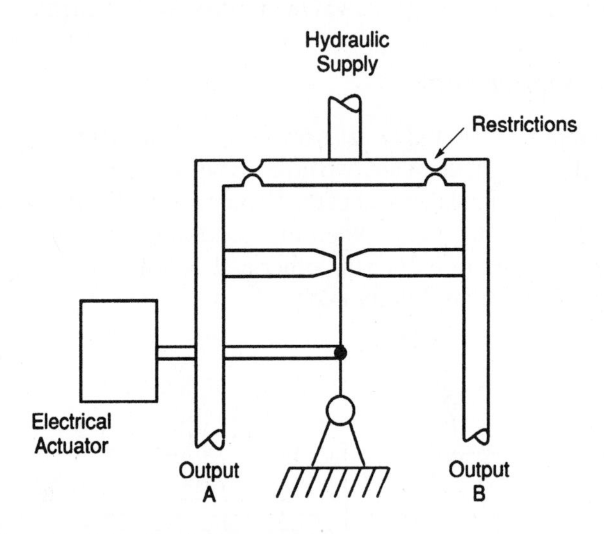

Figure 8-20 A schematic of a hydraulic flapper valve

Hydraulic Spool Valves

The hydraulic spool valve can be employed as a relay to provide control and energy to a control valve or final element. This type of actuator uses a three-way or four-way hydraulic valve, a pilot operated valve, or a servovalve to supply and control the energy to operate the final elements, Figure 8-21.

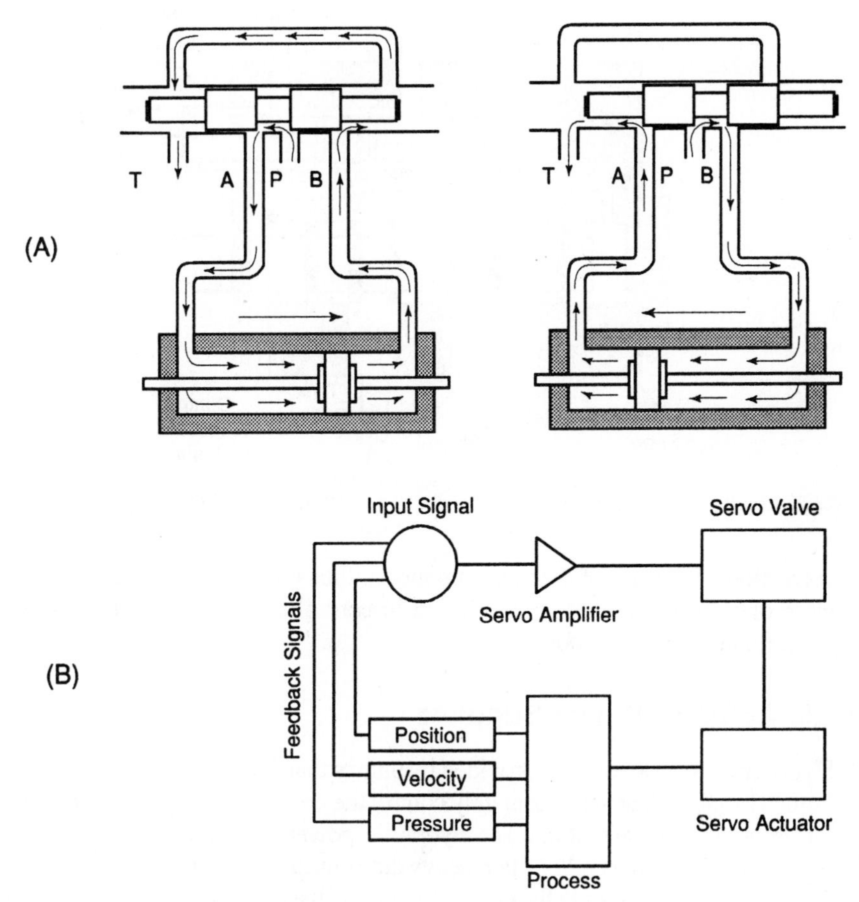

Figure 8-21 (A) Three-way and four-way spool valve actuators; (B) Torque motor and pilot spool valve hydraulic actuators

Of the number of valves and combinations that are available, the hydraulic spool valve is used most often in the control of large and powerful final elements. These valves are designed for either a single-stage operation or a two-stage application. These are powered by a small solenoid or hydraulic pilot to provide control over the position of the directional control valve.

The Hydraulic Servovalve - Another two-stage device, the servovalve, is widely used in precision location. A signal is sent to the torque motor which moves the pilot hydraulic spool valve controlling the master hydraulic cylinder, Figure 8-22. The push rod of the master cylinder has a yoke attached to activate a linear potentiometer. This potentiometer's change in voltage supplies a feedback signal, indicating the position of the master cylinder's push rod. If there is a difference between the potentiometer's voltage and the set-point voltage, a

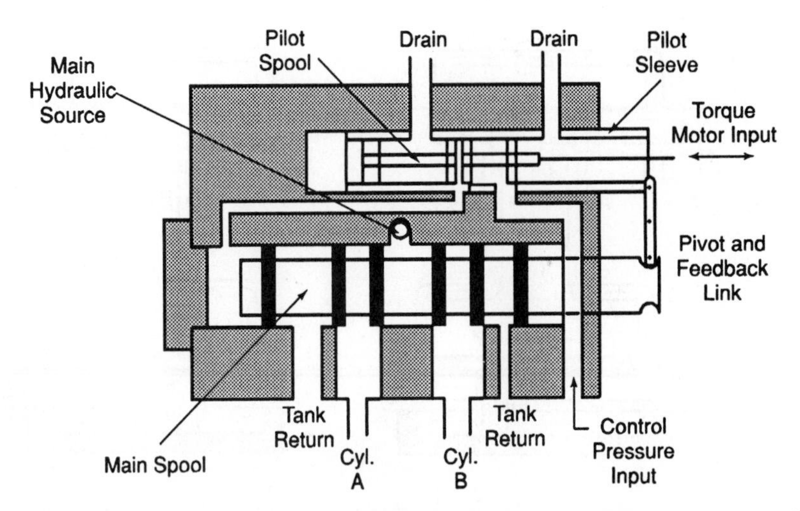

Figure 8-22 Hydraulic servovalve (yoke, linear feedback potentiometer, and torque motor not illustrated)

correction signal is sent to the torque motor, thus completing a feedback loop. This class of actuator has broad application in controlling machines of all types, from aircraft auto pilots to robots.

Hydraulic Actuator Power Supplies

Hydraulic actuator power is provided by pumps that deliver high pressure oil or other liquids to valves that very accurately control the powerful movements of a number of different types of actuators. The hydraulic power supply consists of a reservoir, hydraulic oil, strainer and/or filter, heavy duty pump, electric motor, and accumulator. These components supply the energy to relief valves, directional control valves, flow control valves, and *check valves*, resulting in controlled oil reaching an actuator, Figure 8-23. The actuator usually is a double acting cylinder. Hydraulics supplied to machine cells frequently will be applied to servovalves to precisely position very heavy loads, including heavy machinery and robots.

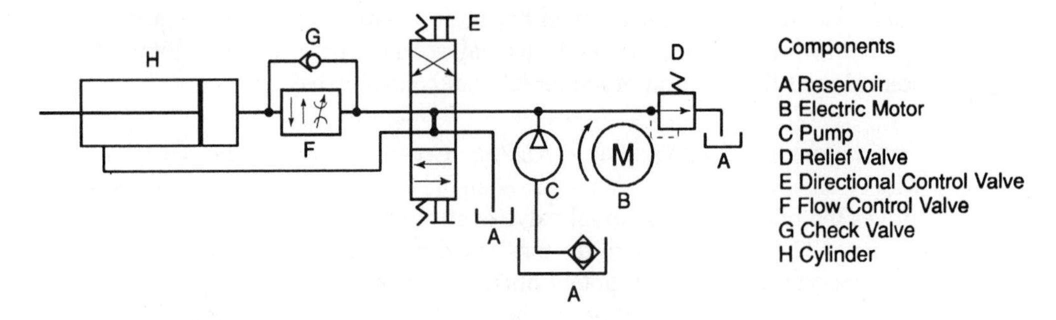

Figure 8-23 Hydraulic power unit and circuit

Hydraulic Components

A number of variation of pumps are used to provide the hydraulic power to meet the demands of the installation. One approach is to use a fixed- or variable-displacement pump and an accumulator. The accumulator is charged by the pump and, upon reaching the required pressure, the pump is discharged back to the tank. The pressure is maintained on the system by the accumulator. A relief valve keeps the pressure in the system at a predetermined maximum and minimum operating range. In this type of power supply, the accumulator must be sized to meet the fluid requirement of the circuits of the system, Figure 8-24.

Hydraulic Reservoir - The reservoir has important contributions to make to the power system in addition to holding a supply of oil, Figure 8-25. It is rugged and built of steel plate and is the mounting base for the motor and pumps of the system. This tank is designed large enough to hold all the oil in the system plus additional space for future circuits' needs. During operation of the pump, the level of the oil changes in the reservoir. This change requires that a breather be attached to the tank so that atmospheric pressure will exist in the tank at all times. Because one of the requirements of the reservoir is heat dissipation, the tank is designed so that air can circulate around the tank on all sides, including the bottom. On both ends of the tank, clean out plates are mounted. On one end, a fluid level sight glass and an oil filling opening are available.

The hydraulic oil in the system is compounded to meet the many demands of many factors. High temperature dissipation, pressures, anti-foam characteristics, viscosity stability, oxidation resistance, fire resistance, and lubricity must all be considered. These oils may be compounded from different base materials, but

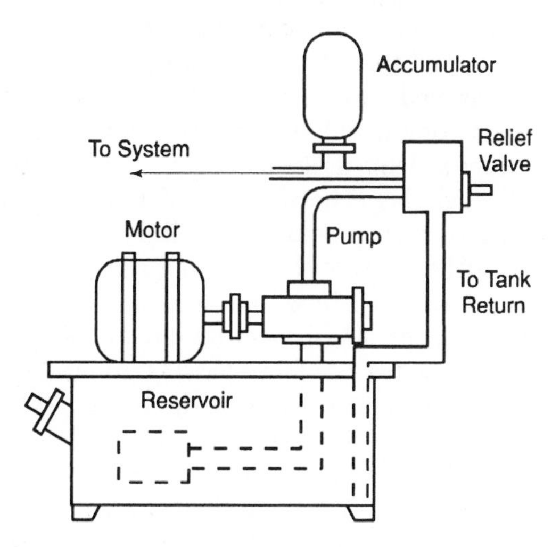

Figure 8-24 A main hydraulic power supply for actuators

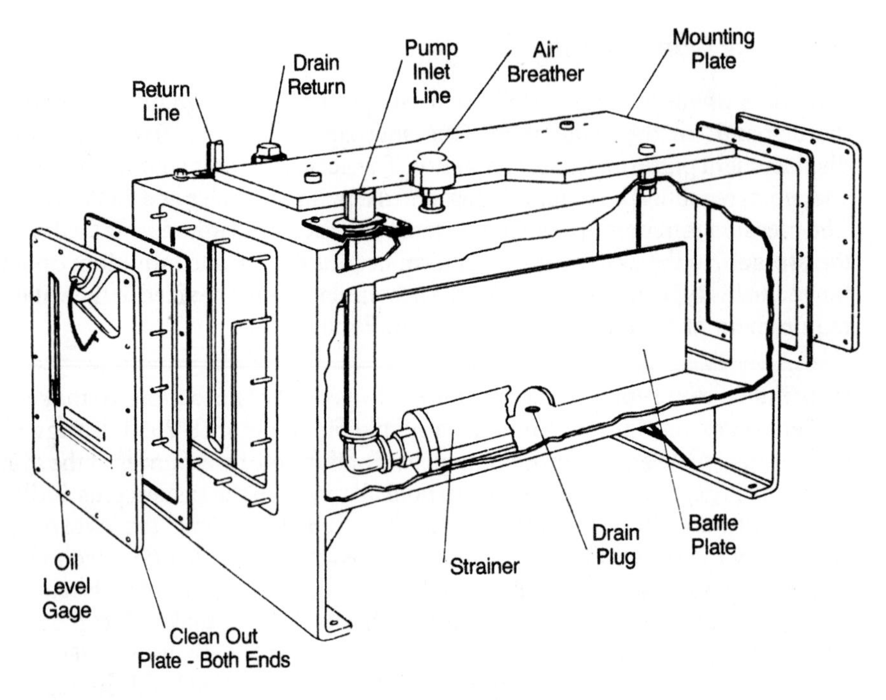

Figure 8-25 An industrial hydraulic reservoir

petroleum-based materials are most often used. A clean, high-quality hydraulic fluid is as important a part of the system and should be given as much care in maintenance as the pumps, motors, valves, tubing, and actuators in the system.

Strainers, filters, and magnetic plugs are the components that keep the hydraulic oil clean and free of foreign particles. Strainers are fine-wire mesh screens wrapped around a metal frame. The strainer protects the intake of the pump from large particles, without creating a large pressure drop upon entering the pump. The filter's elements are made from materials such as wood cellulose, plastic, and activated clay. These filter materials remove fine particles in the hydraulic fluids. To control pressure drop, filters are classified as a full-flow and proportional-flow filters. In a full-flow filter, all the oil passes through the filter of the system. A proportional-flow filter is in parallel with the returning flow and filters a ratio of the fluid passing through the circuit. Using this method of filtering, all the fluid over a longer period of time passes through the filter. The proportional filter greatly reduces the pressure drop in the system. The full-flow filter offers greater resistance to flow when it becomes dirty. It is evident that in a hydraulic system the filters will require scheduled maintenance to avoid equipment damage. In addition to aid the filters, a magnetic plug is mounted in the reservoir to remove iron or steel particles from the hydraulic fluids. These plugs are located at a low point inside the reservoir.

Hydraulic Accumulator - Another component of a hydraulic power supply is the accumulator. This device stores fluid under pressure and *cushions* the shock waves that can appear in hydraulics circuits. They are constructed in different types, but a frequently used accumulator in industry is the pneumatic or gas-charged accumulator. A flexible bladder is contained within an elongated spherical shell and is inflated with air. The hydraulic fluid applies its pressure against the inflated bladder that changes its volume from time to time to stabilize the hydraulic pressure within the system, Figure 8-26.

Another type of accumulator is the spherical-diaphragm accumulator. This accumulator has a diaphragm across the center of the sphere or elongated sphere with compressed gas above the diaphragm and the oil of the system below the diaphragm. The gas, being compressible, changes its volume with a change of pressure on the lower diaphragm's surface. This flexibility allows the accumulator to absorb shocks in the hydraulic system and also provides extra fluid back to the system when a momentarily high demand of energy is required by the system.

Hydraulic Pumps

Hydraulic pumps convert mechanical energy into a hydraulic flow. This is accomplished by displacing a volume of fluid into a confined space and thus generating a pressure. These pumps are classified into two broad categories: positive displacement pumps and nonpositive displacement pumps.

Positive displacement pumps employed in manufacturing can be rotary-type pumps designed to use gears, vanes, or pistons. The gear pump consists of two gears running together in a snug housing. The oil enters the space between the teeth and is carried around the outside between the case and within the gear tooth space. Upon reaching the other side of the gear, the oil is expelled because the meshing gears seal the oil off from totally traveling around the gear. A number of

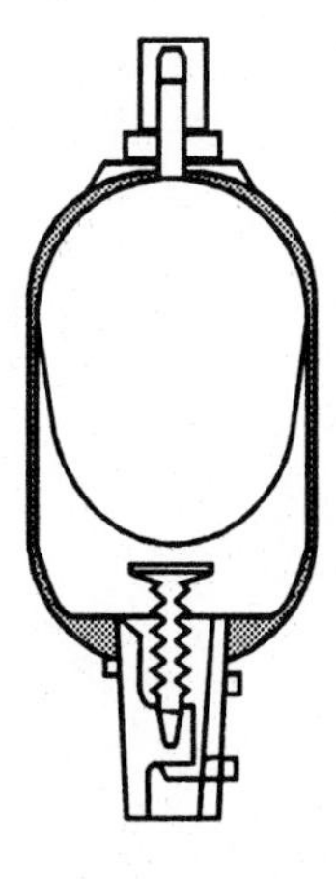

Figure 8-26 Cross section of a bladder accumulator

variations use the concepts of the gear pump, employing separators and external or internal gear applications.

Another design of hydraulic pumps is based upon the concept of an offset cylindrical rotor containing a series of movable vanes in radial slots in the rotor. The oil enters the pump when the vanes are extended, providing the maximum space between the vanes, rotor, and housing. As the offset rotor turns, the ends of the vanes are cammed towards the center of the rotor by the housing wall, thus reducing the volume containing the oil. This forces the oil out the outlet of the pump, Figure 8-27. When the vane has passed the outlet, the volume again starts to expand and refill the space from the inlet side of the pump. With simple modifications and porting, these pumps can be designed as hydraulic motors or actuators.

Hydraulic Rotary Pumps

Hydraulic rotary pumps exist in a number of engineering designs. The basic design principles are piston motors, gear motors, and vane motors.

Hydraulic Piston Pump - Piston pumps are of two configurations: axial-piston and radial-piston construction. The axial-piston motor usually has seven to nine pistons running in a cylinder — not unlike the cylinder of a revolver. The pistons receive high-pressure hydraulic fluid from a valve plate. Slots in the valve plate allow the hydraulic oil to enter the cylinder and sequentially push the pistons down. The pistons are attached to a rod with a ball end connection that is mounted on a cam plate set at an angle to the centerline of the cylinder so that the cylinder

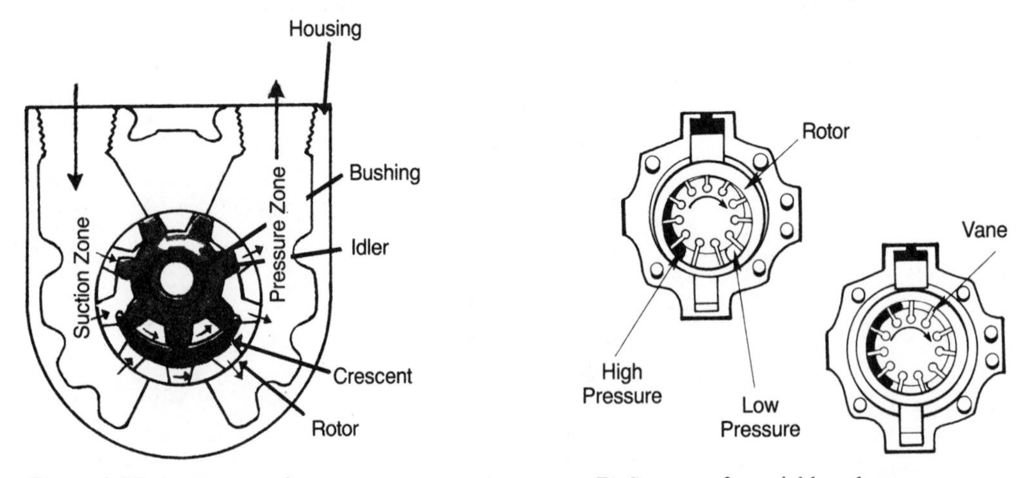

Figure 8-27 (A) Concept of a crescent gear pump or motor; (B) Concept of a variable volume vane pump or motor

and the cam plate are on a bent axis. The energy imparted to the cam plate causes it and the cylinder to rotate. The valve plate is stationary, resulting in one half of the cylinders to be filling with oil (causing the rotation) and the other half of the cylinders to be discharging oil to the reservoirs. The power can be taken from the cylinder barrel or the cam plate to the shaft. The addition of a differential gear arrangement supplies low speeds and very high torque. This design of pump will produce torques to 17,500 lb/in from pressures of 5,000 psi to maximum speed of 4,500 rpm, Figure 8-28.

With porting changes and other slight modifications, this type of hydraulic pump can be converted into a motor. The axial-piston design is widely used in aircraft, manufacturing machinery, and instrumentation in which high torque and control are desired, controlling movement with great smoothness and accuracy.

Hydraulic Radial-piston Pump - Hydraulic radial-piston pumps are designed with the pistons radiating out from the drive shaft. They may be arrayed in different ways. In a typical radial-piston pump, the pistons radiate out from the drive shaft. The pintle element is stationary in the center of the unit and separates the high-pressure hydraulic oil going into the pump from the discharge oil leaving the pump. The rotary barrel, carrying the piston, rotates with the drive shaft. The rotation is produced by the piston's ends pushing against a stationary reactor ring. The center of the rotary barrel is offset (eccentric) to the centerline of the reactor ring. This offset causes the pistons to stroke back and forth as they are rotated around by the pressure entering the cylinders timed by the pintle's cylinder-filling position, Figure 8-29. With porting changes, this type of pump becomes a motor

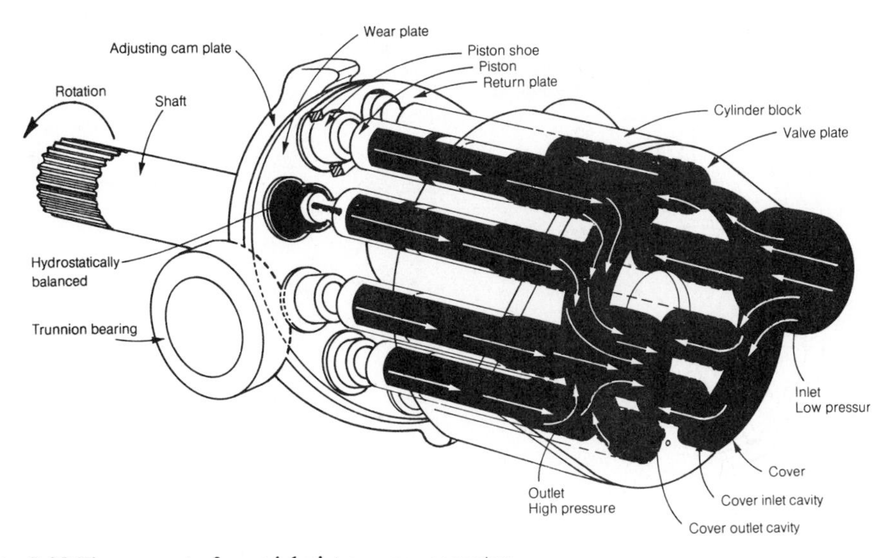

Figure 8-28 The concept of an axial-piston pump or motor

and will deliver the energy of a rotary barrel motor of torques of 46,000 lb/in, from pressures of 5,000 psi with speeds to 2,000 rpm.

Hydraulic Gear Motor - Gear-type motors also come in a number of configurations. Gear-on-gear, gear-within-gear, differential gear, roller-gerotor, and crescent gear are all variations of gear-type motors. For this text, only the gear-on-gear motor application will be discussed.

Gear-on-gear motors are commonly used. They are constructed with a pair of matched spur or helical gears enclosed in a case. The oil pressure enters the space between the gears' teeth and the outside case, thus rotating the gears. The oil leaves the case on the side opposite the entrance. The two gears are always in mesh with each other and provide the seal at the center of the motor. These motors deliver torques of about 6,000 lb/in, with inlet pressures up to 3,000 psi and provide 3,000 rpm, Figure 8-30. Again, these hydraulic devices can be designed for pumping or actuator applications.

Hydraulic Vane Motors - Vane motors are designed with a slotted rotor mounted eccentrically within a circular cam ring. Vanes are free to slide in and out in the rotor slots and seal their outer edge against the cam ring. The vanes are spring loaded so that they will remain against the cam ring regardless of their rotary eccentric positions. Oil under high pressure enters the motor and exerts force on the surface of the vanes, causing the vanes and rotor to turn. The oil is carried between the vanes and is discharged at the outlet of the motor. There are a number of different configurations that are designed with additional ports increasing the

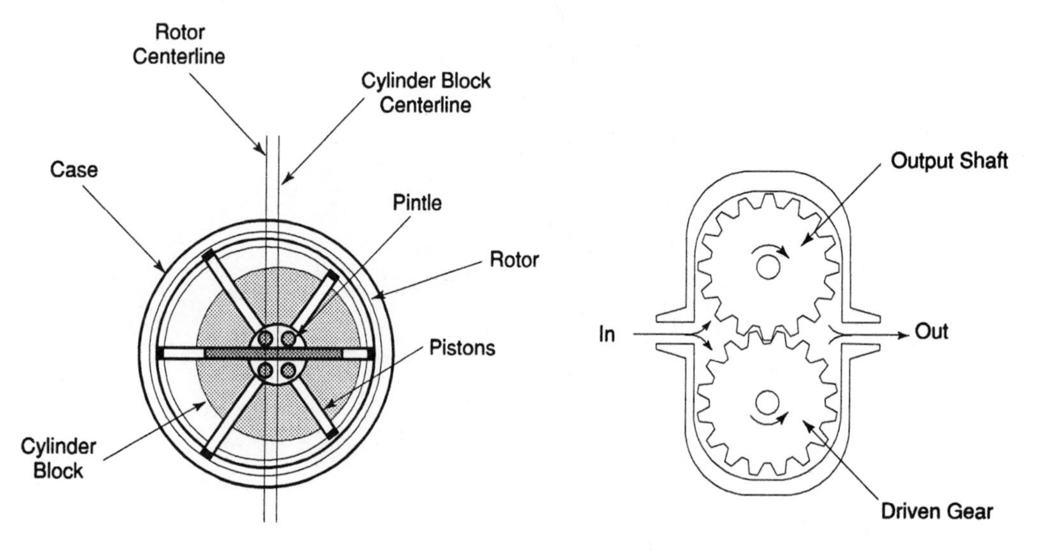

Figure 8-29 Radial-piston pump or motor

Figure 8-30 Gear-on-gear motor or pump

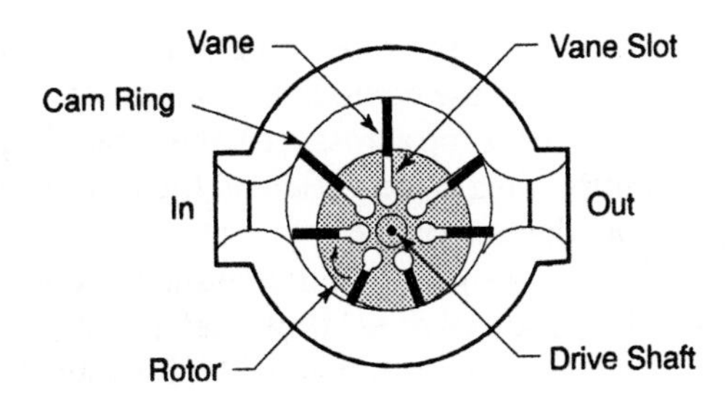

Figure 8-31 Basic vane motor or pump

torque provided by the motor. Vane motors with two ports provide about 4,000 lb/in, with pressures of about 2,500 psi, with a speed of about 4,000 rpm, Figure 8-31. With modifications and a power supply, a vane motor can be designed as a pump.

FINAL CONTROL ELEMENT — CONTROL VALVES

It has been said long ago that the control valve is to fluid process control what the motor is to the mechanical control processes. Fluids under manufacturing processes are primarily controlled by valves. Water, steam, petroleum, chemicals, sewage, gases of all types, and much food production are controlled by valves.

A valve is a device that is employed to restrict the flow of fluids. It is designed to control, regulate, or stop the flow of fluids in processing whether they are liquids, gases, or slurries.

A valve is frequently used as the final control element in an industrial control system. A fluid is most likely the controlled medium. The control valves are designed primarily to throttle the fluids that are providing the energy or materials in a production system. There are a number of popular valve types; each has its own characteristics to meet the needs of a process control function. Some of the common types of control valves are these: ball valves, butterfly valves, gate valves, globe valves, needle valves, pinch valves, plug valves, rotary stem valves, and digital valves. The choice of which valve to use is evaluated against the manufacturing process requirement. One of the most important considerations in the selection of a valve type is the valve sizing for the specific application of the control valve.

Control Valve Sizing

The determination of control valve size is normally based upon a series of flow engineering calculations. These calculations determine the size of the control valve required for a specific application. Final sizing is dependent on good analysis of the data from the specific placement of the control valve. In addition, control valve manufacturers provide computerized control valve analysis for valve sizing information. These computer programs include in the programming such parameters as pressure drop across the valve, specific gravity of the fluid, the fluid's flowing quantity, the flow coefficient, the fluid's critical flow factor, and the valve's cavitation characteristics.

A method of comparing valves to determine the correct one for a particular service is to study the various valves' flow coefficient (C_v). The flow coefficient is the number of US gallons per minute of 60°F water that will flow through a valve, with 1psi pressure drop across the valve. The results of the valve selected should provide a valve of the flow capacity necessary to meet the control requirements.

Control Valve Characteristics

A flow characteristic curve of a valve can be obtained by the shaping of the inner valve or ports of the valve and the manipulation of the valve's opening and closing movements. The characteristic curves are classified as 1) quick opening, 2) linear, 3) modified linear, and 4) equal percentage.

Quick Opening Valves

A quick opening disc or plug valve is often called a *poppet* valve and maintains a straight-line relation between flow and lift through about 70 percent of the stem travel, Figure 8-32. However, the disc is designed for full capacity with a short

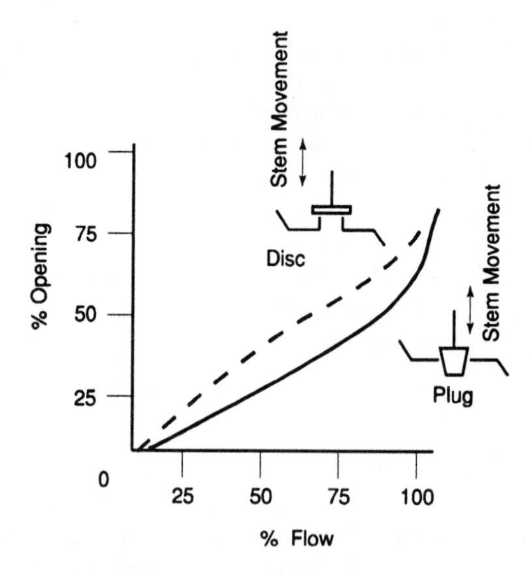

Figure 8-32 Quick opening disc valve and plug valve linear flow characteristics

stem travel. With a stem movement of about one-fourth the diameter of its seat, a full open flow is provided.

Linear Opening Valves

The linear flow characteristic, also seen in Figure 8-32, is obtained with the use of a shaped plug. When compared to the quick opening valves, the main difference is that the plug valve requires greater stem travel to produce the desired flow rate. This characteristic is useful because it can provide greater accuracy over the flow rate, Figure 8-32.

Modified Linear Valves

A modified linear flow characteristic is achieved by altering a plug in the form of a V (called a "V-port") or modified into a "parabolic plug," Figure 8-33. For the application of these flow characteristics, a larger stem travel is required. The sensitivity is decreased at both the high and low flow. This is an advantage for low flow, where a large valve movement results in a small flow change. But it is a disadvantage at the high end of the range, where a small valve movement results in a large flow change.

Equal Percentage Valves

The equal percentage or logarithmic flow characteristic is designed to produce a constant rate of change in flow for a unit of change in the lift of the control valve stem. The V-port or the parabolic plug is machined to a shape so that good control is obtained over a wide range of flow and pressure changes.

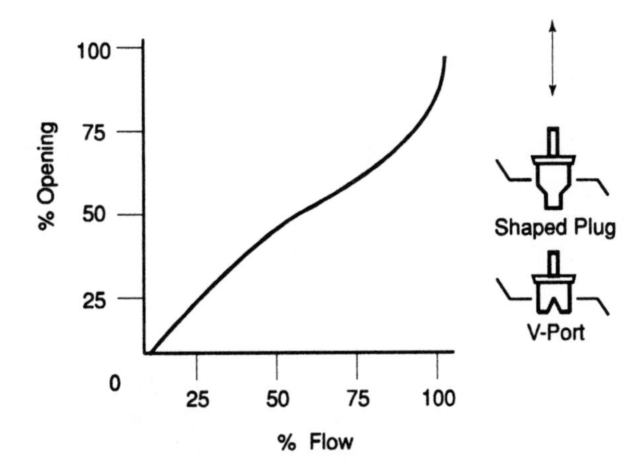

Figure 8-33 A "V-port" or "parabolic plug" provides a modified linear flow characteristic. This delivers better control over low flows.

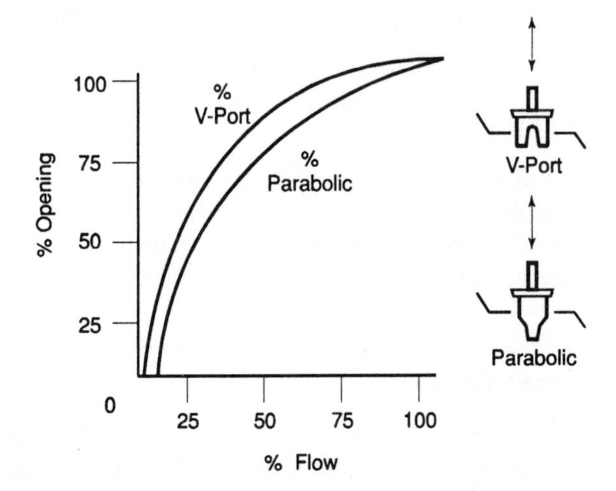

Figure 8-34 The equal percentage or logarithmic valve characteristics provide control over a wide range of flow requirements.

The flow within various chemical and petrochemical industries may fluctuate in a nonlinear manner, thus requiring a control valve with a demand for a very large range of flow. The equal percentage control valve makes this possible with one size of control valve, Figure 8-34.

Types of Control Valves

Manufacturers of control valves have designed and engineered many different configurations for the controlling of fluids. Some basic types of valves are presented below.

Ball-type Control Valve

The ball-type control valve has a spherical plug with a hole or cutaway section through the sphere that controls the flow of fluid through the valve body. A one-quarter turn of the ball will fully open or close the valve. With an accurate actuator, a precise control of flow can be achieved, Figure 8-35.

Butterfly-type Control Valve

The butterfly-type control valve is designed like a damper or carburetor throttle valve. It consists of a disc placed across the flow in a pipe. A one-quarter turn will completely open or close the valve. Turning the valve between these extremes causes a throttling of the flow in the valve. This type of valve is used in large diameter pipes containing low pressures, Figure 8-36.

Figure 8-35 Ball-type control valve—designed V 250 ball 4"-24" hydraulic *(Courtesy of Fisher Controls International Inc.)*

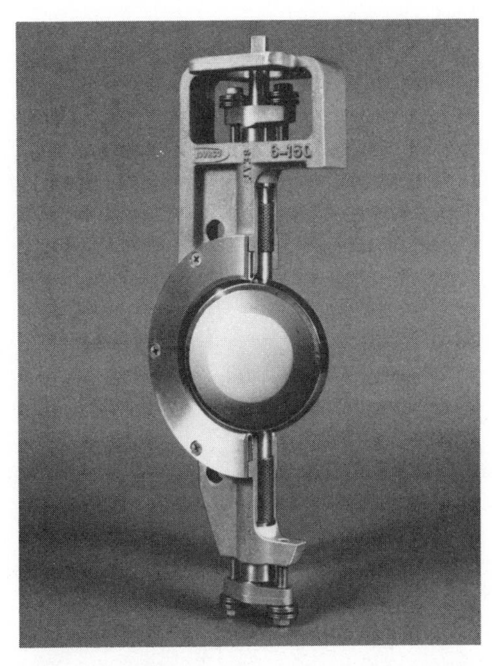

Figure 8-36 Butterfly-type valve *(Courtesy of The Durion Co. Inc.)*

Gate Valve

The basic gate valve has a sliding wedge or a knife-shaped gate to reduce the flow area of the valve, Figure 8-37. This type of valve will throttle flows, but it is not ordinarily considered a control valve. There are other variations of gates used for meeting special needs of industry—for instance, a guillotine gate used in the paper pulp industry; a V-insert; or positioned disc gates for other applications, Figure 8-37.

Globe-type Valve

The globe-type valve changes the flow direction and amount by a plug that is positioned in guides to provide the required flow. There are many variations on the globe body form, each seeking to solve a control problem. Some are single-seated valves, double-seated valves, cage valves, small-flow valves, and many other variations. This type of valve is operated by a plug on the end of a reciprocating stem that connects to an actuator, Figure 8-38A.

A rotary globe valve is a spherically shaped plug mounted eccentrically that rotates to align with a valve seat ring. The valve segment is activated by a rotary stem that provides the controlling position and thus the flow area. This rotary

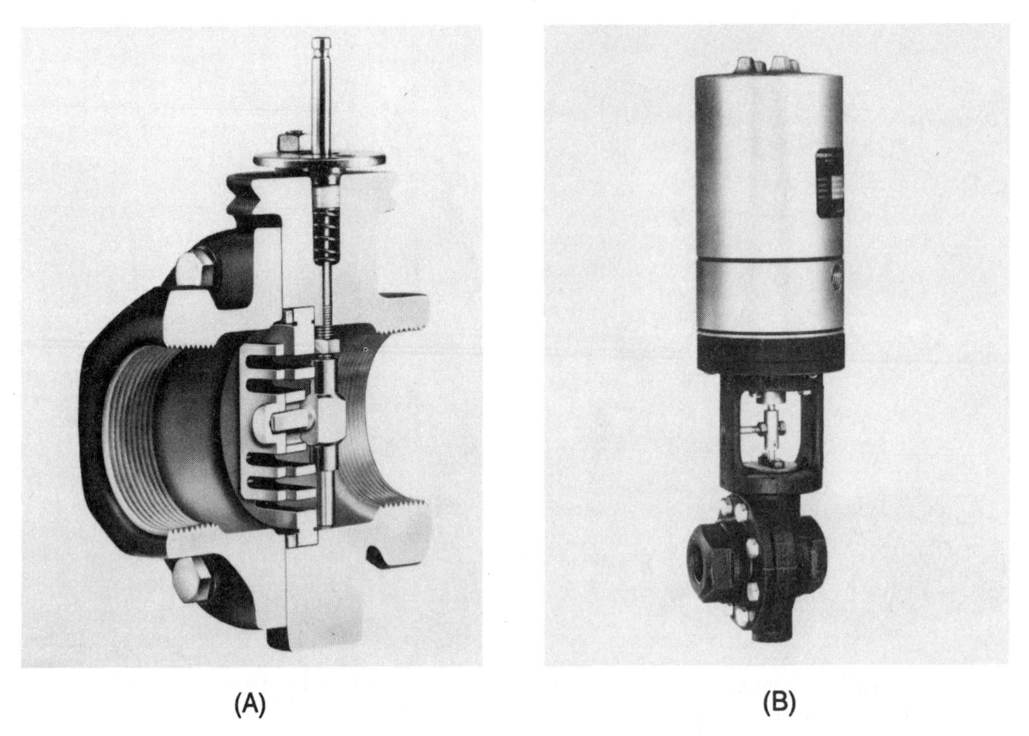

Figure 8-37 (A) Internal construction of a sliding gate valve; (B) Sliding gate valve final process control element, 4-20 mA microprocessor controlled *(Courtesy of Jordan Valve Co.)*

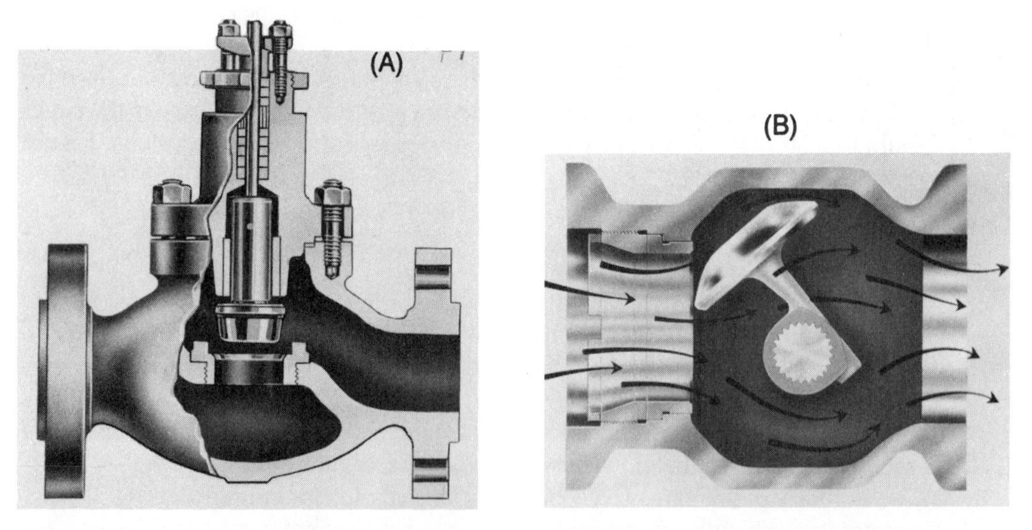

Figure 8-38 (A) A single-ported top guided control valve; (B) A shaft-to-plug eccentric rotation control, with a few degrees of rotation *(Courtesy of Dresser-Masoneilan)*

globe valve has become more popular than the globe plug valve because of its increased rangeability, Figure 8-38B.

Needle Valve

The needle valve is given this name because of the needle-shaped plug that is extended into a seat to provide very sensitive control over the flow area of fluids through a seat. These valves are used in laboratories and commercial plants that apply extremely small flows such as reactive chemicals. Other concepts of small-flow control valves are designed using flat bottomed, V-slotted, or other configurations of plugs that provide variable control over small volumes of fluids, Figure 8-39.

Pinch Valve

The pinch valve is designed to control the flow of a fluid by applying pressure to a flexible material to restrict the flow area. The area is controlled by the compression of an elastomeric tube or sleeve that is closed by mechanical, pneumatic, or hydraulic pressure from an actuator. The walls of the sleeve are smooth. The squeezing action of the sleeve reduces the area, thus restricting the flow.

These pinch valves are successful in controlling slurries, gravel, or minerals in a solution or viscous materials as long as the material does not chemically attach to the sleeve. The sleeve will form itself around gravel or lumps of material and

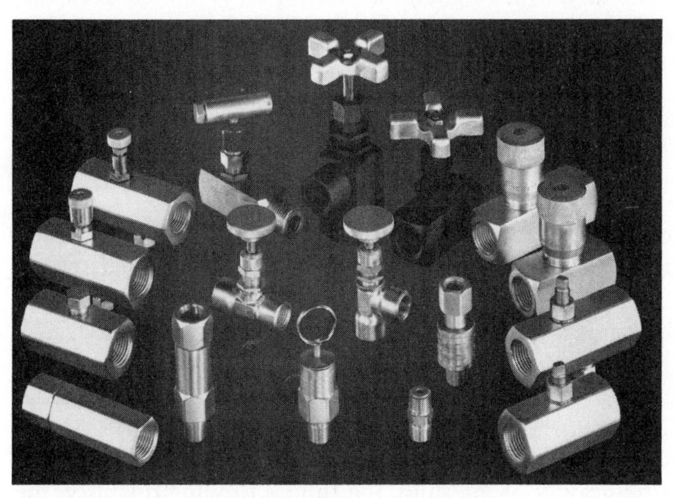

(A)

(B)

Figure 8-39 (A) A collection of needle valves; (B) A cross section of a Series 2000 needle valve *(Courtesy of Rego Co.)*

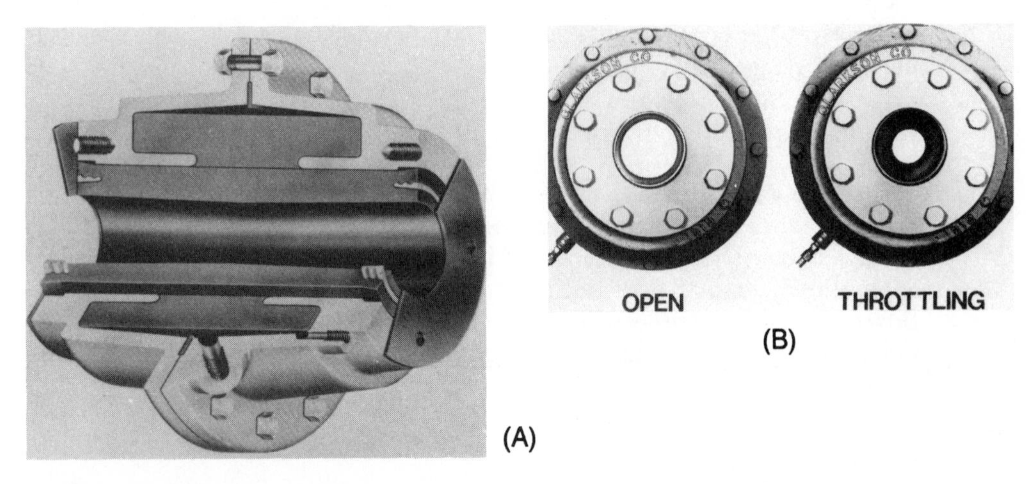

Figure 8-40 A pinch valve demonstration—a shutter closure: (A) A throttling control valve that constricts the flow with a flexible sleeve a pinch valve demonstrating a shutter closure, Series C; (B) End views of a valve showing open and constricted openings *(Courtesy of Clarkson Co.)*

provide a flow restriction during shutdown. This makes the valve very adaptable, Figure 8-40.

Digital Control Valve

Digital control valves are designed with a set of flow passages that are in a binary progression. They are activated with a signal to a series of on/off solenoid valves. The channels are arranged in a binary sequence of 1, 2, 4, 8, etc., to provide a variable flow by opening and closing various solenoids.

SUMMARY/FACTS

- A final control actuator is the motion converter that receives the error signal and provides the change in the system. It may be a diaphragm, switch, solenoid, motor, etc.
- Control actuators may be designed utilizing solid-state electronic, electrical, hydraulic, pneumatic, or mechanical principles.
- A control valve is frequently used as the final control element in an industrial control system.
- The sizing of control valves is concerned with parameters such as pressure drop across the valve, specific gravity of the fluid, the fluid's flowing quantity, the flow coefficient, the fluid's critical flow factor, and the valve's cavitation characteristics.

- A control valve's flow characteristics curves are classified as 1) quick opening, 2) linear, 3) modified linear, and 4) equal percentage.
- The equal percentage or logarithmic flow characteristic is designed to produce a constant rate of change in flow for a unit of change in the lift of the control valve stem.
- The flow within various chemical and petrochemical industries may fluctuate in a nonlinear manner, thus requiring a control valve with a demand for a very large range of flow. The equal percentage control valve makes this possible with one size of control valve.

REVIEW QUESTIONS

1. The unit in the system that receives a signal from a controller and converts that error signal into an energy change for a motion converter is known as what?
2. What is a solenoid?
3. Solenoids can be designed so that they operate sets of electrical contacts. In such an application, they are called __________.
4. What devices use a light-emitting diode to trigger a light-sensitive transistor to complete the required circuit?
5. An example of a rotary motion solenoid or torque motor is an analog device called a __________________.
6. What actuators apply the principles of electromagnetic attraction and repulsion of magnetic poles to supply torque to the armature and shaft?
7. In the shunt field motor, what is the relationship of the field winding to the armature winding?
8. In the induction motor, the stator is connected to an AC source that induces very large current in what?
9. What type of motor is thus described: "It has a rotor made of permanent magnets. The permanent magnets' poles are attracted to the field poles and the movement is correlated with the field rotation of the stator"?
10. What is the SCR or thyristor?
11. Give examples of areas where thyristors may be used.
12. How is the rotation in a stepper motor achieved?
13. Describe the relationship of the torque to the current supplied in brushless DC motors.
14. The activator frequently used to receive a pneumatic control signal for a control valve is a __________________.

15. Describe the process hydraulic pumps use in converting mechanical energy into a hydraulic flow.
16. What type of delivery of hydraulic oil does the yoke-type bent-axis pump produce?
17. What is the name of the component of hydraulic power that stores fluid under pressure and cushions the shock waves that can appear in hydraulic circuits?
18. What motor has pistons running in a cylinder similar to the cylinder of a revolver and receives high-pressure hydraulic fluid from a valve plate?
19. Gear-on-gear motors are constructed with a pair of matched spur or helical gears enclosed in a case. Where does the oil pressure enter in these motors?
20. What is the device that is designed to restrict or throttle fluids in a production system?
21. How are control valve sizes normally determined?
22. What is the value of a flow coefficient?
23. What is a quick opening disc or plug valve often called?
24. How is a modified linear flow characteristic achieved?
25. What is the purpose of an equal percentage or logarithmic flow valve?
26. Describe the function of a globe-type valve.
27. What is used in pinch valves to control the fluid flow?
28. What type of valves are designed with a set of flow passages in binary progression, activated with a signal to a series of on/off solenoid valves?

9

AUTOMATED MATERIAL HANDLING

OBJECTIVES

Upon completing this chapter, you will be able to discuss, present, and apply these material handling concepts:

- Automated material-handling machinery
- Handling bulk materials
- Handling piece parts
- Assembly and part handling
- Automated quality and position control
- Finished product-handling instrumentation

INTRODUCTION

Automated material handling is done in all segments of manufacturing. From the time manufacturing materials arrive on the receiving dock for transportation into storage, from the time of storage into and through processing operations and assembly, from the time of the final inspection, packaging, and storage to the shipping area — all production stages require material-handling machinery.

AUTOMATED MATERIAL-HANDLING MACHINERY

Material handling includes receiving materials in a manufacturing plant and then moving the materials inside the plant. In addition, there is a series of stages of material movement that occur within automated machines. Machine movements include manipulating and orientating component parts into and out of the various production machines. Another important requirement is the removal of waste materials (chips, cuttings, shavings, and scrap) away from the machine to a disposal area. An important concern of instrumentation is the synchronization and control of these material-handling sequences with machine tools and other production equipment.

HANDLING BULK MATERIALS

Bulk materials can be moved by any of a number of machines to provide a direction of material movement. The movement may be horizontal, vertical, at an angle, or a combination of these.

Roller and Wheel Conveyor

A machine very frequently used for material movement is the conveyor. There are a number of conveyor types. Each one can be altered into a number of variations to fill the specific production requirements. The major type — the roller and wheel conveyor — is used in manufacturing to transport boxed and tote-tray-contained materials to the point of need. This type of conveyor can operate with the aid of gravity or with power depending on the production facility requirements.

These conveyors are controlled with circuitry employing microswitches, mercury switches, magnetically actuated reed switches, photoelectric devices, and phototransistors. Other control instrumentation includes inductive, magnetic, or capacitive infrared and proximity switches; revolution counters; tachometers; and area or full bin sensors. These sensors can be applied in various ways with solid-state integrated electronics to control and deliver the amount and position of material to the production machinery, Figure 9-1.

The Flat Belt — Trough Belt and Slat Conveyors

Belt conveyors are employed in a number of different configurations. Flat belts are used to move boxed and contained parts. They also are used as assembly lines, packaging lines, inspection lines, as well as in other applications. Flat belts are also used in conveying small piece parts in tote trays, boxes, or single components. On some types of assembly or testing work, a flat belt is located at the rear of the workstation or in a comfortable reaching position above the workstation

Figure 9-1 Roller and flat belt transporting conveyors *(Courtesy of IBM Information Products Division)*

table. In many cases, the workstation operator will remove the part from the belt and add parts or perform operations on the assembly and then return that part back to the belt. The belt can run continuously at a slow speed, or the belt can be controlled by switches used by operators at multiple points along the belt. In other applications, the belt can be controlled by microswitches, phototransistors, or optic or proximity switches interacting with the products on the belt, Figure 9-2.

Trough conveyor belts provide for the movement of prodigious amounts of bulk raw materials. Materials such as ore, gravel, coal, cement, and sand are transported great distances. These belts are supported by rollers in a shallow, open U shape. The rollers are spaced about six to ten feet apart. This equipment is used to transport loose bulky materials.

Slat conveyors perform similar work to that done by flat belts but they carry heavier loads. They can be made of steel interlocking slats for heavy work or for work done in hot areas such as in heat treating or brazing furnaces.

Screw and Bucket Conveyors

The Archimedes' screw is used as an elevator to transport granular materials such as plastic pellets, scrap, grain, and like materials. The screw will elevate material through angles from horizontal to nearly vertical into a holding container, Figure 9-3A and B.

Bucket elevators are made of a series of buckets mounted on endless chains or belting. The elevator has head and tail pulleys or sprockets as required, one at either end of the bucket conveyor. This type of conveyor is employed in a vertical position so that it can be kept as short as possible. Material that it can handle are powders, granular materials, and gravel.

These conveyors are controlled by weighing the amount of product received in a bin or hopper by using a strain gage or load cell. The conveyor can also be

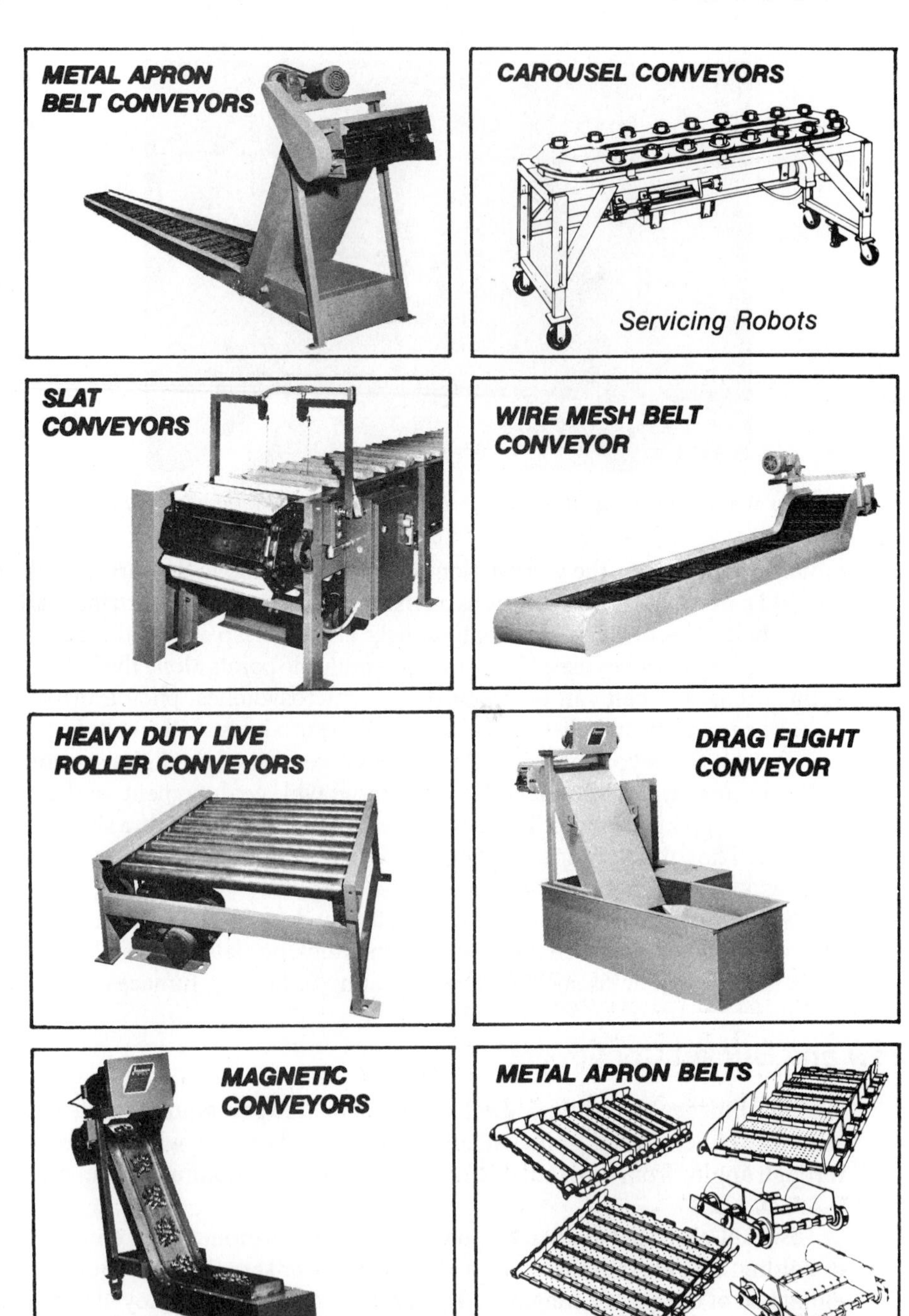

Figure 9-2A A group of different types of flat conveyors *(Courtesy of Jorgensen Conveyors Inc.)*

Figure 9-2B A drag flight conveyor *(Courtesy of Jorgensen Conveyors Inc.)*

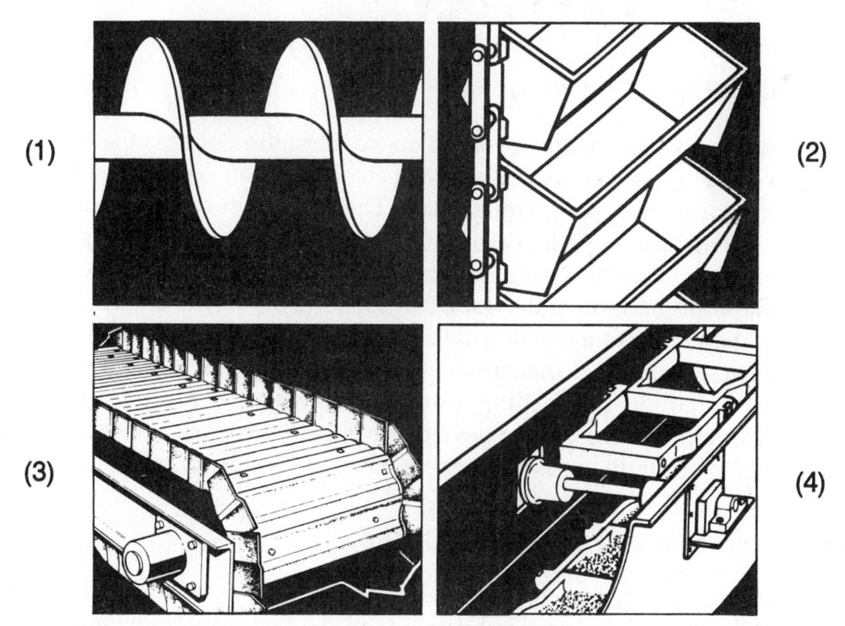

Figure 9-3A (1) Screw conveyor; (2) Bucket conveyor; (3) Slat conveyor; and (4) Drag conveyor *(Courtesy of Jorgensen Conveyors Inc.)*

Figure 9-3B Screw conveyor *(Courtesy of Jervis B. Webb Co.)*

controlled with the application of a number of dry product level sensors: rotating paddle-type level switch; ultra sonic, gamma ray, proximity, or other level sensors.

Fastener and Parts Conveyor

The fastener and parts conveyor is a conveyor with steel pans that are hinged on cross shafts. These cross shafts have a steel wheel on each end of the shaft that is linked together forming a roller chain. The sides of the conveyor are constructed with overlapping side wings that fit flush against the pan providing a sealed belt that can move material (parts and fasteners) flat, at an angle, or through a curved arc. The configuration of the track that supports the rollers determines the angle of elevation or contour position of the conveyor. A flange may be added to the pan making it possible to elevate products at a steep angle into an overhead hopper or part orientation and feeding bowls. The conveyors are controlled with microswitches, photo-optical devices, or phototransistors; or the hopper can be weighed with strain gages to measure when the hopper is full, Figure 9-4.

The Pan, Spiral, and Orientation Chute

Chutes are made from metal formed into shapes of a pan or trough. They transport materials or parts to a location by gravity feed. Chutes are frequently an interlink between two belts or between a belt and a hopper or other container. Chutes can

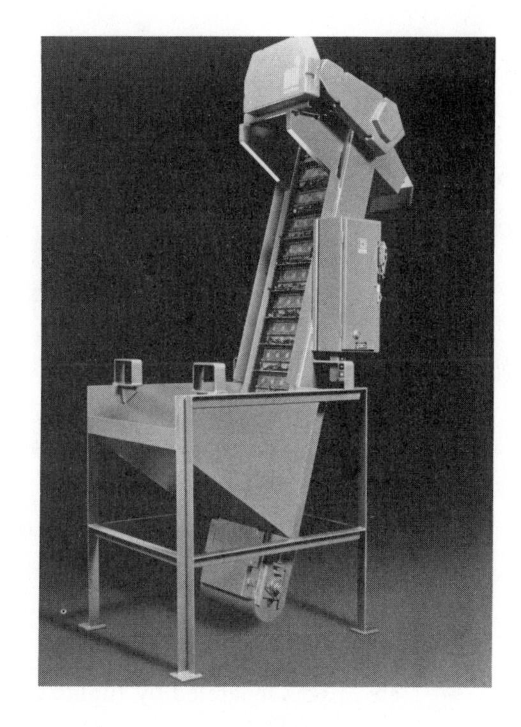

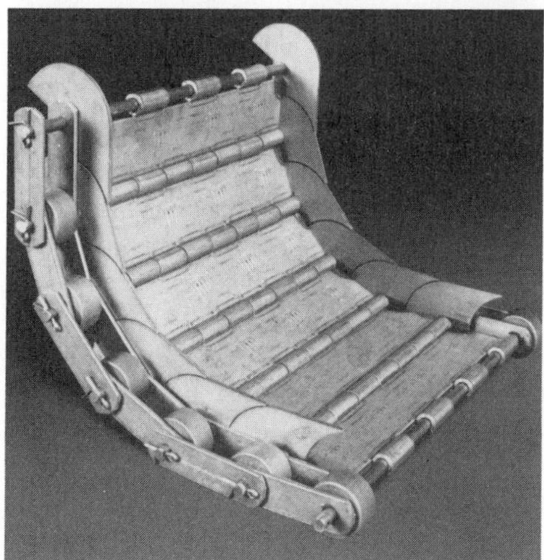

Figure 9-4 Fastener and parts conveyor *(Courtesy of Mayfran International)*

be made moveable so that material may be switched from one belt to another or from one bin to another, Figure 9-5.

A chute can be designed so that is can position parts as they are delivered to additional processing equipment. In this case, the chute is made of welded steel bars and strips that curve to orient the parts as they slide through a caged area.

The control of the chute's position is indicated by microswitches or mercury switches. Position and warning lights can be provided for a visual readout. The chute may be repositioned with a solenoid, electric motors with a gear train, pneumatic cylinder, or a mechanical lever system.

(A) (B)

Figure 9-5 (A)Screw conveyor and pan-type chute; (B) Spiral chute *(Courtesy of Jervis B. Webb Co.)*

The Drag Conveyor

Drag conveyors are built with a rectangular slide as a base. Sliding inside the base are chains separated by flights or bars that contact the product to be moved or elevated. This type of conveyor can move boxed materials up inclines or move loose materials horizontally between the flights. In large manufacturing facilities, waste and scrap can be removed from machines by having these conveyors installed in tunnels under the base of the machine, Figure 9-6.

This type of conveyor can run continuously in a high production area, or it can be activated upon demand. In either case it is controlled by the switching of an electric motor. If automatic control is required, it can be accomplished with a phototransistor circuit, a time interval circuit, or by counting the cycles of the machines generating the product to be moved by the conveyor.

Vacuum or Pneumatic Conveyors

Vacuum or pneumatic conveyors are used to move large amounts of light powdered or granular materials through tubes. These installations function on the concept of a vacuum cleaner. The product is sucked up a tube and deposited in a storage bin.

The plastic industry uses this elevation concept to move and store pelletized raw materials from railroad boxcars to various storage areas in the plant.

These installations are controlled by electrical switches that start the pneumatic pumps and are started on demand by the operator. The level of product in a bin is controlled by a number of dry level indication devices, such as photoelectric, microwave strain gages, ultrasonic or proximity sensors, and other level sensors.

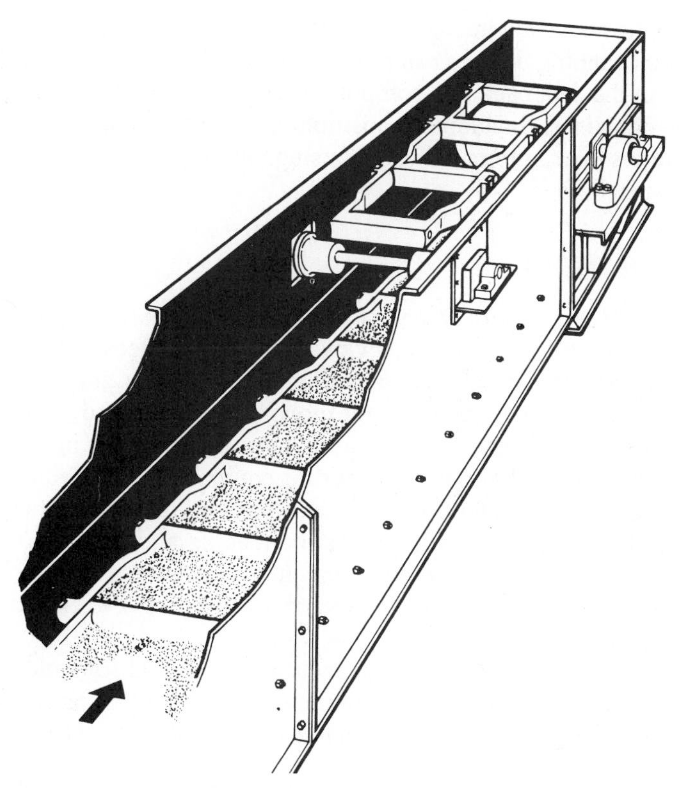

Figure 9-6 Drag conveyor *(Courtesy of Jervis B. Webb Co.)*

HANDLING PIECE PARTS

The handling of individual parts or components is required within the manufacturing operation. Because the range of the configurations and characteristics of the piece parts is so broad, varied solutions are applied.

The Tote Tray

Tote trays are used to move masses of very small parts and small components around various areas of the manufacturing or assembly floor. They are plastic, formed containers that are light and strong and can be stacked when they are empty. These trays can be carried, placed on belts, stacked in carriers, and moved or placed on conveyors for processing for long distance moves within a plant. If the system is automated, the tote trays may carry a sense marking or bar code indicating the content of the tray and then move around the plant or inventory system automatically by a conveyor system.

In the above automated transportation system, magnetically sensitive heads or laser scanning devices can be used to identify the content of the tote tray and transmit the data back to a computer for control and dispatching of the material in the container. In addition, positioning devices can be employed to place the containers in an automated warehousing unit or place them in the proper queues for dispatching.

Monorail Conveyors

Monorail (or trolley) *conveyors* are hung from the ceiling or a beam and are employed to transport heavy, awkward, or large pieces of material, Figure 9-7. They also transport baskets containing assemblies or small parts. This type of conveyor has advantages in that the monorail can be routed around the facility and even out into adjacent storage buildings. Because the moving materials are elevated above the floor, the transporting equipment or material does not consume valuable floor space. The monorail can also move materials on an incline between floors of a building. This improves the loading and moving of materials to an assembly line. The monorail is frequently scheduled and loaded by a computer program so that the correct materials arrive at the proper workstation exactly when they are needed.

Monorails are controlled by electrical relays and solenoids that operate the mechanical track switches. The monorail itself usually runs continuously during the shift. The loading can be manual or automated. The total loading and

Figure 9-7 (A) Monorail material-transfer and fixture-loading system *(Courtesy of Fibro Inc.)*; (B) Monorail from engine build up to load point *(Courtesy of Allied Uniking Corp. Inc.)*

workstation switching control can be handled by local switching or by the computer.

The Railed Slide

Railed slides are used to transport heavy single-piece parts, such as engine blocks or like products, as they move through the manufacturing processes. These railed slides usually are bars with a round cross section. However, they may be in different shapes that allow the workpiece to slide in contact with the rails. These rails can be inclined so that gravity provides the moving energy; or they can be moved by a pneumatic or hydraulic shuttle. When power is necessary to slide the parts, mechanical stroke handling equipment is provided by gravity-activated pushing pawls that contact the workpiece. The stroke is supplied by pushing pawls mounted on a shuttle bar. The shuttle bar is reciprocated back and forth by a cylinder. The pushing pawls retract and slide under the workpiece on the return stroke of the cylinder and again reposition behind the next part to be pushed. This type of heavy part movement is controlled with microswitches, phototransistors, or proximity devices.

Car-on-track Conveyor System

The car-on-track conveyor system operates on a support structure of two hardened wear bars (rails). The cars are driven by a spinning tube and friction drive that provide movement of the cars. A modular design system facilitates the assembly of many different configurations as well as installation for reliable expansion or relocation of the conveyor systems. The control systems are primarily electromechanical and any remote control requirements are easily located. The car-on-track will interface with robotics, hard automation tooling, fixtures, and the strokes of mechanical handling equipment, Figure 9-8.

Vibratory Hoppers

Vibratory part handlers are designed to perform three functions simultaneously. They elevate, orient, and feed the part. They are constructed with a spirally inclined track around the inside perimeter of a bowl on which the parts move up by electromagnetic vibration. The inertia in the part causes the part to move from the bottom of the bowl up the spiral incline plane to the top channel. In traveling, the parts encounter various orienting devices that position the parts or cause them to fall back into the base of the bowl if they are out of orientation.

These part handlers can move pieces made of metal, wood, glass, plastic, rubber, and ceramic without damaging the product. The variety of part shapes that can be processed is almost limitless and depends on the creativity of the engineer designing the baffles, slots, wipers, and cutouts in the track to orient the part, Figure 9-9.

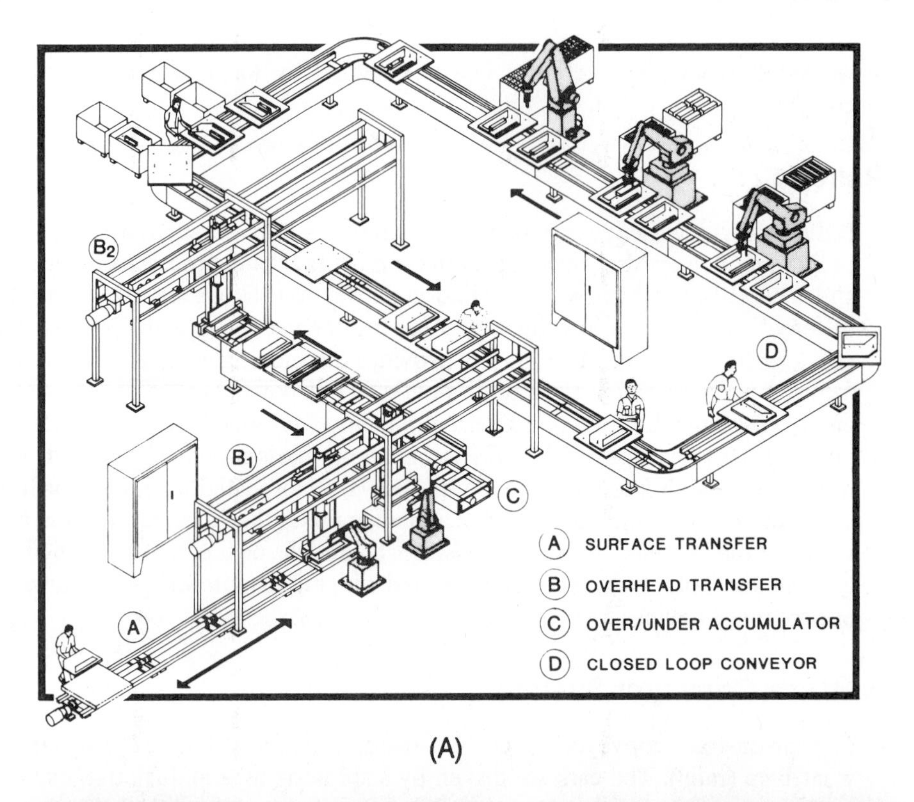

(A)

(B)

Figure 9-8 (A) A car-on-track conveyor system; (B) A conveying system under construction

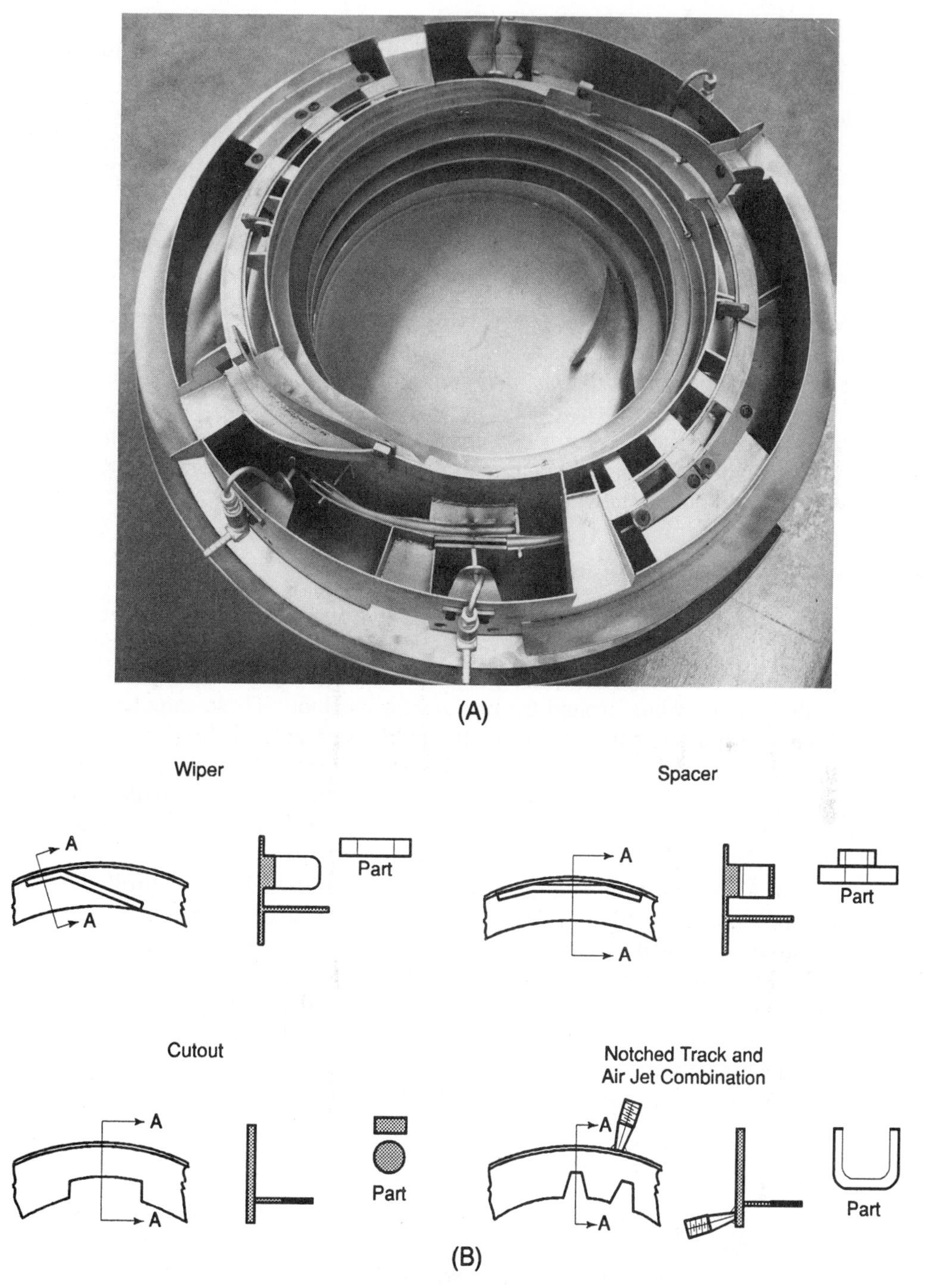

Figure 9-9 (A) Vibratory elevators and part orientator; (B) Orienting devices used in part feeders

This type of machine often is left running continuously during the operation of a plant shift. When required, it can be controlled by a weight measurement of the product. When a batch of parts reaches a predetermined weight, the vibrating hopper can be turned off with such devices as strain gages or load cells. It can be controlled by the count of the part by employing phototransistors or photoresistors and circuitry. Or it can be controlled by the level of the part by installing proximity sensors of the inductive, capacitive, or magnetic types.

Wheeled Carts

Moving materials and large piece parts about a factory floor can be performed by wheeled carts attached to an endless chain moving under the floor. A bar from the cart descends down into a channel and is hooked into the endless chain. This temporary connection provides the motive power. The carts or parts are coded so that a computer can direct the cart around the manufacturing floor to the proper machine where the cart will be removed from the motive power and the desired work carried out on the workpiece.

Automated Guided Vehicles

Large and palletized parts can be transported by means of automated guided vehicles (AGV) that move along paths prescribed by a guide wire embedded in the production floor. These wire-guided carts are utilized to convey materials to production machines around the manufacturing floor. These carts have swivel wheels so that the cart can follow the guide wire around the floor. They can transport material from a storage and retrieval system and carry it to various workstations. The guided carts can also be equipped with a pallet-carrying capacity so they can be used to shuttle large parts from one machine area to another, Figure 9-10.

An AGV's position within a plant can be programmed and controlled by the system's computer in conjunction with the other flexible production centers. These devices entail a host of sensor and control loops that are necessary to move and position the AGV to the desired machine.

Laser-guided vehicles that scan wall-mounted retroreflective beacons to locate the vehicle's position are under development. Upon locating three beacons, this machine orients itself by using the coordinates of the beacons stored in its computer memory to maintain its path to its commanded location, Figure 9-11.

The Robots

Robots are very successful in material handling of small specialized parts, such as in the rapid placing of component parts in circuit boards or feeding parts into *flexible cell* automated machining centers. Robot loading and unloading of all manner of machines with parts under manufacture, as well as the assembly of

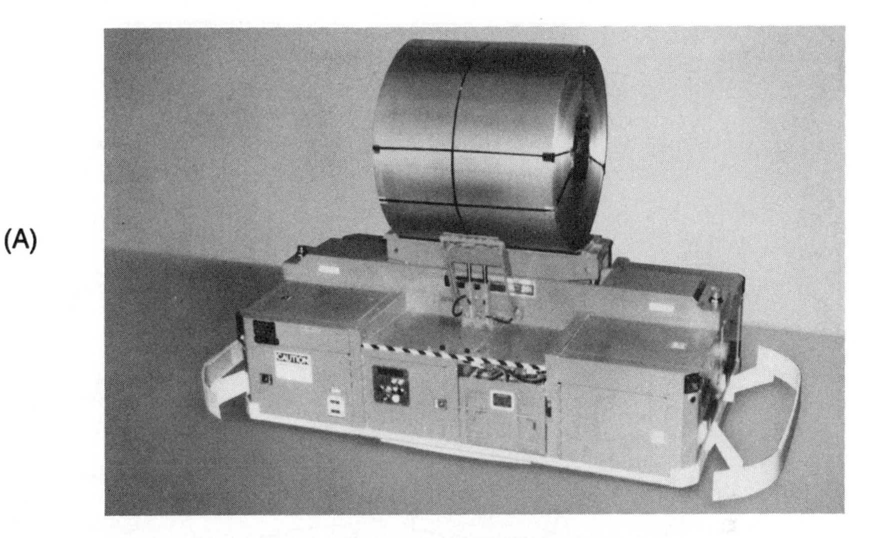

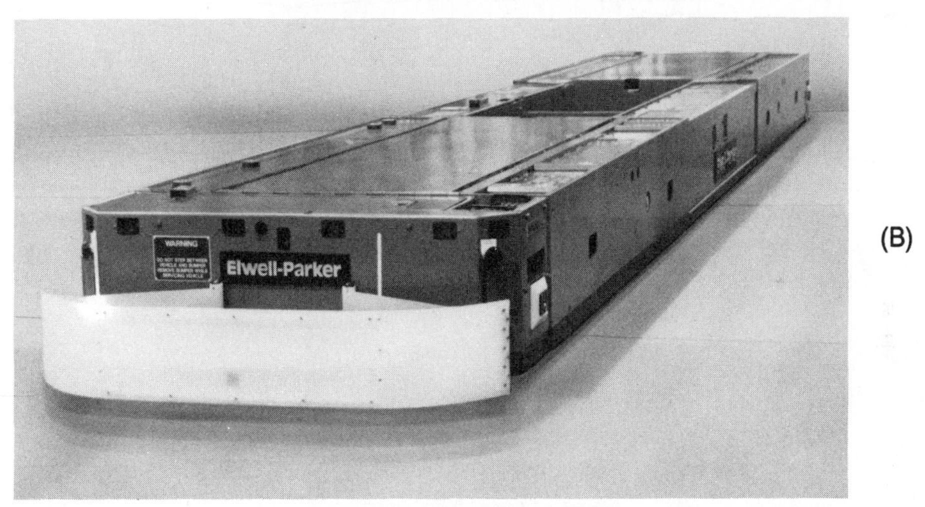

Figure 9-10 (A) A 72 volt AGV designed with a hydraulic lift mechanism for picking up and transporting a 72 inch diameter steel coil; (B) An AGV designed to transfer a 95,000 pound mainframe assembly weldment through the fabrication process. *(Courtesy of Elewell-Parker Industrial Trucks. Engineering, On Board/Off Board Vehicle Control—Control by Engineering Co. of Harbor Springs, Michigan)*

products comprise a quickly expanding field in flexible manufacturing. Pick and place robots have expanded very rapidly and will find many more applications, Figure 9-12A (page 351).

Robots are not generally used where assembly is in a low-volume range because of the cost of the robot. In high-volume manufacturing, a robot may work a single assembly workstation or two robot arms may work interactively on the

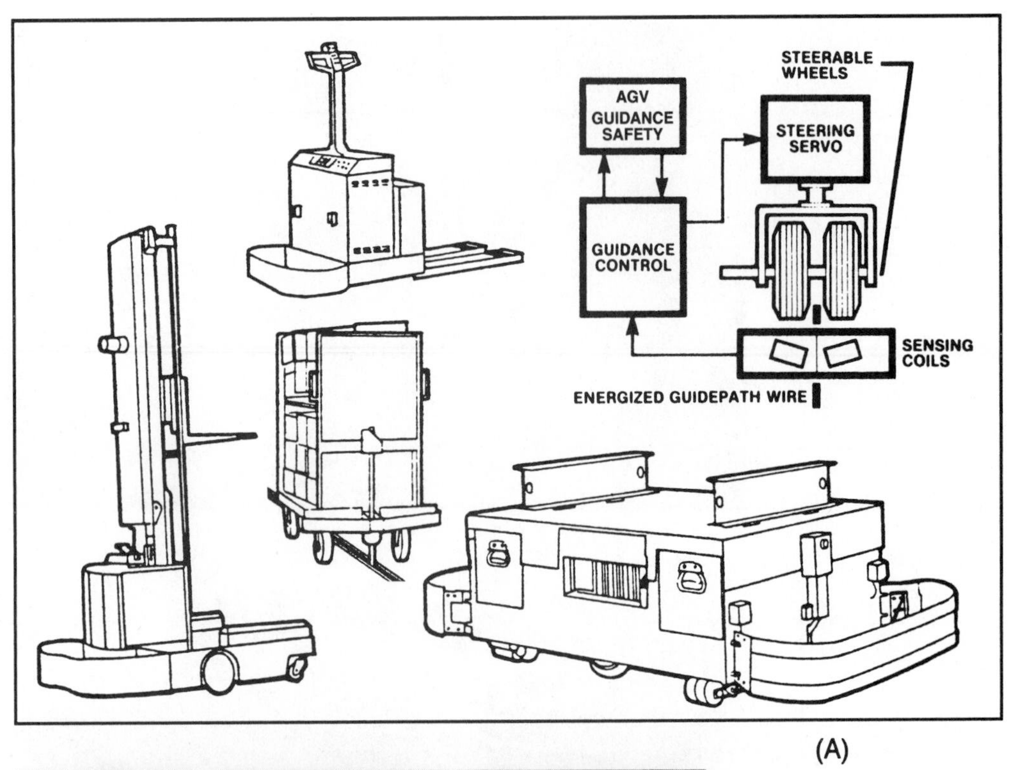

(A)

(B)

Figure 9-11 (A) Automated guided vehicle's configuration and control system; (B) AGVs transporting bulk products *(Courtesy of Eaton-Kenway Inc.)*

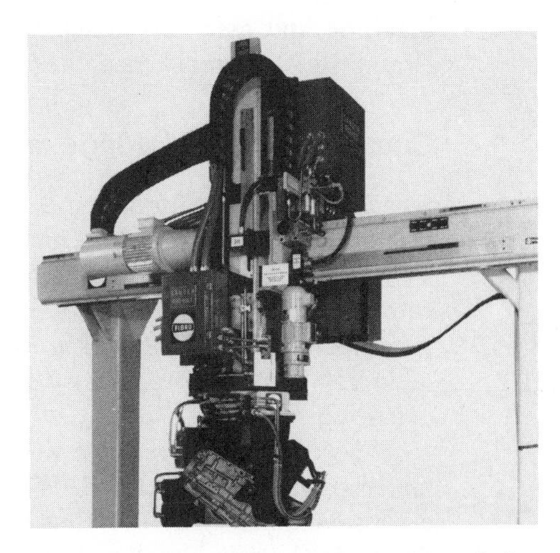

Figure 9-12 (A) A large pick and place robot racking system for automotive floor pans; (B) A transmission case machine loader robot *(Courtesy of Fibro Inc.)*

same station. Part feeders for the robots can be manually loaded magazines, or robots may include part feeders like those employed for machine assembly if the cost warrants, Figure 9-12B. The control of these machines entails a great amount of instrumentation with the applications of most classes of actuators.

ASSEMBLY AND PART HANDLING

Assembly has always been a vexing problem in manufacturing and methods of improving this condition have been studied vigorously. With the introduction of automated assembly machines, considerable improvement has been realized. Assembly is represented on three levels: manual assembly, a special purpose machine assembly, and programmable machine assembly.

Material Handling in Manual Assembly

Material handling for hand assembly employs a number of techniques previously discussed. Usually an individual working at a workstation simply does the moving of the object. In an assembly line, the material is usually handled by a conveyor.

Material Handling by Assembly Machines

Special purpose machine assembly is used in large batch or continuous manufacturing. The materials enter one of three general types of configuration material handling and assembly machines. The carousel, oval or elongated, and the straight-line assembly machines.

Carousel or Circular Indexing Machine

One of the configurations is the carousel or dial configuration. In this mode, the piece parts enter the machine and proceed by indexing from station to station in a circular path. At each station, tools holding fixtures and jigs are mounted on the carousel frame and the work is performed manually on the piece parts or subassembly. When the work is completed, the tools retract and the total indexing table rotates, carrying the workpieces to the next workstation for the next operation, Figure 9-13.

The tooling, fixtures, and workpieces will vary to accommodate the product material being manufactured. When additional work tools are required, a second carousel is added above the first, thus allowing tools to work on the product from above and below as well as from the front of the workstation.

The Oval or Elongated Circle Machine

When additional workstations are required for an assembly machine, the machine can be expanded to achieve additional workstations by changing the configuration from a circular path to an elongated circle or oval. This specialized machine functions as a carousel machine, but it supplies additional workstations for more assembly operations.

Figure 9-13 Carousel or circular indexing assembly machine *(Courtesy of Kingsbury Machine Tool Corporation, Assembly Machine Division)*

Straight-line Assembly Machine

Another configuration of assembly and part-handling machines is in the adaption of Henry Ford's assembly line. In this application, the workstations are placed in a line and transfer equipment is placed between the machines. This assembly machine is indexed with electrical controls so the whole machine is controlled as a single machine. The workpiece travels through the machine with part of the work accomplished at each station. When the workpiece is assembled with a number of smaller parts, it may be mounted on a pallet and the pallet and workpiece progress through the assembly process.

With mass production assembly, the assembly machines will be served with parts and fasteners delivered by vibratory bowls. These bowls orient the parts for automated assembly and will elevate and deliver the part or fastener through a slide or track to the point of insertion into the assembly.

The instrumentation installed to control these types of machines varies from applications of the microswitch to the control of a total unit by a programmable microprocessor or a small computer, Figure 9-14.

Figure 9-14 An in-line "quick change" automatic assembly system *(Courtesy of Schunk Automation Systems)*

AUTOMATED QUALITY AND POSITION CONTROL

Automated quality control can be an integral part of automated and semiautomated assembly lines. These machines inspect the assembled part and, if passed, the part is moved on to the next station or rejected for correction. In special cases, coordinate measuring machines have been programmed and interfaced so that they take advantage of CAD/CAM design geometry to check complex geometries. At this level of quality control evaluation of assembly machine products, considerable instrumentation and computer applications are required.

In automated quality control, the material handling sensing instrumentation is grouped into two broad categories: contact and noncontact. In the contact mode, the workpiece to be sensed is physically intercepted with a touch, force, or slip sensing device. In the case of the noncontact mode, the part or product is not contacted but is sensed with a proximity or vision sensing device.

Contact Position Control

The contact sensor often is a microswitch placed on the side of a conveyor belt to stop the belt when the force of contact with an object activates the switch. This contact occurs when the object is in an approximate position for transfer onto an assembly machine. The microswitch also activates the next operation to start the assembly machine's transfer equipment. Another contact sensor is a pneumatic valve, with a small roller that projects out into the conveyor belt and is activated by the touch of material travelling on the conveyor belt. The roller cam actuator provides a pneumatic limit switch signal to three- or four-way control valves, indicating that material is at hand for assembly, Figure 9-15.

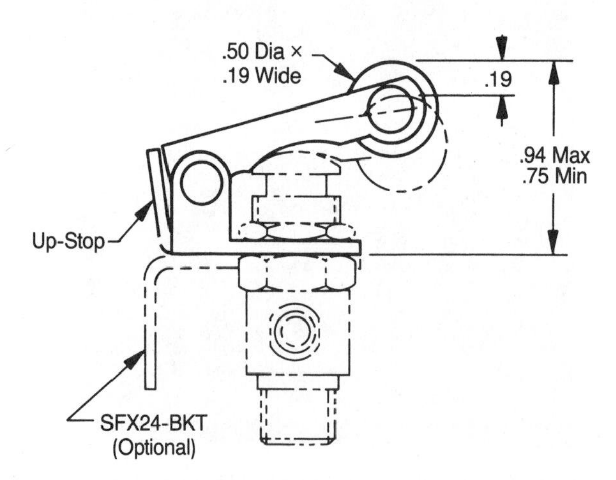

Figure 9-15 A pneumatic limit switch *(Courtesy of Gagne Associates Inc.)*

Robot grippers frequently handle materials in assembly operations. To ensure that the piece parts do not move or slip away from the holding devices, the holding pressure of the gripper may be sensed. This sensing is provided by flexible elastomer pads that compress when a gripper closes on a piece part. The compression of the elastomeric pads provides a piezoelectric pressure signal that can be set into the operational sequence to start the operation when a sufficient pressure is indicated. One of the advantages of this type of switch is that there is no contact bounce, Figure 9-16.

Noncontact Position Control

A noncontact sensor used frequently is one of many proximity sensors. These sensors sense a change in capacitance, inductance, light reflection, and an electromagnetic field. The electromagnetic field sensor utilizes the Hall effect to actuate a circuit when the magnetic field of the piece part causes a voltage change in a semiconductor material, Figure 9-17.

This type of position control is widely used in assembly as well as in production manufacturing applications. They will provide a signal for the object or tool that approaches its desired position. Ultrasonics, magnetic field, and air stream sensors are among a series of sensors applied for positioning control.

FINISHED PRODUCT HANDLING INSTRUMENTATION

The instrumentation for the material handling of the finished product includes the instrumentation necessary for final testing, storage, packaging, and shipping.

Final testing of a product requires an extremely broad selection of instruments to measure the product as it is tested in an operational mode. A hot test duplicates the user's environment as closely as possible. The tested product will be loaded to operational conditions while operating on the expected power or fuel. During the hot test, final adjustments may be completed and sealed. In the manufacture of electronic products, the burn-in phase may be part of this testing procedure.

Testing requires instruments from simple hand tools and meters to coordinate measuring machines or a programmed computer that tests each component and

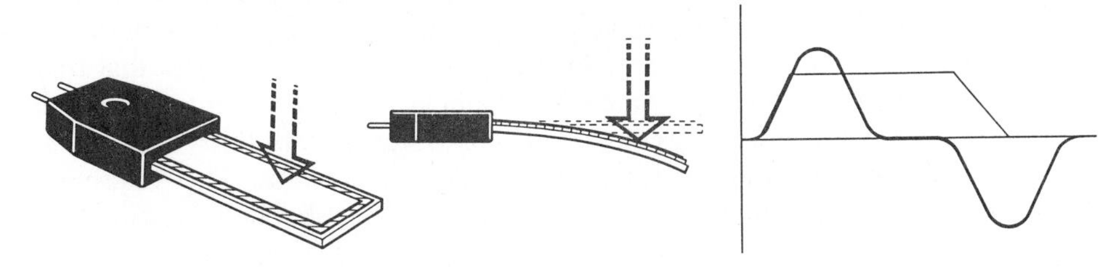

Figure 9-16 Piezo film flexture switch *(Courtesy of Pennwalt Corp.)*

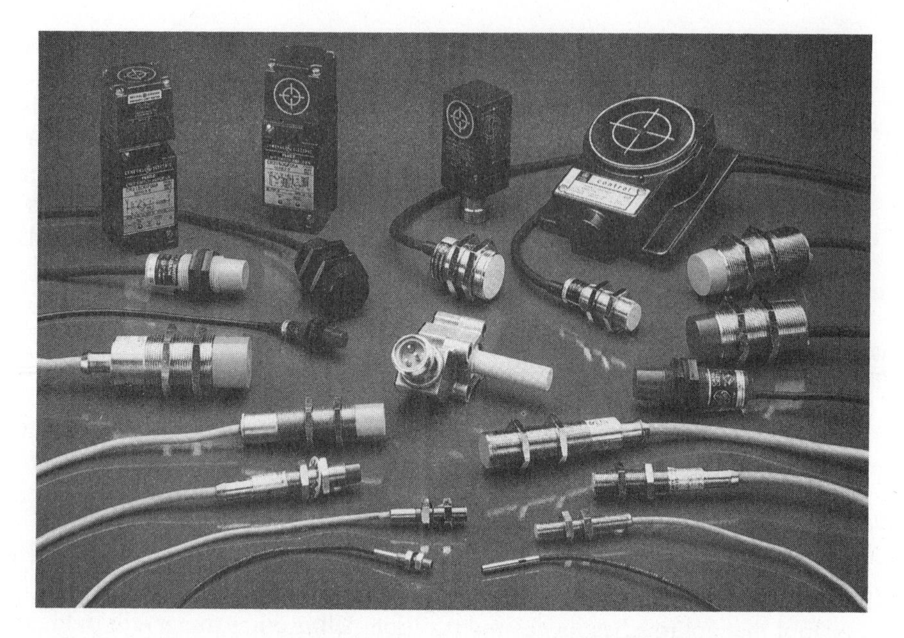

Figure 9-17 Noncontact proximity sensors *(Courtesy of General Electric Company)*

circuit in a logical step-by-step procedure. Quality control in final testing can make use of the CAD/CAM data base by interfacing the data base to a program coordinate measuring machine, allowing the product geometry to be verified before releasing the product.

With the application of state-of-the-art sensors coupled with solid-state electronics, a host of outstandingly accurate instruments have been built. An automatic test system using an instrument controller computer can be used to make final testing total and automated.

High-density Material Handling and Storage

Material handling for products going to or coming off the production line can be stored in a high-density, multilevel, automated storage system. This type of automated storage requires the use of computers to inventory all materials coming in or going out of the system as well as to remember the location of all the items in the system. Frequently, these storage systems will have a robot that travels down the center of the storage area. The robot is a narrow rectangular container that can deliver materials and component bins to storage cells thirty feet above the floor and down the row of cell areas one hundred feet or more. These storage cell areas are often racks in which the bins are stored. The robot delivers bins to both sides of the alley.

This high-density storage area is serviced by conveyors and transfer devices, all controlled by a computer. The computer inventories the materials and components as they are delivered by conveyor from the receiving dock. This system is used to supply materials for a total assembly plant, Figure 9-18.

Another application of material handling may occur at the end of the assembly. For instance, in the manufacture of personal computers, the line conveyor continues and serpentines around high above the work floor carrying the assembled computer. During this transfer, the computers are on and the burn-in stage of testing is performed at the same time as the transfer is done. During the burn-in stage, the computers are all visible from the assembly floor, which results in an operational inspection.

Other products such as disc drives emerge from a clean room assembly and are transferred by a robot into an automated storage area for a burn-in period. At a later time, the robot will deliver the drives to a conveyor that brings them to final quality control, testing, and on to packaging.

Figure 9-18 High density, multilevel automated storage and retrieval system. This system contains 7,000 tons of inventory. *(Courtesy of Nucor Corporation)*

SUMMARY/FACTS

- Material handling includes receiving materials in a manufacturing plant and also the movement of materials within the plant and within various manufacturing processes.
- Bulk materials can be moved by a number of different designs of machines to provide direction of material movement. The material may be horizontal, vertical, at an angle, or a combination of these directions.
- Rubber conveyor belts for handling bulk materials are generally of two types -- a flat conveyor belt and a trough conveyor belt.
- Monorail or trolley conveyors are hung from the ceiling and may be employed to transport heavy, awkward, large pieces of material or baskets containing assemblies of small parts.
- The car-on-track used in manufacturing can transport workpieces, pallets, fixtures, robots, tool trays, etc., and will move products vertically and horizontally as well as provide buffer storage.
- Vibratory part feeders elevate and orient small parts for automated assembly.
- The automated guided vehicle is often used to transport heavy workpieces or palleted workpieces around a computer integrated manufacturing plant.
- One of the configurations of material handling in an assembly machine is the carousel or dial configuration. In this mode, the piece parts enter the circular path and are indexed from station to station. Work (assembly) is performed at each station.
- Another configuration of material handling in a manufacturing machine is the in-line indexing or transfer machine. In this mode, the piece parts pass through in a straight line, being indexed from station to station with work performed at each station.
- In automated quality control, the sensing instrumentation is grouped into two broad categories: contact sensing and noncontact sensing. In contact sensing, there is a physical intercept with the workpiece. In noncontact sensing, the workpiece is sensed by a laser, vision, or proximity devices.
- A final test of a product is performed in an operational mode. A hot test duplicates the user's environment as closely as possible.
- High-density storage and retrieval systems are serviced by robots, conveyors, and other transfer devices.

REVIEW QUESTIONS

1. Material handling is primarily concerned with receiving materials into a manufacturing plant and then moving the materials within the plant. In what other instances is material handling a function?
2. What are the two main types of conveyors?

3. Which type of conveyor is used to carry heavier loads?
4. Describe a bucket elevator.
5. What is a shaker elevator?
6. By what physical phenomenon are materials or parts transported in chutes?
7. A conveyor with chains separated by flights or bars sliding within the base is a ________________.
8. Moving large amounts of light powdered or granular materials through tubes can easily be done by which conveyors?
9. What are light and strong plastic formed containers that can move masses of very small parts and small components around various manufacturing or assembly areas called?
10. A means of transporting heavy, awkward, or large pieces of material is that of a conveyor suspended from the ceiling. What is that conveyor called?
11. How is the flat belt conveyor utilized in assembly or testing work?
12. What three functions are performed by vibratory part handlers simultaneously?
13. What is the method of conveyance in which pushing pawls are used to effect the transport of heavy single-piece parts?
14. Which transporters are attached to an endless chain moving in a channel in the floor?
15. How can an AGV (Automatic Guided Vehicle) be guided without visible external control?
16. How is material usually handled in manual assembly?
17. What are the three main configurations used in material handling and assembly machines?
18. Describe the role of the robot in assembly in high-volume manufacturing.
19. How can automated quality control be an integral part of automated or semiautomated assembly lines?
20. What is the variable that is sensed when robot grippers are used to handle materials in assembly operations?
21. How is this sensing accomplished?
22. What data base can the coordinate measuring machine employ to verify the geometry of the product?
23. Describe the use of a robot in an automated storage system.
24. How can an operational inspection be accomplished simultaneously with transfer of the product?

10

INSTRUMENTATION APPLIED TO MANUFACTURING MACHINERY

OBJECTIVES

Upon completing this chapter, you will be able to discuss and apply these instrumentation concepts:

- Mechanical instrumentation
- Instrumentation used in mass production transfer lines
- Instrumentation by numerical control
- Instrumentation applied to computer numerical control

INTRODUCTION

Kinematics is the study of motion. The interaction between motion and machine elements is the basis of mechanical instrumentation.

MECHANICAL INSTRUMENTATION

Mechanical instrumentation is the oldest form of instrumentation and has been evolving since the Middle Ages. Mechanical instrumentation is considered as analog control in that the work it performs or controls is continuous in output, as contrasted to that of digital control.

Mechanisms

Mechanisms are assemblies of machine elements. The machine elements include rigid links, cams, belts, pulleys, gears, chains, machine frames, and foundations. These mechanisms are motion convertors that provide movement for mechanical instruments and actuator equipment to bring about control of the manufacturing variable.

Linkage Systems

To understand some of the basic concepts of mechanism, consideration must be given to the four-bar linkage system. The link is a rigid member that transmits energy to another member of the system. This interaction produces a motion in the second bar and the joint between them. The four-bar linkage system is the basis for many motion converters and machines, such as in reciprocating engines or automobile front end suspensions.

When a machine member is anchored in a frame, base, earth, or foundation of a machine, that member is defined as a constrained member. The engine block of an automotive engine provides that constraint for many moving elements in the engine that are necessary for the generation of relative movements. In a kinematic diagram, the constrained element is indicated by short diagonal lines, Figure 10-1.

The moveable elements theroretically have six degrees of freedom, the directions or ways in which the moveable elements in a space system can move. But they are usually free to only move in certain directions and planes. The connection between two elements—the joint—is the place where movement

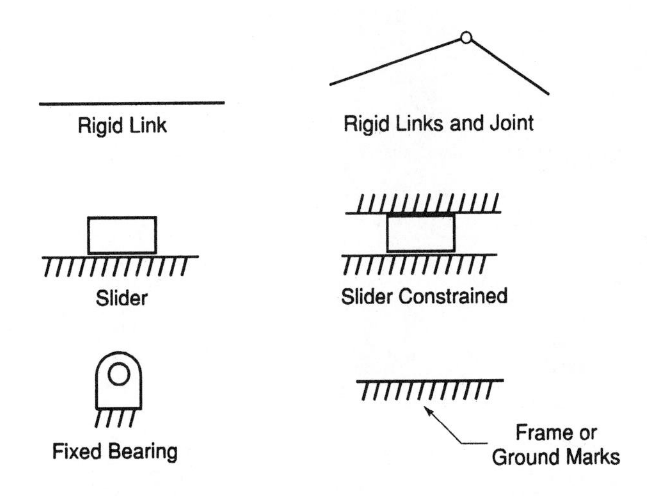

Figure 10-1 Basic kinematic symbols

takes place and is shown on the kinematic diagrams as a small circle. In special cases, such as a ball joint, the connection has additional degrees of freedom and is classified as a space mechanism because it may have up to five degrees of freedom. A common turning pair, such as a connecting rod, has only two degrees of freedom, clockwise and counterclockwise rotation. In any case, a machine will have at least one degree of constraint to provide a resistance that the mechanism will bear against.

Inversions - A four-bar link system can generate a number of motions. Rigid bars connected at their ends with a turning pair, such as a pin or roller joint, can provide a number of mechanisms. The movements produced by anchoring different members progressively around the system can produce different motions: rotary-to-rotary, rotary-to-oscillatory, and oscillatory-to-rotary motions. These different motions are called inversions. The study of inversions has helped discover different applications for machines. Different inversions are obtained by constraining different members of the machines, Figure 10-2. The process is known as *kinematic inversion.*

The slider crank mechanism is one arrangement of a four-link system. It is very useful because it is the basis of a common mechanical reciprocating instrument, Figure 10-3.

This mechanism has two strokes of the slider or piston (up and down) with each rotation of the crank pin or crankshaft. In Figure 10-4, the slider crank is the kinematic linkage of many other machines. The illustrations demonstrate that inversions result as the various elements are constrained. The study of inversions

(A)

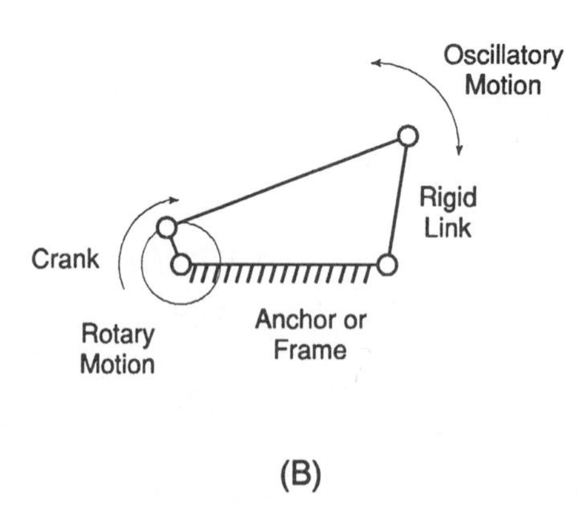

(B)

Figure 10-2 (A) Four-bar linkage, rotary-to-oscillatory motions; (B) A kinematic diagram of the four-bar linkage

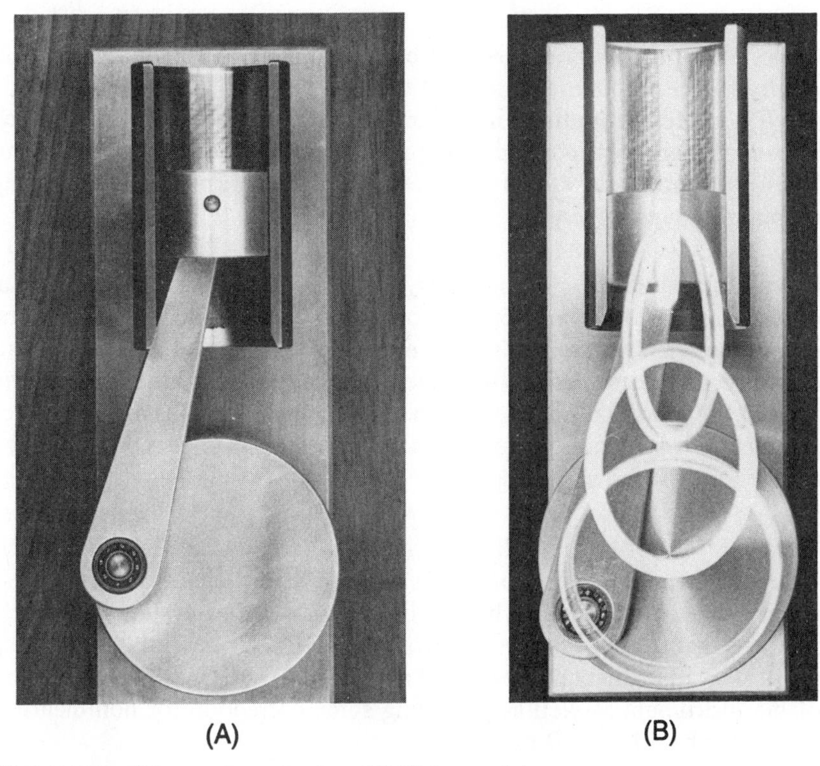

Figure 10-3 (A) The slider crank mechanism; (B) Slider crank traces

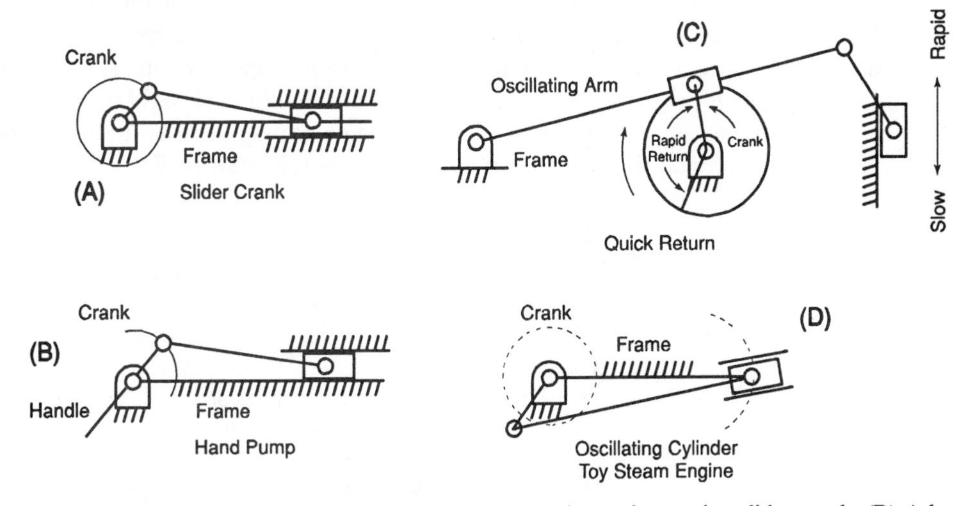

Figure 10-4 Slider crank mechanism inversions; (A) A reciprocating engine slider crank; (B) A hand-powered water pump; (C) A modified Whitworth to a metal working shaper quick-return mechanism; (D) A toy steam engine cylinder oscillates to open and close the steam ports (*From* Mechanisms, *by Virgil M. Faires [co author Robert McArdie Keown], 5th ed., copyright 1960 by McGraw Hill. Used with permission.*)

is important in understanding the movements in mechanical instruments. This exploration process is carried out by systematically anchoring each link and observing the output movements of the system.

It is evident from these inversion illustrations that a number of mechanical movements are incorporated in the mechanical portions of many instruments. These inversions allow them to deliver a characteristic movement or to adjust the output of the instrument.

Oscillatory Motions - An oscillatory motion is the result of the action of a rotary crank pin linked to an anchored pivot bearing. When the throw of the crank pin is very short, it is referred to as an eccentric and its connecting rod or link is spoken of as an eccentric rod. When this system is connected to another link and the crank rotated, the new link oscillates. This link is called a "rocker arm." The eccentric produces a rotary motion, and an oscillatory motion is generated by the rocker arm, Figure 10-5.

Parallel Motions - Parallel motions are frequently used in mechanical instrumentation movements in connection with sensors and gages. These movements are transferred by bar linkages and lever systems, such as in the linkages that transmit movement from the end of a bourdon tube to the pointer shaft. The adjustment of the bar linkages' lengths aids in the setting of the range of the instrument as well as providing some correction for nonlinearity of the bourdon tube, Figure 10-6.

Straight-line Motion - Mechanical instruments in some applications require straight-line mechanisms. They are employed where a moving link must have its end member trace a straight line when going through its movements. A straight-line movement is possible with a Scott-Russell mechanism. A straight motion can be transferred to an element such as a bellows or pointer to keep the element vertical or horizontal throughout the distance moved, Figure 10-7.

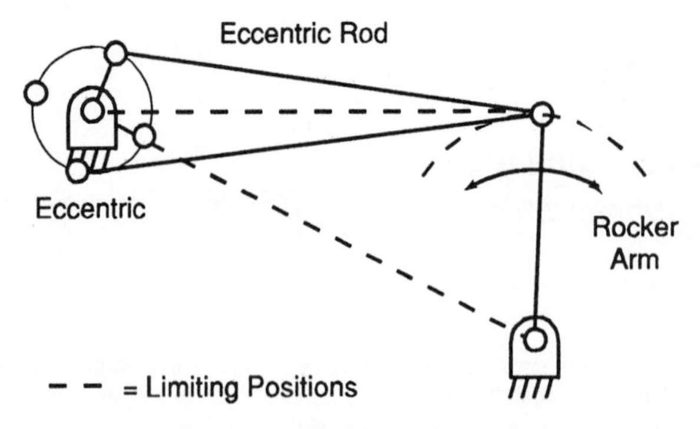

Figure 10-5 An eccentric and the oscillatory motion of a rocker arm

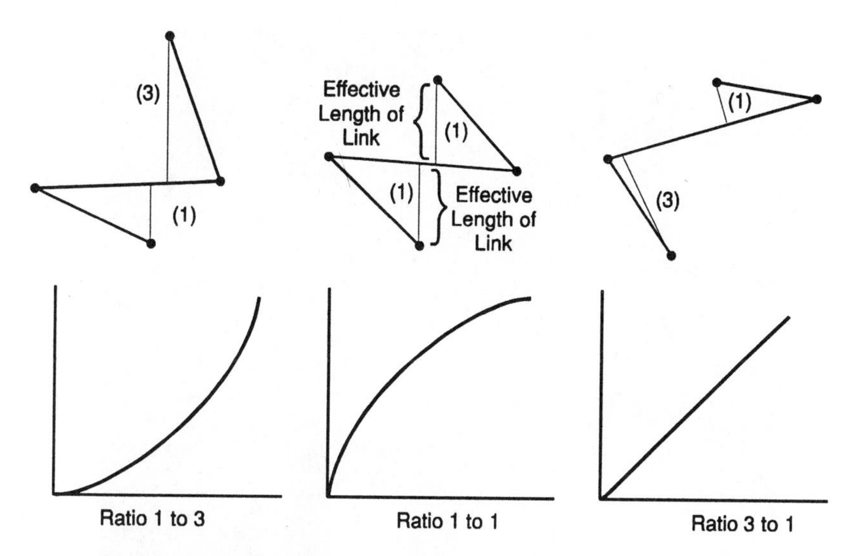

Figure 10-6 Instrumentation parallel-motion linkage. Kinematic correction of nonlinear systems by adjusting the effective lengths of the links in the system.

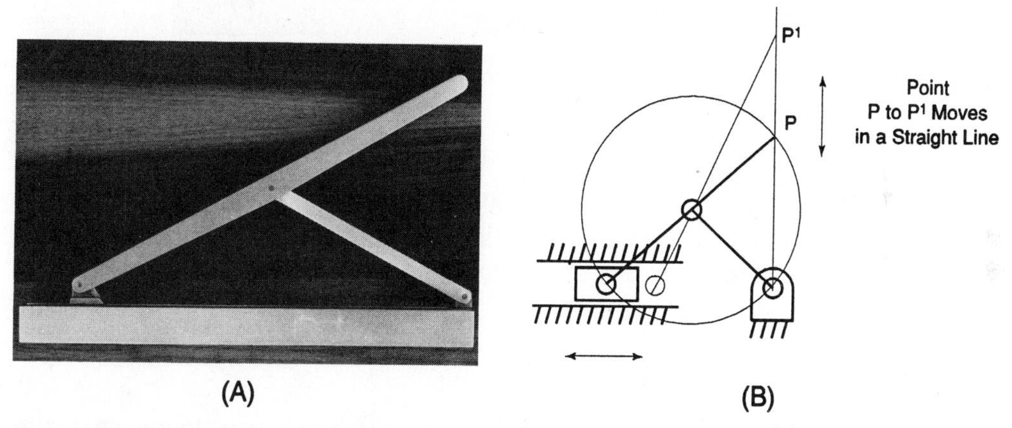

Figure 10-7 (A) Scott-Russell linkage; (B) Point P moves in a vertical straight line

Rotary Motion Speed Control - The fly-ball governor is designed from one of the early mechanical instruments of speed control. It is a feedback device that senses the speed of a steam engine or factory mainline shaft. The fly-ball governor is one form of tachometer with its output connected to a steam control valve and/or to an indicator showing revolutions per minute.

The fly-ball governor is made up of balls on the end of linkages that are spun around by a belt or gear linkage from the engine or shaft being sensed. The balls fly out from the center shaft as the speed picks up and then stabilize in that rotating position. The balls are a centrifugal sensor that, through a linkage, transmits their

Figure 10-8 The fly-ball governor. A change in linkage-effective-length used to sense speed

position to a yoke at the base of the assembly. The yoke's movement in turn provides a setting for a control valve or other controlling device. The fly-ball governor's output is not completely linear at all speeds, so linkages and springs may be added to the linkage system to make corrections, Figure 10-8.

Rotary Motion Convertors

Rotary motion convertors are mechanisms that change rotary motion into reciprocal motion.

Cams - The cam is a mechanism that performs work or controls data in a continuous fashion. An instrumentation characteristic of the cam is that its surface can act as a data storage and retrieval system. This is evident when a cam follower has a defined position for each degree of rotation on the cam. To acquire certain data for instrument control, it requires the rotation of the cam to a predetermined position for the cam follower. The cam's surface represents a step-less change of information. In this fashion, the cam's surface possesses a data memory. The angular rotation of the cam provides an accurate movement of the cam follower that represents the analog data from that point on the cam, Figure 10-9.

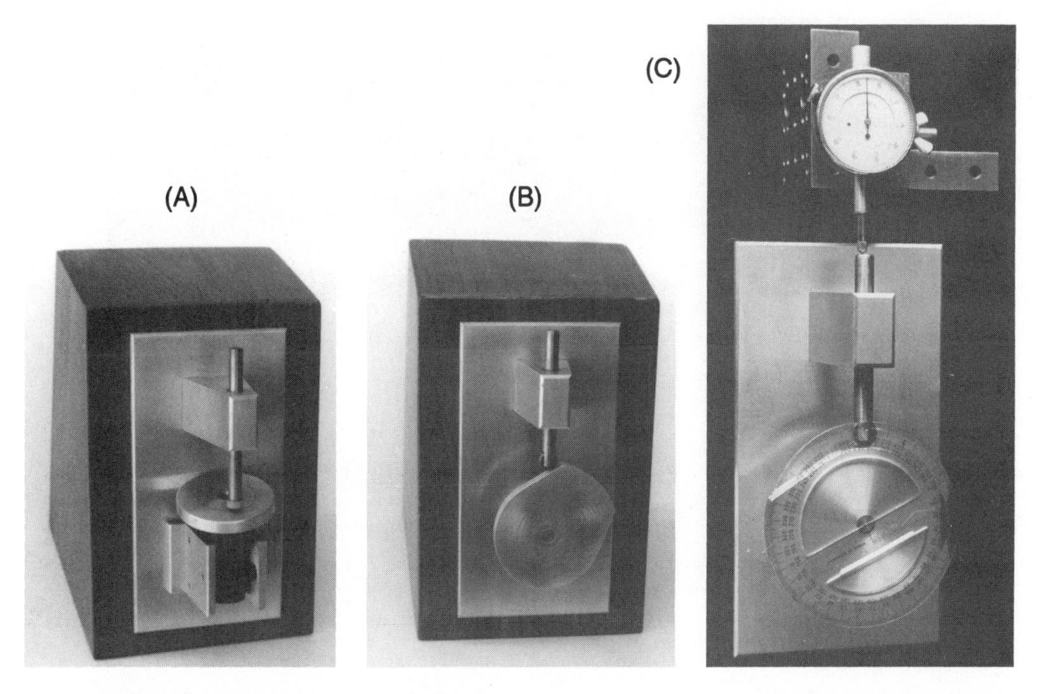

Figure 10-9 Various types of cams: (A) Wobble plate cam (harmonic motion); (B) Plate cam; (C) Graphic analysis (displacement versus cam rotation)

When a cam is continuously rotated, the displacement of the cam follower performs a timing function as well as a change in physical distance controlled by the cam's speed of rotation.

A three-dimensional cam provides data from three sources and integrates the information into a single point on a cam's surface. Each time the cam is returned to that same position, the identical data is retrieved by the cam follower. Each position point on the cam is a memory of the analog data. The position of the cam follower may be used to set the position of another device to a precision setting very rapidly.

Cams are instruments of analog data that provide control over mechanical amplifiers (gears and lever systems, hydraulic systems, and the like). The rotation of the cam supplies the displacement or the amount of movement to control the work. By rotating the cam to a new position, a new data point is provided on the cam for the cam follower. For example, the displacement of the cam follower could be the input to a hydraulic servovalve that causes the hydraulic system to deliver power and make a change in a large system. The movement of a cam may be used to control many systems, from the steering of a ship to the positioning of space antennae, Figure 10-10.

The inverse cam is a special form of cam in that the functions of the parts are reversed. The body with a cam path is the driven member while the cam follower

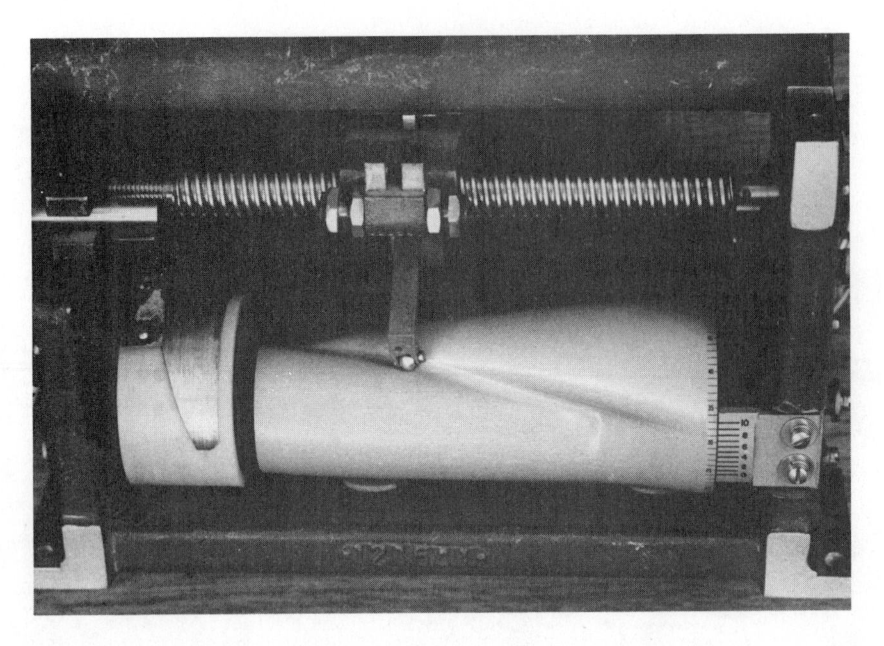

Figure 10-10 A three-dimensional cam

or roller becomes the mechanism driver. An example of an inverse positive cam action is the Scotch Yoke. This positive action cam has been used for centuries to generate a reciprocating motion. A roller is mounted on the end of a crank pin that rotates continuously through three hundred and sixty degrees. The camming surface or face is a groove in which the roller rides. As the crank pin and roller rotate, the roller and cam cause the cammed surface to rise and fall, thus converting a rotary motion to a reciprocating motion. This motion converter has been used to pump water out of mines since the sixteenth century. It also has such applications as powering the needle movements on a sewing machine. This mechanism can be applied to instrumentation by using it as an analog memory device. Rotating the crank pin through a specific number of degrees with today's stepping motor causes the yoke shaft to be displaced a precise distance that in turn opens or closes the stem of a control valve providing control over the flow of fluids, Figure 10-11.

Ratchet - The ratchet is a common mechanism used to produce an intermittent motion used for activating various types of machine feed devices. The ratchet consists of a notched wheel (the ratchet wheel) mounted on a shaft to be indexed or held and a pawl that can either supply power through a lever to turn the ratchet wheel or be a fixed pawl that will hold the ratchet from rotating. The driving pawl may be activated by a variety of crank pins, cams, or eccentrics, Figure 10-12.

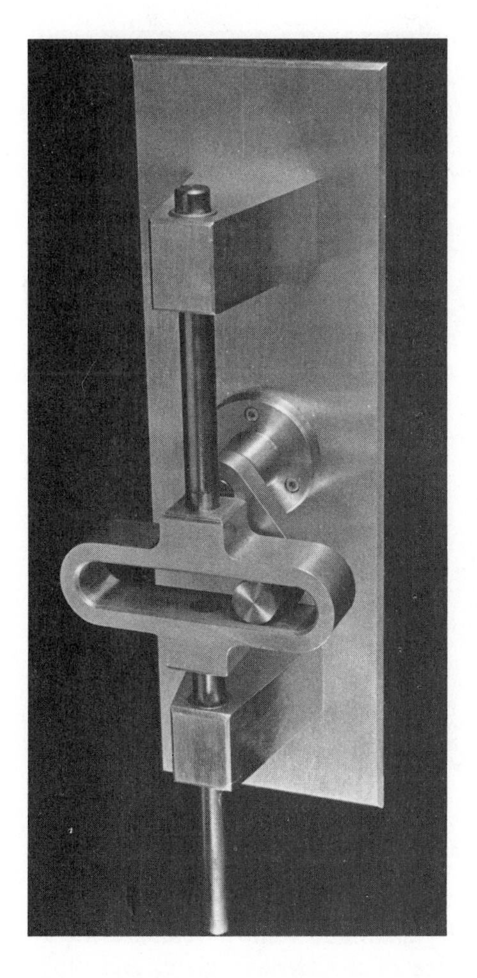

Figure 10-11 An inverted cam—the Scotch Yoke

(A)

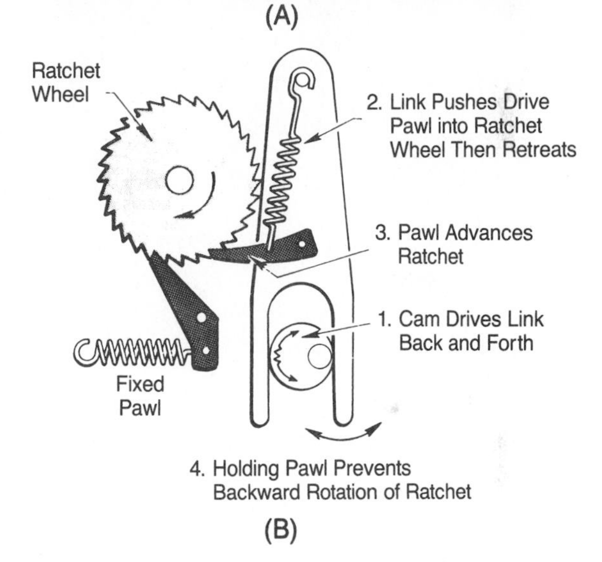

(B)

Figure 10-12 A ratchet wheel with a driving pawl

Ratchets have a number of variations and are supplied to operational intermittent drives to adjust the setting of mechanical instruments or to hold spring-wound mechanisms.

Geneva - The Geneva mechanism provides an intermittent rotation to a shaft. The crank pin enters the camming slot and cams the star wheel around until the crank pin leaves the slot. At this time, a circular section of the crank pin carrier slides against the curved area between the camming slots. The locks the star wheel until a complete rotation of the crank pin has occurred and the pin has entered the next slot. The movement of the Geneva output shaft is timed by the rotational speed of the crank pin and the number of slots in the Geneva plate. The plates are

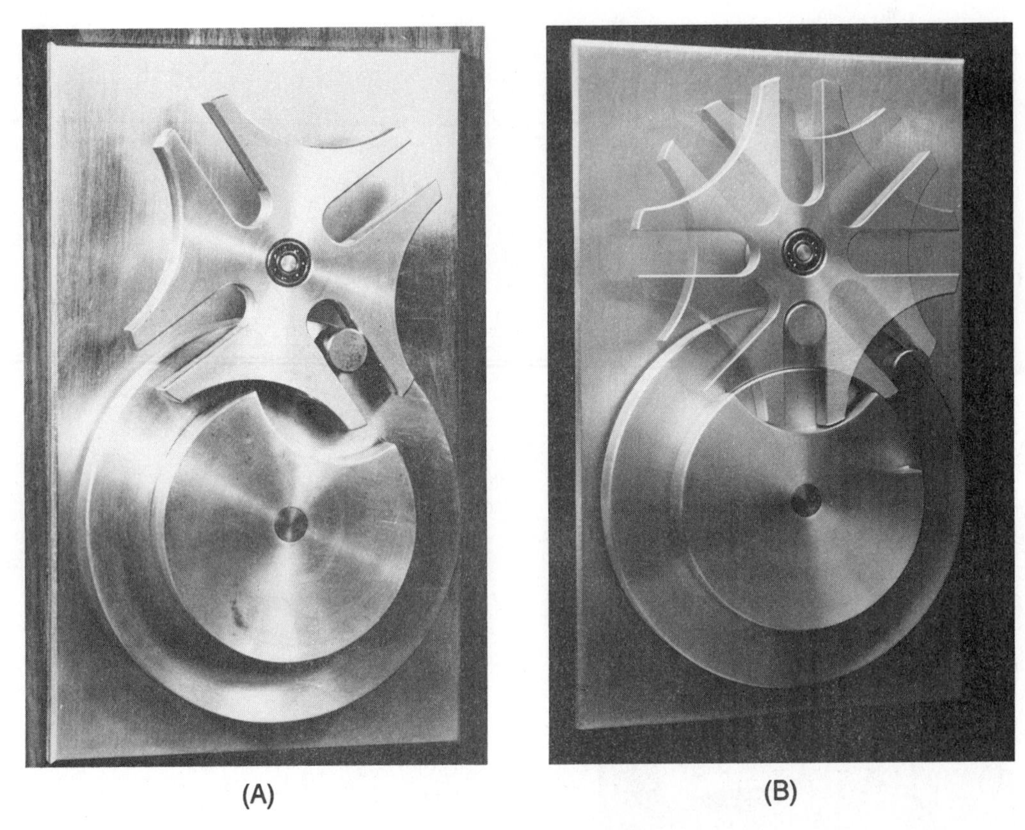

Figure 10-13 (A) The Geneva mechanism; (B) A Geneva phantom view

designed with a minimum of three slots. The maximum number of slots is set by the required size of the pin roller, the size of the Geneva plate, and other dimensions. Many Genevas have four, five, or six slots, depending upon the angle of movement and the output that is required, Figure 10-13.

This mechanism is frequently used as an indexing mechanism for moving tools, conveyors, feeding devices, motion picture projectors, or for other machines and intermittent or timed instruments.

Ball Screws - Ball screws are a combination of antifriction nuts and screws designed to deliver accurate movements with precision and with a minimum amount of force. The balls roll in helical races ground into both the screws and the nut. These parts are preloaded upon assembly to eliminate any backlash within the ball screw parts. The nut assembly is equipped with a tubular assembly that returns the balls back to the front of the nut helix while the screw is turning. This return device provides a constant supply of balls to the nut as it travels down the length of the screw. The nut may receive rotation perpendicular to the screw axis with a worm for thrust loads. More frequently, the drive is parallel to the screw, turning the screw within the nut, Figure 10-14. Applications of the ball screw are found in precision lead screws for milling machines, lathes, and measuring machines.

With the application of ball screws and precision turning devices, such as stepping motors, numerical control machines, and robots, the precision movements necessary for automated manufacturing are available. These mechanical instruments do not make control decisions, but do provide the necessary actuator movements.

Gears

Gears and gearing are machine elements that transmit power or movement from one shaft to another. There are many types and sizes of gears to perform these functions, Figure 10-15.

Spur Gears - The most common type of gear is the spur gear. The spur gear is employed when two shafts are parallel to each other. The tooth elements of the gear are form cut on the outside of a right cylinder. The shape of most spur gear teeth uses the involute curve to form the profile of the gear tooth, so that the two gear surfaces will roll together with their pitch circles tangent to one another in the process of transmitting power or movement, Figure 10-16.

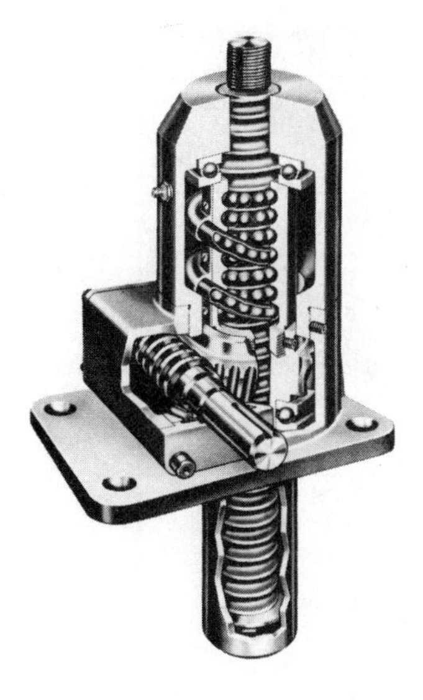

Figure 10-14 Worm-gear-driven precision leveling device ball screw *(Courtesy of Philadelphia Gear Corp.)*

Figure 10-15 An assortment of gear types *(Courtesy of Overton Gear & Tool Corp.)*

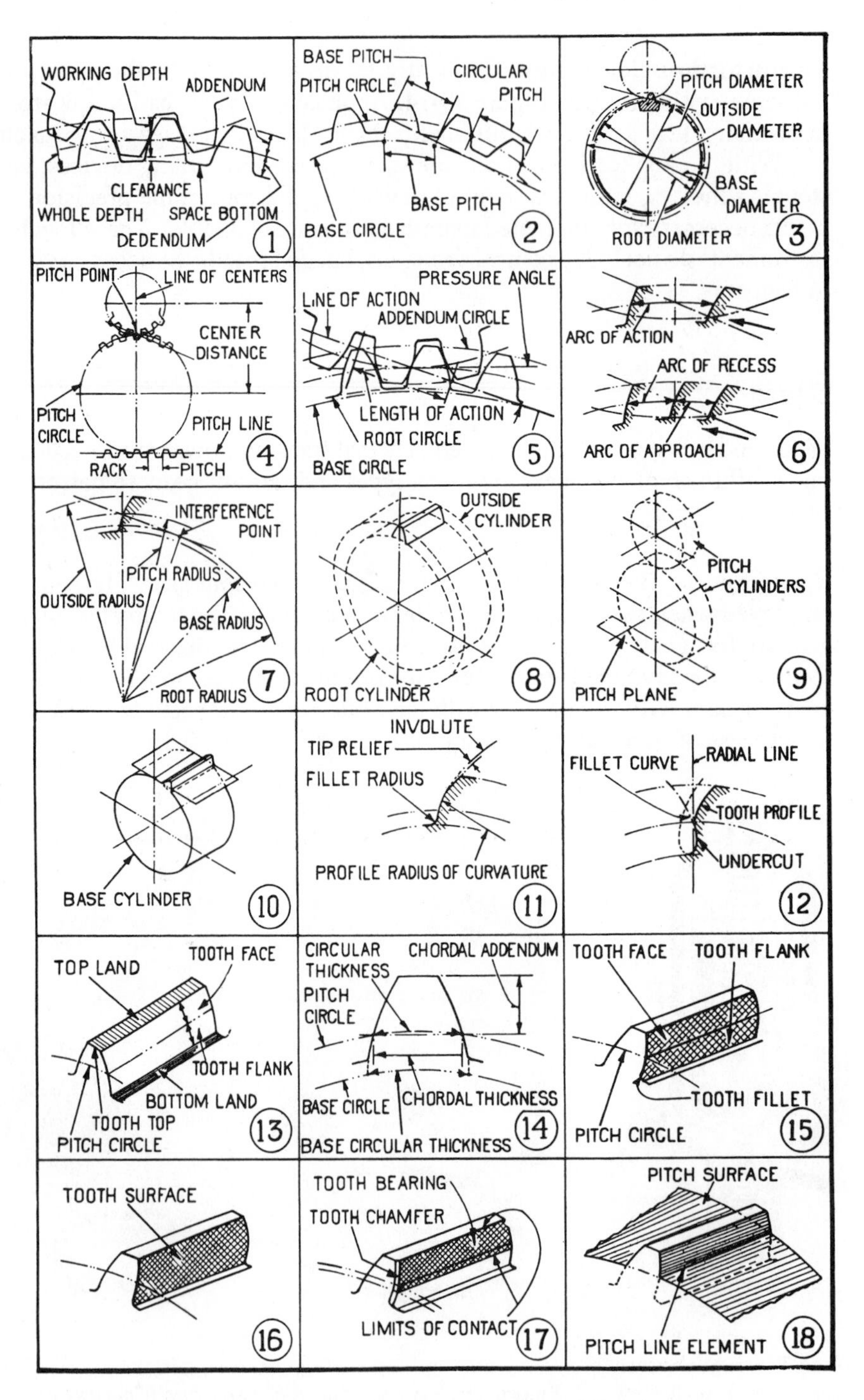

Figure 10-16 Gear nomenclature *(Courtesy of Overton Gear & Tool Corp.)*

Pinion Gears and Sector Gears - A fundamental interest in gearing is the production of a ratio of rotation between the two shafts. The velocity ratio may be calculated in terms of revolutions per minute, or more practically by the ratio of the number of teeth on the gears. In mechanical instrumentation, the movements can be small. Thus, the driving gear may be only a portion of a revolution. The driven gear is a small pinion gear whose shaft carries a pointer. With this application in mind, the larger gear may be a sector gear which is only a portion of a gear. A mechanical bourdon tube pressure instrument uses this concept.

Noncircular Gears - Nonlinear motion can be obtained with the use of *noncircular* gears. They can offer a motion obtainable from a cam and most of the rotation motions available from linkages. They provide the usual gear properties of high operation efficiency and accuracy, while avoiding the pressure-angle and lost-motion problems associated with cams. Noncircular gears give kinematically exact solutions and can be designed to provide virtually any continuous mathematical function that does not require reversed motion, Figure 10-17. Noncircular gears are used in printing presses, packaging machines, conveyors, pumps, as well as precision instruments requiring nonlinear motions.

Figure 10-17A Noncircular gears *(Courtesy of Cunningham Industries Inc.)*

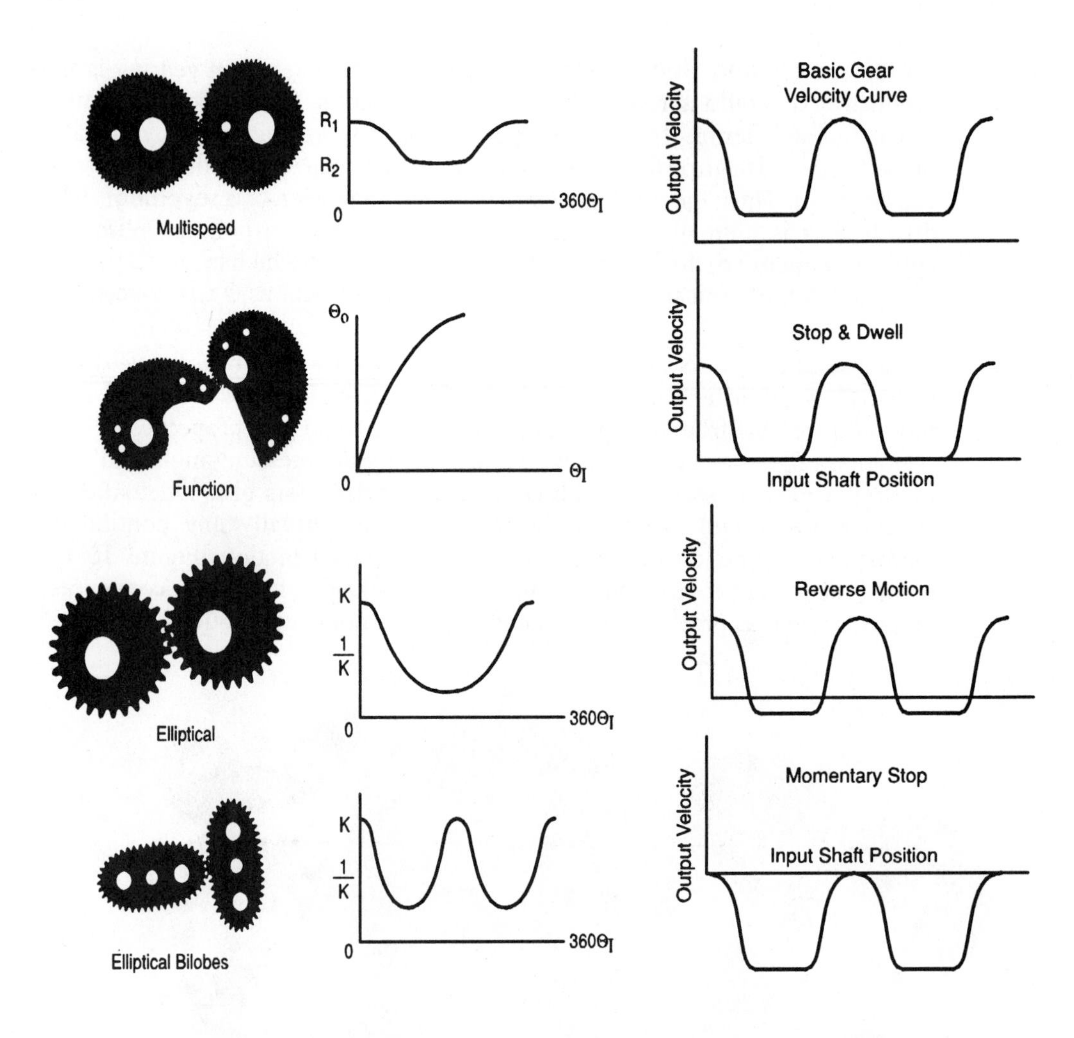

Figure 10-17B Noncircular gear output velocities *(Courtesy of Cunningham Industries Inc.)*

Flexible Connectors

Flexible connectors used in instrumentation are belts, chains, and cables that transmit power or movement over distances that would be impractical for gearing because of the environmental or alignment problems.

Timing Belts - Belts come in a number of configurations, but the belt frequently used in instrumentation is the positive drive or timing belt. Timing belts are flat belts with teeth molded into the contacting surface so that no slippage occurs between the belt and notched pulley. These belts are manufactured from neoprene and nylon to produce a very reliable connector. The belts maintain the ratio between pulleys at slow or high speeds and wear exceedingly well, Figure 10-18. They are used in instrumentation drive systems as well as for position data transferring devices.

Positive belts are also made with reinforced plastic side lugs that mesh directly with spur gears. One gear may have a central groove cut into it so that the belt runs in the groove, or two gears may be placed side-by-side so that the belt runs in the space between them and the belt center will clear the gear surface, Figure 10-19.

Bead Chain Drives - Bead chain drives are available for light instrumentation drives. The sprockets for this chain have indentations or pockets that correspond

Figure 10-18 Synchronous drives—an assortment of pulleys and timing belts *(Courtesy of T.B. Wood's Sons Co.)*

Figure 10-19 A demonstration board of cable chain belts *(Courtesy of W.B. Berg Exhibit Photograph)*

Figure 10-20 A complex instrument dial bead chain drive *(Courtesy of Voland Corp.)*

to the teeth of a gear and the beads of a beaded chain fit into these pockets. With the use of idler sprockets, the shafts of the sprockets need not be in the same plane, so a number of different configurations of drives can result. A precise ratio is maintained by the relationship between the number of pockets of the driving sprocket to those of the driven sprocket, Figure 10-20.

Roller Chain Drives - Roller chain drives are used on heavier duty high-speed long-distance drives. They have the same ratio characteristics as gears except that the center distances between shafts are not critical and can be long distances. The roller chain is made up from parts including rollers, pins, pin links, bushings, and connecting plates. These chains engage the sprocket for about one hundred and twenty degrees of rotation. The tension on the chain is regulated by either an adjustable idler sprocket or by adjusting the distance between the sprocket shafts. In the case where greater power transmission is required, multiple chains are installed.

Friction Drives

Friction drives come in a number of configurations. They are an assemblage of wheels and balls that transmit motion or data.

Wheel and Disc Drive - A wheel and disc drive transmits light loads and supplies a forward or reverse direction of rotation at a variable speed. A small drive wheel contacts a rubber-like covered disc to provide friction between the two. As the

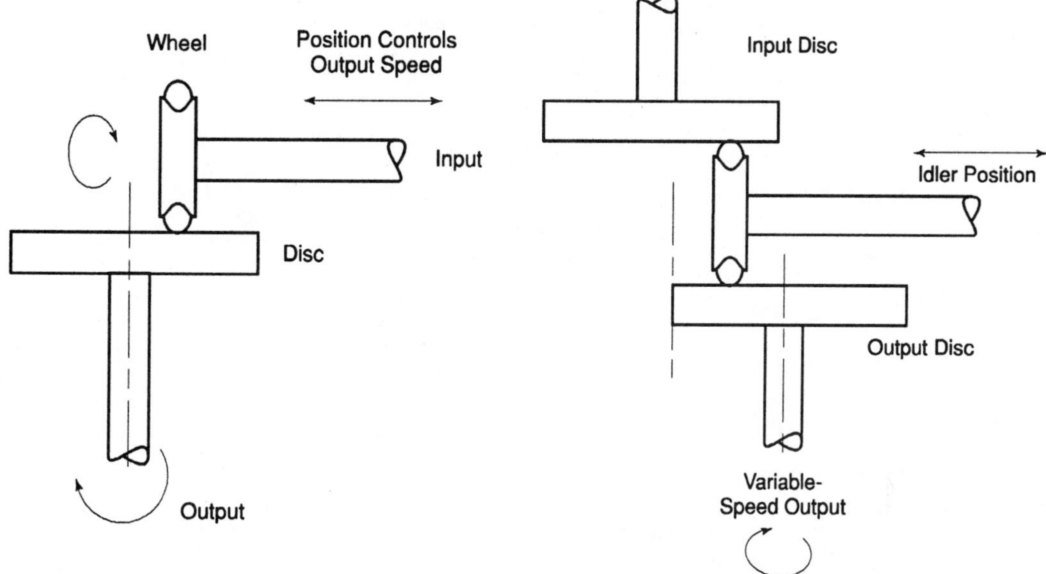

Figure 10-21 A wheel and disc forward and reverse drive

Figure 10-22 Two-disc variable-speed drive

small driver wheel approaches the center of the disc, the output rotation of the disc, is higher until the center is reached. Then the output becomes zero. Upon passing beyond the center, the disc reverses its direction of rotation and again goes through a variable-speed output, Figure 10-21.

Two-disc Variable-speed Drive - A variation of the wheel and disc drive is made by adding a second disc above the wheel. This second disc is offset by the radius of the disc. The wheel between the two discs becomes an idler, but it controls the speed by shifting its position in or out between the center lines of the two discs. This shifting varies the output speed of rotation, Figure 10-22.

Ball and Disc Drive or Integrator - A ball and disc variable-speed drive is made with two caged balls in rolling contact with each other, the input disc, and the output shaft. This was a form of an early *analog computer*. This mechanism is designed to run continuously while the ball cage is shifted in position to change ratios between the input and output shafts. When the caged balls are in the center of the disc, no output is delivered. As the cage is moved to either side of the center point, a different rotational rate is delivered. When the caged balls go beyond the center point, the direction of the output is reversed. Because the radius on the disc can be changed, this mechanism is used as an analog multiplier by changing the ratios of the shafts. In addition, the position of the caged ball can represent a data input and the disc rotation a second data input. The mechanism can be used as an

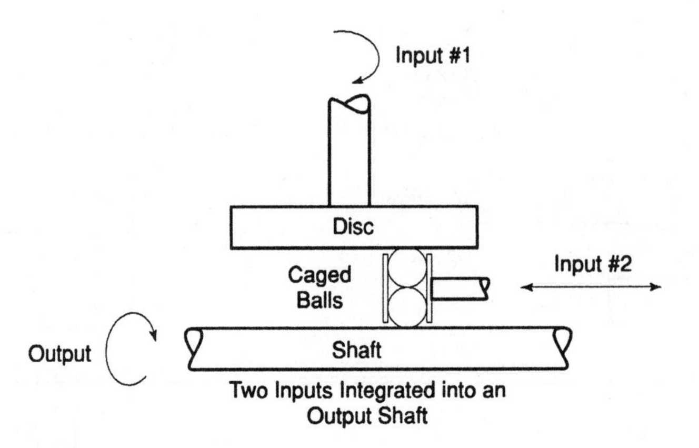

Figure 10-23 Ball and disc drive or integrator

integrator of two data sources, resulting in solution given on a single output shaft of rotation rate, Figure 10-23.

TRANSFER-LINE PRODUCTION INSTRUMENTATION

Instrumentation in mass production transfer lines occurs at different levels of sophistication. The contrast from the microswitch to the laser scanners and coordinate measuring machines interfaced with a CAD/CAM system is a large range. Regardless of the type of instrumentation, the functions of counting, sequencing, verifying, and inspecting the products of the manufacturing transfer lines is the responsibility of control and is an integral part of the manufacturing process.

Counters

Among the oldest and most reliable instruments of the production line are counters or registers that are designed to accumulate products or cycles and report the result. These instruments function as mechanical, electrical, or electronic counters.

Mechanical Counters

Mechanical counters display a row of numbers (or in some cases letters) in a small window to the viewer, Figure 10-24. These are small alphanumeric readouts that display the count at that moment. These are called digital readouts because they display a number for each unit.

Figure 10-24 (A) Stroke-mechanical counter *(Courtesy of ENM Company)*; (B) Rotary-mechanical counter *(Courtesy of Redington Counters Inc.)*

For many applications, the alphanumeric display has an advantage because it eliminates the viewer's error that is due to parallax, i.e., the apparent displacement of the observed object due to a change in the position of the observer. Also, the viewer does not have to estimate fractional movements of a pointer. The information is read to the nearest digit directly.

Mechanical counters consist of a series of wheels, each coupled to the next wheel with a ten to one reduction gear. The viewing edge of the wheel is divided and numbered from 0 through 9 with each number equally spaced on the wheel. When the instrument receives ten strokes or a turn of the unit wheel, it advances the adjacent wheel one-tenth of a revolution. When a wheel completes one revolution and the wheel reads zero, the wheel to its left will move one-tenth of a revolution or one number. This device operates as an automobile odometer, a digital clock, or a water meter.

Electrical Counters

Electrical counters are an application of a small switch that is activated by a light spring steel arm or feeler that contacts the product as it moves by the sensing unit. The closed switch provides an electrical current to a counting mechanism constructed from a coil to energize an electromagnet. This in turn operates a ratchet that moves the counting mechanism one unit on the counting wheel. The counting wheel operates at the ten to one ratio indicated previously.

A linear-measure distance counter is a rotary contact wheel that rolls on the product. Inside the device is a cam that activates a switch each time the measuring wheel makes one revolution and sends a pulse of electricity to the counter solenoid that ratchets the counter ahead one measured unit.

Electrical counters have an advantage in that they may be placed remotely from the actual production area by employing extension wires to the counter.

Proximity Counters

Proximity counters do not physically contact the workpiece. They employ a number of different electrical principles. Counters for various industrial applications use sensors with different electrical characteristics such as light, induction, reluctance, magnetism, and capacitance, Figure 10-25.

Photoelectric devices extend the counting function by being able to locate products and measure distances as well. They use a light beam for sensing. A break in the light beam is caused by objects momentarily interfering with the light. This is used to control the movement of a product on a conveyor belt, counting any solid object passing a point. The light beam is also used to count rotations of a measuring wheel. The measuring is done by a wheel that rolls on the product as it moves. The wheel has a hole or notch machined into it for the light to pass through at each revolution of the wheel. When the light passes through the hole in the wheel, the beam sends a signal that activates an electrical pulse representing the distance of the circumference of the wheel that has been traveled on the product. The number of pulses multiplied by the circumference of the wheel equals the measurement.

The light-emitting diode (LED) is one of the most utilized of the proximity sensors. It can function with a narrow band of infrared light. This makes it very

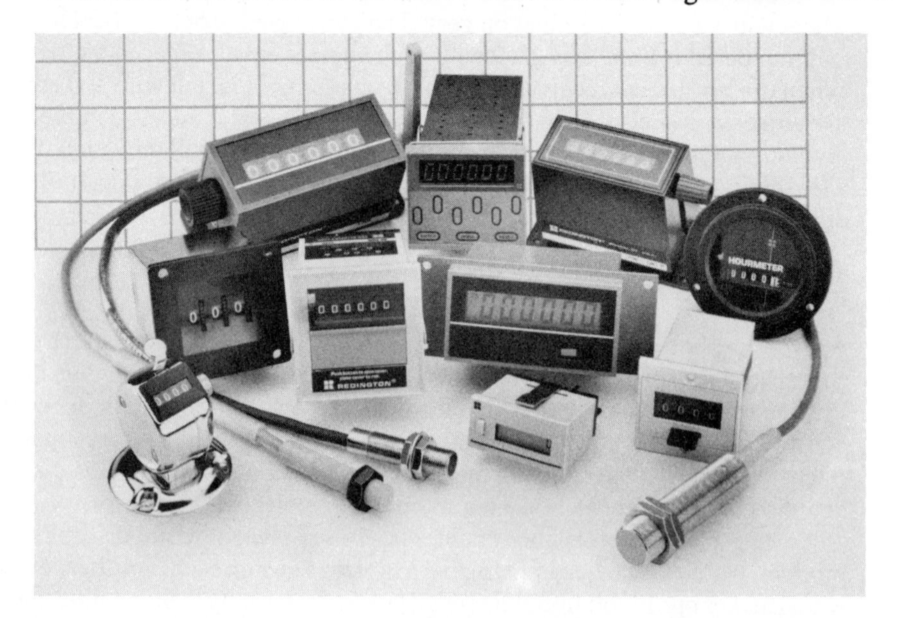

Figure 10-25 Electronic proximity sensors and counters *(Courtesy of Redington Counters Inc.)*

popular in industrial environments because it does not respond to ambient light. The diodes are small and can be placed in nearly any position. This light beam is received by a phototransistor that conducts when the LED beam reaches the transistor, Figure 10-26. In the case where the transistor is receiving a continuous signal and the application requires a reciprocal signal, the output may be changed by placing an inverter after the transistor.

Fiber Optic Counters

Fiber optics are used to sense objects in a different way. A bundle of optical fibers are combined with photoelectric switching devices. The fiber optics are used to sense the presence of products in a pattern. They have several sensing inputs that recognize the pattern. These applications check for the number of parts in a package to ascertain if all of the packaging cavities have been filled, Figure 10-27.

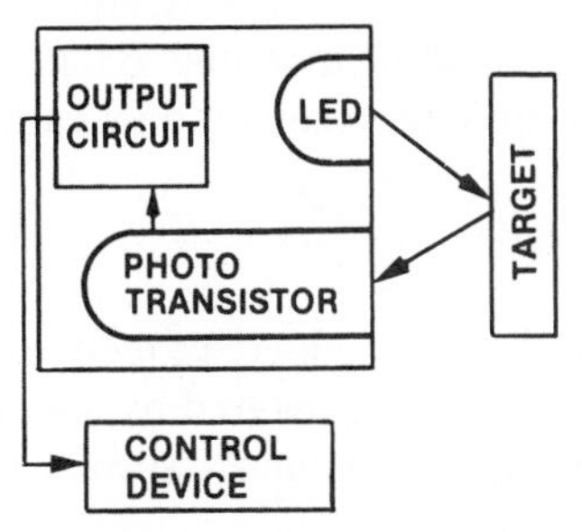

Figure 10-26 Principle of light-emitting diode digital counter *(Courtesy of General Electric)*

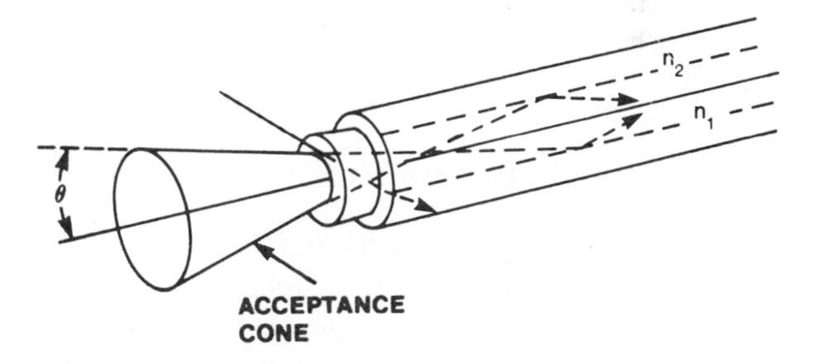

Figure 10-27A Fiber-optic principles

Figure 10-27B Fiber-optic digital interface *(Courtesy of EOTec Corporation)*

Bar Code - Technology has advanced to where counting has broadened into the identification of parts and products. Light sources of light-emitting diode (LED) and helium-neon (He-Ne) laser tubes are applied to bar code readers. Bar code scanners and decoders are combined into a single unit. The scanner illuminates the bar code symbol, and the light is reflected from a carbon printed line back to a photodetector. In turn, the photodetector produces an analog signal that is changed into a digital signal by an analog-to-digital converter. The digital signal is sent to a microprocessor for decoding, Figure 10-28. The bar code readers interface with a computer and perform functions such as counting, part identification, inventory control, tool management, work-in-process, tracking assembly, and shipping data and control.

Sequencing

The sequence of operations enters into most manufacturing as to what is performed first, second, third, and how much time each operation will require. An example of sequencing is the timer on an automatic washing machine. Each cycle is performed in a sequence and with its length of time represented on a cam.

Cam Sequence Timer

Sequencing exists on a number of levels of technology. One level of technology is in the use of microswitches contacting a series of cams powered by a geared synchronous motor. The cams are contoured to provide the sequence and duration that is applied to the operation, Figure 10-29.

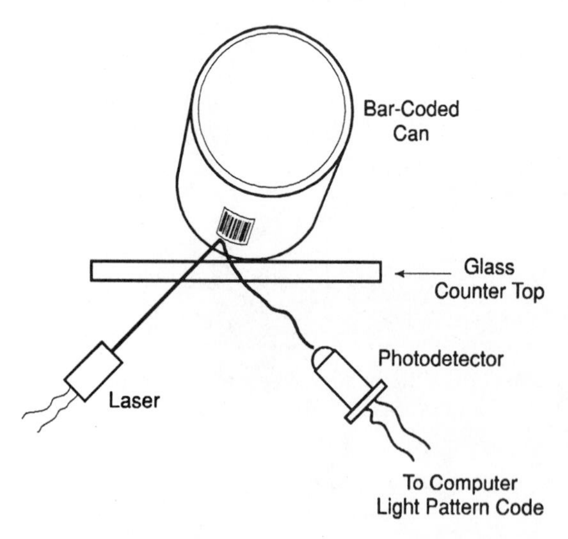

Figure 10-28 The light and dark patterns represent ones and zeroes, a code for the computer.

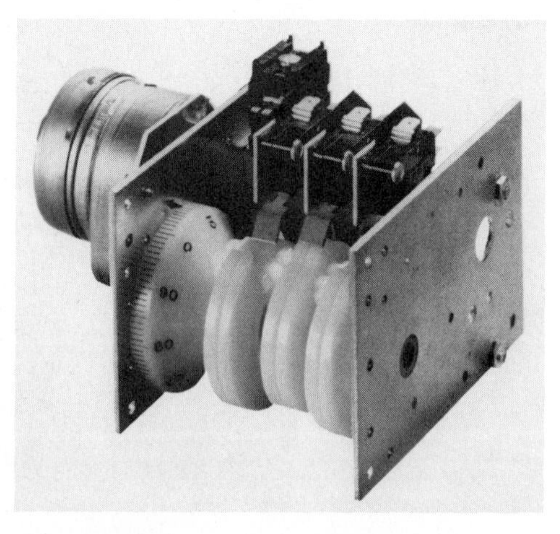

Figure 10-29 A cam-type sequence timer. A precision switch cam programmer *(Courtesy of Automatic Timing and Control Company)*

Relays

Relays wired in ladder-logic sequence is a system for controlling large and complicated production machines. This system consists of large panels of relays wired in a logic that supply the correct signal at the proper time. The sequence is driven by the completion of one event triggering the start of the next event. A ladder diagram shows that these sequenced events are connected across the vertical parts of a ladder, each sequenced event being placed on a rung of the ladder. The event on the rung must be completed before the event on the next lower rung is started; thus the sequence of operations is controlled. These systems electrically control solenoids, start motors, control valves, and other relays and indicator lamps performing their functions in the correct time and sequence.

Numerical Control

Numerical control is a system that provides for the sequencing of operations in manufacturing by having the necessary data and logic programmed onto a punched or magnetic tape. Traditionally, numerical control is associated with positioning control. However, sequencing is a function of this system and applications of numerical control are used to control the sequencing.

Programmable Control

Another level of technology is the application of a programmable controller. The programmable controller is an instrument that has the capability of changing the sequence of the operations. The logic of sequencing grew out of the earlier relay ladder logic in which the relay panel was permanently wired so that specific sequences were followed. With the programmable controller, the sequence can be modified or restated without major hardware modifications. The logic is similar enough that factory maintenance personnel or electricians can successfully change the sequence.

Industrial Computer Control

As factories automate, they are moving toward sequencing operations with industrial computers. One of the major functions of a computer program is the sequencing of data. These machines are easy to interface with a programmable controller or digital-data-collecting instruments. This interface is because of the similarity of the in and out data transmission capability of the computer to the other devices. The computer has provided a capability and flexibility for decisions in sequencing the operations of a plant, including the control of machine operations, pallets, jigs, fixtures, as well as storage and material-handling equipment.

Inspection

Mass production transfer lines require the sequencing of real time inspection between operations. This is done with gaging or probing heads to assure that tools such as drills, taps, or chips have not been broken and remain in the workpiece. In addition, alignment and positioning of work or pallets are checked for the correct position and sequence for the next operation. Other inspection data is acquired and recorded. The size and location of the machining performed by the transfer machine as well as its production output are recorded, Figure 10-30. Warning systems are activated if the inspections are not within the parameters set forth by production requirements.

Figure 10-30 The spindle probe simulates the coordinate measuring machine's functions as well as in process inspection for automatic tool breakage and tool wear monitoring *(Courtesy of Kearney & Trecker Corporation)*

Machine Control

Machine control instrumentation has progressed through a number of levels of development. Early transfer machine control was performed by cams and by microswitches. Later, timers and relays were used. Still later, numerical control was applied. Very highly mechanized transfer machine controls apply feed-forward instrumentation. This allows sensing instruments to acquire machinability data from power consumption and other data from within the transfer machine. This data is supplied to a computer. It is processed by a program that provides correction to the machine while the machine is in operation. This is adaptive control, such as when variations in the hardness of materials may require a change of the cutting tool's RPM, speed, feed, or depth of cut.

This instrumentation is also applied to the selection of machines or operations depending on the type of work presented to the system, Figure 10-31. An example of this is the machining of transmission cases in which two different styles are processed on the same transfer machine within a single batch of parts. An identifying boss, code, or other characteristic is a requirement on each transmission case so that each style of case is identified upon entering the machine. This characteristic is sensed, and the correct program and tools for that case are called up. The tool lengths are checked for that particular part so that the tool performs to specifications.

Figure 10-31 Indexing transfer line machine tool center *(Courtesy of Kingsbury Machine Tool Corporation)*

MANUFACTURING INSTRUMENTATION WITH NUMERICAL CONTROL

The control of machines requires a method of communication between the engineer and the machine. Instructions that the machine can understand and function upon must be precise.

The Binary-coded-decimal Code

As seen earlier, an on/off condition can be related to any number of electromechanical devices. A natural system of communication to a machine has developed in the form of the binary code. The code can represent numbers, letters, and special instructions that position and perform accurate work on the machine, Figure 10-32.

The NC data can be manually programmed directly into a tape code or it can be programmed with a computer aided design station. The data is either punched into a paper or mylar tape, recorded on magnetic tape, or entered directly into a mainframe computer data base. This coding consists of step-by-step detailed instructions for the machine to follow related to positioning the workpiece or tool. It controls the selection of the tools and the rate and direction that the work or tools will move with respect to each other. The program will also provide additional miscellaneous functions, such as changing the tool, starting or stopping the spindle, starting or stopping the coolant, setting the cutting speeds, rewinding the tape, as well as starting or stopping the machine. Special preparatory functions, such as dwell, drill cycle, tapping cycle, boring cycle, circular interpolation, linear interpolation, and many other such functions, are coded into a machining program.

Other uses of NC data are found in applications of flame cutting, bending, welding, punching, riveting, filament windings, textile weaving, and the control of robotic positions.

The Cartesian Coordinate System

To communicate part geometry and positions in space to the machine it is necessary to provide a structure for accurately locating positions. The system used is the rectangular or cartesian coordinate system. This system is described as two planes at ninety degrees to one another with equal base scales. This is the X and Y axes concept used to locate positions. In three-dimensional programming, a Z (or third) axis is added that is perpendicular to both of the other axes. With the third axis, positions are located in space within the extent of the axes. An additional concept is applied to the machines and workpieces. In this concept, rotation around the X axis is designated as A and can be either plus A or minus A depending on the direction of rotation (clockwise or counterclockwise). The rotation around axis Y is called B and rotation around axis Z is referred to as C,

	MAG TAPE CODING							(ISO) ASCII									EIA								
CHARACTER	1	2	3	4	5	6	7	1	2	3	●	4	5	6	7	8	1	2	3	●	4	5	6	7	8
0	—			—							o		O	O						o			O		
1	—						—	O			o		O	O		O	O			o					
2		—					—		O		o		O	O		O		O		o					
3	—	—						O	O		o		O	O			O	O		o		O			
4			—				—			O	o		O	O		O			O	o					
5	—		—					O		O	o		O	O			O		O	o		O			
6		—	—						O	O	o		O	O				O	O	o		O			
7	—	—	—				—	O	O	O	o		O	O		O	O	O	O	o					
8				—			—				o	O	O	O		O				o	O				
9	—			—				O			o	O	O	O			O			o	O	O			
a	—				—	—	—	O			o				O		O			o			O	O	
b		—			—	—	—		O		o				O			O		o			O	O	
c	—	—			—	—		O	O		o				O	O	O	O		o		O	O	O	
d			—		—	—	—			O	o				O				O	o			O	O	
e	—		—		—	—		O		O	o				O	O	O		O	o		O	O	O	
f		—	—		—	—			O	O	o				O	O		O	O	o		O	O	O	
g	—	—	—		—	—	—	O	O	O	o				O		O	O	O	o			O	O	
h				—	—	—	—				o	O			O					o	O		O	O	
i	—			—	—	—		O			o	O			O	O	O			o	O	O	O	O	
j	—					—			O		o	O			O	O	O			o		O		O	
k		—				—		O	O		o	O			O			O		o		O		O	
l	—	—				—	—	O			o	O			O	O	O	O		o				O	
m			—			—		O		O	o	O			O				O	o		O		O	
n	—		—			—	—		O	O	o	O			O		O		O	o				O	
o		—	—			—	—	O	O	O	o	O			O	O		O	O	o				O	
p	—	—	—			—					o		O		O		O	O	O	o		O		O	
q				—		—		O			o		O		O	O				o	O	O		O	
r	—			—		—	—		O		o		O		O	O	O			o	O			O	
s		—			—			O	O		o		O	O	O			O		o		O	O		
t	—	—			—		—			O	o		O		O	O	O	O		o			O		
u			—		—			O		O	o		O		O				O	o		O	O		
v	—		—		—		—		O	O	o		O		O		O		O	o			O		
w		—	—		—		—	O	O	O	o		O		O	O		O	O	o			O		
x	—	—	—		—						o	O	O		O	O	O	O	O	o		O	O		
y				—	—			O			o	O	O		O					o	O	O	O		
z	—			—	—		—		O		o	O	O		O		O			o	O		O		
+								O	O		o	O		O						o		O	O	O	
—						—	—	O		O	o	O		O						o				O	
/	—				—			O	O	O	o			O		O	O			o		O	O		
.	—	—		—	—	—	—		O	O	o	O		O			O	O		o	O		O	O	
,	—	—		—	—					O	o	O		O		O	O	O		o	O	O	O		
%			—	—	—		—	O		O	o			O		O	O	O		o	O				
&					—	—					o									o					
(											o	O		O					NOT	o	ASSIGNED				
)											o	O		O		O			NOT	o	ASSIGNED				
:									O		o	O	O	O					NOT	o	ASSIGNED				
CR								O		O	o	O				O				o					O
DELETE								O	O	O	o	O	O	O	O	O	O	O	O	o	O	O	O	O	
SPACE											o			O		C				o		O			
=											o									o					
TAB								O			o	O						O	O	o	O	O	O		
											o									o					

Figure 10-32 Magnetic tape coding; ASCII (ISO); and EIA punch code tape

Figure 10-33. With these six movements, positions, shapes, and rotations can be defined within the coordinate system's space.

The Cylindrical Coordinate System

Another system referred to in NC machining and robotics is the cylindrical coordinate system. In this system the X, Y, Z coordinates are used, but the point of origin is placed at the center of a theoretical cylinder. The point in space is defined in terms of a radius (R) rotated through an angle (A) rising to a height of (Z). A position is indicated as PT1 as (R, A, Z) with the arc of the radius starting from the X axis, Figure 10-34.

Tape Controlled Machine Program

Tape controlled machines are numerical controlled machines that receive their coded data for position, feed, speed, tool change, and a complete system of preparatory functions and miscellaneous functions from the punched or magnetic tape. The information is punched into a one inch paper or mylar tape with eight binary coded data channels available down the length of the tape. There is a space between the third and fourth channels for sprocket holes that drive the tape through the tape reader but provide no data information. The tape is read by the

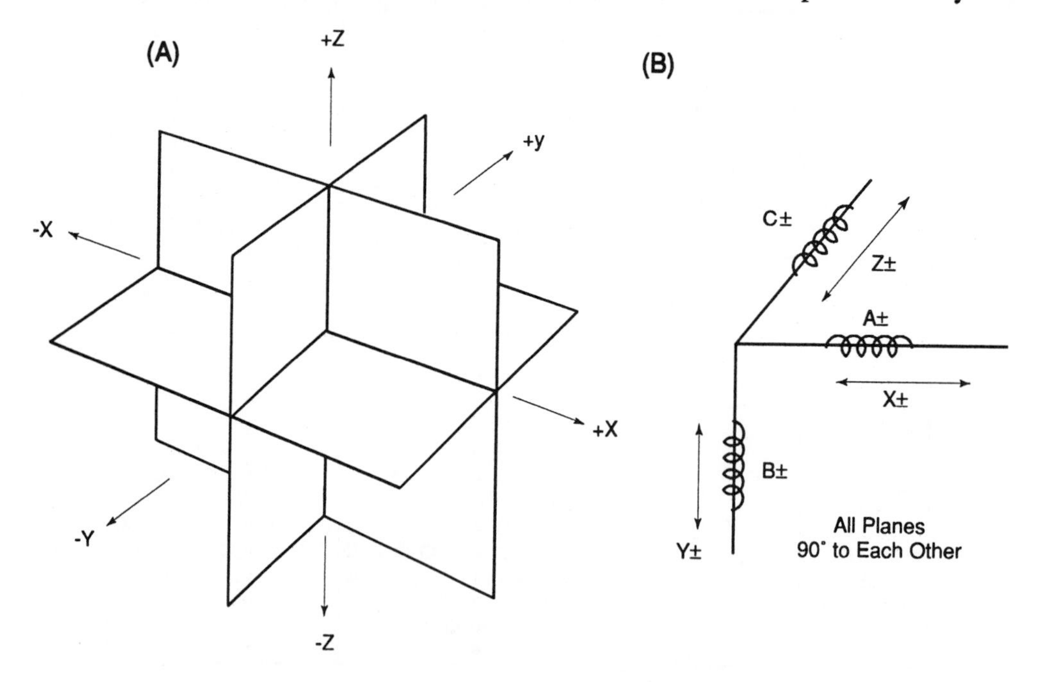

Figure 10-33 (A) Cartesian coordinate system; (B) Diagram of X, Y, Z, and A, B, C movements

photoelectric tape reader providing information to the machine a block at a time. A block of data is a discrete position that is stored in the memory of the tape control unit until the command is received to move the machine to the next position.

This program represents the positions and functions for tools, robots, or machines. The program is the key because it provides flexibility in that the device or machine receiving the positioning data may be changed or adjusted as requirements change. This flexibility is achieved through the application of the programming data needed to control the machine.

The numerical control machine uses either incremental or absolute positioning. Incremental positioning means that each position movement is programmed from the last position. In the absolute positioning system, the machine location is given in relation to the zero point or the origin position of the machine, Figure 10-35.

The microcomputer has provided additional flexibility by supplying a floppy disc drive for the storing of dozens of programs that can be recalled or modified to meet the objective of a newer program. These microcomputers can be coupled with a tape punch and a portable hard copy printer to deliver rapidly produced reprogrammed tapes that provide machine flexibility.

Numerical control machines are programmed using three basic systems: point-to-point, straight-line, and contouring. In the point-to-point system, the tool, robot, or machine moves to the next position for an operation by the most direct route from one coordinate position to the next coordinate position. The point-to-point program is used for drill press types of work or is applied to a pick and place robot operation.

In straight-line control, the tool is moved parallel to one of the three axes at a controlled feed rate. The axis perpendicular to the first or second axis position

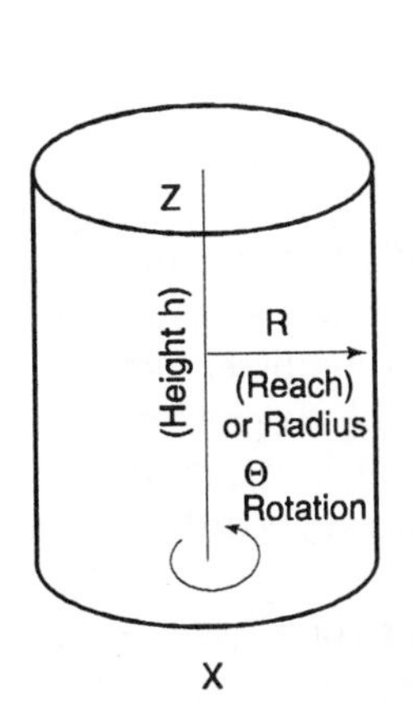

Figure 10-34 Cylindrical coordinate system

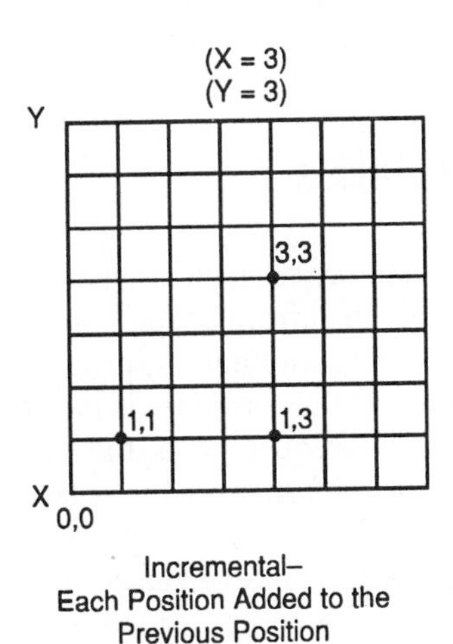

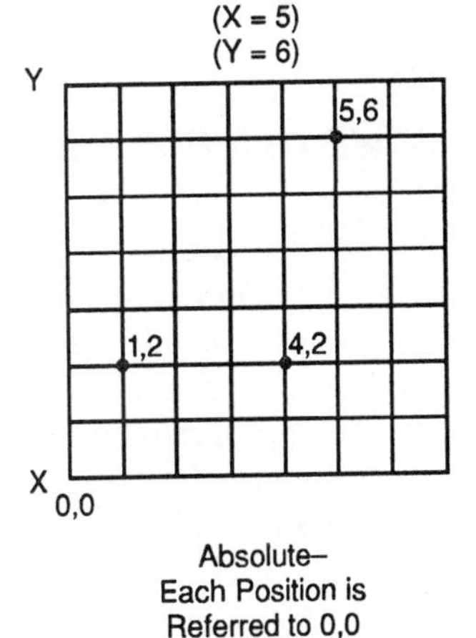

Figure 10-35 (A) Incremental programming; (B) Absolute programming

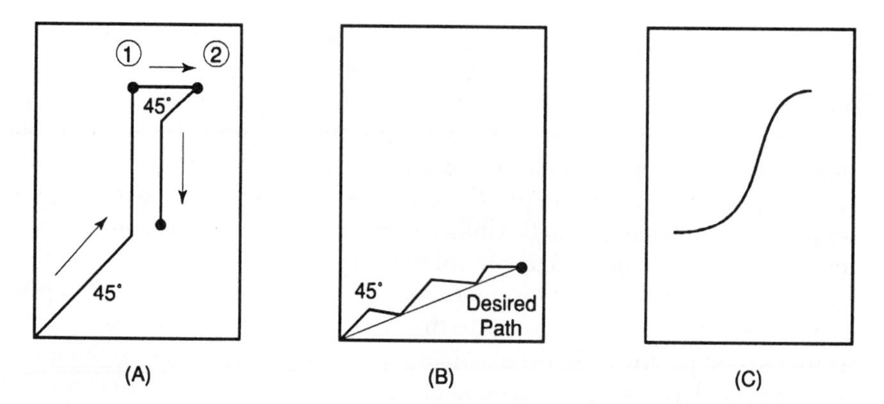

Figure 10-36 Three programming system of numerical control: (A) When the programmed motion is simultaneous, the two movements are equal, resulting in a 45 degree movement until one movement is complete; (B) On a slant-line movement, a small variation will result. On straight-axis tool paths, there is no error; (C) Interpolating the approaching coordinate points continuously provides a smooth contoured form

represents the depth of cut. This produces slots or other rectangularly shaped work. A large number of milling and machining operations employ this programming system. Applications similar to robot seam welding also use straight-line control.

In the contouring system, two or three axes are positioned simultaneously. When two axes are moved at different rates, cam-like shapes may be generated. With three axes moving at different rates, the surface machined results in a sculptured form with changing contours, Figure 10-36. This system of control is used for machining complex curving surfaces or positions as are performed by spray painting robots.

COMPUTER CONTROLLED MACHINES

The computer controlled machines receive instruction directly from the memory system of the computer rather than from a tape. Tapes do wear out with repeated use and must be replaced. With a computer, the program is in a computer memory. The program will last as long as the computer is in service. The program may be reprogrammed or repeated as frequently as the information is required. The computer memory may be loaded from a magnetic tape, from a keyboard, or from a microcomputer.

The movements of the machines can be programmed on a coordinate system by programers or by the use of a computer aided design system. In the CAD system, the machine control unit receives the programs from a computer memory or directly from the CAD station's main computer memory.

A computer controlled system can handle a great deal of data rapidly and accurately. Because of the computer's ability to calculate machine positions very rapidly, the distances between the positions of the machine can be very short. The computer interpolates a series of points to control the movement of the tool path. This is accomplished by a programming feature—applying algebraic formulae to the data to produce circular arcs. The radius of the arc and its length and feed rate are provided, and the computer generates a series of points necessary to control the velocity of the machine's slides. The slides are constantly changing at the rate and position indicated for the arc's production. Using this circular capability, parabolic curved figures with differing dimensions are available.

In addition, linear interpolation with the appropriate dimensional information and with the correct feed rates can produce straight-line motion of the tool path. In some cases, the straight-line paths require driving slides at different speeds to produce an oblique straight-line movement.

Computer numerical control equipment with continuous path capability can generate a continuously controlled motion or tool path, Figure 10-37. This is accomplished by the computer interpolating a series of points to control the movements simultaneously on multiple axes.

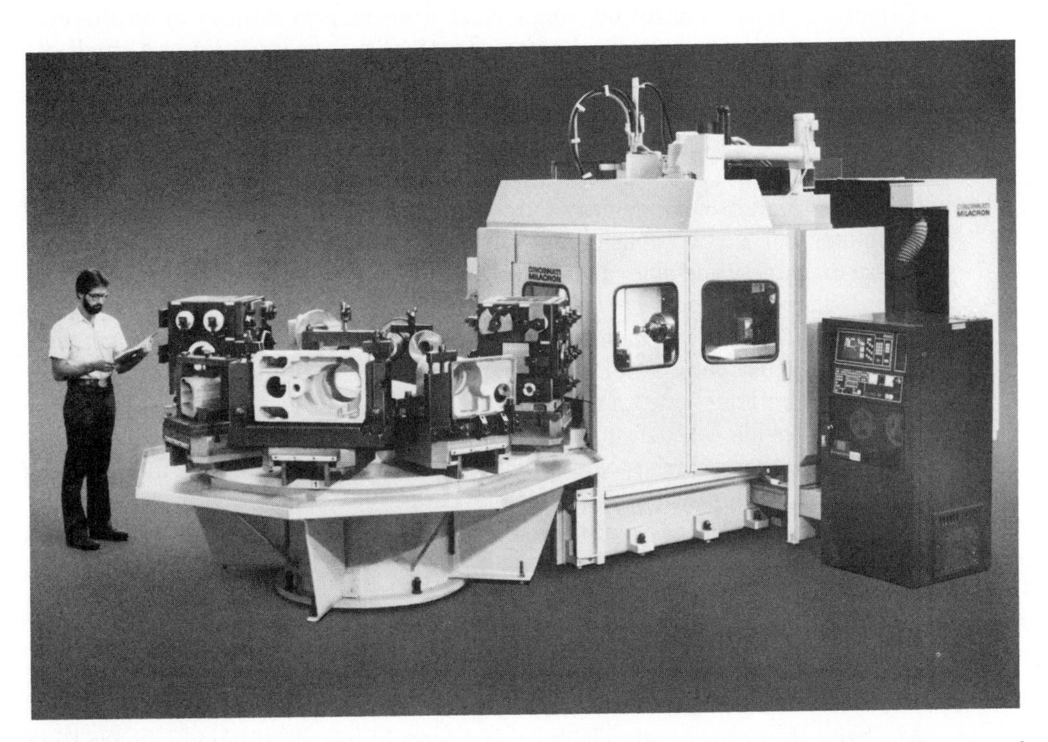

Figure 10-37 CNC machining center with a series of eight pallet automatic work changers *(Courtesy of Cincinnati Milacron)*

Adaptive Control

Tool temperature, torque, RPM, and other sensing data may be entered as part of an input to the computer in an effort to optimize the working parameters. Adjusting the machinability factors and reducing the travel time between air spaces and surfaces to be cut when machining metals are examples of optimization. The computer takes these data inputs from the sensors and evaluates the processes' effectiveness. As a result, the computer program is modified and changes are made in the work under production. In adaptive control, the computer program adjusts the processes so that an excellent product is delivered. The adaptive control requires considerable software and interaction with the data base.

Additional information can be in the form of process monitoring and reporting for engineering, management, and quality control. They may also be statistically collected, calculated, graphed, and/or reported.

There are many programming considerations necessary for NC. These are treated in specialized textbooks.

Direct Numerical Control

Direct numerical control machines receive instruction directly from the bulk memory of a large centralized computer rather than from tapes or a small on-station computer. One of the chief advantages of this system is the elimination of the punched tape and the tape reader that frequently was the cause of errors. This system today is recognized as a higher level of control in automated manufacturing.

A large number of machines and controllers are connected directly to a central computer that can be a considerable distance away from the manufacturing area. These machines time-share the central computer which stores all part configurations that each individual machine can request. The instructions from the central computer are stored in a memory buffer in the machine. The position movements are given to the controller of the machine one block at a time. Some applications utilize two buffers—one for receiving the data from the central computer and the other to provide the data to the machine. This special machine control unit employs software which provides flexibility for the unit for circular interpolation and other specialized operations for tool paths.

This system allows real-time management supervision of the production processes at all time.

SUMMARY/FACTS

- Mechanical instrumentation is the oldest form of instrumentation and has been evolving since the Middle Ages.

- Mechanical instrumentation is considered as analog control in that the work it performs or controls is continuous in output, as contrasted to that of digital control.
- When a machine member is anchored in a frame, base, earth, or foundation of a machine, that member is defined as a constrained member.
- The moveable elements of a machine theroretically have six degrees of freedom or movement. In reality, there is at least one degree of constraint so that the machine remains in tact.
- The study of inversions is obtained by systematically constraining one member at a time and observing the motion of the resulting outcome.
- Parallel-motion devices are valuable mechanical devices because they can be used to correct for nonlinear instrument elements.
- An instrumentation characteristic of the cam is that its surface can act as a data storage and retrieval system.
- The Geneva mechanism delivers an accurately displaced and timed movement. This movement can be employed to drive indexing devices.
- Noncircular gears are used to obtain a nonlinear motion. They can provide virtually any continuous mathematical function that does not require reversed motion.
- Counters are among the oldest and most reliable instruments on the production line. Counters or registers are designed to accumulate product amounts and cycles and report the resulting count.
- Sequencing exists on a number of levels of technology. The order and the length of time in which operations occur are essential in certain production processes.
- A programmable controller is an instrument that is capable of having its sequence and lengths of time reprogrammed. The logic of sequencing grew out of the earlier relay ladder logic in which the relay panel was permanently wired so that specific sequences were followed. With the programmable controller, the sequences can be modified.
- Advanced manufacturing can apply feed-forward instrumentation. This mode of control allows sensing of machinability data from power consumption and other data from within the production process. This data is processed through an ongoing computer program. The output of the computer makes corrections in the manufacturing while the work is in progress. This is referred to as adaptive control.
- Numerical control is a tape control system employed for engineers to communicate with a machine. The communication is carried out through a binary code that represents numbers, letters, and special instructions that position and control the machine.
- Computer numerical control machines receive instruction directly from the memory system of a computer rather than from a tape. The computer's memory can be loaded from a keyboard, a magnetic tape, a microcomputer, or a computer aided design or computer aided manufacturing system.
- Direct numerical control machines receive their control instructions directly from the bulk memory of a large centralized computer.

REVIEW QUESTIONS

1. What is a mechanism?
2. What is the function of a mechanism?
3. In a linkage system, what do we call a member that is anchored to a frame, base, or foundation of a machine?
4. Theoretically, how many degrees of freedom does a moveable element have?
5. Name the degrees of freedom as found in a common turning pair.
6. What are the motions produced by anchoring different members progressively around a system called?
7. What are the forms of some of these motions?
8. Describe the motion of the slider crank mechanism.
9. What motion results from the action of a rotary crank pin linked to anchored pivot bearing?
10. What is an eccentric rod?
11. When the eccentric rod and eccentric are connected to another link and the crank rotated, the new link oscillates. What is that link called?
12. The motions that result from such linkages as those that transmit motion from the end of a bourdon tube to the pointer shaft are called ____________.
13. What movement is transferred to an element or pointer to keep it vertical or horizontal throughout the distance moved?
14. What is the action of the balls in a fly-ball governor?
15. Describe how a cam's surface can act as a data storage and retrieval system for instrumentation.
16. When a cam is continuously rotated, the displacement of the cam follower performs two functions. Name them.
17. In what way does the data provided by a three-dimensional cam from three sources become effective?
18. What does the rotation of the cam provide?
19. In the inverse cam, as the crank pin and roller rotate, the roller and cam cause the cammed surface to rise and fall. This converts a rotary motion to a ____________ motion.
20. What is a common mechanism used to produce an intermittent motion for activating various types of machine feed devices?
21. What mechanism provides a locked intermittent rotation to a shaft?
22. How is backlash prevented within ball screw parts?
23. Describe the contour of the teeth of a spur gear.

24. Production of a ratio of rotation between two shafts is the fundamental concern of ___________.
25. Why can a sector gear be used in such instruments as the mechanical bourdon tube pressure instrument?
26. What is the function of noncircular gears?
27. In certain environmental or alignment problems, when it is impractical to use gearing for instrumentation, another means is used to transmit power or movement over the distances. Name that means and give some examples of it.
28. What is a timing belt?
29. Describe the output of a ball and disc drive integrator.
30. Name three functions of photoelectric devices.
31. Why is the LED (light-emitting diode) one of the most utilized in industrial environments.
32. Describe how a bar code reader works.
33. How has counting broadened into identification of parts and products?
34. What is involved in the sequencing of operations?
35. In numerical control, the sequencing is performed by having the necessary data and logic programmed onto ________________.
36. What is the chief advantage of the programmable controller in manufacturing?
37. What is the advantage of feed-forward instrumentation?
38. The natural system of communication between the engineer and the machine has developed in the form of ___________.
39. To communicate part geometry and positions in space to a machine, a system for accurately locating positions is used. What is that system called?
40. How does the polar or cylindrical coordinate system differ from the cartesian?
41. When the tool is moved parallel to one of the three axes at a controlled feed rate, what name is given to that program in a numerical control machine?
42. Of what value is the programming feature of circular arcs in a computer controlled machine system?
43. Optimization of the working parameters by computer programming is accomplished with the application of __________ control.
44. Direct numerical control systems are so effective because individual machines and controllers are able to time-share what?

11

STRATEGIES OF PRODUCTION AUTOMATION

OBJECTIVES

Upon completing this chapter, you will be able to discuss and apply these manufacturing concepts:

- Manufacturing cells
- Flexibility in manufacturing
- The classification of robots
- The control of robots
- Industrial power supplies

INTRODUCTION

Flexible manufacturing systems interlink the technologies of material shaping, forming, and assembly with information processing into an efficient productive unit. Flexible manufacturing objectives are to reduce material inventories, reduce setup time, increase capacity utilization of the facility, improve the quality consistency, and provide greater responsiveness to market products.

The building blocks for flexible manufacturing systems are the manufacturing cells. These cells can consist of two or more computer linked numerical control machine tools; with their material handling equipment, they form a flexible manufacturing system.

MANUFACTURING CELLS

A manufacturing cell is a group of various machines and material-handling equipment that receive materials or parts and perform a series of operations upon the workpiece. These operations include machining, assembling, or testing. These units are not a total production line but rather a subsystem within a larger system. Control is accomplished by a large variety of sensors. They provide data to a programmable controller or a cell computer. The cell is designed to be an unmanned or a lightly manned work station. This results in a reduction in the possibilities of human error and produces more predictability and reliability in the manufactured product, Figure 11-1.

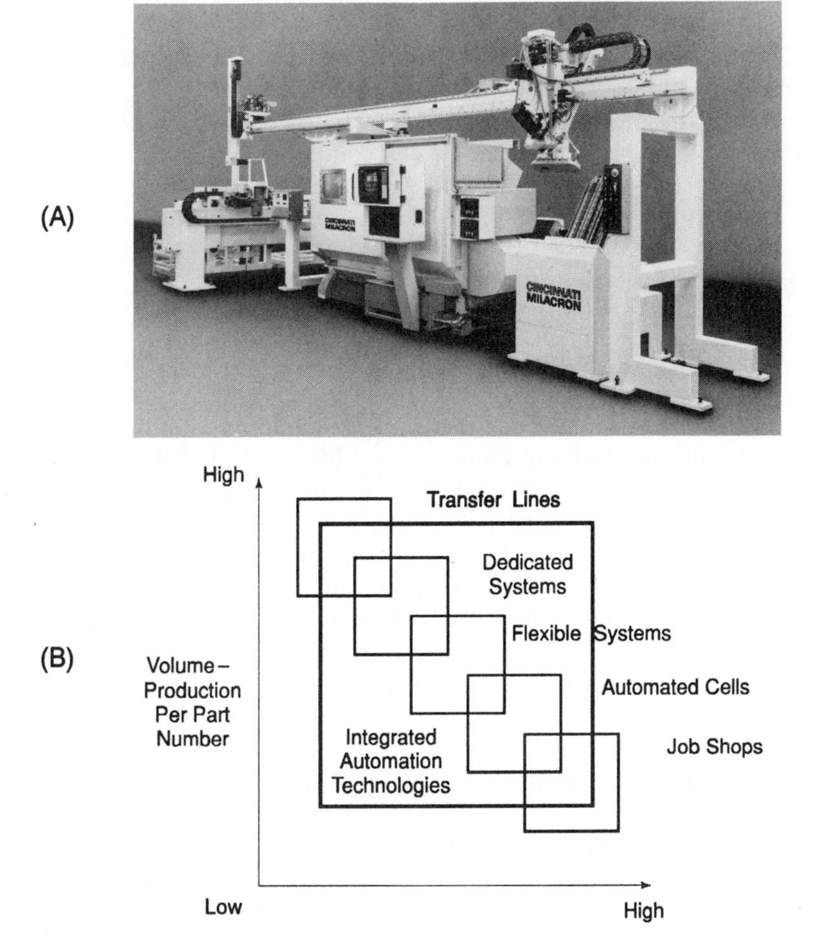

Figure 11-1 Variety—Part Numbers per System (A) A flexible production CNC turning cell; (B) Volume-production per part number verses variety-part numbers per system *(Courtesy of Cincinnati Milacron)*

A Series of Cells

A flexible manufacturing system contains a series of manufacturing cells. Manufacturing cells often begin as automated islands within a manufacturing plant. Because of the cost of the cells, the plant may be able acquire them only over a period of time. The first ones acquired thus appear as isolated islands within the plant. A typical manufacturing cell may consist of two turning centers and an industrial robot for automatic work handling or any number of combinations of tools or machines.

The instrumentation of the manufacturing cell applies to the various sensors previously mentioned that control position, motion, rate, velocity, count, temperature, and other manufacturing variables. These sensors send data to the controller or cell computer.

The cell computer is a device with information processing capability that coordinates multiple stations, machines, or unit operations. Programmable controllers are taking on more information processing capability, but they do not have the flexibility that users demand of a cell computer.

FLEXIBLE MANUFACTURING

The flexible manufacturing system provides changes through the control system so that parts can be produced with differing configurations within the same machine. These stand-alone cells may have several machines that are controlled by microcomputers and can run unattended for a period of time. The cells can contain automatic tool control and material-handling devices such as automatic pallet handling or robot loading of machines. The system can deliver almost any machining operation as well as include other manufacturing operations. These operations may include, CNC punching, nibbling, tapping, milling, forming, bending, pressing, and laser control for high-speed cutting of unusual shapes, Figure 11-2. The FMS may also include a coordinate measuring machine in the system that communicates with a CAD/CAM system. This system provides a program for real-time manufacturing corrections.

Numerical Control in Flexible Manufacturing

Flexible manufacturing systems composed of a number of work cell machining centers are able to apply many of the advantages of transfer machine automation. But they have the added advantage of computer numerical control to give the cells flexibility for the manufacture of a family of products, Figure 11-3. These systems can perform different categories of work within the part family because of

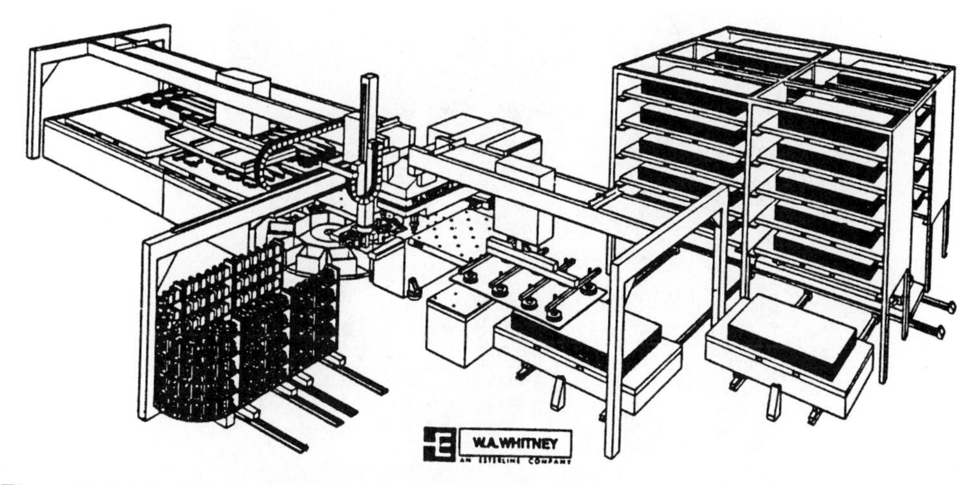

Figure 11-2 A flexible manufacturing cell *(Courtesy of W.A. Whitney Corp.)*

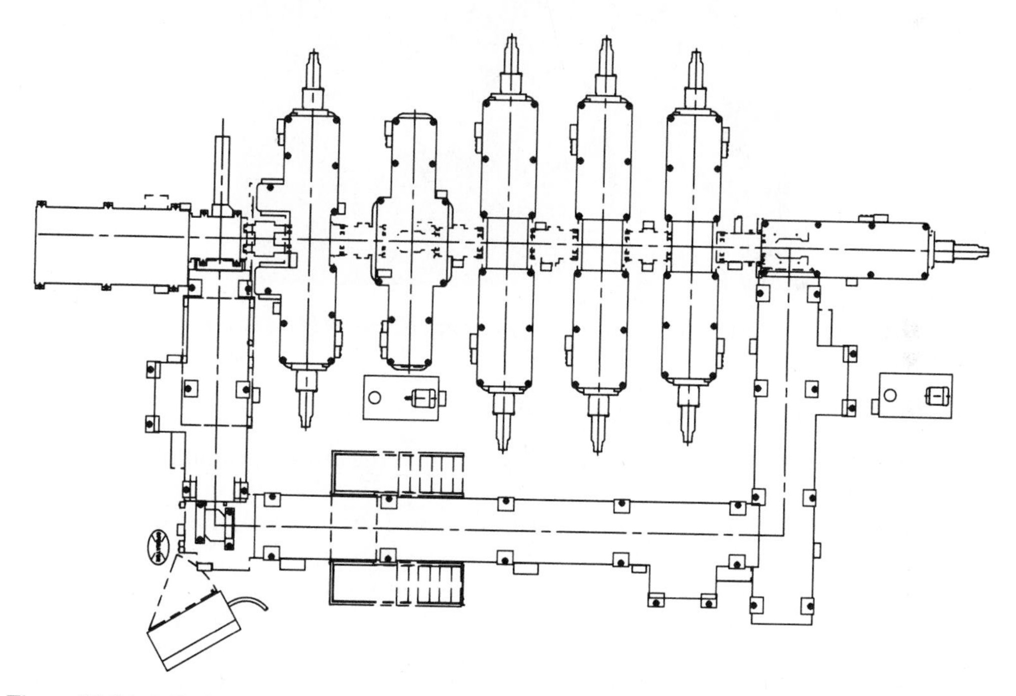

Figure 11-3A A flexible manufacturing system with a ten-station palletized transfer line

changeable programming. Because of this reprogramming, they are capable of manufacturing entirely different family parts without expensive and extensive retooling or the necessity of buying new machinery or transferring equipment.

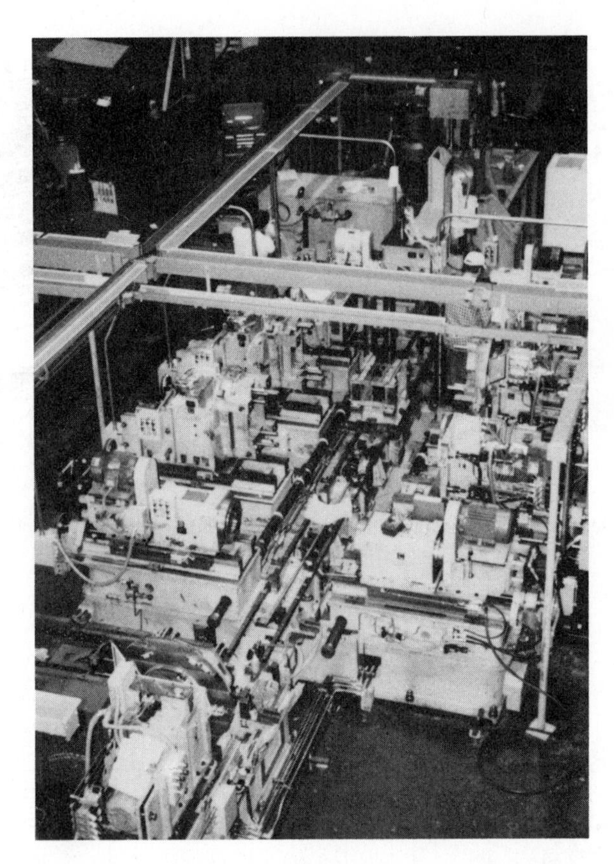

Figure 11-3B Layout of the ten-station palletized transfer line *(Courtesy of NATCO Inc.)*

Manufacturing Automation Protocol

As stated, the flexible manufacturing system is the result of combining a number of cells or modular machining or assembly systems that perform multiple tasks, Figure 11-4. Early problems were encountered because a number of these early manufacturing control interfacing units were incompatible. There was a great need to link together the controllers of all transfer devices, machines, and their numerical control programs, as well as the tool changers, pallet changers, head changers, fixtures, and robotic machines operating with computer numerical control (CNC). All of these units needed to be interfaced with the host computers within the multilevel communication network. All of this equipment within the flexible manufacturing system must interact. The trend has been that sensory-interactive machines were built by different vendors, and so linking communication between the various pieces of control equipment was a difficult and major technical requirement. These developments made it necessary to standardize a communications protocol. To solve this problem, the *Manufacturing Automation*

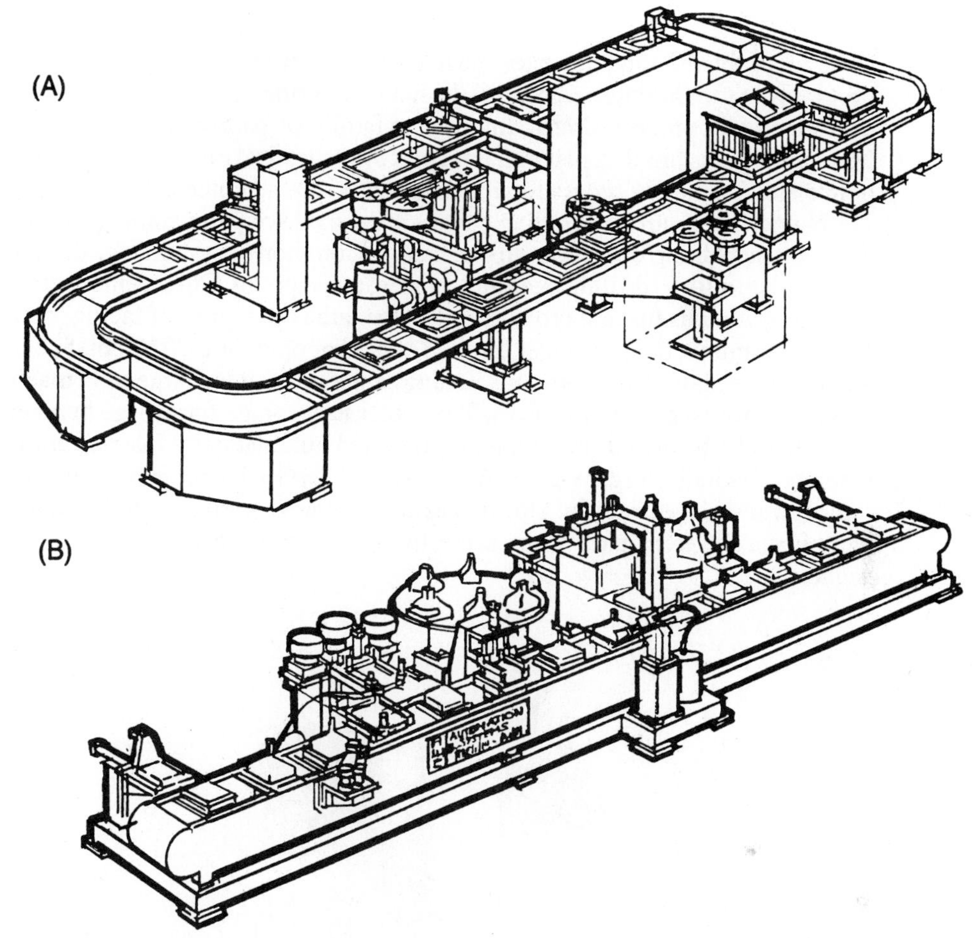

Figure 11-4 A flexible machining system layout. (A) Nonsynchronous assembly; (B) Synchronous precision assembly *(Courtesy of R.W.C. Inc.)*

Protocol (MAP) was developed. MAP created a communication network by providing a broadband backbone network that serves a series of carrier band subnets. The standardization of flexible manufacturing communication was very important, because as the number of interfaces in a flexible manufacturing system grew, the communication problems expanded exponentially.

Numerical Control Machining Centers

The numerical control machining centers are reprogrammable manually or by computer aided manufacturing systems. With the computational capacity of graphic engineering displays and with the use of sophisticated applications of software, the programmer has information made available to him/her for reprogramming. This data is easily transferred from data file to data file until the correct program is obtained and down loaded to the manufacturing cell controller. These programs in turn initiate and control the manufacturing operation. The machines can effectively process almost any part or family of parts at the required rate. Machine slides, spindles, material transport systems, feed rates, and start and end points are completely programmable and sent to the machines.

Numerical control machining centers are capable of performing a number of different machining operations in one center. Common operations for the center are contour milling, drilling, reaming, boring, and tapping. Other centers provide turning operations that are provided by all the lathe operations of facing, boring, drilling, reaming, tapping, taper turning, and other operations. The chief characteristic of the numerical control machining center is that it is a large rigid machine that is controlled on multiple axes. The center has storage for a large number of tools that can be called up by the program and automatically inserted into the machine's spindle, Figure 11-5. At the end of the tool's program, the tool is removed and placed back into the magazine for storage until required by the next part. For a different part, these same tools may be used in a reprogrammed sequence.

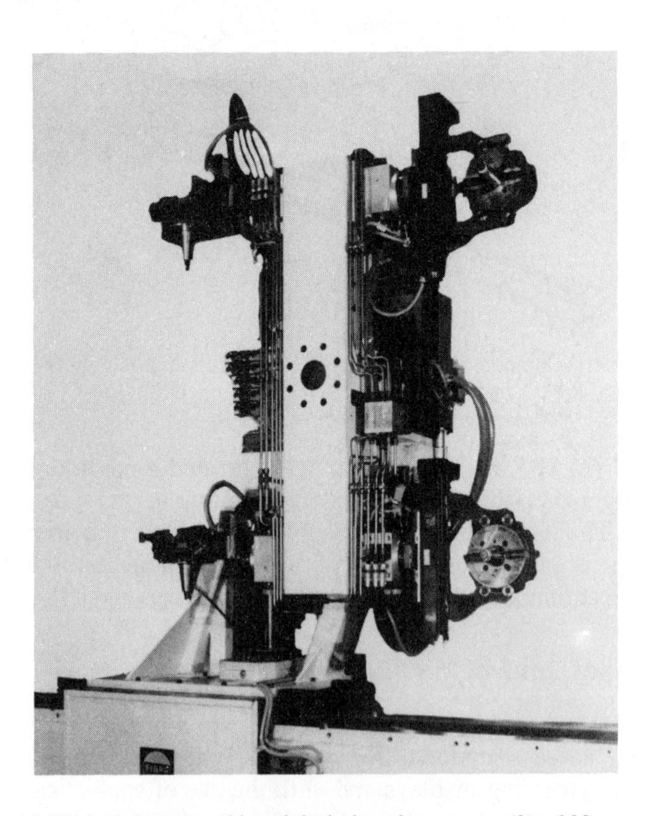

Figure 11-5 Double spindle tool changer—this unit is designed to remove (2) x 200 pound tools and rotate new tools into position *(Courtesy of Grantry Unit by Fibro Inc.)*

Modular Machining Centers

The modular or building block design for machining centers are precision machine tools designed to meet the diverse and exacting needs of high-volume manufacturing. Machining center modules may be built to different specifications. The configurations of the machine are designed so that the various machine modules may be placed into position for a manufacturing run. The same machine may be rebuilt with the modules in different positions for a different product, Figure 11-6. The building modules are: bases, slides, milling spindles, drill and tapping, and power units.

(A)

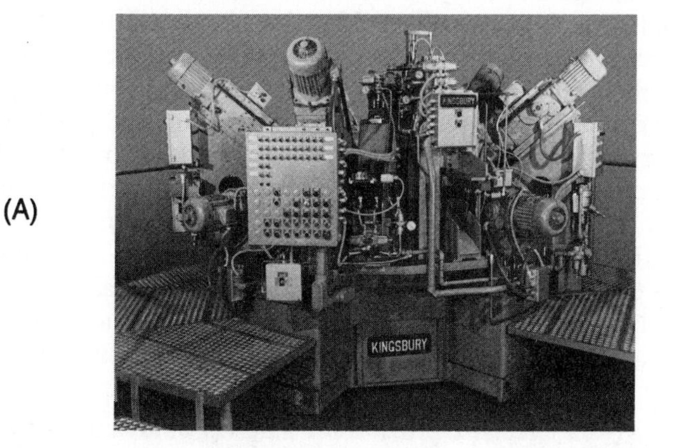

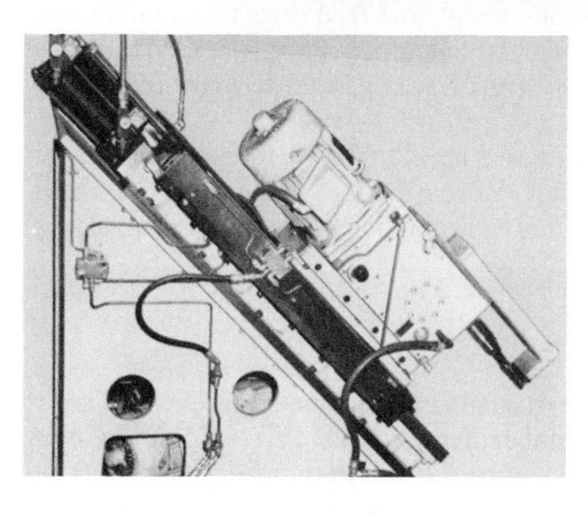

(B)

Figure 11-6 (A) A fifteen-spindle automatic indexing horizontal rotary modular machine; (B) Modular attachments for index base units *(Courtesy of Kingsbury Machine Tool Corporation)*

Modular Assembly Machines

Modular assembly machines have technologies similar to modular machining centers in that they can be rotary (carousel), straight-line, or race track (oval) in configuration. However, the modular machine assembles parts into components and components into products by mechanical device applications. Modular machines assemble batches of a product with flexibility obtained with the aid of a programmable controller. The flexibility is obtained by being able to reprogram to change the drives of stepping motors, to close solenoids, and to resynchronize part feeders and tools.

These mechanical assembly machines are designed to automatically perform assembly and fastening operations. Utilizing the flexibility of the modular assembly machines, several variations (families of the product) can be assembled. Machine assembly operations require part storage, orientation of the part, feeding of the part to the machine, and locating the part automatically in the machine.

Fastening of parts in assembly may include the operations of pressing, swaging, staking, clinching, riveting, spot welding, brazing, soldering, and bonding with adhesives. It also includes mechanical fastening by drilling and tapping holes for automatic screw driving and automatic nut running. These machines are capable of printing or labelling instructions, fastening by molding plastic around inserted components, casting molten metal into or around existing components, as well as performing physical and electrical inspections.

Part Feeders - Part feeders provide for a series of functions for automatic assembly machines. They elevate, orient, and feed parts to assembly machines. They feed parts from a bowl with slide chutes for a continuous feed to the assembly machine, such as screw and nut drivers. These part feeders also position parts for pick-and-place robots, Figure 11-7.

In some systems, component parts are delivered to the assembly machines by a spiral-shaped buffer storage rack. Storing the parts in the rack allows an interval of time for any irregularity in the consumption rates which can vary between assembly machines.

Robot Part Handlers

Robots are frequently used as part handlers in automated assembly. These robots are capable of feeding a large number of machines in different configurations. The largest group of part handler robots are classified as nonservo devices. The others are servorobots. The servorobots can be programmed as continuous-path robots.

Nonservo Robots - The early robots were called pick-and-place or bang-bang robots. They are open loop devices meaning that there is no feedback information to the controller that would provide for a position correction. These robots are powered by hydraulic, pneumatic, or electric actuators with stops at the end of the

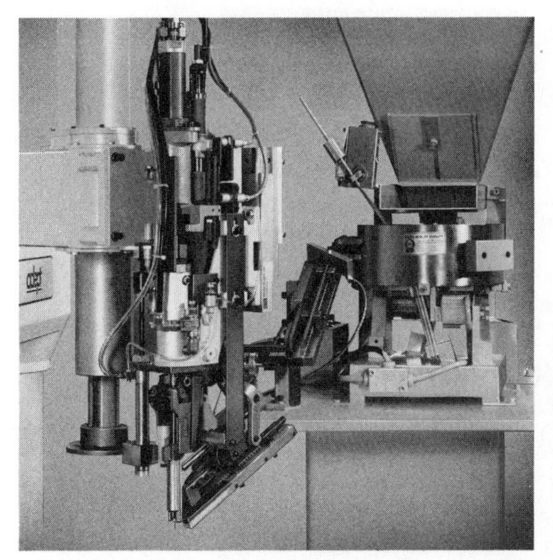

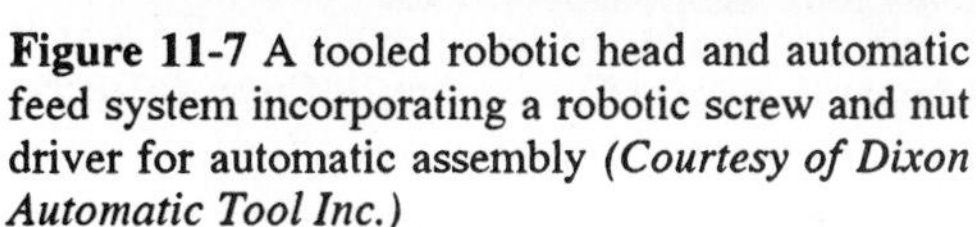

Figure 11-7 A tooled robotic head and automatic feed system incorporating a robotic screw and nut driver for automatic assembly *(Courtesy of Dixon Automatic Tool Inc.)*

Figure 11-8 A nonservo robot *(Courtesy of I. S. I. Manufacturing Inc.)*

machine's arm travel. This bang-bang robot received its name because it is controlled by limit switches that are placed at the end of the robot's desired position. The limit switch signals the sequencer to index to the next sequence. This in turn sends a command to a valve or relay controlling the next sequence. The sequencer provides the capability for a considerable number of motions in its program, but the robot arm will always move to the end point of each axis. Therefore there are only two positions on each axis. The arm of the robot stops abruptly. It bangs. To reduce the shock of stopping, a special valve or shock absorbers are added to the unit. The programming is done by adjusting the end stops for each axis and setting the desired sequences of movements for a material-handling operation. The *nonservo robot* is relatively high-speed, low in cost, and is applied in many small part moving applications, Figure 11-8. It is a very reliable robot. It does a wide range of work, from loading circuit boards to placing sheet metal parts into assembly machines.

Servo Robots - The *servo robot* is the robot most people think of when they think of a robot. It can be a continuous-path robot. When the robot is activated, the controller will address the memory location of its controller for the first command position. The feedback from the arm will be compared with the command position and any difference becomes an error signal indicating that the arm is not in the correct location. The error signal is amplified and transmitted to the servovalves on each actuator to position the arm into the correct position.

Figure 11-9 A servo robot tool changer *(Courtesy of Kearney & Trecker Corporation, A Cross & Trecker Company)*

As the robot arm is positioned by the actuator, its position is fed back to the controller by these devices: encoders, resolvers, potentiometers, tachometers, or velocity sensors. In some cases, servovalves control the hydraulic cylinders that provide the movement of the robot. In other systems, the movements are controlled by electrical circuits and the movement provided by electrical motors. These motors are under constant control to adjust the robot arm to a position in which the feedback signal and the control signal are equal, Figure 11-9.

Industrial Robots - Industrial robots are applied to different functions in manufacturing. Robots in the role of part handlers are manipulators that have great freedom through programming, Figure 11-10. But robots are also used as flexible power tools to perform complex work movements such as welding, spray painting, drilling, machining, and cutting shapes.

ROBOT CLASSIFICATIONS

Robots are classified according to different criteria by engineers based upon developmental history, analysis, or functions. Early robotics was merely the application of automatic slides, cams, and microswitches. Later, automatically controlled pneumatic or hydraulic cylinders sequenced to perform work or to locate workpieces were applied. Still later, with the application of numerical and electronic controls, programmable robotic devices appeared. Now, with the

(A)

(B)

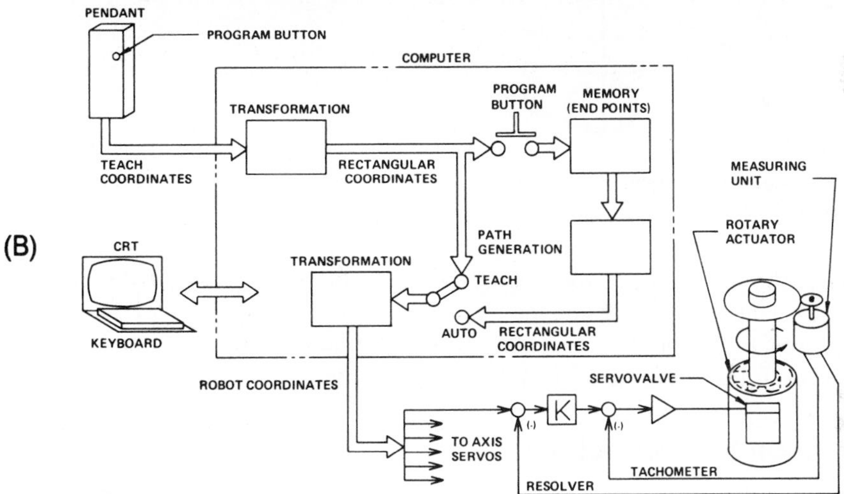

Figure 11-10 Two-machine/robot machining cell. (A) A Milacron grantry robot moves prismatic and cylindrical parts between horizontal turning centers; (B) Computer control system *(Courtesy of Cincinnati Milacron)*

addition of computer controlled multiaxis flexibility, with tactical, visual, and laser sensing coupled with learning abilities, the robot has become a key industrial tool.

The analysis of kinematic space machines and their degrees of freedom has provided an engineering rationale for the study of robots. This analysis can represent up to six degrees of freedom and is useful in the programming of movements of the robot. In relating to robots, the axes of motion are described as these: the vertical traverse, or the up-and-down movement of the robot arm or body; the radial traverse, or the extension or retraction of the robot arm; the rotational traverse, or the rotation of the robot body clockwise or counterclock-

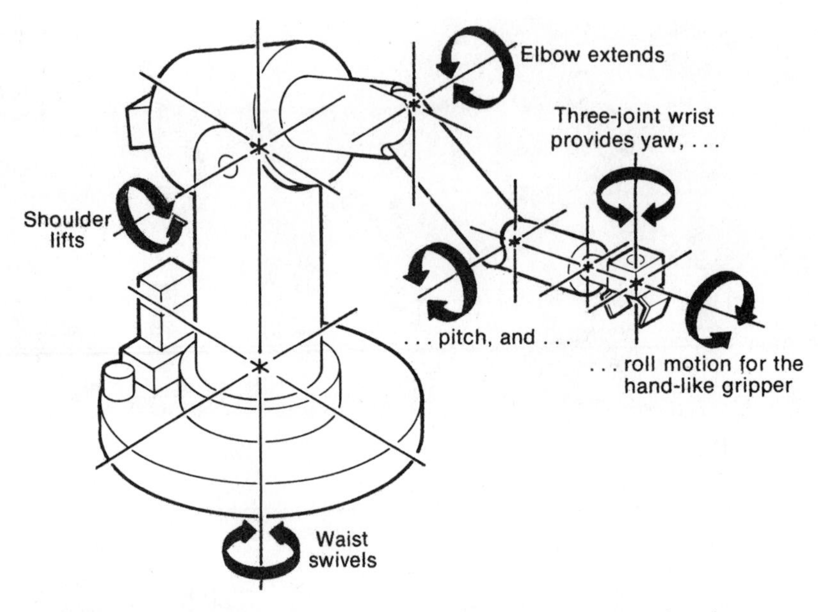

Figure 11-11 The anthropomorphic arm of the robot is programmed for a series of motion sequences in one or more bearing joints to achieve six directions of movement or freedom

wise; the wrist roll, or rotation of the wrist; the wrist pitch, or the up-or-down movement of the wrist; and the wrist yaw, or the right-or-left swivel of the wrist, Figure 11-11. The movements of the robot shoulder, arm, and wrist result in the successful positioning of the tool or gripper that is capable of performing work.

Classification by Technology

As previously stated, robots are classified by various criteria. One of these methods is by the level of technology used in their construction. These are referred to as low technology, medium technology, and high technology.

Low-technology Robots

The low-technology robot includes the pick-and-place or the bang-bang robot. It is a nonservo robot with a limited sequence designed to move light objects. It usually is activated by a pneumatic drive system.

Medium-technology Robots

The medium-technology robot is a servo-controlled robot operating with a programmed controller. The location and positioning of the robot arm vastly increases the accuracy. The robot can spot weld, unload die casting machines, and perform other similar functions as well as handle materials up to one hundred and fifty pounds in weight. It is activated and controlled by hydraulic systems.

High-technology Robots

A high-technology robot functions with a servo-controlled system and an extensive feedback system. These systems are usually hydraulic/electronic or electric/electronic systems. As the hydraulic/electronic systems move the actuators to new positions, various feedback instrumentation devices (such as encoders, potentiometers, resolvers, tachometers, or velocity sensors) return position data back to the controller. The controller in turn compares the input position to the robot's output position and provides an error signal that corrects the robot's position. The high-technology robots supply maximum flexibility through programming the axes of the manipulator to any position within the travel of the arm. The motions are smoothly executed with control of the speed with acceleration and deceleration velocity control. The robots are designed to handle massive weights of the manufacturing cell.

The electric/electronic robots are also high-technology robots that employ either AC or DC electric motors in their drive units. This drive train contains gear reduction sections to generate sufficient torque for heavy loads. The control instrumentation is computer activated for all axes.

Classification by Lifting Capacity

Sometimes robots are classified by their lifting capacity, the terms used are simply these: light duty, capable of working parts or tools up to 15 pounds; medium duty, with capacities of fifteen to fifty pounds; and heavy duty, with working capacities of fifty pounds or more.

Robot Configurations

Another classification is that of the configuration of the robot. Four basic configurations or variations of robots are presently available: a jointed-arm configuration, a spherical configuration, a cylindrical configuration, and a rectangular configuration. Each robot configuration will generate its own work envelope or the space that can be reached by the working arm, Figure 11-12.

Jointed-arm Robots

The jointed-arm robot's three major axes are determined by the robot's arm sweep, shoulder swivel, and elbow extension. The rotary motion is obtained by rotating the arm sweep in a horizontal plane about the base. The shoulder swivel motion is a rotation perpendicular to the horizontal base plane, similar to the movement of a man's shoulder. The elbow extension has a similar movement but

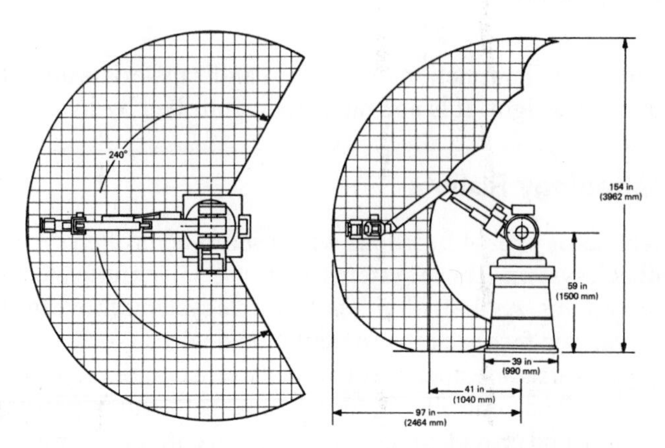

Figure 11-12 A jointed-arm work envelope *(Courtesy of Cincinnati Milacron)*

at an extended position, it is more like a man's elbow. This configuration of robot is built to machine tool quality standards. It contains combined hydraulic/ electrical power units. The robot's extended arm working area is the partial section of a sphere. This working area is the work envelope. The work envelope is determined by plotting the position of the tool, work, or gripper when it is in the extended position and moved through all its directions, Figure 11-13.

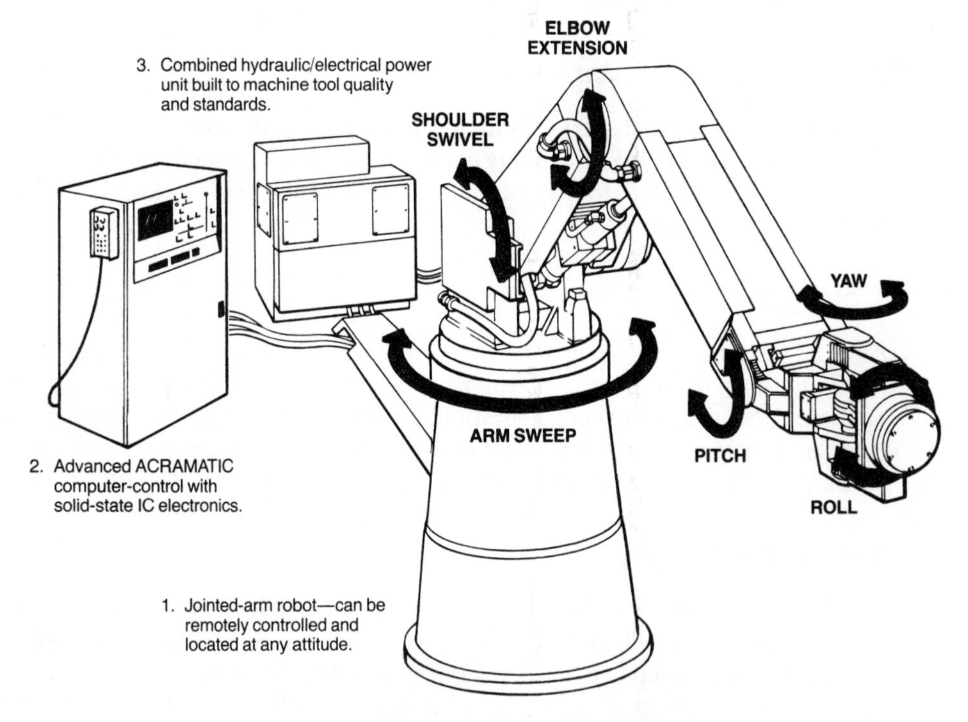

Figure 11-13 A jointed-arm robot *(Courtesy of Cincinnati Milacron Industrial Robot Division)*

Spherical Robots

The spherical configuration robot is commonly called a straight-arm robot. The vertical motion, in the shape of an arc, is obtained by pivoting the manipulator on its base. The radial motion is obtained by rotating the arm in a horizontal plane about the base. The radial motion is also in the form of an arc. This system provides a work envelope that is a partial sphere.

Cylindrical Robots

The cylindrical robot configuration achieves vertical movement by moving up and down on a vertical column. The radial motion is provided by moving a horizontal shaft in or out, and the rotary motion by rotating the robot in a plane around the base of the machine. This provides a work envelope that is a section of a cylinder.

Rectangular Robots

Rectangular robots move in two planes perpendicular to each other. The movements are like the Cartesian coordinates system used in numerical control. The work envelope is that of a rectangle. To provide additional flexibility, these robots can be combined with a rail or track system either on the floor or overhead to provide an additional length of travel. The track is frequently a part of a manufacturing cell's transfer system. This subset of the rectangular robot is referred to as a gantry configuration robot. In many cases, these gantry or shorter distance transport crane-like structures are incorporated into the machine system. The basic robot configurations are represented in Figure 11-14.

Robot Grippers and End Effectors

The robot end effector is a gripper (or hand). It may be a closing device that holds the work, or it may be an end-of-arm tool holder. Gripper end effectors place parts into machines. They can load machining centers and presses. They can unload pallets and die castings from machines. End effectors with tool holding capability perform work such as welding, assembly, painting, grinding, and cutting. In the latter case, the end effector holds a tool that performs work on the workpiece directed through its operation by the robot's control system, Figure 11-15 (page 414).

Gripper Sensors

Tactile sensors are added to grippers to improve the robot's ability to perform pressure-sensitive tasks of all types. They control the amount of pressure exerted so that fragile products can be handled without crushing. They also prevent

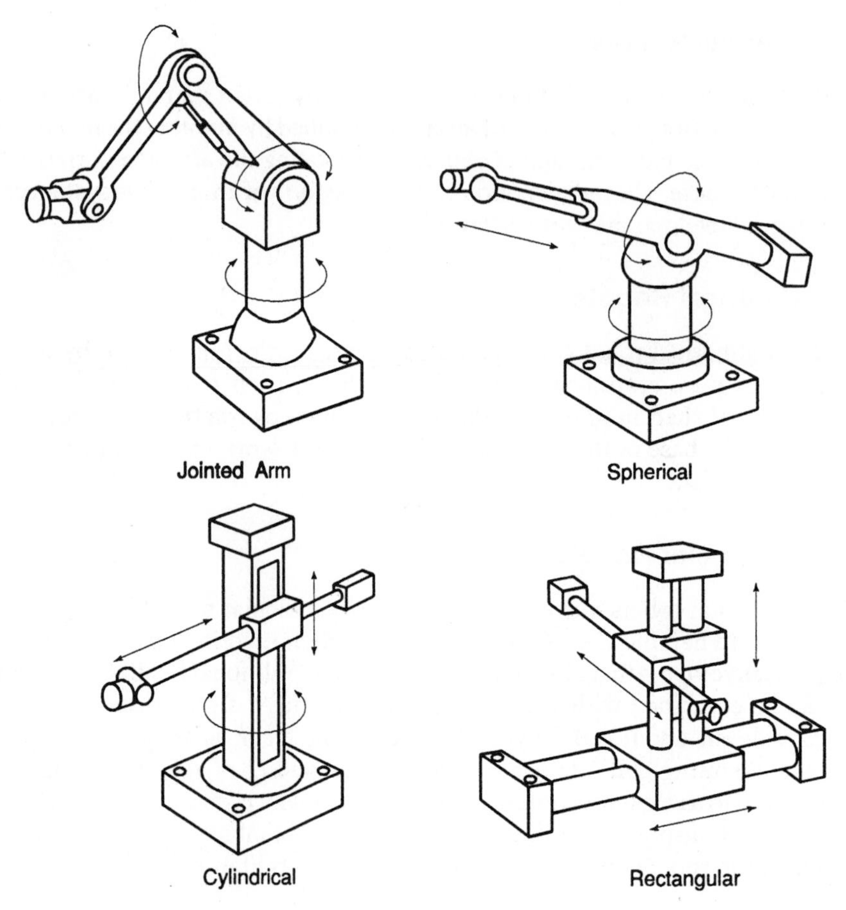

Figure 11-14 Basic robot configurations

slippage of the object held. This type of sensor manufacturing employs elastomers, piezoelectric materials, optical fibers, and strain gage technology within silicon chips, Figure 11-16. Pressure-sensitive assembly operations, such as assembling carburetors and instruments, are performed with sensitive grippers.

CONTROLLERS FOR ROBOTS

Controllers are designed to fit the level of robot technology.

Low-technology Control

Low-technology control functions are performed by hard stops and microswitches or timed with clock- or position-driven cams to provide a sequence to pneumatic

(A)

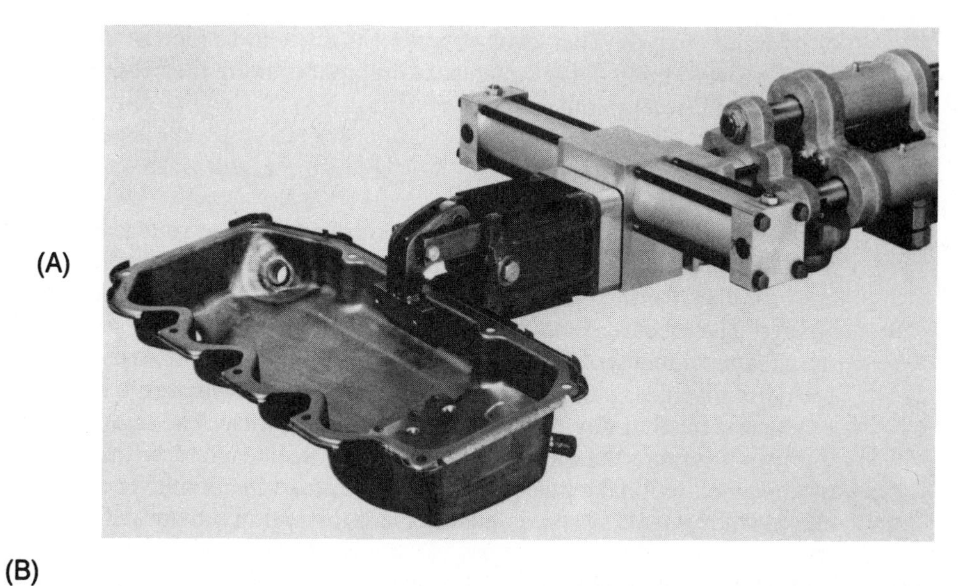

(B)

Figure 11-15 (A) Typical robot end effector; (B) End effector configurations *(Courtesy of I. S. I. Manufacturing Inc.)*

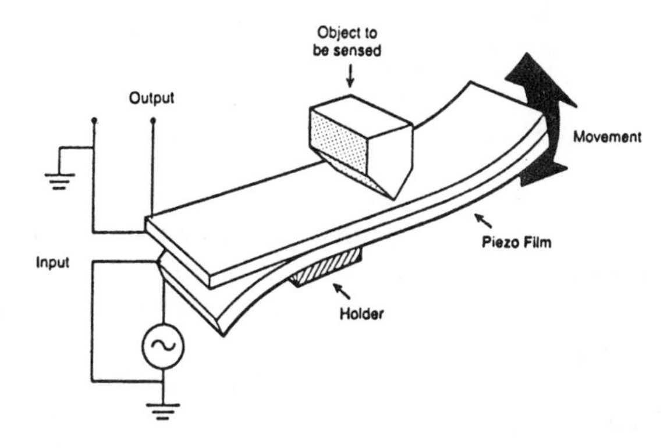

Figure 11-16 Bimorph design—Piezo film applied to robotic tactile sensor for assembly applications *(Courtesy of Pennwalt Corp.)*

or hydraulic valves. This level of control is difficult to reprogram. Screws and slides provide limited adjustments, but reprogramming the robot requires reorganizing and resetting the control system.

Medium-technology Control

Medium-technology control provides control over two to four axes to make this level of robot applicable in many more operations. The controller has a microprocessor with a memory. Thus it can more easily be reprogrammed with a software change. The robot receives feedback from sensors that supply servo control over the axes' movements of the manipulator. This machine has more memory than the previous machines. Therefore it can be programmed to service a number of production machines with a load and unload capability. The microprocessor for this machine limits the speed at which the robot will respond, so that only one axis of movement will take place at a time. Because of this, time is consumed while all axis movements are completed to place the part in the correct position.

High-technology Control

A high-technology robot controller requires a minicomputer with a considerable amount of memory because it is called upon to provide program activity supporting up to six axes of movement of a manipulator simultaneously. In addition, it may hold complete programs for each model of product that is presented by an assembly line or manufacturing cell. To provide additional memory, co-microprocessor systems are added to the robot control system, thus shortening the data execution time necessary to position the end effector.

The axes of the robot are controlled separately by a control loop from the microprocessor system that activates the drive element of that portion of the robot arm. This control loop may activate a pneumatic cylinder, hydraulic motor or cylinder, stepping motor, or DC servomotor, depending upon the arm's design. These joints can be driven directly or through chains, cables, bands, gears, or lead screws to complete the movement and position of the end effector.

Devices that Sense a Robot's Position

To ensure that the robot's position has been reached, a feedback system is included that provides data that can be compared to the reference. Any error is transmitted back to the microprocessor controller for change, Figure 11-17.

In the control of a high-technology robot, the data coming to the robot is a program received from a mainframe computer or a minicomputer that controls the production activity at the manufacturing cell. This data enters the system of the cell at the controller reference input port. The information at the reference point is compared with the real position of the robot element of the arm. If a difference is found, the controller provides an error signal that is transmitted to a digital-to-

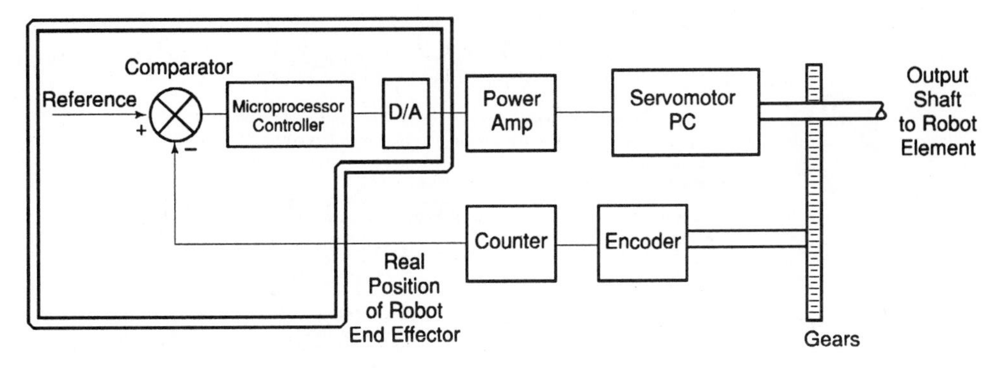

Figure 11-17 A block diagram of a single axis of a robot control

analog converter that supplies an analog signal to a power amplifier with adequate energy to power a DC servomotor. The motor output shaft provides the mechanical energy to move the robot's arm element to the new position.

Encoders

To make certain that the element has reached the new position, an encoder is attached to the robot arm element. The encoders are of two types: incremental or absolute. The incremental encoders provide the logic states of 0 and 1 alternately

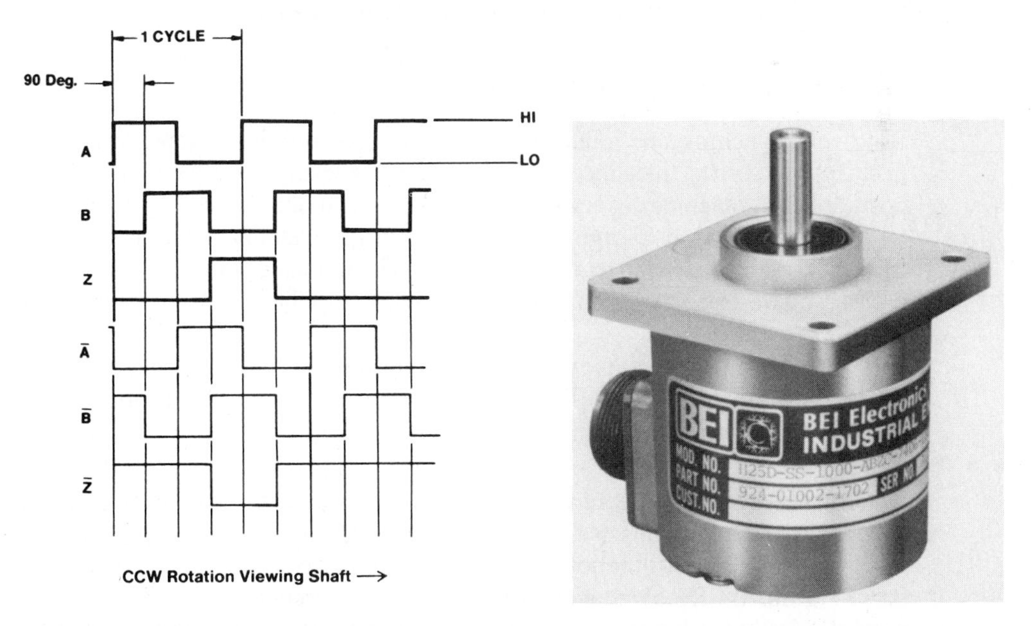

Figure 11-18A Encoder provides position data actuator back to a controller *(Courtesy of BEI Motion Systems Co. Industrial Encoder Division)*

ENCODER OPTIONS

8-BIT ABSOLUTE

8-BIT ABSOLUTE POSITION ENCODER

The basic unit provides a gray code output and an interrogation input line. This interrogate line is useful when the outputs of several encoders are tied in parallel to one processing unit. Optional features include a gray to natural binary code converter, an output latch (memory), and a variety of output I.C. configurations.

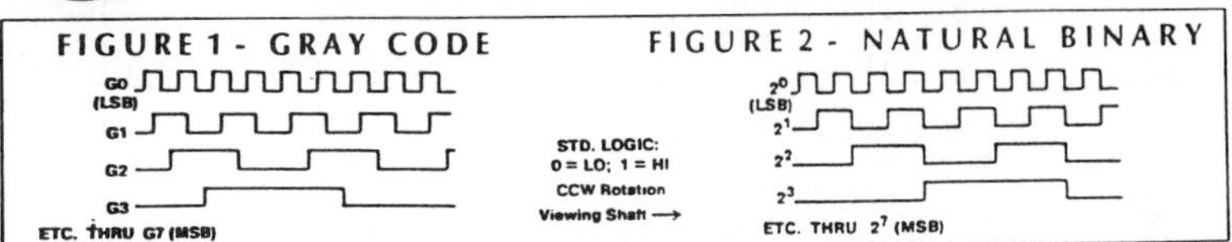

Figure 11-18B The digital codes for an eight-bit absolute position encoder *(Courtesy of BEI Motion Systems Co. Industrial Encoder Division)*

for each successive cycle of resolution or counts per turn. Absolute encoders provide a unique binary word for each position, resulting in a signal in a binary or gray code depending upon the instrument chosen. The gray code is used as an error minimizing system because there is only a one bit change at a time from one number to the next, Figure 11-18. This output is sent to a counter or an interrogate line. The encoder receives an analog position and reports a digital output signal. The counter accumulates the encoder's pulses that are sampled in a constant time interval. The microprocessor is applied to the comparator of the individual control loop and again any additional corrections of position are made.

Resolvers, Synchros, and Inductosyns

Resolvers, synchros, and inductosyns are rotational and linear transformer-type position-measuring transducers. They provide accurate feedback signals for positioning of machines operating in an industrial environment. These resolver-to-digital and synchro-to-digital position-measuring systems appear as small AC motors from one half inch to about four inches in diameter. An AC input voltage with a range of frequencies up to a few kilohertz is supplied to the rotor windings.

Resolvers - In resolvers, two stator fixed coils are located ninety degrees apart to provide the output voltage. The output voltage depends upon the position of the resolver's rotor.

Synchros - Synchros are similar in construction to resolvers, except that they have three stator windings that are each one hundred and twenty degrees apart and connected in a Y-configuration. The voltage is induced across any two synchros' stator terminals. Resolver and synchro-to-digital converters transform these voltage relationships into digital signals that represent the actual position of the actuator.

Inductosyns - Inductosyns measure linear positions directly. They are constructed as two flat, rectangular, magnetically coupled parts. One part of the scale is fixed to the bed of the reference device and the other scale is fastened to the part that is to be measured. The base materials have a bonded printed-circuit track of a continuous rectangular waveform pattern each with the same cyclic pitch. A small air gap exists between the slider, which is about four inches long, and the fixed scale, which is a series of ten inch sections. The scale is supplied an AC voltage. An output voltage is induced into the slider, proportional to the slider spacing within the pitch of the printed waveform pattern of the scale. The stator voltages are supplied to the inductosyn-to-digital converter whose output corresponds to the displacement of the activator. The output is supplied to a microprosessor, Figure 11-19.

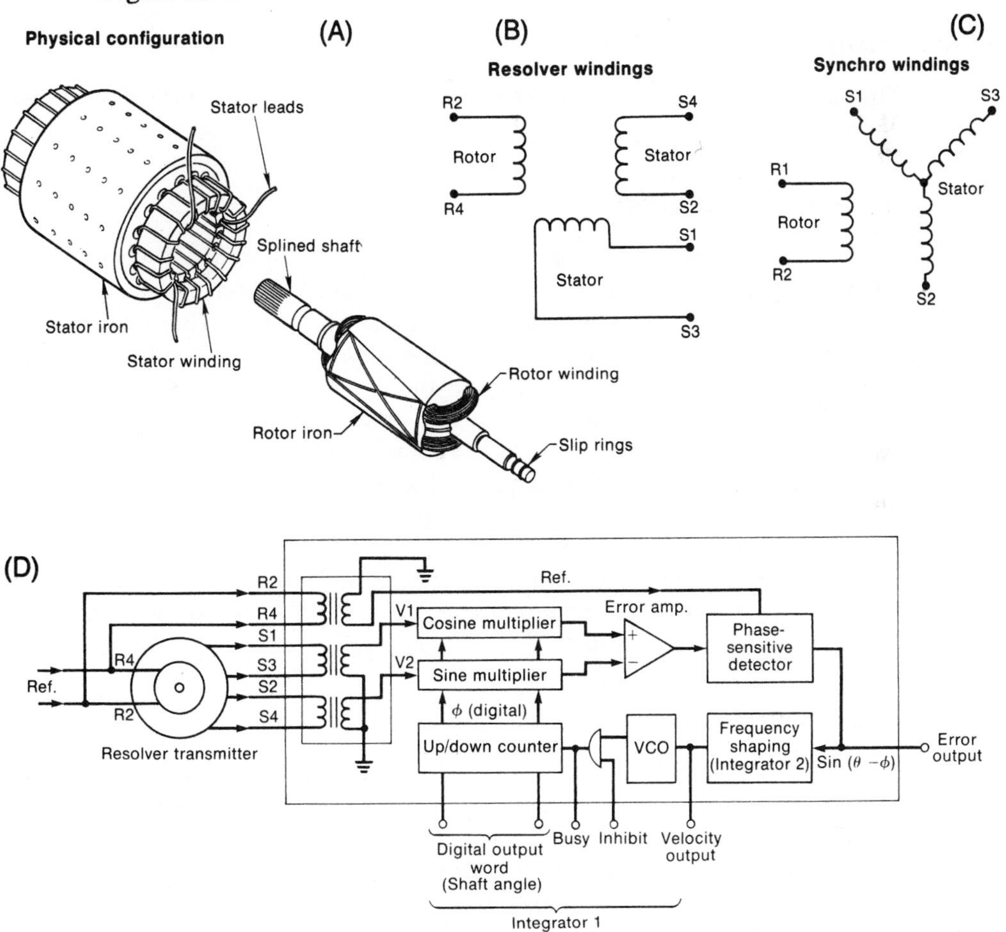

Figure 11-19 Diagram of resolvers, synchros, and industosyns: (A) Physical configurations; (B) Resolver winding; (C) Synchro winding; (D) Inductosyn configuration *(Courtesy of Analog Devices Inc.)*

Tachometers

High-technology robots are frequently driven by DC brushless motors or with AC motors. Within the various movements of the manipulator, the torque will change as angles and loads change, resulting in variations in velocity of the parts of the robot. A tachometer is used as a velocity feedback device or a rate-of-motion sensor that is applied to reduce nonlinearity of voltage caused by the effect of load and torque. The tachometer is frequently a small DC generator that provides the signal entered into the control loop so that the motor will overcome the torque and keep the motor speed constant. A controlled velocity in axes' movements of a robot is an important requirement for a work cell as the end effector approaches its programmed point.

Robot Program-controllers

The movements of robots are effected by a number of position control devices that are related to their level of technology.

Sequence Controllers

Step-sequence programs are employed in low-technology robots where movements are repeated in a series or steps. This type of sequenced program is found in nonservo or pick-and-place robots or similar operations. The program may consist of a series of adjustable limit switches and/or adjustable mechanical stops that program a series of movements in hydraulic or pneumatic cylinders. When one movement is made and reaches a stop or switch, the second movement is activated and started, moving until its end point is reached. The machine continues to the next step until the whole sequence has been completed. At this point, the machine stops and waits until the first step in the sequence is activated and the process repeats itself.

Sequence controllers are reliable, accurate, and very successful for a large range of work. But their programming lacks flexibility and changing their operations is time consuming.

Sequential Switching with Logic Circuits

The electronic controls of manufacturing cell robots are based on the microprocessor. Electronic sequential logic circuits perform the functions of timing, sequencing, and storage of data. Sequencing of operations is controlled by clock signals. The data is stored as binary code in flip-flops and changed by logic input and output signals. The two most commonly used devices in this function of sequencing are counters and shift registers for storing the data as it is scanned. These logic circuits are available in integrated circuits and are used to produce decision and switching operations.

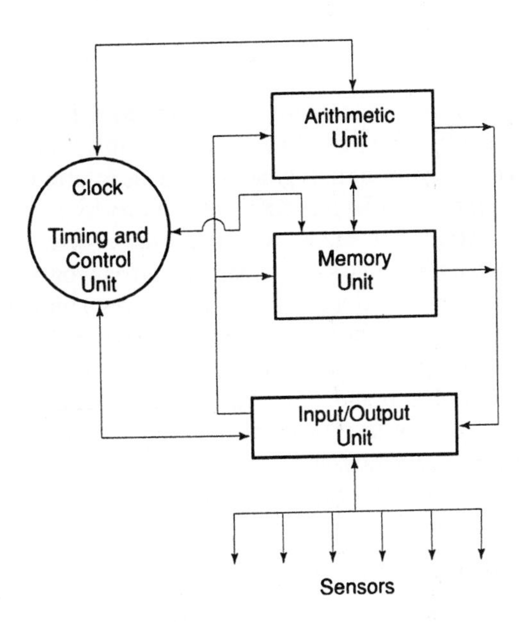

Figure 11-20 Elements of a digital system

In instrumentation and digital control circuits, there are times when events or spaces between events are counted. The timing can be performed by counting pulses of the logic device receiving accurate pulses. This is usually a monostable multivibrator or clock. These pulses are applied to the variable being controlled as the signal is modified by the sensor. The counting of these pulses provides a time element and value when a change of signal is observed. A microprocessor can scan a thousand transducers while synchronized with the clock looking for pulses. The value of these pulses is compared to each set-point and the microprocessor determines, by a voltage value, if it is on the set-point or how far it is off the set-point, Figure 11-20. The microprocessor provides a signal to switch things on or off according to the rate of change and in a prescribed sequence.

Microprocessor Controllers

A microprocessor on a chip (or "smart chip") is an integrated circuit with all of the central processor unit functions included within it. It has a main storage including a random access memory (RAM) that can be read from or written into, a read only memory (ROM), an arithmetic logic unit (ALU) for performing arithmetic operations, and special register groups (a clock for timing signal, a computer controls unit, and an instruction set). Input and output integrated circuits complete the microprocessor. The microprocessor may be thought of as a programmable building block that can be self-contained or applied in groups to supply the sophistication necessary to carry out the control requirements. The smart chips usually control a single control loop. In this case, the microprocessors can be on a single chip.

High-technology manufacturing cells can contain microprocessor-based servosystems for controlling positional data of the machinery and/or robots. The servosystem has its position signals generated in the microprocessor controller by a program. Or the signal may be placed in the memory by a teaching pendant. Each servo axis has a system that responds to the microprocessor's memory circuit. Each circuit has separate registers or storage locations for the servo position data coming from the programmed memory. A second register contains the feedback data or the actual position on the axis location from a resolver. These two registers are scanned by the microprocessor and compared for any difference in positional locations between the command and actual positions of the axis. Any difference between the two registers is sent to a third register (a velocity register) that commands the velocity and movement of the axis to the new position by providing controlled voltage to the actuator, a servomotor, Figure 11-21.

Minicomputer Controllers

A minicomputer is a small computer that applies a microprocessor as a CPU. It contains all of the functions of a large-frame computer. These devices have less memory capacity and are slower than the large computer, but they are less expensive. They provide a practical capacity for process monitoring and control of most manufacturing cells in a flexible manufacturing system. In a manufacturing facility employing computer aided manufacturing, the minicomputer may be found at the machine level and again at the supervisory level. It is referred to

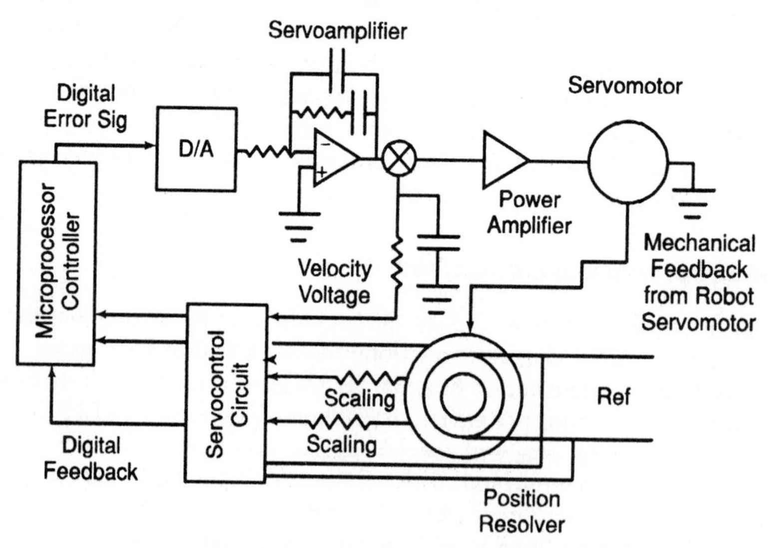

Figure 11-21 Microprocessor control of a robot's single axis position

as a "satellite computer." This minicomputer is usually linked to a series of smaller microprocessors at the manufacturing cell level. The satellite computer is connected into the plant large-frame computer and provides data of plant operation, engineering, management, business related costs, and supervisory functions. This computer supplies the data for management reports at the corporate level and executive decision making.

POWER SUPPLY

Power is needed to supply the energy necessary for flexible manufacturing systems. The power to operate a manufacturing cell requires electrical power, pneumatic power, and hydraulic power. All forms of power may not be needed in one facility but very frequently two of the three are used within the production.

Electrical Energy

The most common form of energy consumed within plants is electrical energy. Electrical energy is also the source of pneumatic and hydraulic energy delivered to the machines.

Alternating current of 60 cycles per second is delivered to production facilities at a number of nominal voltages: 120, 120/208, 240, 480, 600, and 2400 volts. Some of these voltages are delivered in single phase, while the higher voltages are delivered in a three-phased power. Voltages of 120 are used on motors from 1 to 15 horsepower; 208, 240, 480, and 600 volts are used on motors from 1 to 200 horsepower, and 2400 volts are used on those that are 444 horsepower and larger. Industrial motors of 5000 horsepower are available as prime sources of mechanical energy for large manufacturing machinery.

The power supplies available for manufacturing cells and their robots require either alternating current or direct current, depending on the actuator and control applied.

Unregulated Power Supplies

The power supplied by the electrical companies is considered as unregulated power. In most industrial applications for transformers, motors, and appliances, the supplied AC power is satisfactory. However, during the working day, load changes occur within the power utility's system. These load changes are corrected, but for specialized equipment, the variation may cause difficulty and require additional control.

Half-Wave Rectifiers - Direct current power supplies frequently utilize unregulated alternating power supplies as a source of energy for rectification. One

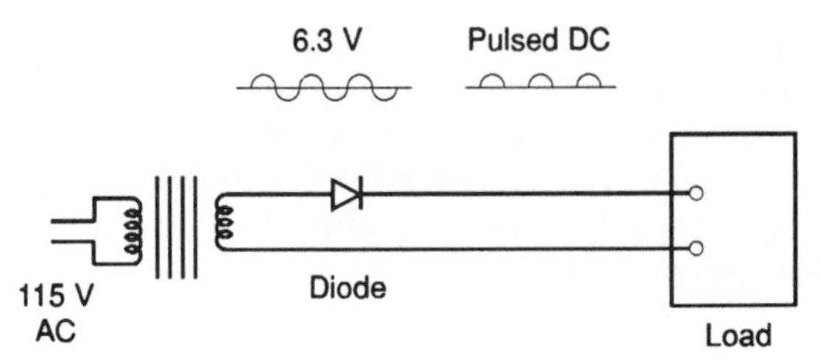

Figure 11-22 Schematic of a half-wave DC rectifier

method of producing direct current is to eliminate one negative half-cycle of an alternating sign wave of an AC current. This is provided by adding a diode between a transformer's secondary winding and the load. This diode application is referred to as a half-wave rectifier and is an adequate DC power supply for many applications, Figure 11-22.

Full-Wave Rectifiers - A full-wave diode rectifier uses both half-cycles of an AC voltage supply to make more current available. The secondary of a transformer is connected to opposite corners of four diodes positioned on a shape similar to a baseball diamond. The diodes are placed between the bases on the diamond. Diodes one and four are forward biased on the positive half-cycle and conduct current to the load. On the negative half-cycle, diodes three and two are forward biased and conduct the current. As a result both the positive half-cycle and the negative half-cycle of the alternating current sine wave provide a positive voltage to the load, Figure 11-23. The full-wave rectifier will supply double the voltage and current of the half-wave rectifier to DC equipment.

Silicon-Controlled Rectifiers - Equipment that requires high currents, heavy loads, and directionalized motor circuits employ silicon-controlled rectifiers. This device may be called an SCR or a thyristor. The heavy currents produced by a DC servoamplifier circuit are built to power and control DC motors. The SCR is triggered or turned on by a pulse to its gate and the current will flow. The current will continue to flow after the pulse to the gate is no longer there. So to stop the current flow, it is necessary to disrupt it with another electronic switch. The SCR can be controlled by pulses from microprocessors to provide direction and distance positioning of machinery and robots. Figure 11-24.

Regulated Power Supplies

Integrated circuit devices with digital logic control are very sensitive and require voltage-regulated power supplies. Regulators have been designed to provide this precise stable control. They also provide overload protection, maximum load current adjustment, and thermal shutdown if the device is overheated. These characteristics provide protection to the circuits that are serviced.

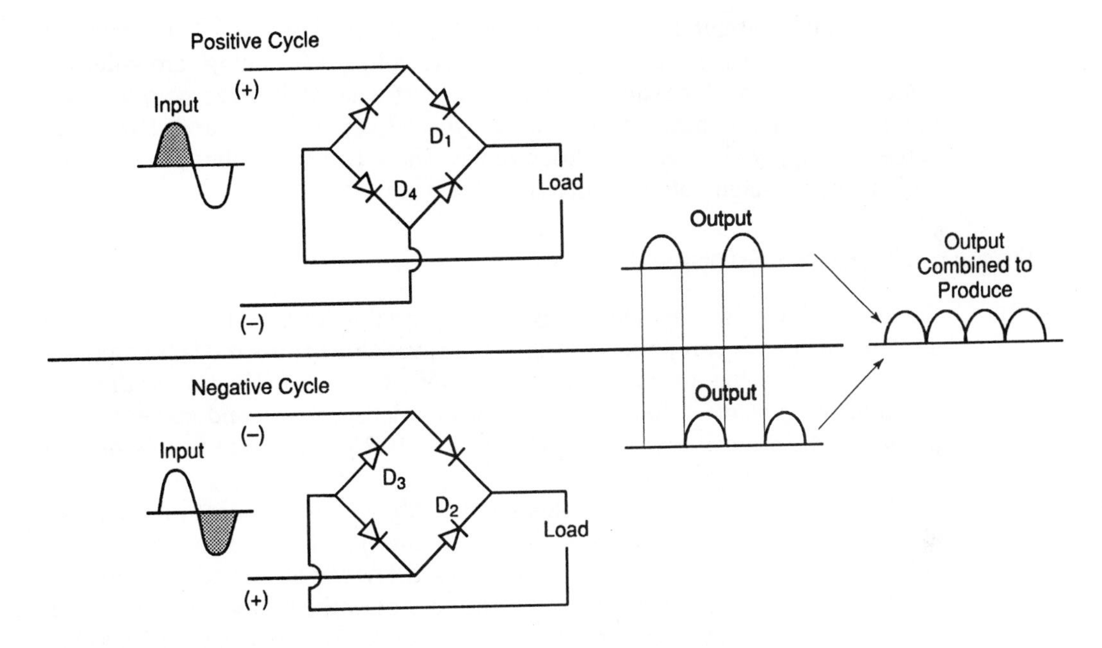

Figure 11-23 The concept of a full-wave rectifier

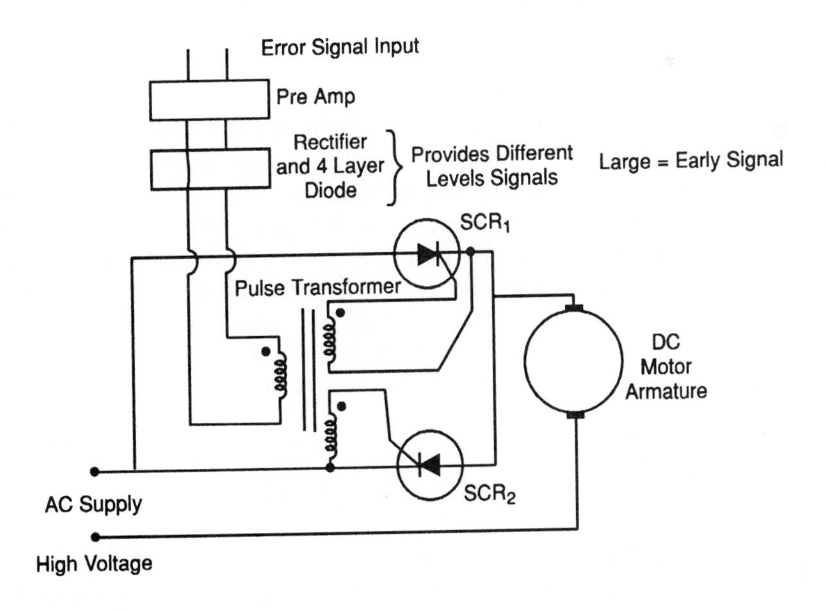

Figure 11-24 SCR application to power rectification and control for DC motors

Integrated circuit regulators are capable of supplying different types of voltage outputs. They can supply both positive voltages (meaning zero volts to a selected positive voltage) and negative voltages (meaning from zero voltage to a minus voltage). A common supply of this type delivers +18 volts and -18 volts as is used to supply electronic control circuits. These power supplies are also sold with a fixed voltage output or an adjustable output.

Pneumatic Power Supply

Pneumatic power is delivered through a compressible fluid, usually compressed air. The medium is compressed by any of the available commercial air compressors. Frequently the air compressor is located at a central location within the manufacturing plant -- often in the maintenance department -- and serves a large portion of the production facility. The air must be clean and free from moisture and the pressure regulated.

Pneumatic motors and cylinders require lubrication. To meet this requirement, a lubricating unit is placed in the air line just prior to the valves and cylinders. At the manufacturing cell machines, the power is applied to pneumatic motors or cylinders to perform work. The control of the motors or cylinders is provided by pneumatic directional control valves. These valves are of a number of types but frequently are solenoid-activated spool valves. The valves can be plumbed in a number of configurations to provide the various actuator functions required. This type of pneumatic power can be applied to double-acting pneumatic cylinders for the opening and closing of robot grippers.

Hydraulic Power Supply

Hydraulic power is provided by pumps that deliver high-pressure oil or other liquids to valves that very accurately control the powerful movements of a number of different types of actuators. The hydraulic power supply consists of a reservoir, hydraulic oil, strainer and/or filter, heavy duty pump, electric motor, and accumulator. These components supply the energy to relief valves, directional control valves, *flow control valves,* and check valves resulting in controlled oil reaching an actuator. The actuator usually is a double-acting cylinder. Hydraulics supplied to machine cells frequently will be applied to servovalves to precisely position very heavy loads, including heavy machinery and robots.

APPLICATIONS OF ROBOTS IN MANUFACTURING

Manufacturing time studies have indicated that an area where improvement could be made in manufacturing is in the reduction of the time consumed between processes during manufacturing. Parts would accumulate at machines waiting for processing, resulting in large inventories of parts waiting to be processed. In hard automation, this problem was resolved with the building of transfer machines to lines between the processing machines.

Robot Orientations

With flexible manufacturing, the robot is applied in many ways to transport material, load machines, and unload machines. The work must reach the envelope of the robot, whether it be by monorail, conveyor, or other work-transporting system. The transporter system also must orient the workpiece so that the robot gripper can contact the work in a consistent and reliable position. The control of the robot is set in action to move the work to the proper machine and position for loading. If the robot is in the center of a group of machines, a cylindrical coordinate robot can be programmed to supply the next available machine.

For some time, robots have been mounted on the floor, on the side of a machine, or on an overhead frame to carry out their work. In an effort to more efficiently utilize robots' capabilities on larger products or to interact with more stations, robots are being mounted on transporters.

If the manufacturing cell is in a straight-line configuration, the robot can transport materials to the machines by being mounted on an overhead monorail, gantry, traverse tables, or track for linear travel, Figure 11-25.

Figure 11-25 A gantry-mounted robot *(Courtesy of Grob Systems Inc.)*

Figure 11-26 Automated guided vehicle *(Courtesy of Control Engineering Affiliate of the Jervis B. Webb Co.)*

The Automated Guided Vehicle

Another transporting device is the automated guided vehicle (AGV), Figure 11-26. Flexible manufacturing systems are making use of AGV systems to get the right parts and subassemblies to the right place at the right time. These mobile robots are controlled by a number of methods: tow chains below the floor, cables or wires in the floor, a system of lasers and reflectors, and direct computer control integrated in the central control system. These transporters are providing flexibility in increasingly larger work cells or small workpiece manufacturing.

Fabrication Robots

Fabrication of products with the use of robots is a rapidly growing area of industry. Fabrication includes all the processes that have been performed in the past by manual operation. The running of many types of machines, arc welding, spot welding, spray coating, applying of adhesives and sealers, operation of die casting machines, drilling, light machining or grinding, electrical-wire-harness building, or the assembly of a vast array of products are all being performed by robots.

The instrumentation on these machines employs many types of sensors and digital feedback loops that are discussed in earlier chapters on control. The innovation of fabrication with the use of robots has been a function of interfacing with the computer to provide flexibility.

Fixed Robotic Assembly

By applying robotics to the assembly of products, the cost of the product is decreased because of less machine downtime and increased production rates. In most cases, the products selected are small and physically designed for robotic assembly.

In a system where mass volumes of assemblies are produced, fixed automation is a realistic choice. This requires assembly fixtures, part orienters, and part feeders, as well as part positioners and programmed tools. These requirements are best utilized in a product with a stable production output over a considerable time period.

Examples of successful fixed automated assembly machines are those that are programmed to place electronic components into circuit boards at high speeds. The electronic parts are first oriented in a machine that sequences and places the individual parts on a continuous tape that feeds the sequenced parts into the assembly machine. This type of assembly machine indexes the empty boards under the part transfer units. The board is released to a positioning unit and in turn is moved in the X and Y axis positions to receive the parts. The programmed table and board is positioned under the sequenced part, and the transfer is completed, Figure 11-27.

Figure 11-27 Circuit board assembly robot 20/20 high-density placement system *(Courtesy of EPE Technology)*

Flexible Robotic Assembly

A flexible automation assembly operation is one in which the machines can be reprogrammed to assemble a different product once the batch has been completed. One of the keys to flexible assembly is the part design. The end effectors of the robot require a surface that makes it possible for the part to be picked up and moved into its assembly location. To do this, the surface that is contacted by the robot is designed onto the part so that the gripper has a consistent surface to grasp. The orientation and location of the new assembly part then becomes a software or program change to assemble the new parts into a different product.

Vision Controllers - A different application of automated flexible assembly is to orient the parts by utilizing a vision controller. The scrambled parts are fed past a recognition device that scans the parts and collects image data. The vision controller determines the part orientation by comparing data from the required orientation previously taught the to the controller. If the vision controller recognizes a part as being in the correct position it will pass it into the assembly system. In other cases, it will be returned back into the reorientation process. The chief advantage of a vision controller is that at the completion of a batch assembly, the orientation system can be reprogrammed by reinstructing the vision controller for a new part, Figure 11-28.

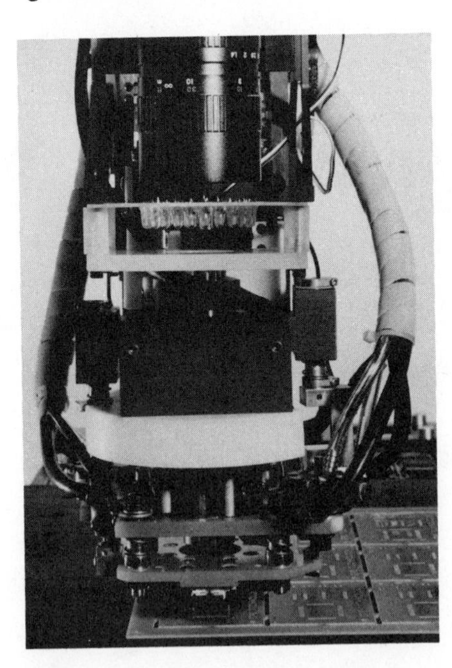

Figure 11-28 20/20 placement head reflowing fine pital device with hot head. The hot head allows users to reflow a soldered device *(Courtesy of EPE Technology)*

Redesigning Parts - Another principle applied in flexible automation assembly is in the redesigning of parts. Combining functions and providing for nondirectional shaped parts that can be placed into the assembly with any directional orientation are design considerations to be made. These changes result in products with fewer parts to be assembled. The combining of two or more parts into one redesigned part with gripper contact points becomes a major feature of reprogrammable assembly.

Robotic Inspection and Measurement

The coordinate measuring machine (CMM) was first developed as a manual inspection tool. It is now found as a part of a flexible manufacturing system. The coordinate measuring machine can reduce inspection time and costs dramatically, especially when applied to complex part measuring. Because of how rapidly and accurately they work, the CMMs have changed a concept of quality control from defect detection to defect prevention. The CMMs use computers and digital technology in taking measurement data, thus making feedback data available to the manufacturing production computer. The work of the CMM is performed by moving probes of the machine in three-dimensional space. The machines can store a vast number of calibrated probe-tip positions that can be compared with the measured probe-tip positions. These measuring machines are designed to function in one of three modes of operation: by the operator moving the probe from point to point; by using a manually controlled power-driven probe; or by a direct computer controlled and driven probe.

An advantage of a computer-controlled coordinate measuring machine is the ability of software to be applied to the system. The manufacturer of the CMMs can supply measuring programs, or the programs can be written specifically to perform the subroutine commands necessary for the exploitation of the machines for automatic feedback control.

The coordinate measuring machines exist in a number of configurations: cantilever, bridge, column, horizontal arm, and gantry. These constructions provide a structure for movement of the probe in a three-dimensional rectilinear or polar coordinate system of measurement. In each case, they provide an accurate measurement and recording of the spacial coordinate location of the probe on the features being measured, Figure 11-29.

The industry has recently developed a new technology of noncontact probing (with the use of lasers or video) to be used on fragile materials that cannot be touched. The flexibility in this measurement and control is available with the application of sophisticated software.

Auxiliary Equipment

Other common applications of flexible manufacturing are in the control of conveyors, automatic guided vehicles, positioners, feeders, and safety zone

Figure 11-29 A gantry CMM universal measuring machine *(Courtesy of Carl Zeiss IMT Division)*

shields and fences. Each installation has its own unique group of auxiliary equipment to be controlled.

The computer has provided a means of scheduling and controlling the auxiliary equipment that services a flexible manufacturing system. In addition, it has provided a method of modifying operations in real time with the opportunity of program changes to bring about coordination, accuracy, and efficiency.

Conveyors are started and stopped either by optical sensors or weight depending on the types of materials transported. Automatic guided vehicle codes are read and the product is shuttled to the correct processing area, the tools and programs retrieved, and the work performed. Part positioners, orienters, and feeders are controlled to provide the necessary fasteners or subassemblies for completion of the product. These are sensed by optical, proximity, mechanical, and other sensors for automatic control.

Safety zones, shields, light curtains, and fences are necessary around robot-supported and other flexible manufacturing areas, Figure 11-30. Safety devices activated by infrared beams warn of danger when the safety zone is invaded. Automatically controlled equipment in many cases stands and waits for its operational instructions. In this period of time, there may be a great temptation for personnel to perform servicing, cleaning functions, or other operations in support of the equipment. This is done because automated machines operate without warning, thus the need for safety devices to be installed in hazardous areas.

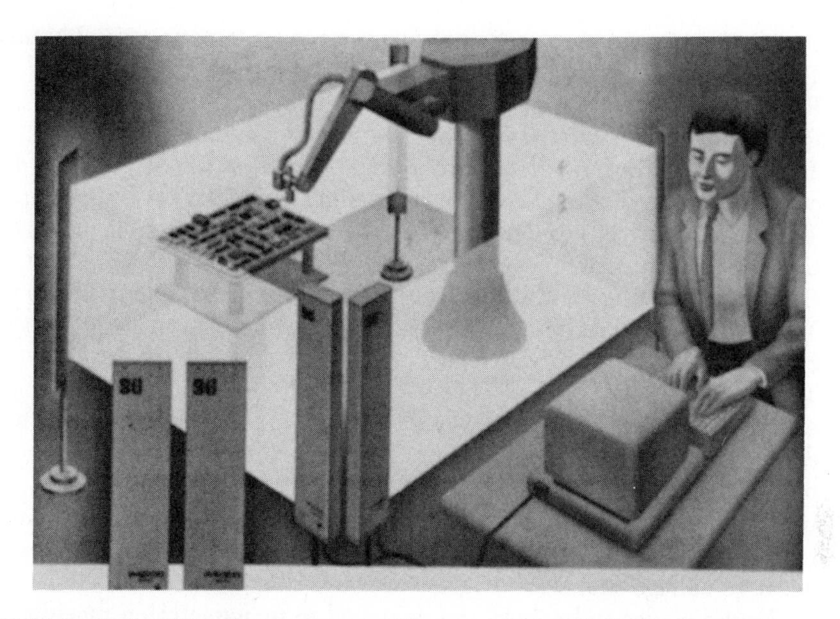

Figure 11-30 A robotic field perimeter protector *(Courtesy of Scientific Technology Inc.)*

Robots in manufacturing are applied to many manufacturing production processes. In the transporting processes, they move materials, they load and unload machines, and they work in palletizing and packaging applications. In the fabricating processes, they do spray coating; they grind, polish and drill; they do light machining, metal flame cutting, metal laser cutting, spot welding, and arc welding. In assembly processes, they combine parts and subassemblies into a product. In quality control, they inspect and test in-process and completed products.

The common ingredient in all these processes is the instrumentation applied and controlled by a computer technology.

SUMMARY/FACTS

- A flexible manufacturing system consists of a series of automated cells that are capable of being programmed for batch production for a family of parts.
- A manufacturing system made up of sensory-interactive machines requires a standardized communications protocol. The protocol widely used is the Manufacturing Automation Protocol or MAP.
- Mechanical assembly machines are designed to perform assembly and fastening operations.

- Part feeders provide for a series of functions for automatic assembly machines. They elevate, orient, and feed parts to the assembly machines.
- Robots are frequently used as part handlers in automated assembly.
- Robots are classified into two groups -- nonservo and servo robots.
- The four basic configurations of robots are: jointed-arm configuration, spherical configuration, cylindrical configuration, and rectangular configuration.
- The robot end effector is a gripper. It may be a closing device that holds the work or an end or arm tool holder.
- One method of transmitting the position of a robot arm is to employ an encoder. The encoder is attached to a robot element and a position signal is sent to its controller.
- The most common form of energy consumed within manufacturing plants is electrical energy. Electrical energy is also the source for pneumatic and hydraulic energy delivered to the equipment. This energy is considered as unregulated power.
- Integrated circuit devices and digital logic circuits are very voltage-sensitive and require a regulated power supply. In addition, they may require both positive and negative voltage sources.
- Hydraulic power and control are delivered to manufacturing machines with a large group of hydraulic components consisting of a reservoir, pumps, valves, accumulators, and cylinders.
- The automated guided vehicle is a mobile robot that transports large or heavy parts under manufacture that may contain self-power and control.
- Robotic inspection and measurement is carried out with a coordinate measuring machine. The CMMs use computer digital technology and software in taking measurement data from the workpiece, thus making feedback data available to the manufacturing computer.
- Safety zones, shields, light curtains, and fences are necessary around robot-supported and other flexible manufacturing areas.

REVIEW QUESTIONS

1. What term is used to describe a system of manufacturing that uses a series of automated cells capable of being programmed for batch production of a family of parts?
2. Why is the fact that reprogramming is possible so important in manufacturing?
3. What is the meaning of MAP?
4. What is the chief characteristic of the numerical control machining center?
5. What are three functions of part feeders?
6. Early robots, such as pick-and-place or bang-bang robots, are considered open loop devices. How can this be a disadvantage?

7. What condition indicates an error signal in a servorobot?
8. The economic cost of a flexible manufacturing system limits its growth. What is a typical method of its development?
9. Name some features that are now applied to robots to make them a key industrial tool.
10. List the four basic configurations or variations of robots available.
11. How is the work envelope of a jointed-arm robot determined?
12. How does the end effector of a robot perform such work as welding or assembly?
13. What are the two types of encoders attached to the robot arm element?
14. List some linear and rotational transformer-type position-measuring transducers that provide accurate feedback signals for positioning machine operations.
15. What device is applied to reduce nonlinearity of voltage caused by the effect of load and torque?
16. A controller that consists of a series of adjustable limit switches and/or adjustable mechanical stops that program a series of movements in hydraulic or pneumatic cylinders is called a ____________ .
17. A microprocessor compares the value of _______ representing the set-point and those from the sensor and then provides a signal to switch on or off an actuator according to a prescribed sequence.
18. What is a satellite computer?
19. Integrated circuit devices with digital logic control are very sensitive and require voltage ____________ power supplies.
20. List some of the means a robot can be mounted on to transport materials in a straight-line manufacturing cell.
21. What are mobile robots that transport heavy workpieces to manufacturing machines and around the facility?
22. Where can the software applied to a computer-controlled coordinate measuring machine be obtained?
23. Name some types of safety measures used in robot-supported and other flexible manufacturing areas.

12

A STRATEGY OF AUTOMATION WITH COMPUTERS

OBJECTIVES

Upon completing this chapter, you will be able to explain and apply the principles and concepts of Computer Integrated Manufacturing:

- Interfacing with other manufacturing systems
- Integrating production control
- Integrated material-handling control
- Integrated inspection control
- Computer-aided quality assurance
- Data base management systems
- Interchangeable modular components
- Software systems
- Proximity data acquisition
- Alternate branch sequencing and programming
- The automated factory

INTRODUCTION

The computer is a masterful instrument that the men and women in the manufacturing world are learning to exploit. The more production problems confront the computer, the more applications and solutions are sought through the use of the computer.

COMPUTER INTEGRATED MANUFACTURING

In the evolving manufacturing system, the computer becomes the primary instrument that utilizes sensing, decision, and control to implement production. Added to this manufacturing function is its analysis of economic conditions and business decisions. With this combination an integrated system evolves.

The Scope of Computer Integrated Manufacturing

The first considerations of computer integrated manufacturing (CIM) is to include a study of the total manufacturing enterprise: strategic planning, financing, marketing, design, product engineering, manufacturing planning, and control of the production floor.

Computer integrated manufacturing is more than a technical commitment to manufacturing. Rather, it is a philosophical dedication to manufacturing development and continued manufacturing improvement. This is a very long-term commitment to excellence required by all levels of corporate management, finance, and production.

The computer networks also link interactions with the management elements of Manufacturing Resource Planning (MRP II): simulation and optimization, management graphics and decision support systems, shop-floor data-collection systems, quality control, cost control, and inventory control. Computer networks provide the necessary communication between these various hierarchical computer levels within the corporation.

The production objective of CIM is to provide for the interaction of flexible manufacturing systems with data collection, decision, and control. The computer integrated manufacturing system at the operational floor level links three areas of automation: group technology, robotics, and material handling.

The computer network processes information concerning programs for cell computers, programmable controllers, robotic systems, material-handling systems, material processing and assembly, inspection, and testing with correction.

Building a CIM organization for many companies is a calculated step-by-step movement with the interfacing of the various islands of information and automation. The Computer Aided Design and Computer Aided Manufacturing (CAD/CAM) islands frequently are the first to be interfaced with the market management to assure that production is working and solid business decisions have been made. The Manufacturing Requirements Planning (MRP II), process planning and shop floor control are also integrated. The flexible manufacturing systems of cells are linked with the tools under direct computer control for production in a CIM system.

Computer integrated manufacturing includes interaction of all the data and expertise of a total manufacturing organization. It makes use of communication to integrate all levels of the enterprise to support the plant floor operations.

This strategy of manufacturing has been made possible by the application of the computer to solve problems with a very large number of variables. The understanding of many variables through computer analysis has made and continues to make great changes in the financing and reorganizing of corporations of all types. The necessity of being competitive in world markets has forced the use of computer technologies in the development of successful strategies by discovering areas of manufacturing that have become too high-priced or are losing value in world competition. The computer can supply information as to the time when a series of management decisions should be made. To do this, the right questions must be asked and the proper data supplied to the computer. This information will provide alternative strategies, so that business and manufacturing decisions can be successfully and economically made.

Interfacing with Other Machine Systems (MAP)

In order to interface customers' desires with manufacturing machine systems, communication within the total manufacturing organization is critical. The difficulty has been concerning the compatibility of various parts of the system. Various vendors' equipment requires a common dialect of communications so they can interface with each other's manufacturing systems. These requirements for Manufacturing Automation Protocol (MAP) are important standards for the specification of interconnections between intelligent networks. A MAP network architecture allows a fully integrated system with communication from the consumer to indirectly control the manufacture of an individualized product. The customer can directly order various option specifications to be included on his or her purchase.

The production and control communication is received and distributed throughout a smart factory by using a broadband MAP architecture. The broadband distribution system has few distance limitations and is excellent in supporting a network for large facilities. A broadband communication system is less susceptible to ambient factory electronic noise than many systems.

Local Area Network (LAN)

Flexible subsystems supply necessary communication by a local area network (LAN). LAN supplies the production information for flexible manufacturing. These subsystems improve job flow and decrease production downtime.

When machines built by different vendors are built to MAP and LAN standards, they provide the method and procedures for multilevel communication between the customer, computers, engineering, controllers, manufacturing systems, quality control, testing, workstations, and management for the total factory. The application of the LAN to a purchase order from a distant source results in a completely assembled product with the options ordered to industry and customer specifications. This flexibility by the network provides for everchanging market and customer requirements as well as helps meet manufacturing goals, Figure 12-1.

Integrated Machine Control

Interfacing provides access to the mainframe digital computer. The various programs to be executed are stored in the computer. The program to be transmitted to the tools or machines is identified and sent. This supplies and controls the positioning of the machines or tools and is the source for supplying commands to

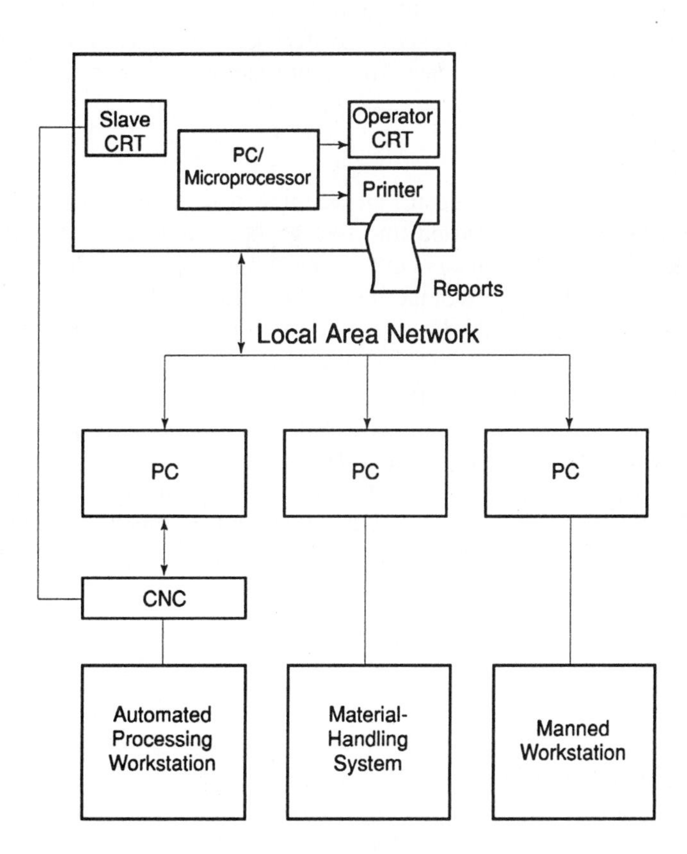

Figure 12-1 An automated manufacturing controller applying a local area network

carry out the operations of the machines. Interfacing also includes data feedback signals indicating that operations are in progress, completed, or stopped because of an error.

Integrated Production Control

Interactive data is supplied by computer monitoring of a number of manufacturing cells. This processing data has made it possible to integrate the total production floor. Mainframe computers can contain the production information from each process as the manufacturing takes place. The number of units produced per hour, the number of parts that performed during the test, the number of parts that failed during the test, and the location in the process where the failures occurred are all a part of the data captured. These and other data are entered automatically into the line computer and transferred to the supervisory computers. While this type of data is continuously arriving, many real-time decisions concerning production rates, scheduling, and pacing of production lines can be intelligently made.

This data can be used to program the retrieving of parts from an automated robot storage system and to pace the materials entering a multiple production line. The data can be used as a basis for input to an inventory planning system.

Sensor Data

The sensors applied in securing machine and robot data are built into the equipment and are classified into two groups: contact sensors and noncontact sensors. The contact sensors contain the subgroups of tactile sensors and force-torque sensors. Examples of tactile sensors, those that touch the work, are probes or microswitches that send a signal or control a circuit by opening or closing to indicate a position or a count. Force-torque sensors measure reactions to forces and movements. These types of sensors are strain gages or piezoelectric gages that are cemented to the various components of the robot gripper. These sensors measure the force on grippers and the wrist.

The noncontact sensors do not touch the work and include proximity sensors based on the concepts of capacitance, inductance, ultrasonics, magnetics, air jets, or vision. With these types of sensors, the robot gripper or machine part's position can be sensed as it approaches its final position.

Machinability Sensing

The use of adaptive control instrumentation is increasing in metal cutting operations because of the interaction between the machines and the computer. The computer makes data available that anticipates a potential change in machinability characteristics. This is predictive of changes in the work material as it occurs. When a hard spot is encountered in the metal, the machine adjusts. Adaptive control instrumentation measures the deflection of the cutting tool spindle or the power consumed by the cutting tool. Adaptive control provides adjustments of the speed and feed rate to the cutting process to maintain the desired horsepower delivered to the machine's spindle thus increasing the cutting tool's life.

The power consumed is measured by monitoring voltage, current, and the power factor of the load drawn from the power supply. The computer is programmed with data parameters such as machine overload limit, time delay limit, machine spindle *surge* time delay, RPM, vibration limits, torque, and various other machinability material parameters. These and other parameters are monitored in a real-time mode by the controlling computer.

The ability to adapt cutting tool conditions in a matter of milliseconds and change cutting speed and feed to optimize tool cutting increases productivity. The wear and breakage of tools are minimized. The time the tools are "cutting air" (the space between two surfaces to be cut) is reduced by increasing the feed until a new cutting surface appears. Corrections for variations in hardness or machinability of materials are carried out.

Integrated circuit technology offers sensors for vibration, torque, voltage, and amperage that become part of the tool itself and provide data for control of processes that were not available a few years ago.

Integrated Material-Handling Control

The integration of material-handling systems with the production facilities is a reality with the application of computer control. Whether the materials are to be moved, lifted, transferred, or stored, computer techniques provide a vast improvement in the efficiency of material handling.

The reason for an integrated system is to gain control over a complicated and separated system. Efficiency is increased as planned control is realized. The computer delivers this control, efficiency, and accuracy to the production plan. Instruments that detect jams, count articles, read bar codes, sort materials, sense changes, inspect products, provide routing, stop conveyors, limit motion, trip alarms, position workpieces, provide time intervals, or start equipment are all sensing instruments. These devices control and monitor the functions of integrated material handling with the aid of the computer.

Automated Storage and Retrieval System

With an interactive data facility, an automated storage and retrieval system becomes a reality. An automated storage and retrieval system incorporates high-rise storage cells that are loaded and inventoried with the use of a robot travelling in narrow interfacing aisles, Figure 12-2. The raw material and components for products, after leaving the storage and retrieval area, are transported within the plant, depending upon their configuration and weight, by such equipment as conveyors, monorails, or automated guided vehicles.

Figure 12-2 An automated storage and retrieval system *(Courtesy of Eaton-Kenway, Inc.)*

Automated miniload storage and retrieval systems are widely used in manufacturing systems where the parts produced or components consumed can be moved in tote trays. Electronic and computer manufacturers employ these systems of material handling because the parts are light-weight, and a large variety of parts are required. The automated storage facility is interlinked with the manufacturing facilities by means on many automated roller conveyors. Computer and computer-memory disc-drive manufacturers use a highly elevated section of their conveyor systems as a test (burn-in) area.

Monorail and Automated Guided Vehicles

Automated monorails frequently are suspended from the workarea's ceiling and serpentine from workstation to workstation. Because the monorail system is elevated, it saves floor space. The system's individually controlled carriers can be started, stopped, switched, and rerouted independently to provide the material to its processing or assembly equipment. It can be integrated into the total system by interfacing with the computer.

Automated guided vehicles are used in flexible manufacturing and computer integrated manufacturing systems because they are able to get heavy and awkward parts and subassemblies to the correct place at the right time. Their function is to transport large workpieces and fixtures to machine centers, Figure 12-3. The carts

Figure 12-3 A typical automated guided vehicle and point-to-point path *(Courtesy of Eaton-Kenway Inc.)*

can transport heavy workpieces mounted on machine-tool pallets and transfer them directly to in-line machine ways. They can rotate a pallet into a working position on a machine. Frequently the AGV is designed to follow a single wire imbedded in the concrete plant floor. Other methods of guidance employ lasers and reflectors mounted in crucial positions, computer controlled dead reckoning, ultrasonic vision, and—the latest—direct computer control.

Integrated Inspection Control

The application of the computer to measure products under manufacture and report these measurements to the system's minicomputer will generate a real-time measuring system. The inspection data in the computer is compared to the data being measured, and corrective signals are returned to the manufacturing machines. The machines are all using the same data base and therefore an integrated network is established.

Coordinate Measuring Machine (CMM)

The coordinate measuring machine is a rigid and stable measuring device. It is a stress-relieved steel structure with a granite base. The moveable parts of the machine are constructed with air bearings for smooth and error reducing movements. The machine is essentially a means for moving a probe or part within a polar or rectilinear X, Y, Z coordinate system. The machine accurately records the spatial coordinate positions of the probe on the workpiece at each location. The machine is controlled by programmable software that automatically interpolates the measuring probe tip's position during the measurement. The program regulates the probe's attitude or length and corrects each axis for inaccuracies in pitch, roll, yaw, scale, straightness, and squareness to the other axes. The physical designs of these machines are varied to provide the best geometry for the parts being inspected. The configurations of the machines are based upon the same principles as machine tools. They employ the construction concepts of the cantilever, bridge, column, horizontal arm, and gantry, Figure 12-4.

Programming the Coordinate Measuring Machine - The availability of commercial software for coordinate measuring machines delivers great power and flexibility to these machines. These programs can be put into action with a terminal keyboard command. Commercial measuring programs are available for nearly all the common geometric characteristics of figures and alignments as well as the geometric tolerance elements. They measure the elements of form and position as well as tolerances of flatness, straightness, roundness, cylindricity, and conicity. They also supply the attitude tolerances for parallelism, squareness, and angularity. The operation of the CMM data can be accessed through a manufacturing shop floor PC terminal that provides flexible correction and control.

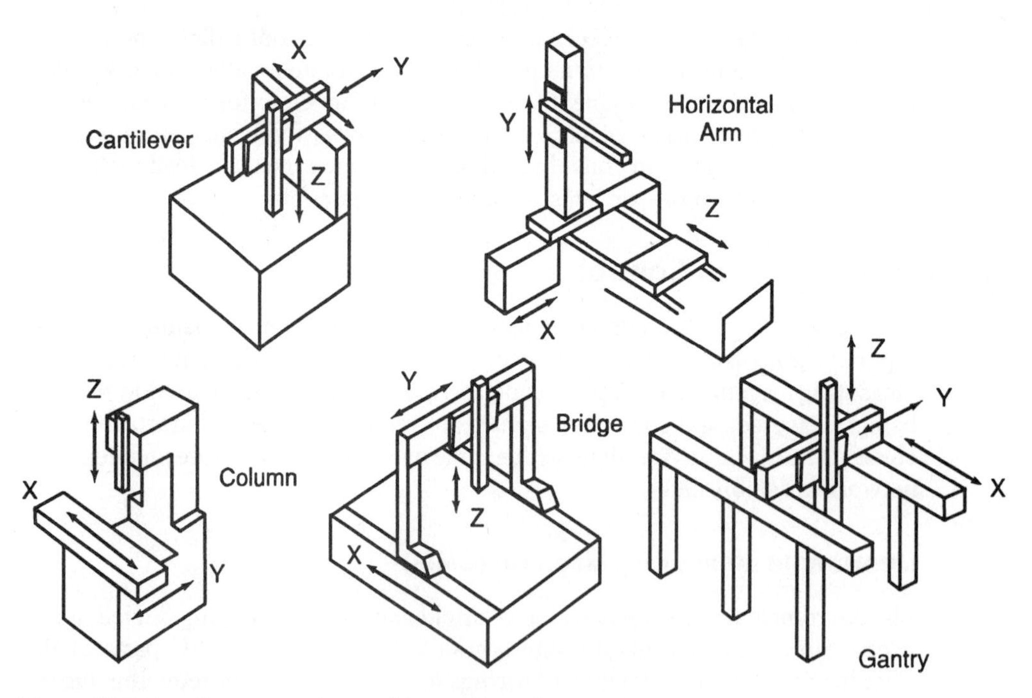

Figure 12-4 Coordinate measuring machine configurations

Figure 12-5 Coordinate measuring machine *(Courtesy of Kearney & Trecker)*

The integration of inspection modules is a critical design parameter in manufacturing systems. The coordinate measuring machine is the most flexible form of automated inspection, Figure 12-5.

Computer-Aided Quality Assurance

In computer integrated manufacturing, a communication system with a data base containing part geometry files is required. With the link between this primary data source and the manufacturing floor, a computer-aided quality assurance system can provide feedback to manufacturing process control. The primary instrument of quality assurance is the coordinate measuring machine.

The CMM takes the part geometry data from the data base and converts it into Cartesian, polar, and in special cases, spherical coordinate systems. The probe examines the corresponding point on the workpiece and the CMM calculates the deviation of the measurement from the nominal dimension stored in the data base. A position that is out of tolerance is flagged for corrective action. Software provides a means for evaluating these geometric tolerance conditions. When inspection is augmented with commercially available software programs that are entered into the CMM system, complex and detailed measurements can be quickly and accurately provided.

Figure 12-6 Video-measuring system Q-See 400 *(Courtesy of Optical Gaging Products Inc.)*

Electro-optical Inspection - Electro-optical inspection systems provide a noncontact system measurement and inspection. Small precision parts that are too fragile to be touched by a probe are examined by a laser beam or a machine vision camera. In some systems, the microprocessor digitizes a vidicon camera picture. The CMM locates the feature to be measured, and measures it relative to a master grid. The grid information is compared with the nominal dimension for analysis, Figure 12-6.

Trend Analysis - The computer-aided quality assurance system provides feedback to the processing machines. The data-process trend-analysis capabilities of the coordinate measuring machine provide information for keeping the manufacturing systems under control. The trend-analysis data is sent to the mainframe computer for the generation of management reports. This requires a computer link and hardware that interact with all of the company's other organizations. Management, engineering, planning, designing, production, marketing, and outside suppliers use this trend-analysis data.

Data Base Management Systems

Computer aided design systems have changed the design process by exponentially expanding the data base holdings. That expansion requires an organized method of management for data storage and control. The difficulty of keeping track of computer aided design and computer aided manufacturing information that is revised or passed from one computer to another (or from one network to another) has become a problem. These are functions of the data base management systems (DBMS).

These management programs are developed to do the job of data base configuration management. This is the result of the vast amount of information associated with engineering and manufacturing applications. These data management systems' software provides data management and data process management. For example, the data management system allows engineers working on different projects to use a common data file and it protects files from unauthorized updates. The data process management prescribes design methodologies, provides tools and data files, and controls access to files. It also maintains an audit trail of the products in the system.

Large mainframe computer manufacturers and private companies lead in providing data base management programs to handle data in a format that links common networks that interrelate. In integrating factory operations, hierarchical networks handle local data needs with interdepartmental data movement interrelated to other networks by data base management software, Figure 12-7.

Computer integrated manufacturing systems are aided with DBMS by controlling engineering change orders and having them automatically move through the system. The system can provide security so that only authorized users can have the data available. Customers or vendors can be linked to this system to communicate electronically with the company for product specifications or interdependent designs.

Interchangeable Modular Components for Machinery

Manufacturing machines have evolved from individual machine tools and assembly machines dedicated to a single product. When a new product or major model change was presented, the manufacturing facility for the production of the new product required the rebuilding of the production system. A system developed gradually whereby the tool sections were modularized.

Units Assembled into Different Configurations

Modules are self-contained functioning parts of a tool system that can be assembled in a number of configurations. This is the concept of standardized building-block construction. This design allowed a modular component such as a drill unit to be placed within a manufacturing machine in almost any position required by the fixture and workpiece. These modules can be mounted vertically, horizontally, or at any angle necessary to perform drilling, reaming, counter boring, spot facing, and other rotary tool operations.

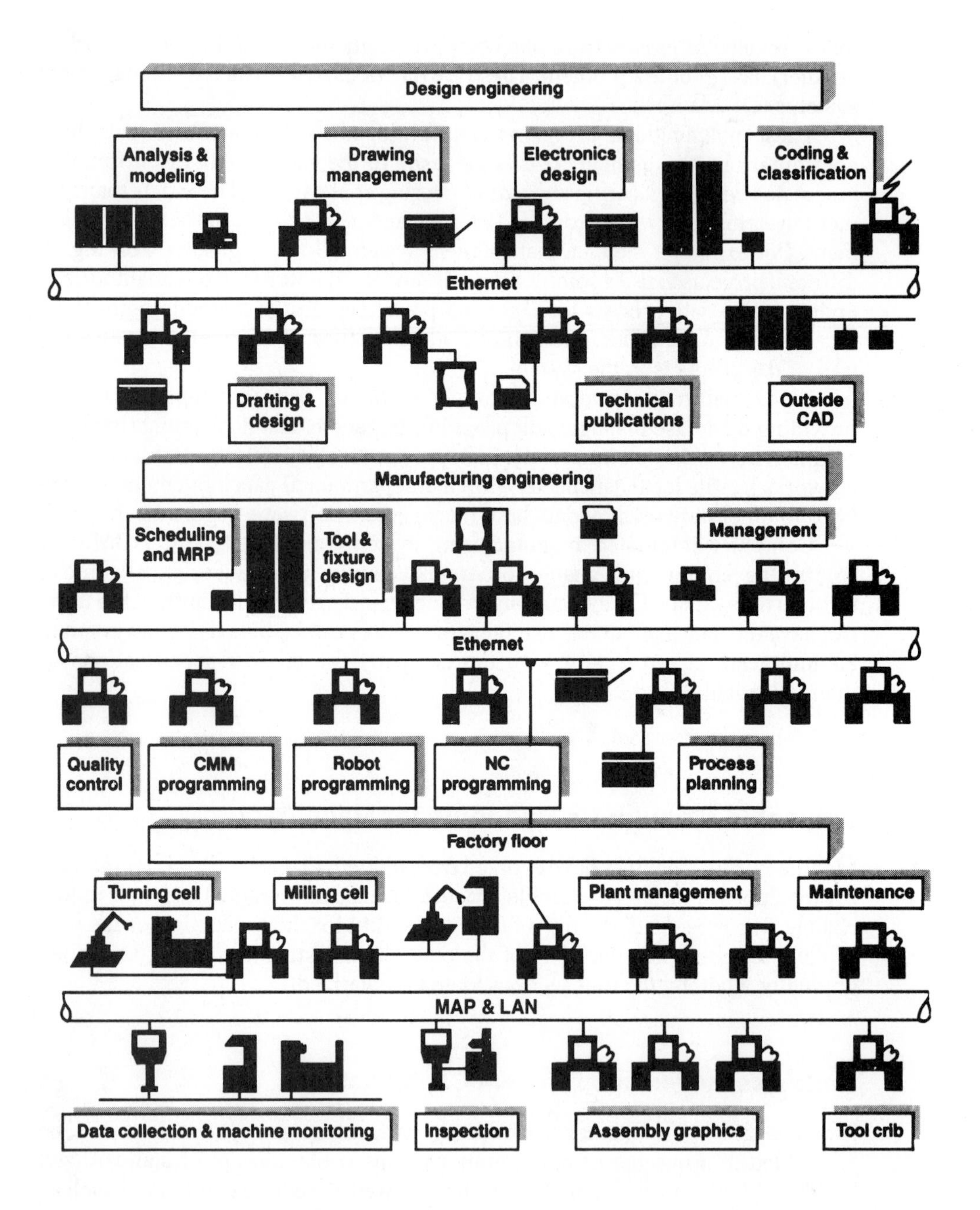

Figure 12-7 Hierarchical communication in computer integrated manufacturing provides a data base management system

Figure 12-8 A modular computer controlled manufacturing system—guards removed *(Courtesy of Kingsbury Machine Tool Corporation)*

Units Build Different Machines - *Modular tools* are designed that can perform all the machine functions, thus providing the opportunity to reconfigure the modular units into tools that can be applied to the manufacture of different products.

Modular machines are commonly applied to machine tool operations, forming operations, and welding operations. The concept can be adapted to almost any operation that is subjected to an end-of-batch machine change.

With the request of greater flexibility for multiaxis machines in the transfer machines, Figure 12-8, the machine builders have included computer numerical control to provide reprogrammability. Flexible controls require CNC, a minicomputer, or an industrialized personal computer.

Software Systems

The primary functions in a manufacturing system are to control, sequence, and change or correct the parameters of that production system. These functions are performed most efficiently with software.

Machine Control Systems

Programmable logic controllers (PLCs) have gradually replaced the relay switching controls of sequencing and timing in machine tools and transfer machines of mass production. Increasing the memory capacity of the programmable controller has increased and broadened its applications. The microprocessor chips that supply the memory capacity have been so sufficiently reduced in cost that they are economically practical for use in instrumentation. This improvement of the controller has broadened its responsibilities to include data acquisition and storage, arithmetic calculations, and signal conversion. They network with other systems. With the application of microprocessor technology, the physical size of the equipment is reduced. The programmable controller is placed near the production machines, Figure 12-9.

A reason for the programmable controllers' success is that they are designed to function in the industrial environments of temperature variations, electronic noise, vibration, and rugged use within manufacturing. Another reason for success is the ease with which programs can be modified or changed. A portion of the program can be modified in a real-time mode and the corrections in the process observed directly, providing convincing confidence as to the control adjustment.

The PLC is utilized where repetitive, regular, and sequential work is found, as in long-running processes. These controllers are employed in transfer lines and industrial operations of other mass produced products. They are available for networking and can provide color graphics at the station's terminal.

Figure 12-9 Programmable controller monitors robot assembly of stepping devices for household appliances *(Courtesy of Siemans Capital Corporation)*

Sensor Reprogrammed in Real Time to New Parameters - Programmable controllers are not subject to wear as relay control panels are. Thus they are reliable and cost effective. The PLCs can be programmed by electricians and technicians that are familiar with relay-ladder language. These controls supply color graphics of the process under control and diagnostics for troubleshooting. In addition, they also can include repair manuals, tool and preventative maintenance charts, and part inventory for equipment within the PLC's memory. The programmable controller offers centralized shop-floor control of automated manufacturing cells and systems, Figure 12-10.

Requirements for Sensor Changes - A change in the raw materials entering the production system (such as shape, composition, position, location, size, or consistency) requires an adjustment in the calibration of the sensors. Software reprogramming can provide the means of making these sensor changes. This flexibility, provided by real-time programming, results in greater product reliability.

Multiple Sensor Integration - In addition to product and production change software, programming is required to receive and integrate multiple sensor instrumentation. Sensor information can be received from different machines with a family of sensors that determine such information as position and

Figure 12-10 Programmable controller *(Courtesy of Cincinnati Milacron)*

Figure 12-11 Gripper head sensing switch indicates robot gripper position *(Courtesy of I.S.I. Manufacturing Inc.)*

orientation of the cutting tool, robot positions and gripper sensing, and the orientation of workpieces entering the feed system of the machine. Tactile sensing will provide control over the force or pressure necessary to hold the workpiece. This sensing prevents the slippage of workpieces during manipulation and transportation, Figure 12-11. The newest field is vision sensing, used to recognize and select products that are workpieces for a particular workstation.

Proximity Data Acquisition

Noncontact sensing provides proximity data acquisition. These sensors or switches deliver a signal without any moving parts or physical contact with the measured variable being detected. The proximity sensor can be used to provide data as a workpiece approaches a predetermined positioning or location. These sensors can be applied to a broad area of applications, from indicating a jetliner's flap positions to the location of a robot arm.

These instruments are designed by employing inductive, magnetic, ultrasonic, and capacitive principles. By including custom microprocessor chips in the sensor, it can measure time intervals between changes, determine rates at which objects are moving, and accumulate counting data of products. They can also send updates to a monitoring computer.

Data Trend Information

In addition to sensor on/off control, the proximity sensors are capable of measuring time intervals between movements of parts or equipment that provide rate data. This makes trend data available. The trend data can be used to provide an indication of a workpiece or equipment as it approaches the desired position. By having trend data available as an object approaches, deceleration corrections can be activated. Information relative to approaching a manufacturing position is important in the controlled movement of heavy products. This deceleration-rate data becomes very important as heavy transported workpieces approach a holding fixture. Without proper deceleration, the workpiece will crash into the fixture and damage the machine.

Changeover to New Products

With the changeover to different models or completely different products, the noncontact sensor does not become obsolete. If its responsibility was to perform such tasks as to locate holes, position robot arms, monitor tool wear, or fill cartons, the sensor can be readjusted to perform a similar function on the next product. Remounting, orientating, locating, and recalibrating may be required for it to become part of a functioning system.

With the application of computer aided manufacturing and robotics, these proximity sensors need to become more flexible. When coupled with the computer, trend data can be calculated in real time and control signals instantly will respond to the conditions at the point of production. At this diagnostic and correction level of sophistication in manufacturing, it will still require a supervisor to oversee the operation of these machine complexes.

Alternate Branch Sequencing and Programming

Computer control in an integrated manufacturing factory delivers an overall view of the manufacturing process. The present operations are displayed and monitored. In the event a difficulty occurs in the manufacturing process, the display can provide an alarm or warning that a problem has arisen.

An Optimal Selection

Computer-based controllers display operating data and fault flags or messages to the control CRT. With the examination of the production problem, the operator can question the computer terminal as to what possible alternatives are available. These various options are evaluated and the best solution to restore productivity to the manufacturing system is selected.

Study of the Slow Process or Machine

In a manufacturing plant, failures of machines or necessity of repairs are inevitable at some time. These disruptions result in slowdowns. A strategy often employed is to work around the machine or process that is causing the slowdown. In some cases, a control parameter such as motor speed may need adjusting. Or a minor program change may clear a shutdown interlock and reset the process.

In serious long-term problems, the system may need to be studied and a procedure worked out to resequence the operations. These program modifications may provide relief until new capacity or machines can be added to solve the problem. The essential information needed for the machine or systems diagnosis comes from the instrumentation. This instrumentation consists of sensors at critical points that report to the controlling computer.

Part and Work Progress Reporting - In computerized and monitored electronic assembly plants, thousands of parts are obtained, distributed, and assembled rapidly and efficiently. The processes include automated insertion, lead trimming, soldering, cleaning, test, and inspection, Figure 12-12. These parts are often tiny

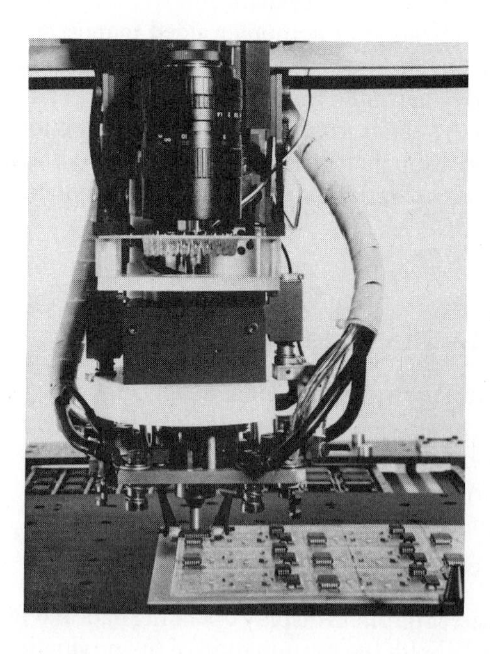

Figure 12-12 Automatic insert machine placement head—picking for placing surface mountings *(Courtesy of EPE Technology)*

and fragile. They are monitored and tracked so that no delay occurs at any point in production. The computer system can quickly provide the information necessary for the best utilization of materials, equipment, and personnel. The supervising engineer can also use the data acquired in the assembly operations to provide a standard work-time data base. The system can be used to monitor each workarea with the objective of switching personnel or equipment to areas where shortages may have occurred.

Schedule Comparison and Work Progress - The microcomputer is employed as part of a data acquisition link. On electronic component assembly lines sensors perform counts, tests, and inspections on continuous operations. The data are displayed at the end of the workarea for the local supervisor and also in the production and planning control offices. The work progress is reviewed and compared to the planned schedule. If deviations exist, the individual units can be supervised and corrections made or alternative actions started.

Data about work progress for manufacturing is available at any time the information is desired. This can be used to determine if the work schedule is being maintained or where work interactions are causing a delay, Figure 12-13. The work progress data may provide feedback for future production planning and control.

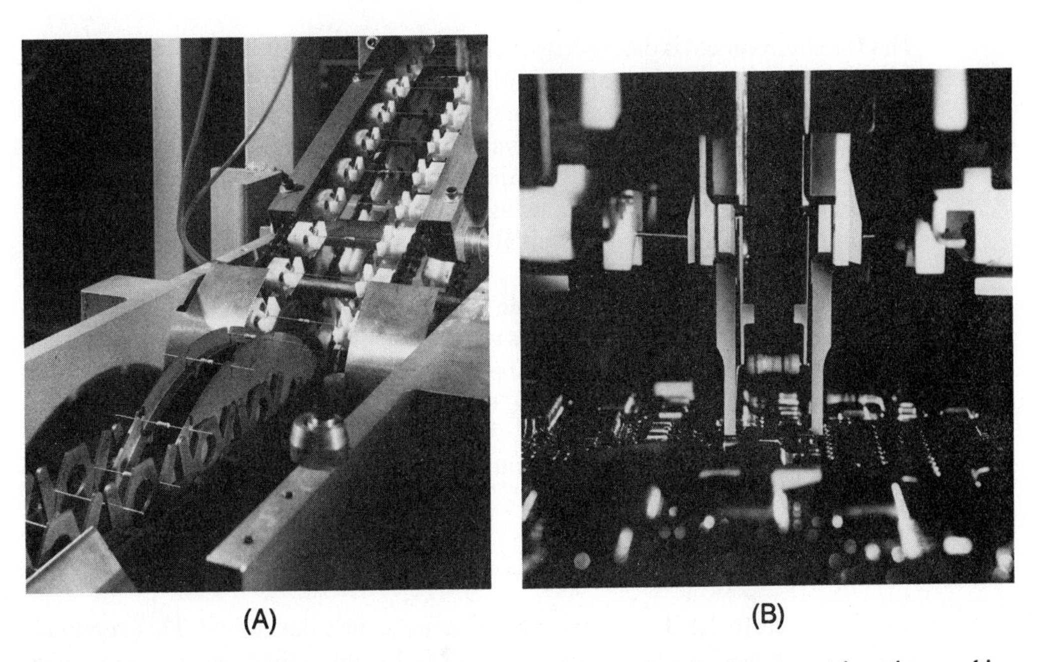

(A) (B)

Figure 12-13 (A) Electronic component sequencing and transfer; (B) Component inserting machine *(Courtesy of Universal Instruments Corporation)*

IDENTIFICATION OF INVENTORY AND STORAGE

In manufacturing, the inventory falls into different categories: raw materials, components, work in progress, and finished products. When products enter the manufacturing system, they require some form of accounting and storage until needed. These products represent an investment cost that should not be out of synchronization with the capacity of the manufacturing plant.

Production Schedule and Inventory Rationalized

The production schedule controls the required inventory and the necessary storage of the materials. Through the years, the difficult task of balancing inventory levels with their costs and production needs has provided any number of theories and plans. One such plan is designed to produce the desired unit in the quantities of units needed and at the time needed. This plan is known as "Just In Time" (JIT) production.

With the application of computer production data to inventory control, a system of inventory management and control has been offered to improve the problem of material flow.

Just in Time Inventory Control

The JIT environment is designed to reduce the amount of inventory that is required for manufacturing by providing a rationalized inventory program. JIT intends to deliver materials and parts to the manufacturing location at the time that they are needed. JIT's object is to reduce waste. Anything that adds cost or does not increase the product's value is challenged. Work-in-progress, scrap, rework, inspection, work orders, warehousing, and in-plant storage are to be kept to a minimum. These objectives are not always evident in existing inventory control systems.

One type of JIT software module functions with a post-deduct, production-reporting system. This system allows workers on the plant floor to count parts and production at either the end of the line or at intermediate points of manufacture. At a predetermined consumed level, the computer signals a part's report, and a replenishment request is triggered. This system permits an overview of the inventory movement within the manufacturing facility for the management and control of production, Figure 12-14.

Continuous Flow Manufacturing (CFM)

Improvements to the JIT operation have been recently developed. This version is called continuous flow manufacturing (CFM). In this application nonvalue added activities are eliminated. Items such as receiving raw material, inspection, putting

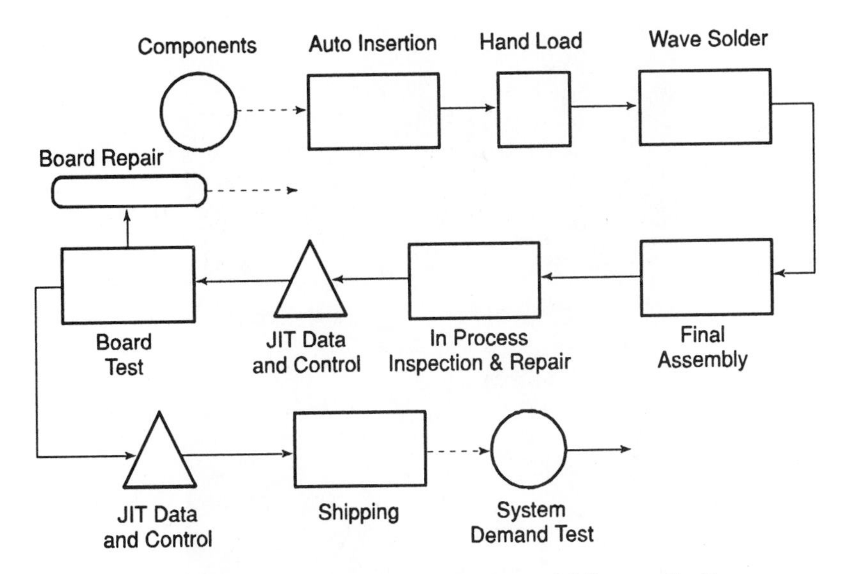

Figure 12-14 Block diagram of a just in time circuit board material flow application

into stock, issue to production, stock on line, packaging, and shipping are incorporated into the production line process. This requires an extensive educational program for the work force so that they manage their own quality program. The goal is for the work force to manage their own jobs properly and deliver a zero defect system.

Physical Plant Records and Inventory

Tools, machines, and supplies represent a large financial investment. In the management of these resources, an organization is required to prevent loss, damage, or theft. The instrument that has developed into a successful deterrent to loss is the computer.

Tool Management

Production tools, cutting tools, specialized tooling, jigs and fixtures, gages, control panels, instruments, calibration measuring and testing tools, as well as numerous hand tools require a monitoring and checking system. A computer aided tool-inventory system provides a real-time check for all tools under its control, Figure 12-15. The system may be operated from a traditional tool crib, a tool magazine at a workstation, or an external tool center with an automated guided vehicle tool-transportation system or other delivery systems. The computer terminal presents a tool data system containing dimensional and tool holder data as well as other pertinent information on the inventory of each tool. The physical

Figure 12-15 Tool management gantry robot *(Courtesy of Fibro Inc.)*

condition of the tool (such as dull, broken, damaged, or missing) can be noted so the computer program can issue a purchase order for replacement.

Machine Inventories

Large production machines are not frequently removed from the premises, but they do require servicing. A service record is kept so that they are maintained in the best working order possible. The machines need complete records, such as the date they entered service, various serial numbers and model numbers, and the date and by whom they received servicing. In addition, service repair manuals and special service center telephone numbers for diagnostic services and machine repair parts can be entered into the computer.

Special Materials

In manufacturing inventory, special care is required of certain materials used in the production processes. Items such as precious metal, dangerous chemicals, nuclear radioactive material, and dangerous genetic materials are to be inventoried and safety protection provided. The computer can list, acknowledge, and log all materials requiring special handing.

Packaging and Shipping

In computer integrated manufacturing, the area of packaging and preparing for shipping is highly automated. The finished products are delivered to the packaging area on conveyors or pallets to be readied for processing for shipment.

Selection of Carrier

The selection of the carrier or shipping method rests on a number of factors. The size and amount of the products being shipped, the distance to be shipped, the fragility of the product being shipped, the time available for travelling, the reliability of the carrier in meeting the delivery schedule, and the cost of the shipment are all factors to be considered. Rough handling of a shipment, causing damage to the product on delivery, cannot be tolerated. There may be special requirements in addition to the ones listed above, such as temperature sensitivity. Some products can be exposed to only certain temperature ranges during shipment. Temperatures above or below certain points may render the product unserviceable. The requirements involved in the selection and scheduling of the carrier rely on the computer.

The common bulk carriers are the truck, train, plane, and ship. The carrier selected will depend upon the characteristics of the product being transported. The carrier available for most deliveries is the truck. Trucks are used for both short and long hauls. They can be used to pick up the product and deliver it to another carrier. Trucks are very versatile in the loads they can transport. In some cases, special trucks are built to deliver unique products—bulk-steel-hauling trucks are an example. A chief advantage of the truck is that its scheduling is very flexible, and all that is needed is a primitive road. The other bulk carriers are important but they lack the flexibility of the truck.

Selection of Packaging for Shipping

Shipping containers have been the subject of much research in the last few years. The application of polymer materials to shipping containers has made remarkable improvements. Polymer products are applied in many creative ways in packaging materials: as sheets for moisture and vapor proofing, with air bubbles for shock and thermal resistance; foam for form fitting shock and thermal protection; and popcorn-shaped kernels to settle around irregular shaped fragile objects and instruments. These types of materials are usually housed with a paper carton that is wrapped with a plastic sheet as a moisture barrier. For overseas shipment, items are carefully orientated so that the labels are facing the outside and stacked on a pallet. The whole load is placed on a turntable and the cartons on the pallet are wrapped with a continuous sheet of plastic.

Smaller items may be placed in different types of envelopes, ranging from insulated bags, sealed plastic envelopes, and tubes. Small, very fragile parts can be shipped by placing their connectors or projections into foam plastic blocks and shipping the component and block as one piece.

Very heavy and large items are mounted on wooden skids and a crate is built around the product for shipping. The product may have parts waterproofed with Cosmoline or covered with grease and wrapped with paper held in place with a steel band. The outside of the crate can be covered with wood and reinforced with a series of steel bands around the total crate.

The selection of the type of packaging relates to the requirement of the product. It is necessary to protect the product so that the buyer will receive it without any damage. In some cases, the cost of the shipping container will be a major cost of the item.

Control of Packaging Machines

Often high-production products are packaged by machines. The products arrive in the packaging area on a belt, roller conveyor, or chute and sometimes are inspected. In the case of small items, the parts may be placed in kits and sealed in a plastic bag. Larger items are placed in container boxes with the necessary kits of small parts and documentation added to the box. The instrumentation for controlling the packaging processes are photo-optical devices that detect the presence of a part, count parts, inspect the parts for some characteristics, position the parts for packaging, or sort the packages according to destination. These devices are also employed to start and stop the equipment.

Filling machines are employed to package powders, liquids, granules, or solid products in plastic bags, wrappers, tubs, bottles, tubes, barrels, cans, or boxes. The container depends upon the nature of the product being packaged. In each case, the product is either counted, timed, weighed, or the volume or level measured to meet the standard printed on the specifications or label.

Labeling

The computer provides the information for labeling most products because the data base in the accounting and billing process has generated the data from the original purchase order. The bill of lading, packing list, and labels are printed at the packaging and shipping department of the organization. Documents, such as assembly manuals and parts lists, are placed in their container and inserted into the box, carton, container, or crate for shipping.

At this time, any notice for special handling may be attached. Special products that must remain in vertical positions will have instructions like "This End Up" on the package surface, or "Instruments—Do Not Drop," etc.

Dispatching

Dispatching in shipping is concerned with the identification of the area where the product is to be picked up by the transportation company. The computer can supply the most direct and available transportation method for the product's destination and assign it to the factory's pick-up point. It may be the factory shipping dock or at another designated point in the local area. In each case, the product is moved to the pick-up area of the plant for the carrier.

THE AUTOMATED FACTORY

The automated factory is the result of combining the concepts of computer integrated manufacturing, flexible machining systems, manufacturing cells, computer-based robots, material-handling systems, and management systems.

The major corporate executive level responsibilities for the factory decisions are in the areas of manufacturing management, strategic planning, financing, and marketing. Manufacturing management and engineering have a responsibility to provide leadership for automation throughout the factory, including material handling, manufacturing processes, assembly, inspection, testing, and the physical plant.

Because of the extensive costs involved in the designing and construction of the automated factory, certain considerations should be clear in the minds of the total leadership of the corporation. Such considerations may be these: to study the philosophies and methods used by other successful organizations that have been leaders in automation use; to study and define corporate strategies and principles; to define a five to ten year plan of objectives and then to start the study of specific manufacturing goals. It is important to gain a holistic view of what the manufacturing operations will entail and then start with a first goal that will be successful with unquestioned support. Management should study the manufacturing system for activities whose values added are marginal or redundant. Operations that do not result in more or improved products should be minimized or eliminated. Management should study available information network systems and select a system with a standardized protocol that will respond to the total factory needs for a reasonable time in the future. Management should study the reactions of working people; they will provide the success of the manufacturing program. It is important to start their retraining as soon as possible and include them whenever possible in the planning and organization of objectives and goals.

Automated Factory Machine Development *

In the past, individual machine tools were considered as the basis for all machining operations. They had to be very general in their design to provide the necessary flexibility that was required in job shops. With the growth in production,

* Adapted from K.T. Swasey - *KT's World of Advanced Manufacturing Technologies*

flexibility in operator control and the machines were reduced. The high-volume dedicated transfer machines delivered the highest production output, but their flexibility or ability to provide products with different requirements was limited. So the high production resulted in a reduction in flexibility.

To gain back flexibility in manufacturing machines, the application of numerical control machines was introduced. The concept that a machine could be programmed to provide this flexibility resulted in a nearly universal machine tool that could be programmed to manufacture almost any shaped part and be economically manufactured. This innovation was of great importance to both small and large machining companies.

With the coupling of the computer directly to the machine control, another phase occurred in manufacturing. The application of the computer took different forms, such as Computer Numerical Control (CNC) and Direct Numerical Control (DNC). Computer numerical control added faster supervision and control to the machining processes. But the system did not deliver the information needed by supervisory personnel for the decision making required for computer integrated manufacturing (CIM). Direct numerical control (DNC) was designed to operate through a network to provide control for a number of machines with a central computer. The latter development was necessary for computer integrated manufacturing to become a reality.

Automated Factory Machining Modules

Modularization of machine tools is a concept for increasing flexibility in a high-production manufacturing system. Modules that have been developed for advanced manufacturing technologies include these: NC milling, turning, machining center, manufacturing center, head indexer, head changer, and others.

Numerical Control Milling

The numerical control milling machine is frequently an individual machine that is controlled by a magnetic tape or more commonly by a one-inch mylar or paper tape. The codes necessary for movement of the various axes and functions are punched into the tape and are read by a machine control unit. These machines can be reprogrammed by changing the tape by punching a new command onto the new tape. NC is very successful because of its accuracy and its ability to exactly repeat the same program many times.

The Turning Module

Under supervisory computer control, the use of the turning module on palletized workpieces requires a different application from that of traditional lathe turning. When the work remains on a pallet, the module may take one of two forms—either that of pallet rotation or that of tool rotation. A pallet with a workpiece mounted can be moved into a moving cutter head that will deliver lathe-type turned surfaces, Figure 12-16.

By keeping the workpiece on the pallets, a number of savings are realized. There is an elimination of setup, a reduction of labor because of the lack of loading and unloading functions, workpiece integrity between machines and after the first piece, and setup time is eliminated. In addition, quality is monitored because of the application of a supervisory computer control system.

The Boring Module

Boring operations are needed to produce holes of various sizes that must be located correctly and to an accurate size. Some of the holes may be large and also require a considerable reach as well as being concentric with the center line. The palletized workpiece is moved into position and the boring head moves forward and machines the proper holes and surfaces to the required geometrical interrelationships. The boring module is also shown in Figure 12-16.

Figure 12-16 A boring module is on the left and a turning module is on the right (a facing operation). *(Courtesy of Kearney & Trecker Corporation, A Cross & Trecker Company)*

The Standard Machining Center Module

The standard machining center provides part manufacturing with a flexibility of products but lower volume. These are often the requirements of the companies' customers. This module frequently is the main module for the machining of medium-size workpieces. A vast number and configuration of parts are produced in a machining center, Figure 12-17.

The Manufacturing Center Module

The manufacturing module is similar to the machining center, but it is designed for higher production with continuous operation and little operator intervention. The *manufacturing center module* has in addition a multipallet storage magazine. This provides five additional stations for workpieces, with one on the shuttle unit. Six parts can be machined in uninterrupted sequence. The internal shuttle mechanism removes a completed workpiece from the machining area and replaces it with the next part so that machining can be resumed immediately. The automatic sensing and control device is a spindle probe with a tool post that delivers a number of indicating functions. The sensing and control device synchronizes the machine and workpiece automatically. The sensing and control device automatically resets the machine to home position, finds the part before starting the program, and references the machine to a previously machined surface

Figure 12-17 Machining center modules serviced by wire-guided vehicles *(Courtesy of Kearney & Trecker Corporation, A Cross & Trecker Company)*

or target point on the fixture. In addition, it performs in-process inspection by taking dimension data off the part by the sensor and storing it in the computer memory and/or displaying the information. The system will also sense for broken tools or cutting edges and provide a warning if a tool is damaged. The manufacturing center modules utilize adaptive control which senses any variations in cutting conditions and adjusts the feedrate and spindle speed to maintain proper constant load on the cutter, Figure 12-18.

The Head Changer Module

To increase productivity in a manufacturing system, a multiple spindle technology is added to a series of heads. The head changer module is available in a number of sizes and utilizes the concept of interchangeable multiple spindle heads. The various heads relate to the special family of work parts to be machined. The

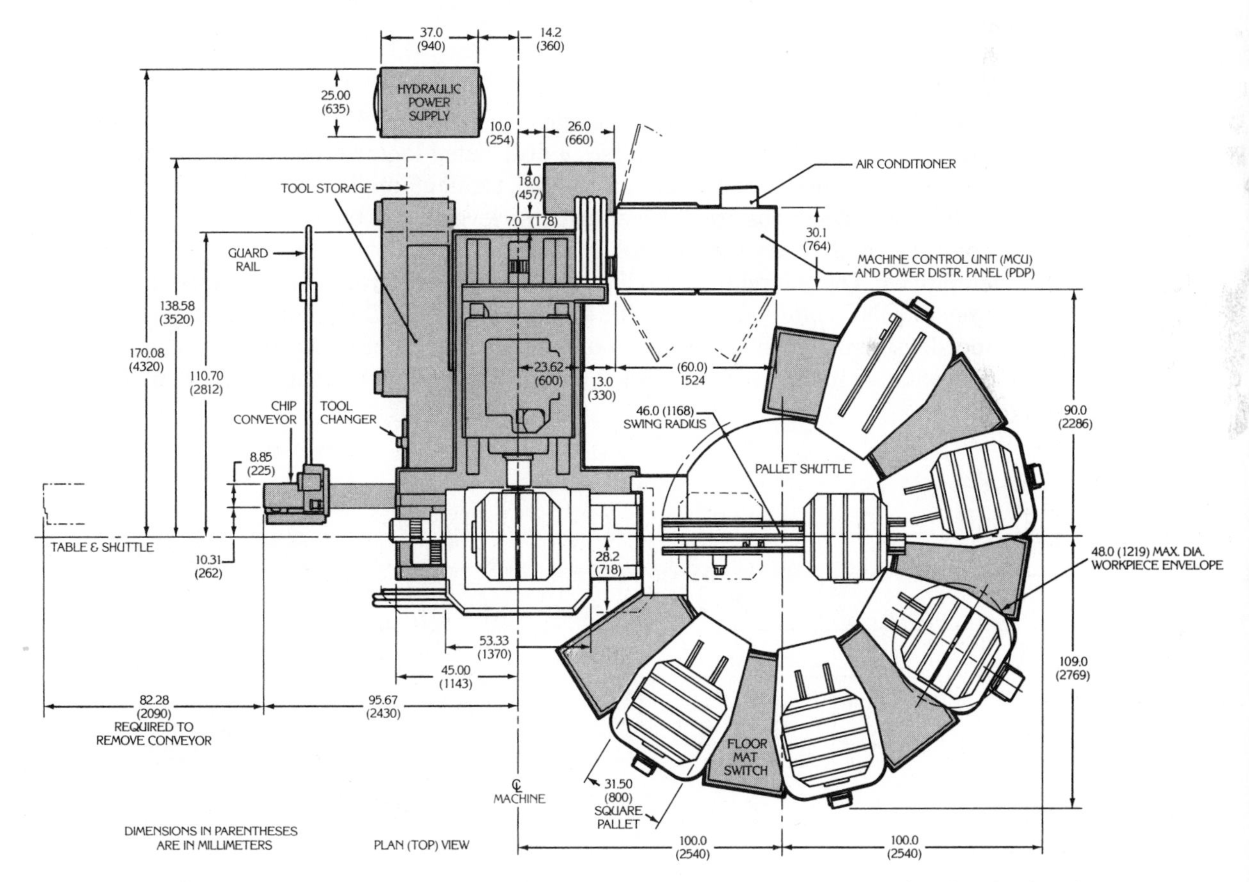

Figure 12-18 A Milwaukee-matic manufacturing center *(Courtesy of Kearney & Trecker Corporation, A Cross & Trecker Company)*

spindles of the heads are available in a number of different configurations. These interchangeable multiple spindle heads are stored and selected as required for the jobs being machined. The typical work performed is drilling, tapping, boring, reaming, or special requirements. The head changeover in many cases is accomplished while the machine is cycling, thus increasing the machine's machining time, Figure 12-19.

The Head Indexer Module

The *head indexer* is a module of a high-production tool. They are efficient metal removing devices because they can provide simultaneous machining. Eight to twelve heads can be mounted to the index table. The head indexers are normally tooled for processing specific families of parts and will accommodate a wide range of applications for workpiece size and operations. The heads can range in size from thirty to sixty inches.

The Control Module

A supervisory computer control module can be designed with different philosophies of control. One system utilizes a distributed logic architecture. This logic enables three modes of operation: manual, semiautomatic, and fully automatic. In this case, decisions are made at machine level and at the supervisory level of control. To activate the control module, a series of software modules are employed. These modules provide such software controls as an operator menu system, DNC communications, file use logging, CAD/CAM computer interface, tool data management, automatic tool setter interface, operator procedure display, preventative maintenance manager, station production queuing, load balancing,

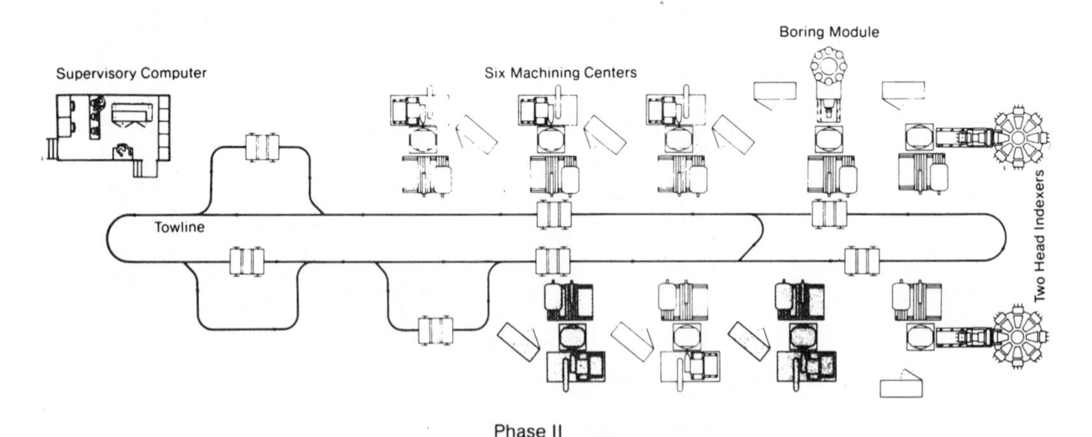

Figure 12-19 The layout of a seven-station random FMS system with the illustration of two head indexers *(Courtesy of Kearney & Trecker Corporation, A Cross & Trecker Company)*

shop-floor graphics display, basic reports, and many other planning and control options, Figure 12-20.

Factory Software

The realization of computer integrated manufacturing is the interaction between production, material handling, and management. The key to this interaction has come to be known as software. Computer integrated manufacturing requires software programs that manage all types of manufacturing processes that include machining, assembly, and inspection. They also manage the production variables for the instrumentation of chemical, electrical, and mechanical controls.

In addition, control management needs software programs that provide scheduling, routing, material handling, automatic tool delivery, station control, fixture management, part serialization, instrumentation and many others.

Software programs are also required for factory planning. Capacity planning programs include such items as these: calendar manager, MPR II, production simulator, production plan implementer, planning reports, and others.

Figure 12-20 A supervisory computer control system control module and a shop data entry unit *(Courtesy of Kearney & Trecker Corporation, A Cross & Trecker Company)*

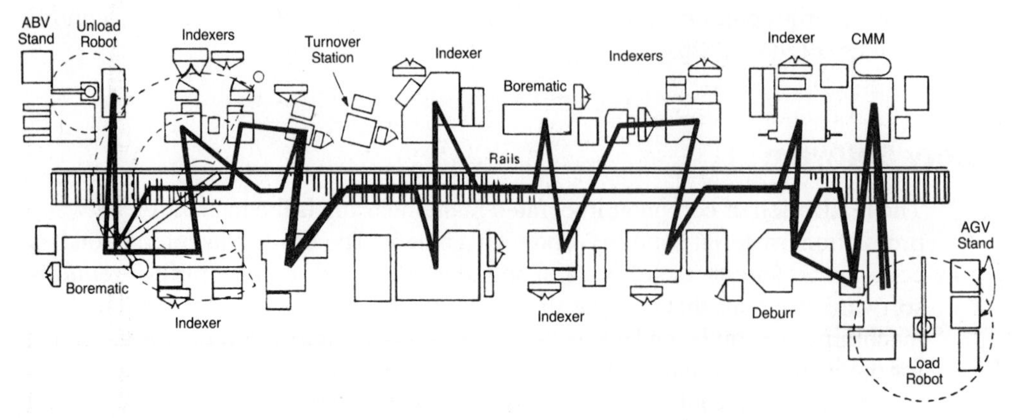

Figure 12-21 Work movement through a computer integrated manufacturing unit attended by robots, AGVs and CMM

As indicated, software programs are very important in the integration of manufacturing and require intensive planning and implementation so that a factory may be totally integrated. A vast amount of truly creative programming work is required to integrate the activities of the shop floor, Figure 12-21.

Automated Factory Instrumentation

Instrumentation in the automated factory is built around the chip microprocessor that contains solid-state sensors and provides a conditioned signal directly to the factory's computer. The computer is the focal instrument in the total control system that accumulates data, makes control decisions, and delivers integrated control to the production processes.

The automated factory is the application of a computer integrated manufacturing system designed for batch manufacturing. Production involves the mixing of materials and chemicals, the distilling of products, and the weaving of textiles. In the metal manufacturing industries, the target of the integrated automated cells will be on the performance of the requirements of casting, forging, milling, turning, drilling, punching, forming, inspection, finishing, heat treating, assembly, and testing.

Production Control

In the factory, the identification and tracking of all moveable machines, automated guided vehicles, pallets, materials, parts, tools, and fixtures are recorded. These items are inventoried, programmed, and controlled by the factory computer. The system employs a binary bar code that can be read with sensors and data supplied to the computer for item identification, classification, and control.

Bar Code Readers - The bar code reader contains a photosensor that reads light and dark bars. The sensor reads patterns of changing light and dark as reflected from a laser beam of the bar code back to the sensor. The light and dark pattern spots become ones and zeros that are buffered into the computer's input port. With the aid of a complicated program, the computer provides alphanumeric information for identification and further processing, whether it be a decision or printed information.

The Tool Center - The small tools and cutters to perform these forming and machining tasks have been kept in tool magazines at the machine center. In the newer automatic factories, the tools are to be kept, maintained, and inventoried in a central automated tool center. A tool transfer system is present to deliver the needed tool just in time to the correctly coded spindle, with the tool's dimensions set for operation—all under the control of the factory control computer. During the shift, a particular tool may be used in a number of different machines. The automated tool center measures all returned tools for wear by applying an automated computer coordinated measuring machine as the tool is returned to its automated storage position or placed in a service or repair storage area. A management report is generated from the coordinate measuring machine data indicating which tools need to be reground, replaced, or are not needed by the system. The automated tool center is serviced by a robot controlled by the factory control computer. All production machines are serviced by the tool control system, Figure 12-22.

Artifical Intelligence

One branch of artificial intelligence (AI) is concerned with the designing and development of automated systems that deliver intelligent processes, products, and actions. Intelligent outcomes are related to human cognitive processes such as written language, speech, perception, vision, the ability to manipulate symbols, and the ability to learn. When these characteristics are applied to manufacturing problems, the science is called "expert systems" (ES).

Expert Systems

Vast amounts of technical knowledge and experience about the manufacturing processes and machinery are lost when a pivotal craftsman or technician retires form the workforce. To reduce this loss of expertise, progressive industries have set about developing a system whereby this resource may be preserved and made available for continued use at a specific workstation or system. In some areas this process may be referred to as "knowledge engineering" (KE).

Figure 12-22 (A) A storage and retrieval tool center; (B) The transportation by AGV to machine tool *(Courtesy of Kearney & Trecker Corporation, A Cross & Trecker Company)*

An example of an expert system would be the providing of a data base that contains the acquired knowledge and experiences of a senior individual performing a specific task. An example of such a task might be vibration analysis on a large manufacturing transfer machine, where the responsibility is to maintain by diagnosing, preventing, or fixing breakdowns of these massive rotating machine systems.

An expert system is a powerful computer-based program that simulates the thinking processes and experiences that a human expert would apply to solving a discrete problem. The expert system captures the knowledge, thinking processes, and procedures as well as the expert's rules for problem solving that had been gained through years of experience on a specific application of work. Those gained-experience rules are referred to as "heuristics."

Once an expert system specialist has interacted with the expert and has programmed his/her knowledge, experience, and thinking processes, the data is constructed into a series of rules for problem solving. These rules and data can be added to the expert system data base. In the future, a new employee may call up the computer data base on the specific machine or problem and receive tested and accurate instruction on the manufacturing system's machinery.

Instrumentation's Future

Within the growth and development of instrumentation and automation for manufacturing, unbelievable changes have taken place in the past score of years. It is interesting to contemplate whether applications of artificial intelligence, expert systems, and superconductors will provide advancements in instrumentation as did the transistor, integrated circuit, microprocessor, and—the Alethia** of all instrumentation—the computer.

SUMMARY/FACTS

- First considerations of computer integrated manufacturing include a study of the total manufacturing enterprise: strategic planning, financing, marketing, design and product engineering, manufacturing planning, and control of the production floor.
- Computer integrated manufacturing is more than a technical commitment to manufacturing. Rather, it is a philosophical dedication to manufacturing development and continued manufacturing improvement. This is a very long-term commitment to excellence required by all levels of corporate management, finance, and production.
- The computer integrated manufacturing system at the operational floor level links three areas of automation: group technology, robotics, and material handling.

** Alethia - The ancient Greek personification of truth

- Computer integrated manufacturing includes interaction of all the data and expertise of a total manufacturing organization. It makes use of communication to integrate all levels of the enterprise to support the plant floor operations.
- When machines built by different vendors are built to MAP and LAN standards, they provide the method and procedures for multilevel communication between the customer, computers, engineering, controllers, manufacturing systems, quality control, testing, workstations, and management for the total factory.
- The coordinate measuring machine is controlled by programmable software that automatically interpolates the measuring probe tip's position during the measurement. The program regulates the probe's attitude or length and corrects each axis for inaccuracies in pitch, roll, yaw, scale, straightness, and squareness to the other axes.
- A data base management system allows engineers working on different projects to use a common data file and protects files from unauthorized updates. Data process management prescribes design methodologies, provides tools and data files, and controls access to files. It also maintains an audit trail of the products in the system.
- Commercial CMM measuring programs are available for nearly all the common geometric characteristics of figures and alignments as well as the geometric tolerance elements. They measure flatness, straightness, roundness, cylindricity, and conicity. They also measure parallelism, squareness, and angularity.
- Electro-optical inspection systems provide a noncontact system measurement and inspection. Small precision parts that are too fragile to be touched by a probe are examined by a laser beam or a machine-vision camera. In some systems, the microprocessor digitizes a vidicon camera picture.
- The programmable logic controller (PLC) is utilized where repetitive, regular, and sequential work is found, as in long-running processes. These controllers are employed in transfer lines and industrial operations of other mass produced products. They are available for networking and can provide color graphics at the station's terminal.
- Computer-based controllers display operating data and fault flags or messages to the control CRT. With the examination of the production problem, the operator can question the computer terminal as to what possible alternatives are available. These various options are evaluated and the best solution to restore productivity to the manufacturing system is selected.
- Just in Time (JIT) inventory control intends to deliver materials and parts to the manufacturing location at the time that they are needed. JIT's object is to reduce waste. Anything that adds cost or does not increase the product's value is challenged. Work-in-progress, scrap, rework, inspection, work orders, warehousing, and in-plant storage are to be kept to a minimum.
- In manufacturing inventory, special care is required of certain materials used in the production processes. Items such as precious metal, dangerous chemicals,

nuclear radio-active material, and dangerous genetic materials are to be inventoried and safety protection provided. The computer can list, acknowledge, and log all materials requiring special handling.

- The instrumentation for controlling the packaging processes are photo-optical devices that detect the presence of a part, count parts, inspect the parts for some characteristics, position the parts for packaging, or sort the packages according to destination. These devices are also employed to start and stop the equipment.
- The automated factory is the result of combining the concepts of computer integrated manufacturing, flexible machining systems, manufacturing cells, computer-based robots, material-handling systems, and including management systems.
- Computer numerical control added faster supervision and control to the machining processes, but the system did not deliver the information needed by supervisory personnel for the decision making required for computer integrated manufacturing (CIM). Direct numerical control (DNC) was designed to operate through a network to provide control for a number of machines with a central computer. The latter development was necessary for computer integrated manufacturing to become a reality.
- The head indexer is a module of a high-production tool. They are efficient metal removing devices because they can provide simultaneous machining. Eight to twelve heads can be mounted to the index table. The head indexers are normally tooled for processing specific families of parts and will accommodate a wide range of applications for workpiece size and operations. The heads can range in size from thirty to sixty inches.
- The expert system captures the knowledge, thinking processes, and procedures as well as the expert's rules for problem solving that had been gained through years of experience on a specific application of work.

REVIEW QUESTIONS

1. What is the production objective of computer integrated manufacturing (CIM)?
2. Computer monitoring of a number of manufacturing cells produces interactive data that is very important because it allows for ______________ time decisions.
3. The use of the computer in adaptive control instrumentation makes data available that anticipates a potential change in machinability characteristics. What variables are measured to predict these changes?
4. High-rise storage cells in automated storage and retrieval systems can be loaded and inventoried by __________.
5. What system of automated manufacturing measurement provides automatic correction with the aid of the system's computer?
6. How is electro-optical inspection done?

7. The difficulty in keeping track of information that is revised or passed from one computer to another has been alleviated with the use of what system?
8. The concept of allowing a modular component to be placed within a manufacturing machine in almost any position required by the fixture and workpiece is known as what?
9. In what type of work are programmable logic controllers most often used?
10. How does a proximity sensor achieve data-trend information and why is that information important?
11. In the event that a difficulty occurs in the manufacturing process, the computer display can provide an alarm or warning that a problem has arisen. What procedures can be followed?
12. In computerized and monitored electronic assembly plants, the computer system can provide the data for a standard work-time data base. How does this increase productivity?
13. Why is the Just in Time (JIT) environment so effective?
14. What is the advantage of a computer-aided tool inventory system?
15. In machine inventories, it is necessary to show complete records as to the date they entered service, serial and model numbers, etc. It is also wise to include other information for repair. What are some of those records?
16. In packaging high-production large items, besides the necessary kits of small parts, another important thing to include in the packing is the ______________.
17. What were some of the reasons the described 1967 concept of a flexible manufacturing system did not find wide acceptance?
18. Automated factory instrumentation is built around the __________ that contains solid-state sensors and provides a conditioned signal directly to the factory's computer.
19. What is the role of the tool-transfer system in automated factory instrumentation?
20. What is the key that manages the interactions with all types of manufacturing processes as well as the control of corporate management planning and decision making?
21. What is one potential scientific application being considered in the future of instrumentation?

APPENDIX A
ACRONYMS

(ADC)—Analog-to-digital converter. An interface device that converts an analog signal into an equivalent digital signal.

(AGV)—Automated guided vehicle

(AI)—Artificial intelligence. A science of making machines intelligent in order to make them more useful and to understand intelligence.

(ALU)—Arithmetic logic unit. A part of the CPU that performs arithmetic and logic operations on input operands under control of the program instructions.

(ASCII)—American Standard Code for Information Interchange. A seven-bit binary code used to represent one hundred twenty-eight letters, numbers, punctuation symbols, and other similar characters.

(BCD)—Binary-coded decimal. A system where decimal digits are represented by the first ten equivalent codes using four binary bits in the 8421 system.

(CAD)—Computer-aided design

(CAD/CAM)—Computer aided design and Computer aided manufacturing

(CFM)—Continuous flow manufacturing

(CIM)—Computer integrated manufacturing

(CMM)—Coordinate measuring machine. A digital inspection tool that measures positions in three-dimensional, space employing software.

(CNC)—Computer numerical control. Computer control over individual machine tools.

(CPU)—Central processing unit. A microprocessor that does the computing.

(CRT)—Cathode ray tube. An electron beam device in a sealed glass envelope, commonly used as the primary display of an oscilloscope or terminal.

(DAC)—Digital-to-analog converter. Interface circuitry that converts a digital signal into a equivalent analog signal.

(DBMS)—Data base management system

(DIP)—Dual-in-line package. A semiconductor package with parallel rows of pins from the two lengthwise edges of the package.

(DNC)—Direct numerical control. Computer control over machine tools and materials handling systems.

(DTL)—Diode transistor logic

(EMF)—Electro motive force. Synonymous with millivolt output.

(EPROM) or **(EROM)**—A programmable, read-only memory. It can be erased by the exposure to ultraviolet light.

(ES)—Expert system. A rule-based system that performs a task that normally takes humans a long time to acquire.

(FET)—Field-effect transistor. A transistor consisting of a source, gate, and drain, whose action depends on the flow of majority carriers past the gate from the source to the drain. The flow is controlled by the transverse electric field under the gate.

(FMS)—Flexible manufacturing systems. A group of processing stations, usually NC machines, connected together by an automated work part handling system.

(HIC)—Hybrid integrated circuit. It consists of an assembly of one or more semiconductor devices and a thin-film integrated circuit on a single substrate, usually of ceramic.

(IC)—Integrated circuit. Electronic circuits containing thousands of devices on a semiconductor material.

(I/O)—Input-output. The section of the system that handles data communications between the CPU and external peripheral devices.

(JIT)—Just-in-time

(KE)—Knowledge engineer. A person who designs and builds expert systems.

(KV)—Kilovolts. A voltage measurement in 1000 volts.

(LAN)—Local area network

(LASER)—Light amplification by stimulated emission of radiation.

(LDR)—Light deoebdebt resistor. A solid state device that lowers its resistance when a high intensity light strikes it.

(LED)—Light-emitting-diode. An activitated diode that emits photons.

(LSI)—Large scale integration. A high-density, monolithic semiconductor integrated circuit on a single silicon chip containing more than the equivalent of 100 logic gates.

(LVDT)—Linear variable differential transformer. A transducer that produces a voltage as a function of core position.

(LVRT)—Linear variable reluctance transducer. A device that uses a variable reluctance to control the setting up of flux in a magnetic circuit.

(MAP)—Manufacturing automation protocol

(MRP II)—Manufacturing resource planning

(N-type)—A semiconductor material in which the carriers are electrons and are therefore negative.

(NBS)—National Bureau of Standards responsible for maintenance of standards in the United States.

(NPN)—A semiconductor material. A structure consisting of a layer of P-type material placed between layers of N-type material, as commonly used in the bipolar type of transistor.

(P-type)—A semiconductor material in which the majority carriers are holes and are therefore positive.

(PC)—Personal computer (or Micro computer). A computer that is small, powerful, user friendly, and inexpensive enough for individual use.

(PID)—Proportional integral derivative control. A control mode that consists of the combination of the proportional, integral, and derivative control modes.

(PLC)—Programmable logic controller

(PM)—Permanent magnet

(PN)—Diode. A crystal diode that will allow current to flow only in one direction.

(PNP)—A semiconductor material. A structure consisting of an N-type region placed between two P-type regions, as commonly used in bipolar transistors.

(PROM)—Programmable read-only memory. A memory chip that cannot be rewritten or altered during operation, but can be reprogrammed off-line by special equipment.

(PSIA)—Pounds/square inch absolute. A pressure measuring instrument with a vacuum applied to the sensing equipment.

(PSIG)—Pounds/square inch gage. A pressure measuring instrument with atmospheric pressure applied to the sensing element.

(RAM)—Random-access memory. A read/write memory having the capability that any location in the memory can be addressed and accessed at the same time.

(RC)—Resistance capacitance. A circuit delivering a time constant.

(RF)—Radio frequency. In instrumentation it may be superimposed on a signal voltage and would be referred to as unwanted noise.

(ROM)—Read-only memory

(RTL)—Resistor-transistor logic

(SCR)—Silicon controlled rectifier. A thyristor.

(SI)—International system of units

(TTL Gates)—Transistor-transistor logic. A family of digital circuits, a bipolar transistor logic family used in digital circuits in integrated circuit technology. Logic gates: AND, OR, NOT, NAND, NOR, exclusive OR, and exclusive NOR.

(UJT)—Unijunction transistor. A very rapid action electronic switch.

GLOSSARY

Absolute pressure—pressure measured relative to a vacuum, usually expressed in pounds per square inch absolute (psia)

Acceleration—the rate of change of velocity (speed), usually expressed in "g" or gravity units

Accumulator—a chamber in which a liquid can be stored under pressure and from which it can be withdrawn upon demand. It is often used to smooth the variation in pressure in a hydraulic system.

Accuracy—the difference between the reading of an instrument and the true value of what is being measured, expressed as a percent of the full instrument's scale

Acidity—a measure of hydrogen ion content of a solution

Actuator—the unit that provides the power or movement to a final control element of a control system. The actuator performs the action indicated by the controller to the final element. Actuators may consist of pneumatic motors, various electric motors, hydraulic cylinders or motors, solenoids, air diaphragms, or cylinders.

Alethia—the ancient Greek personification of truth

Alkalinity—a measure of hydroxyl ion content of a solution

Alumel—an aluminum-nickel alloy used in the negative branch of a type K thermocouple

Ambient condition—the conditions of temperature, pressure, humidity, radiation, etc., existing in the area around the instrument

Ambient temperature—The temperature in the area around an object of concern

Amplifier—a device for increasing the strength of a signal

Analog computer—a device that converts mathematically expressed variables into mechanical equivalents

Armature—a moving member of a device that a magnetic flux reacts upon to produce a torque

Attenuate—to reduce in value, amount, amplitude or severity; weaken. The reciprocal of gain.

Automated guided vehicles—wheeled carts, guided by various means: embedded wires, computer control, or laser triangulation. They carry large or palletized parts to the various workstations.

Automatic controller—a device that measures the value of a variable and compares it to a set-point of selected standard and corrects the system so that it remains within the instrument's limits

Bandwidth—a symmetrical region around the set-point in which proportional control occurs

Barometer—an absolute pressure instrument measuring the local ambient pressure

Base—one of the three semiconductor regions of the bipolar type of transistor (the base, emitter, and collector)

Black body—a theoretical object that radiates the maximum amount of energy at a given temperature and absorbs all the energy incident upon it. The name "blackbody" is used because it is the color that is defined for the total absorption of light.

Calibration—a test during which known values of measure are applied to the transducer and corresponding output values are recorded under the specified test conditions

Cavitation—a condition in which liquid changes its phase, such as from a liquid to a vapor form, causing damage to pump parts and other fluid-handling equipment

Check valve—a valve that permits flow of fluid in only one direction

Cold junction—the reference junction of a thermocouple, usually the lower of the two junctions, thus the name "cold junction"

Constantan—a copper-nickel alloy used as a negative lead in the type E, J and T thermocouples

Control mode—the output form or type of control action used by a controller, such as on/off, proportional, proportional-integral, or derivative control

Chromel—a chromium-nickel alloy that makes up the positive branch of K and E thermocouples

Cryogenics—related to very low temperatures, usually below zero degrees Fahrenheit

Cushion—a device usually built into the ends of a hydraulic cylinder. It restricts the flow of fluid to the outlet port, thereby decelerating the motion of the piston.

Cycle—the completed sequence of a periodic motion that occurs during a period

Cylinder—a linear motion device in which the thrust or force is proportional to the effective cross-sectional area and the hydraulic pressure acting on it. Cylinders may be single acting or double acting.

Damping—the energy-dissipating characteristic which, together with the natural frequency, determines the limit of frequency response

Dead time—the time that elapses while the input to an instrument varies sufficiently to pass through the dead zone and causes the instrument to respond

Derivative action—control action by which the rate of change of the error signal determines the magnitude of the corrective action to be applied. The unit is calibrated in the time units. When subjected to a ramp change, the derivative output precedes the straight proportional action by this time. This is a PID control.

Differential pressure—the pressure difference measured between two pressure sources, usually expressed in pounds per square inch differential (psid)

Directional valve—a valve that selectively directs or prevents fluid flow to desired channels

Displacement—the change in position of a body measured from the point of rest or reference

Distributive control—a series of romote microcontrollers or field controllers are placed throughout the facility near the production process being controlled. These "smart" microprocessors control the operations, but in turn are under control by a control room station operator.

Dither—an alternating signal imposed upon a hydraulic cylinder in order to reduce the effects of a seal's stickability

Doping—the introduction of an impurity into the crystal lattice of a semiconductor to modify its electronic properties (such as adding boron to silicon to make the material more P-type)

Drain—one of the three regions of a unipolar or field-effect transistor (the drain, gate, and source)

Drift—in electronics, a change in a parameter due to a temperature change

Emmissivity—the ratio of energy emitted by an object to the energy emitted by a blackbody at the same temperature. Also, the emissivity of an object depends upon its material and surface texture.

Encoder—delivers information concerning position or count. They are of two types: incremental and absolute

Error—the difference between the actual and the true value, often expressed as a percentage of either span or full-scale value

Excitation—an external electrical voltage applied to a device for its proper operation

Feedback—data about the status of the controlled variable which may be compared with that which is desired, in the interest of making them coincide

Final control element—the component of a control system (such as a valve or motor) that directly regulates the flow of energy or material to the process

Flexible machining cell—a station in a larger automated processing network, in which different types of cells may be linked by material-handling devices

Flexible manufacturing system—the result of combining a number of cells or modular machining systems that perform multiple tasks

Flow control valve—a valve that controls the rate of fluid flow

Flow rate—the number of units of volume of a fluid passing by any given point in one unit of time, such as gallons per minute

Fluidics—the use of flowing gases or liquids to sense, transmit, and control other devices and equipment

Four-way valve—a directional control valve having four flow paths

Frequency—the number of vibrations or cycles in a unit of time, usually expressed in cycles per second or hertz

Gage pressure—pressure measured relative to ambient pressure (psig)

Harmonic—one whose vibration frequency is an integral multiple of that of the fundamental frequency

Head changers—an application of multiple spindle technology. It utilizes the concept of interchangeable multiple spindle heads stored at random and selected as required for the machining operation.

Head indexers—machines tooled for a specific family of parts and are a higher production tool than changers. The machine axes are associated with the workpiece table or the head index table or both. When one cycle is complete, the total head indexes and starts the next cycle.

Homogeneous—the same composition throughout the material

Hot junction—the measuring junction; usually warmer than the reference junction

Hydroscopic—a substance capable of absorbing moisture from the air

Impedance—the total opposition that a circuit offers to the flow of alternating current or any other varying current at a particular frequency. It is a combination of resistance R and reactance X, measured in ohms.

Insulator—a material having high resistance to the flow of electric current; often called a dielectric

Integral control action—the controller's output is proportional to the time integral of the error input. When used in combination with proportional action, it is often called reset action.

Kinematic inversion—anchoring or locking different members of a mechanism and producing a totally new motion, movement, or machine

Kinematics—the study of motion and the elements of machines. It is often called the study of mechanisms.

Linearity—the extent to which a calibration curve approaches a straight line

Load—a change in level of material force, torque, energy, power, or other variables applied or removed from a process or other component system

Machining center modules—stand-along machines such as NC milling, NC turning, vertical lathe, boring modules, etc. This module is the backbone for workpiece processing.

Manufacturing automation protocol—a computer communication network that provides a broadband backbone network that serves a series of carrier band subnets

Manufacturing cell modules—a group of processing modules combined to bring a family of parts to a desired level of completion without leaving the cell. The design of the cell is usually determined by the processing needs of the part family.

Manufacturing center modules—consist of a multi-pallet system working in conjunction with a machining center. It can operate on several types of parts simultaneously, selecting them randomly from a preloaded queue of palletized parts.

Mass—a body that has weight in a gravitational field

Mechanism—an elementary machine, a motion converter such as a lever, gear, cam, etc.

Modular tool—standardized units or sections for easy reconfiguration or flexible rearrangement. This allows flexibility so that the units can be applied to different product configurations.

Modulus of elasticity—the ratio of stress to strain in an elastic material

Monorail conveyor—a monorail or trolley hung from the ceiling that is used to transport heavy, awkward, or large pieces of material

Noise—unwanted electrical signal, usually generated from sources external to the measuring system

Noncircular gears—provide nonlinear motions; virtually any continuous mathematical function that does not require a reversed motion

Non-servo robot—an open loop device, meaning that there is no feedback. They travel to a preplaced stop and then sequence to the next movement.

Offset—the difference between what is received and what is desired

On/off control—a control mode in which the output is either full on or off, depending on whether or not the input is below or above the set-point

Open loop—a control without feedback, such as an automatic washing machine

Orifice—a restriction; the length is short in respect to the cross section dimensions

Overshoot—the control system, in reaching the desired level, goes above the set-point

Phase—the measurement of time change between the input and output signal, usually expressed in degrees of phase angle

Piezoelectric—a transducer utilizing a crystalline material that produces an electrical charge when subjected to a strain

Pilot pressure—an auxiliary pressure used to actuate or control hydraulic components

Poppet—a style of valve that opens and closes quickly and prevents flow when it is against its seat

Pressure reducing valve—a valve that limits the maximum pressure at its outlet regardless of the inlet pressure

Pressure switch—a switch operated by a rise or fall in fluid pressure

Proportional band—refers to the percentage of the controller's span of measurement over which the full travel of the control valve is divided. The reciprocal of gain is expressed as a percentage.

Proportional control—control action in which there is a fixed gain or attenuation between output and input

Proximity—a technique of measuring the distance between the sensing probe tip and a surface such as that of a workpiece

Range—the measured values over which the transducer is intended to measure

Response curve—obtained by applying a step change, either by load or set-point, and plotting the response of the controlled variable with respect to time

Relief valve—a pressure operated valve that bypasses pump delivery to the reservoir, limiting the system pressure to a predetermined maximum value

Repeatability—the maximum error or deviation that can be expected when the same value is input at two different times

Reproducibility—the exactness with which a measurement or other condition can be duplicated over a period of time

Resolution—the smallest detectable change in a measurement

Seismic—the construction of a transducer where the inertia of a mass is used as a base of reference

Self-regulation—the ability of an open-loop process or other device to settle out at some new operating point after a load change has taken place

Sensitivity—the minimum change of input to which the system is capable of responding

Sequence valve—a pressure operated valve that diverts flow to a secondary actuator while holding pressure on the primary actuator at a predetermined minimum value after the primary actuator completes its travel

Servomechanism—an actuator that includes its own automatic feedback control system

Servo robot—when the robot is activated, the controller will address the memory location in its computer for the first command position and a signal will be sent to move the robot to that position

Servo valve—a valve that controls the direction and quantity of fluid flow in proportion to an input signal

Set-point—the instruction given to an automatic controller to determine the point or value at which the controlled variable will stabilize

Solid-state technology—devices and circuits fabricated from solid materials, such as semiconductors, ferrites, or films

Spool valve—a sliding, relieved or channeled cylinder that can be positioned within a ported sleeve that, when positioned, provides for the directional flow of fluids

Strain gage—a series of wire or foil elements in the form of a Wheatstone bridge. As force is applied/relaxed, the components of the bridge change resistance. External voltage excitation is supplied to the bridge, resulting in millivolt output proportional to the changes in force.

Surge—a transient rise in hydraulic pressure in a circuit

Transducer—a device (such as a thermocouple) that converts one form of energy to another and will provide a proportional signal of the variable being measured

Transistor—a semiconductor device that uses a stream of charge carriers to produce active electronic effects

Vacuum—a perfect vacuum is the absence of gaseous fluid

Velocity—the rate of change of displacement, usually expressed in inches per second or millimeters per second, etc.

Wall attachment—a flow held against a surface by differences of speed and/or pressures of the fluid

Wheatstone bridge—an electrical network of fixed resistors

BIBLIOGRAPHY

Books

The AI Business: The Commercial Use of Artificial Intelligence. Edited by P.H. Winston and K.A. Prendergast. Cambridge, MA: The MIT Press, 1988.

Bateson, R.N. *Introduction to Control System Technology.* Columbus, OH: Merrill, 1989.

Faires, V.M. and R.M. Keown. *Mechanism.* New York, NY: McGraw-Hill, 1960.

Humphries, J.T. and L.P. Sheets. *Industrial Electronics.* Boston, MA: Breton, 1986.

Hunter, R.P. *Automated Process Control Systems: Concepts and Hardware.* Englewood Cliffs, NJ: Prentice-Hall, 1987.

Johnson, C.D. *Process Control Instrumentation Technology.* New York, NY: John Wiley & Sons, 1982.

K.T. Swasey. *KT's World of Advanced Manufacturing Technologies.* West Allis, WI: Kearney & Trecker, 1983.

Kirk, F.W. and N.R. Rimboi. *Instrumentation.* Chicago, IL: American Technical Publishers, 1975.

Koren, Y. *Robotics For Engineers.* New York, NY: McGraw-Hill, 1985.

Malcolm, D.R. Jr. *Robotics - An Introduction.* Boston, MA: Breton, 1985.

Malvino, A.P. *Electronic Principles.* New York, NY: McGraw-Hill, 1984.

McWane, J.W. *Introduction to Electronics Technology.* Boston, MA: Breton, 1986.

O'Connor, P.J. *Digital and Microprocessor Technology.* Englewood Cliffs, NJ: Prentice-Hall, 1983.

O'Higgins, P.J. *Basic Instrumentation.* New York, NY: McGraw-Hill, 1966.

Patrick, D.R. and S.W. Fardo. *Industrial Process Control Systems.* Englewood Cliffs, NJ: Prentice-Hall, 1979.

Pippenger, J.J. and G.H. Tyler. *Industrial Hydraulics.* New York, NY: McGraw-Hill, 1962.

Process Instruments and Controls Handbook. Edited by D.M. Considine. New York, NY: McGraw-Hill, 1985.

Magazines

American Machinist, Assembly Engineering, Machine and Tool Blue Book, Manufacturing Engineering, Measurement and Control, Modern

Materials Handling, Production, Production Engineering, Tool Engineering. (Magazine advertisements provide an excellent source of vendor addresses.)

Literature

Manufacturing vendor literature and photographs are the most outstanding source of material and are credited under the illustrations.

INDEX

A,B,C rotation, 386
Absolute encoders, 238–40, 416
Absolute humidity, 100
Absolute pressure, 91–92
Absolute pressure devices, 155
Absolute viscosity, 210
Acceleration, 103, 242
Accelerometers:
  piezoelectric, 245–47
  strain gage, 247–48
  variable capacitance, 244, 245
Accumulator, 321
Accuracy, 3, 14
Acidity, 105, 221–23
AC induction motor, 297
AC motor, 297
AC servomotors, 311–12
Actuators. *See* Control actuators
Adaptive control instrumentation, 392, 438–39
Addition (binary), 53–54
Air gaging, 251, 252
Air purge, 205, 206
Alkalinity, 105, 221–23
Alternating current motors, 304
  induction, 305–06
  synchronous, 306–07
  universal with AC power supply, 307
Aluminum oxide, 215, 217
Ambient condition, 92
Amperage, 31–32
  changes in, 86
Analog control systems, 285
Analog data input, 288
Analog output, 289
Analog-to-digital converter, 13–14, 288
Analytical measuring system, 257
AND gate, 59
Angular momentum mass flowmeter, 171–72
Angular velocity, 228
Archimedes' principle, 204
Archimedes' screw, 337
Arithmetic logic unit (ALU), 419
Armature, 150
Artificial intelligence (AI), 467, 469
Assembly machines, 351–53
Assembly and part handling, 351–53
Astable multivibrator, 73–75
Autocollimation, 255
Automated factory, 8, 25–26, 459
  instrumentation in, 466
  machine development in, 459–60
  machining modules, 460–65
  production control in, 466–67
  software for, 465–66
Automated guided vehicles (AGV), 348–50, 426, 440–41
Automated material handling, 19, 335
  assembly and part handling, 351–53
  bulk materials handling, 336–42
  finished product handling, 355–57
  integrated control of, 439–41
  piece parts handling, 343–51
  quality and position control, 354–55
Automated storage and retrieval, 439–40

Back scattering, 105
Backward gage devices, 155
Ball and disc drive, 377–78
Ball screws, 370–71
Ball-type control valve, 328, 329
Bar code reader, 382, 467
Bare capacitance probe, 190
Barometer, 140
BCD code, 239, 240
Bead chain drives, 375–76
Bellows, 144–45, 275–76, 312, 314
Bending-type accelerometer, 248
Bernoulli's theorem, 163

Bicolor level gage, 185, 186
Bimetallic thermometers, 114–17
Binary code, 386, 387, 416
Binary communication, 51–52
Binary numbering system, 52–53, 55–58
Bistable amplifier, 277–79
Bit, 55
Blackbody, 131
Bluff body, 172
Boole, George, 59
Boolean algebra, 53–55, 87
Boring module, 461
Boundary-layer effect. *See* Coanda effect
Bourdon tube, 11, 143
Boyle's law, 91
Breakdown voltage, 37
Bridge amplifiers, 71
Brushless DC linear motor, 309, 311
Brushless DC motor, 297, 309
Bubbler, 205, 206
Bucket conveyors, 337, 340
Bulk materials handling, 336–42
Buoyancy level sensors, 188–89
Butterfly-type control valve, 328, 329
Byte, 55

Calibration, 15, 16
Cams, 366–68
Cam sequence timer, 382
Capacitance, changes in, 86
Capacitance sensors, 190
Capacitors, 37
  charge storage in, 38
  time constant function, 38–39
  variable, 148
Capacity, 266
Capsules, 146
Carbon composition resistor, 34
Car-on-track conveyor system, 345, 346
Carousel or circular indexing machine, 352
Cartesian coordinate system, 386, 388
Cathode ray tube (CRT), 79
Cesium, 194–95, 206
Charles' law, 139
Chutes, 340–42
Cipolletti weir, 167
Closed loop, 17–18, 36, 64, 265
Coanda effect, 96, 172, 277, 278
Cobalt 60, 194, 206
Cold cathode ionization gage, 154
Cold junction, 125
Colorimetric measurement, 222
Compound motor, 303
Computer Aided Design and Computer Aided Manufacturing (CAD/CAM), 435
Computer aided drafting (CAD), 6–7, 390
Computer aided manufacturing (CAM), 6–7
Computer controlled tool center, 26, 467
Computer integrated manufacturing (CIM), 8, 25
  alternate branch sequencing and programming, 451–53
  artificial intelligence, 467, 469
  automated factory, 459–67
  data base management systems, 444–46
  inspection control, 441–44
  interfacing with other systems, 436–38
  inventory and storage, 454–59
  material-handling control, 439–41
  modular components for machinery, 445, 447
  production control, 438–39
  proximity data acquisition, 450–51
  scope of, 435–36
  software systems, 447–50
Computerized autocollimation, 255
Computer numerical control (CNC), 6, 21–22, 390–92, 398–99, 460
Computers, 6–7
  automation with, 25–26
Concentric orifice plate, 162–63
Conductance probe, 190
Conductivity, 37
  changes in, 85
Conductors, 36–37
Conductor wire, 33
Cone and plate plastometer, 212–13
Contact position control, 354–55
Continuous flow capillary tube, 210
Continuous flow manufacturing (CFM), 454–55
Contouring control, 390
Control, 263
  computer, 290–91
  controller technologies, 273–86
  control loops, 264–65
  direct digital technology, 287–90
  distributive control systems, 292–93
  microcomputer, 286–87
  microprocessor, 286
  modes of, 267–73
  response characteristics, 266–67
Control actuators, 18, 296
  electric, 297–12
  hydraulic, 315–18
  hydraulic power supplies, 318–25
  pneumatic, 312–15

Controllers:
 digital, 13–14, 284–86
 electrical, 12
 electronic, 13
 fluidic, 276–79
 hydraulic, 13, 279–83
 mechanical, 10–11
 pneumatic, 11–12, 274–76
 programmable, 283–84
 robotic, 412–21
Control loop, 16, 264
 closed loop, 17–18, 36, 265
 gain in, 64
 open loop, 17, 264–65
Control module, 464–65
Control valves, 325
 characteristics of, 326–28
 sizing of, 325–26
 types of, 328–32
Conventional current theory, 32
Conveyors:
 bucket, 337, 340
 car-on-track, 345, 346
 chutes, 340–42
 drag, 342, 343
 fastener and parts, 340, 341
 flat belt, 336–37
 monorail, 344–45, 440
 roller and wheel, 336
 screw, 337
 slat, 337
 trough belt, 337
 vacuum or pneumatic, 342
Coordinate measuring machine (CMM), 24, 429, 441
 electro-optical inspection and, 444
 programming, 442–43
 quality assurance and, 443–44
 trend analysis and, 444
Coriolis effect, 94–95, 171
Counters:
 electrical, 379–80
 fiber optic, 381–82
 mechanical, 378–79
 proximity, 380–81
Covalent bonding, 43–44
Crayon temperature indicators, 136
Cryogenic temperature measurement, 139
C-type bourdon tube, 143
Curie point, 153
Cutting air, 439
Cycling, 15–16
Cylindrical coordinate system, 388, 389
Cylindrical sensor, 117–18

Damping, 106
D'Arsonval voltmeter, 40, 300
Data base management systems (DBMS), 444–46
Data trend, 451
DC servomotors, 311
Dead time, 15, 266–67
Deformation, 99, 209
Degrees of freedom, 361, 407
Density, 97–98, 204
Derivative action, 273
Dew point, 100–101, 216
Dew point hygrometer, 216–17
Diaphragm box, 188
Diaphragm metering pump flowmeter, 181–82
Diaphragms, 145–46
Differential pressure gage, 141, 155
Differential pressure instruments, 161–66
Digital controllers, 13–14, 284–86
Digital control valve, 332
Digital data input, 288–89
Digital electronic multimeter, 40–41
Digital motor actuators, 307–12
Digital optical tachometer, 234
Digital readouts, 378
Digital stroboscope tachometer, 232
Digital technology, 51–58
Digital thermometers, 119
Digital-to-analog converter, 13, 289, 290
Diode:
 back biased, 44
 forward biased, 44
 light-emitting, 45
 zener, 45–46
Diode thermometer, 121
Direct current motors:
 compound, 303
 rotation reversal in, 303
 series DC, 301–02
 shunt-wound, 302
 solid-state controls, 304
 speed control in, 303
Direct digital technology, 287–90
Directional control valves, 279–80
Direct numerical control (DNC), 392, 460
Discharge curve, 39
Dispatching, 459
Displacement, 101, 234
Distance, 107
Distributive control systems, 292–93
Dither, 282

Documentation, 291
Doping, 43–44
Doppler ultrasonic flowmeter, 176–77
Drag conveyor, 342, 343
Drift, 15
Driving pawl, 368, 369
Dynamic error, 15–16

Eccentric orifice plate, 163
Elbow flow sensor, 166
Electric actuators, 297–312
Electrical capacitance hygrometer, 217
Electrical controllers, 12, 13
Electrical counters, 379–80
Electrical current, 31–32
Electrical energy, 421
Electrical level sensors, 189–93
Electrical pressure transmission devices, 146–54
Electrical temperature sensing systems, 117–29
Electrolytic hygrometer, 215–16
Electromagnetic flowmeter, 173–75
Electromechanical tachometer, 229
Electron flow theory, 32
Electronics:
  amperage, 31–32
  digital technology, 51–58
  impedance, 40
  inductance, 39
  integrated circuits, 77–79
  measurement of electrical values, 40–43
  op amps, 62–76
  resistance, 32–39
  semiconductor logic, 59–62
  semiconductors, 43–51
  voltage, 30–31
Electro-optical inspection, 444
Emissivity, 131
Encoders, 57–58, 415–16
  absolute, 238–39
  construction, 240
  incremental, 239
  incremental modular, 240, 241
End effectors (grippers), 23, 355, 411–12
End measure, 4
Engineering workstation, 25–26
Equal percentage valves, 327–28
Error signal, 13, 17
Expert systems, 467, 469

Fabrication, 426
Factory automation. *See* Automated factory
Fahrenheit, 112
Fastener and parts conveyor, 340, 341
Fastening, 404
Fiber optic counters, 381–82
Fidelity, 15
Field effect transistor (FET), 48
Filters, 320
Final control element, 18, 325. *See also* Control valves
Finished product handling, 355–57
555 integrated circuit, 73–75
555 timer, 72–73
Fixed resistor, 33–35
Fixed robotic assembly, 427
Flapper, 274, 275
Flat belt conveyor, 336–37
Flexible cell, 348
Flexible connectors:
  bead chain drives, 375–76
  roller chain drives, 376
  timing belts, 374–75
Flexible manufacturing, 7–8, 20, 396
  applications of robots in, 424–31
  Manufacturing Automation Protocol and, 400–01
  manufacturing cells in, 397–98
  modular assembly machines, 404
  modular machining centers, 403
  numerical control in, 398–99
  numerical control machining centers, 401–02
  part feeders, 404
  power supply for, 421–24
  robots in, 404–21
Flexible robotic assembly, 428–29
Flip-flop controller, 276–79
Floats, 188–89
Float viscometer, 211
Flow:
  in confined conduits, 92
  differential pressure instruments, 161–66
  electromagnetic flow instruments, 173–75
  fluidic flow concepts, 96
  laminar, 93
  mass flow, 94–96
  mass flow instruments, 168–72
  in open channels, 94
  oscillatory instruments, 172–73
  positive displacement instruments, 177–82
  pressure drop and, 93
  turbine instruments, 175–76
  turbulent, 93
  ultrasonic instruments, 176–77
  variable area instruments, 166–68

Flow coefficient, 326
Flowmeters:
  electromagnetic, 173–75
  metering pump, 181–82
  oscillatory, 172–73
  positive displacement, 177–81
  turbine, 175–76
  ultrasonic, 176–77
Flow nozzle, 163
Flow rate, 92
Fluid-filled thermometers, 113
Fluidic controllers, 276–79
Fluidic flow, 96
Fluidic flowmeter, 172–73
Fluid processing, 5
Flumes, 94, 167–68
Flux, 235–37, 306
Fly-ball governor, 365–66
Force, 98, 223
Force-torque sensors, 438
Ford, Henry, 4
Franklin, Benjamin, 31
Frequency, changes in, 86
Friction drives:
  ball and disc, 377–78
  two-disc variable-speed, 377
  wheel and disc, 376–77
Full-wave rectifiers, 422, 423

Gage pressure, 92
Gain, 48, 270, 271
  closed loop, 64
  open loop, 64
Gases:
  behavior of, 91
  changes in, 84
Gas-filled thermometers, 113
Gas-purge level sensor, 188
Gate valve, 329, 330
Gear motors, 324
Gears:
  nomenclature, 372
  noncircular, 373, 374
  pinion and sector, 373
  spur, 371
Geiger counter, 105, 194
Geneva mechanism, 369–70
Germanium, 43, 44
Globe-type valve, 329, 331
Gray code, 58, 239, 240, 416
Gripper sensors, 411–12
Gyrator, 39
Gyroscopic (Coriolis) mass flowmeter, 170–71

Half-wave rectifiers, 421–22
Handcrafted production, 3
Hardware, 290
Head changer module, 463–64
Head indexer module, 464
Head loss, 93
Heat transfer, 88–90
Helium-neon laser tube, 382
Heuristics, 469
Hexadecimal code, 57
High-density storage, 356–57
Homogeneous circuit, 126
Hook gage, 184
Hook's law of elasticity, 224
Hot bulb thermometers, 139
Hot cathode ionization gage, 153–54
Hot junction, 125
Hot test, 355
Humidity, 100, 213–14
Hybrid circuits, 197–98
Hybrid control systems, 273, 274
Hydraulic accumulator, 321
Hydraulic actuator:
  controls, 315–18
  power supplies, 318–25
Hydraulic controllers, 13, 279–83
Hydraulic flapper valve, 315, 316
Hydraulic jet pipe valve relay, 315, 316
Hydraulic power supply, 424
Hydraulic pumps, 321
  gear motor, 324
  piston, 322–23
  radial-piston, 323–24
  vane motor, 324–25
Hydraulic reservoir, 319–20
Hydraulic servovalve, 317–18
Hydraulic spool valves, 316–17
Hydrogen ion, 105, 221
Hydrometer, 204–205
Hydrostatic head, 91, 161, 167, 187
Hygrometer, 214–15
  dew point, 216–17
  electrical capacitance, 217
  electrolytic, 215–16
  impedance, 218
  infrared, 220–21
  microwave, 218
  quartz crystal, 221

Hysteresis, 68–69

Impedance, 39–40
Impedance bridge, 86
Impedance hygrometer, 218
Impeller mass flowmeter, 171–72
Inclined manometer, 142–43
Incremental encoder, 239, 415
Incremental modular encoder, 240, 241
Inductance, 39
  changes in, 85
Induction motor, 305–06
Inductive sensors, 241
Inductosyns, 417
Industrial computer control, 383
Industrial Revolution, 3–4
Industrial robots, 406
Infrared hygrometer, 220–21
Infrared thermometers, 133, 135
Input:
  analog, 288
  digital, 288–89
Input/output (I/O), 79
Inspection control, 384, 429
  integrated, 441–44
Instrumentation:
  in automated factory, 466
  computer numerical control and, 25, 390–92
  machine control, 385
  measurement by changes in materials, 83–87
  mechanical, 19, 360–78
  numerical control, 386–90
  transfer lines production, 19, 378–85
Instrument development, 8–9
  control sensing with computers, 10
  electrical instruments, 9–10
  pneumatic instruments, 9
  simple instruments, 9
Instruments:
  calibration of, 16
  contact, 249
  dynamic characteristics of, 15–16
  noncontact, 250
  static characteristics of, 14–15
Insulators, 37
Integrated circuit pressure sensors, 155–56
Integrated circuits (ICS), 77, 197
Interference band, 253, 254
Interferometry measurement, 107, 253, 258
International Practical Temperature Scale, 111
Inventory control:
  continuous flow manufacturing, 454–55
  just in time, 454
  machines, 456
  special materials, 456–57
  tools, 455–56
Inverse cam, 367–68
Inversions, 362–64
Inverting op amp, 65
Inverting zero-crossing voltage detector, 66
Ionization chamber, 206–07
Isotopes, 103, 105

Johansson gage block, 4–5
Joint, 361–62
Just in time (JIT) production, 454

Kinematic inversion, 362
Kinematic linkage systems, 19
Knowledge engineering, 467
Krypton 86, 253

Labeling, 458
Laminar flow, 93
Large-scale integrated circuits (LSI), 10, 77
Laser interferometry, 258
Laser level sensors, 199–200
Law of Intermediate Metals, 126
Level, 97
  buoyancy sensors, 188–89
  electrical sensors, 189–93
  hydrostatic head measurement, 187–88
  laser sensors, 199–200
  load cell as sensor, 194
  nuclear sensors, 194–95
  solid-state sensors, 197–98
  ultrasonic sensors, 195–96
  visual observation of, 183–87
Lever systems, 223–24
Light-dependent resistor (LDR), 191, 193
Light-emitting diode (LED), 45, 232, 380–82
Light wave interference, 253
Linear motion actuators, 297–300
Linear opening valves, 327
Linear potentiometer, 36
Linear variable differential transformer (LVDT), 148–49, 235, 236, 243
Linear variable reluctance transducer (LVRT), 236–37
Linkage systems, 361–66
Liquid column pressure devices, 140–43
Liquid-filled thermometers, 112
Liquid height, 91

Liquids, changes in, 84
Litmus paper test, 222
Load, 88, 266–67
Load cell:
  as level sensor, 193, 194
  pneumatic, 224
  semiconductor strain gage, 227–28
  strain gage, 225–27
Local area network (LAN), 436–37
Logic gates, 59–62, 87
  fluid equivalents, 280

Machinability sensing, 438–39
Machine control, 385
  integrated, 437–38
Machining modules. *See* Modules
Magnetic field, 39
Magnetic induction transducer, 229, 230, 231
Magnetic plugs, 320
Magnetic tape coding, 387
Main frame computer, 10
Management reports, 26
Manometer, 140–42
  inclined, 142–43
  well, 142
Manual assembly, 351
Manufacturing Automation Protocol (MAP), 400–01, 436
Manufacturing cells, 8, 21–22, 396–98
Manufacturing center module, 462–63
Mass flow, 94–96, 168–72
Mass flow meter, 95
Mass production, 4–5
Mass thermal dispersion sensor, 170
Material handling, 19. *See also* Automated material handling
Maudslay, Henry, 4
Measurement:
  of acceleration and vibration, 242–48
  of acidity and alkalinity, 221–23
  by changes in materials, 83–87
  of density and specific gravity, 204–09
  of dimension, 249–60
  of displacement and motion, 235–42
  electrical values, 40–43
  of flow, 161–82
  of humidity and moisture, 214–21
  of pressure, 139–56
  process control instruments, 14–16
  process variables of, 87–107
  of rotational speed, 228–34
  of temperature, 112–39
  of viscosity, 210–13
  of weight and force, 223–28
Mechanical controllers, 10–11
Mechanical counters, 378–79
Mechanical instrumentation, 19, 360–78
Mechanical pressure devices, 143–46
Mechanical temperature sensing systems, 112–17
Mechanisms, 361
  flexible connectors, 374–76
  friction drives, 376–78
  gears, 371–74
  linkage systems, 361–66
  rotary motion converters, 366–71
Mechanized production, 3–4
Memory (computer), 78
Mercury, 112
Metering pumps, 181–82
Microcomputer, 10, 78–79, 255, 286–87
Microprocessor, 77–78, 286
  controllers, 419–20
  rotational speed measurement, 233–34
Microwave hygrometer, 218
Microwave level sensors, 196, 197
Minicomputer, 10
  controllers, 420–21
Modified linear valves, 327
Modular tools, 447
Modules, 403, 404, 445, 447
  boring, 461
  control, 464–65
  head changer, 463–64
  head indexer, 464
  manufacturing center, 462–63
  numerical control milling, 460
  standard machining center, 462
  turning, 460-61
Modulus of elasticity, 143
Monolithic sensor, 156
Monorail conveyors, 344–45, 440
Motion, 235
Motors:
  alternating current, 304–07
  direct current, 301–04
Mueller bridge, 117
Multiple input gates, 62
Multiturn potentiometer, 36

NAND gate, 61
Narrow band pyrometers, 133
National Bureau of Standards, 83

Needle valve, 331
Noise, 15, 68
Noncircular gears, 373, 374
Noncontact position control, 355
Noncontact pulsed infrared sensor, 258, 260
Noninverting zero-crossing voltage detector, 66
Nonservo robots, 404–05
Nonzero-crossing voltage detector, 67
NOR gate, 61
NOT operator, 60
Nozzle, 274, 275
Nuclear level sensors, 194–95
Numerical control instrumentation, 383, 386–90, 398–99
Numerical control machining centers, 401–02
Numerical control milling machine, 460
Nutating disk flowmeter, 178

Observation and comparison, 2
Octal code, 55, 56
Offset, 272
Ohm's law, 12, 42–43
One-shot multivibrator, 75
On/off control, 268
Op amp, 62
    bistable, 277–79
    bridge amplifiers, 71
    construction of, 63–64
    digital-to-analog converter, 289, 290
    555 timer, 72–73
    gain of, 64
    hysteresis added to, 68–69
    inverting, 65
    inverting zero-crossing detector, 66
    noninverting zero-crossing detector, 66
    nonzero-crossing detector, 67
    sawtooth frequency generator, 72
    square wave generator, 73–75
    time delayed switch, 75–76
    unijunction transistor high speed switch, 72
    as voltage comparator, 65
    voltage follower, 69–70
    voltage range detector, 67–68
    voltage-to-current conversion, 70–71
    zero-crossing detector, 66
Open channel flow, 94, 166
Open flow nozzle, 168
Open loop, 17, 64, 264–65
Operational amplifiers. *See* Op amp
Optical/brightness pyrometers, 133
Optical comparator, 251
Optical flats, 253
Optical tooling, 253–58
OR gate, 60
Orifice plate, 161–63
Oscillating-piston flowmeter, 178
Oscillatory flowmeters, 172–73
Oscillatory motions, 364
Output, 189
Oval or elongated circle machine, 352
Oval-shaped gear flowmeter, 180

Packaging, 457–59
Paints, temperature sensitive, 135
Parabolic plug, 327
Parallax, 379
Parallel motions, 364, 365
Parshall flume, 167–68
Part feeders, 404
Part geometry files, 443
Pascal's principles, 140
Pellet temperature indicators, 136
Peltier, Jean, 125
Peltier effect, 126
Peristaltic pump flowmeter, 181
Permanent-magnet stepper motor, 307–308
Phase, 150
pH concentration, 105, 221
    measurement methods, 222–23
Photoelectric level control, 191, 193
Photo-SCR, 50
Photosensor, 231–32
Phototransistor, 232–33
Piece parts handling, 343–51
Piezoelectric accelerometer, 106, 245–47
Piezoelectric crystal transducers, 153
Piezoresistive principle, 154, 227–28
Pinch off voltage, 48
Pinch valve, 331–32
Pinion gears, 373
Piston metering pump flowmeter, 182
Piston pumps, 322–23
Pitot tube, 165
Plastometers, 212–13
Platinum wire, 117
Plumb bob and tape, 183–84
Pneumatic actuators, 312–15
Pneumatic controllers, 11–12, 274–76
Pneumatic diaphragm motors, 312, 313
Pneumatic load cells, 224
Pneumatic mechanical actuators, 312, 314
Pneumatic power supply, 424

Pneumatic rotary motors, 315
Pneumatic technology, 5
Point-to-point control, 389
Poppet valve, 326
Position, 101, 234
    control of, 354–55, 414–18
Positive displacement flowmeters, 177–81
Potential difference, 30–31
Potentiometers, 12, 35–36, 147
Potentiometric sensors, 222–23
Power rating, 33
Power supply, 23–24
    electrical energy, 421
    hydraulic, 424
    pneumatic, 424
    regulated, 422, 424
    unregulated, 421–22
Precision, 3
Precision measuring microscope, 250
Precision quartz accelerometer, 245–47
Pressure drop, 93
Pressure measurement, 90, 139
    absolute pressure, 91–92
    electrical transmission devices, 146–54
    gage pressure, 92
    hydrostatic head, 91
    liquid column devices, 140–43
    mechanical devices, 143–46
    solid-state electronic devices, 154–56
Process control. *See* Control
Process instruments. *See* Instruments
Process variables, 87
    acceleration, 103
    density and specific gravity, 97–98
    distance, 107
    flow, 92–96
    force, weight, stress, and strain, 98–99
    humidity and dew point, 100–01
    level, 97
    pH concentration, 105
    position, rotation, and speed, 101–02
    pressure, 90–92
    temperature, 88–90
    thickness, 103, 105
    vibration, 105–06
    viscosity, 98
Production control, 438–39, 466–67
Production schedule, 454
Programmable control, 383
Programmable logic controller (PLC), 283–84, 447–49
Progress reporting, 452–53
Proportional bandwidth, 270–72
Proportional control, 268, 270–73
Proportional-plus-integral control, 272
Proportional-plus-integral-plus-derivative control, 272–73
Proximity counters, 380–81
Proximity sensors, 241, 258–60, 450–51
PSIA, 140
PSIG, 141
Psychrometer, 214
Pulse frequency tachometer, 229–30
Pulse transducers, 238–40
Pumps, 321–25
Pyometric cones, 136–37

Quality assurance, 443–44
Quality control, 354
Quartz crystal hygrometer, 221
Quartz thermometers, 138
Quick opening valves, 326–27

R,A,Z coordinates, 388
Radial-piston pumps, 323–24
Radiant energy, 129, 131
Radiation density sensor, 206–07
Radiation thermometers, 129, 131–35
Radioactivity, 103, 105
Radium, 194
Railed slides, 345
Random access memory (RAM), 287, 419
Ratchet, 368–69
Ratchet wheel, 368, 369
Ratio pyrometers, 133
RC circuit:
    frequency change and, 86
    high speed switch and, 72
    square wave generator and, 73–75
    time delayed switch and, 75–76
Reaction curve, 267
Read only memory (ROM), 287, 419
Recording labels, temperature sensitive, 135
Redesigning parts, 429
Reference temperatures, 111–12
Reflex glass level sensor, 185
Relative humidity, 100
Relay, 297–300, 383
Reproductibility, 15
Reprogrammable tools, 21
Resistance, 32–33, 266
    changes in, 85

Resistance Thermal Detector (RTD), 117, 170
Resistance thermometer bridge, 120
Resistance thermometers, 117–20
Resistant wire, 33
Resistors:
  fixed, 33–35
  values, 33–35
  variable, 35–36
Resolvers, 237, 238, 416
Response, 15, 18, 266–67
Reverse breakdown, 45
Reynold's number, 93
Rheostat, 148
Rho, 32–33
Robots:
  applications in manufacturing, 24, 424–31
  classifications of, 406–11
  controllers for, 23, 412–14
  cylindrical, 411
  end effectors (grippers), 23, 355, 411–12
  high-technology, 409, 414
  industrial, 406
  jointed-arm, 409–10
  lifting capacity and, 409
  low-technology, 408, 412, 414
  in material handling, 348, 351
  medium-technology, 408–09, 414
  nonservo, 404–05
  position sensors for, 414–18
  program controllers for, 418–21
  rectangular, 411
  servo, 405–06
  spherical, 411
Robot units, 7
Rocker arm, 364
Roller chain drives, 376
Roller and wheel conveyor, 336
Rotary globe valve, 329–31
Rotary motion converters, 19, 366–71
Rotary motion solenoids, 300
Rotary motion speed control, 365–66
Rotary pumps, 322–25
Rotary variable differential transformer, 149–50
Rotating lobe flowmeter, 180
Rotating vane flowmeter, 179
Rotation, 101
Rotational speed, 228
Rotor, 305
RPM, 228

Safety devices, 430
Satellite computer, 421
Sawtooth frequency generator, 72–75
Schedule comparison, 453
Schmitt trigger, 75
Scotch yoke, 368, 369
Scott-Russell mechanism, 364, 365
Sector gears, 373
Seebeck, Thomas J., 124
Segmental orifice plate, 163
Seismic mass, 243
Self-regulation, 267
Semiconductors, 43
  changes in, 87
  diodes, 44–46
  field effect transistor, 48
  principles of, 43–44
  silicon controlled rectifier, 48–50
  thermistor, 121–24
  thyristor, 48
  transistors, 46–47
  voltage divider, 50–51
Sensitivity, 15
Sensors, 15, 110–11
  acceleration and vibration, 242–48
  acidity and alkalinity, 222–23
  contact, 438
  density and specific gravity, 204–09
  dimensional, 249–60
  displacement and motion, 235–42
  flow, 161–82
  humidity and moisture, 214–21
  level, 183–200
  noncontact, 438
  pressure, 139–56
  rotational speed, 228–34
  temperature, 112–39
  viscosity, 210–13
  weight and force, 223–28
Sequence valve, 280–81
Sequencing:
  alternate branch, 451–53
  cam sequence timer, 382
  industrial computer control, 383
  inspection, 384
  logic circuits, 418–19
  numerical control, 383
  programmable control, 383, 418
  relays, 383
Series DC motor, 301–02
  speed control in, 303
Servo robots, 405–06
Servosystem, 420
Servovalve, 18, 281–83

Set-point, 12, 14, 35, 86, 268, 272, 275, 419
Shear rate, 209, 212–13
Shear stress, 209, 212–13
Shipping, 457–59
Shunt-wound motor, 302
  speed control in, 303
Siemens, William, 117
Sight glass, 185
Signal conditioned transducers, 156
Silicon, 43–44, 154
Silicon controlled rectifier (SCR), 48–49, 422, 423
  photo-SCR, 50
Slat conveyor, 337
Slider crank mechanism, 362–63
Slip, 306
Slip tube level sensor, 186–87
Slowdowns, 452
Smart sensors, 10, 77
Software, 290, 447–50, 465–66
Solenoid, 297, 298
Solids, changes in, 84
Solid-state devices, 87
  counting and measuring instruments, 242
  electronic pressure devices, 154–56, 198
  level control sensors, 197–98
  motor controls, 304
  rotational speed, 232–33
Solid state switching, 75–76
Specific gravity, 97–98, 204
Specific humidity, 100
Speed, 102
  control of, 303
Spring systems, 224
Spur gears, 371
Square wave generator, 73–75
Squirrel cage armature, 305
Standardized building-block construction, 445
Standard machining center module, 462
Static error, 14–15
Stator, 237, 305
Steady-state cycling, 15
Stepper motor, 297, 307
Still well, 167
Storage and retrieval. *See* Automated storage and retrieval
Straight-line assembly machine, 353
Straight-line control, 389–90
Straight-line motion, 364
Strain, 99
Strainers, 320
Strain gage, 106, 151–52
  accelerometer, 247–48
  load cell, 225–27
  semiconductor load cell, 227–28
Stress, 98
Subtraction by complement (binary), 54–55
Synchronous motor, 306–07
Synchros, 237–38, 416

Tachometers, 102, 228, 418
  AC-type, 229
  DC-type, 229
  digital optical, 234
  digital stroboscope, 232
  magnetic induction, 230, 231
  photosensor, 231–32
  pulse frequency, 229–30
  solid-state, 232–33
Tactile sensors, 438, 450
Tape controlled machine program, 388–90
Temperature:
  cryogenic measurement, 139
  electrical sensing systems, 117–29
  heat transfer, 88–90
  hot bulb system, 139
  mechanical sensing systems, 112–17
  quartz crystal measurement, 138
  radiation sensing systems, 129–35
  scales, 88
Temperature-compensated monolithic sensor, 156
Temperature sensitive materials, 135–37
Testing, 355–56
Theodolites, 257
Thermal level detectors, 190–91
Thermal mass sensor, 169–70
Thermistor, 121–24
Thermocouples, 124–29
Thermometers:
  bimetallic, 114–17
  diode, 121
  fluid-filled, 113
  gas-filled, 113
  hot bulb, 139
  quartz, 138
  radiation, 129, 131–35
  resistance, 117–20
Thickness, 103, 105
Thyristor, 48
Time delayed switch, 75–76
Timing belts, 374–75
Tool center, 26, 467
Torque motor, 300
Torr, 153–54
Tote trays, 343–44

Transducers, 76, 110–11
  acceleration and vibration, 242–48
  acidity and alkalinity, 222–23
  density and specific gravity, 204–09
  dimensional, 249–60
  flow, 161–82
  humidity and moisture, 214–21
  level, 183–200
  position placement and motion, 235–42
  pressure, 139–56
  rotational speed, 228–34
  temperature, 112–39
  viscosity, 210–13
  weight and force, 223–28
Transfer lag, 267
Transfer machines and lines, 5–6
  instrumentation in, 19, 378–85
Transistors:
  biasing, 47
  field effect, 48
  NPN, 46, 47
  unijunction, 50, 72
Transit time ultrasonic flowmeter, 176–77
Trapezoidal weir, 167
Trend analysis, 444, 451
Triple point, 112
Trough belt conveyor, 337
TTL voltages, 296, 300
Turbine flowmeter, 175–76
Turbulent flow, 93
Turning module, 460–61
Two-color pyrometers, 133
Two-disc variable-speed drive, 377

Ultrasonic density sensor, 208
Ultrasonic flowmeter, 176–77
Ultrasonic level sensors, 195–96
Ultrasonic viscometer, 212
Unijunction transistor, 50
  high speed switch, 72
Universal motors, 301–02
  with AC power, 307
U-tube manometer, 140–42, 187–88

Vacuum or pneumatic conveyors, 342
Validation (program), 291
Values:
  instruments for measurement of, 40–43
  resistors, 33–35
Valve, 325. *See also* Control valves
Vane motors, 324–25
Vapor pressure, changes in, 84–85
Variable capacitance accelerometer, 244, 245
Variable capacitor, 148
Variable reluctance induction transducer, 151
Variable-reluctance stepper, 308–309
Variable reluctance transducer, 150–51
Variable resistors, 35–36
Vena contracta, 161
Venturi tube, 163–64
Vibrating U-tube, 207–08
Vibration, 105–06
Vibration density sensor, 207–08
Vibratory hoppers, 345, 347–48
Viscometer:
  float, 211
  ultrasonic, 212
Viscosity, 98, 209
  absolute, 210
Vision sensing controllers, 428, 450
Visual level observations, 183–87
Voltage, 30–31
  changes in, 85–86
Voltage comparator, 65
Voltage divider, 50–51
Voltage follower, 69–70
Voltage range detector, 67–68
Voltage-to-current conversion, 70–71
Voltmeters, 31
  D'Arsonval, 40, 300
  electronic digital multimeter, 40–41
Vortex precision flowmeter, 173
Vortex shedding flowmeter, 172
V-port, 327

Wall effect. *See* Coanda effect
Weight, 99, 223
Weirs, 94, 166–67
Well manometer, 142
Wheatstone bridge, 12, 71, 117, 191, 209, 247, 248
Wheel and disc drive, 376–77
Wheeled carts, 348
Whitworth, Joseph, 4
Wide band pyrometers, 132
Wire gauges, 33

X,Y,Z coordinates, 257, 386, 388

Zener diode, 45–46
Zener voltage, 45
Zero-crossing voltage detector, 66